Aircraft System Identification

Theory and Practice

Second Edition

Aircraft System Identification

Theory and Practice

Second Edition

Eugene A. Morelli

Research Engineer
NASA Langley Research Center
Hampton, Virginia USA

Vladislav Klein

Professor Emeritus
George Washington University
NASA Langley Research Center
Hampton, Virginia USA

Published by

Sunflyte Enterprises
Williamsburg, Virginia

MATLAB® is a registered trademark of The Mathworks, Inc.

Aircraft System Identification
Theory and Practice
Second Edition

Published 2016

Sunflyte Enterprises
115 Crownpoint Road
Williamsburg, Virginia 23185

ISBN 0-9974306-1-3

For Susan

Table of Contents

Preface

This book is based on the research and teaching activities of the authors, mostly at the NASA Langley Research Center in Hampton, Virginia. It is intended as a resource for researchers and practicing engineers, as well as a textbook for post-graduate and senior-level courses. The reader is assumed to have undergraduate-level familiarity with topics such as differential equations, linear algebra, probability, statistics, and aerodynamics. For readers without this preparation, the material in the appendices should provide adequate background information on the necessary concepts. In writing the book, the goal was to provide a comprehensive treatment of both the theoretical underpinnings and the practical application of aircraft modeling based on experimental data, which is also known as aircraft system identification.

The scope of the book is restricted to fixed-wing aircraft, assumed to be rigid bodies. However, the methods discussed are generally applicable, and can also be applied to flexible vehicles, rotorcraft, and spacecraft, among many other applications. Methods discussed in the book are used routinely for risk reduction during flight envelope expansion of new aircraft and modified configurations, comparison with results from wind tunnel tests and analytic methods such as computational fluid dynamics (CFD), control law design and refinement, dynamic analysis, simulation, flying qualities assessments, accident investigations, and other tasks.

Aircraft system identification is an applied engineering discipline, which means that a complete education in this field of study must include practical hands-on experience. To that end, the book includes a software toolbox, written in MATLAB®, which implements most of the methods discussed in the text. The toolbox is called SIDPAC, which is an acronym for System IDentification Programs for AirCraft. SIDPAC can be obtained by request from NASA Langley at https://software.nasa.gov/software/LAR-16100-1. Full contact information and organizational affiliation will be collected from all requestors, and all requestors must pass NASA security screening before receiving SIDPAC. Chapter 12 provides complete documentation for the software.

SIDPAC can be and should be applied to modeling problems of particular interest to the reader. Example problems using real data from flight tests and wind tunnel experiments are provided in the text. Calculations and plots for these examples (except for the X-29 examples, due to data restrictions) can be regenerated in full detail using the sidpac_text_examples.m script. The reader can study this file to see exactly how each example was done. SIDPAC also includes demonstration scripts, which provide complete examples of how tools in SIDPAC can be used to solve aircraft system identification problems. Each demonstration script has a name that includes the characters _demo.m.

Each directory of files in SIDPAC includes a file named Contents.m, which is a list of the files in the directory, along with a short description of each file.

All SIDPAC routines are written in MATLAB®, so it is possible for the user to study the implementations, and use the software as a basis for writing additional routines to solve related or specific problems. No special MATLAB® toolboxes are necessary to use SIDPAC, just standard MATLAB®.

The software package also includes a nonlinear simulation of the F-16 aircraft, written in MATLAB® and documented in Appendix D. This software can be used to simulate realistic flight test maneuvers to provide data for exercising the SIDPAC tools, and for further research and development. The code includes tools to generate linear models from the F-16 nonlinear simulation. The user can run simulated maneuvers using the F-16 nonlinear simulation or the extracted linear models, then use SIDPAC tools to identify models based on the simulated data. This provides feedback on the modeling success, because (unlike in a real flight test situation) the correct model structure and model parameter values are known.

The authors are grateful to both the NASA Langley Research Center and George Washington University for financial support and their important roles in our careers and at several stages of the book preparation. The authors also thank their colleagues at NASA Langley, particularly Dr. Jared Grauer, Jim Batterson, Dr. Pat Murphy, Kevin Cunningham, Dan Murri, Jay Brandon, John Foster, Carey Buttrill, Dr. Christine Belcastro, Dr. Dave Cox, and many colleagues and graduate students, for their insights and assistance with our research and quest for understanding in this interesting and important area of study. Thanks to Dr. Ravi Prasanth for his careful reading of the manuscript, which resulted in corrections in several equations and a clearer exposition.

Finally, and most importantly, we wish to extend our deep gratitude to our families, for encouragement and support over the long period of time required to produce this book, and indeed over the course of our lives. It is a pleasure to dedicate this book to all of you.

Eugene A. Morelli
Vladislav Klein

November 2016

Chapter 1

Introduction

One of the oldest and most fundamental of all human scientific pursuits is developing mathematical models for physical systems based on imperfect observations or measurements. This activity is known as system identification. This book describes the theory and practice of system identification applied to aircraft, which is also called aircraft system identification.

A good practical definition of system identification is that of Zadeh [1.1]:

> *System identification is the determination, on the basis of observation of input and output, of a system within a specified class of systems to which the system under test is equivalent.*

Implicit in this definition is the practical fact that the mathematical model of a physical system is not unique. In general, the guiding principle for model selection is the parsimony principle, which states that of all models in a specified class that exhibit the desired characteristics, the simplest one should be preferred. There are both theoretical and practical reasons for the parsimony principle, and these will be discussed further throughout the book. The preceding definition also mentions that system identification is based on observations of input and output for the system under test. In practice, these observations are corrupted by measurement noise. This requires the introduction of statistical theory and methods, as will be described in detail. Finally, there must also be some definition of what is meant by the word "equivalent" in the preceding definition. There is more than one way in which a model can be considered equivalent to a system under test. The most common approaches will be presented and explained.

The most important requirement for a mathematical model is that it be useful in some way. Sometimes this means that the model can be used to predict some aspect of the behavior of a physical system, while at other times, just the values and accuracies of parameters in the model provide the desired insight. In any case, the synthesized model must be simple enough to be useful, and at the same time complex enough to capture important dependencies and features embodied in the observations or measurements.

1.1 System Identification Applied to Aircraft

System identification is one of three general problems in aircraft dynamics and control, which can be understood with reference to Fig. 1.1. The three problems are:

1) *Simulation*: Given the input $\boldsymbol{u}$ and the system $\boldsymbol{S}$, find the output $\boldsymbol{y}$.

2) *Control*: Given the system $\boldsymbol{S}$ and the output $\boldsymbol{y}$, find the input $\boldsymbol{u}$.

3) *System Identification*: Given the input $\boldsymbol{u}$ and the output $\boldsymbol{y}$, find the system $\boldsymbol{S}$.

Figure 1.1 **Aircraft dynamics and control**

For many applications, an aircraft can be assumed to be a rigid body, whose motion is governed by the laws of Newtonian physics. System identification can be used to characterize applied forces and moments acting on the aircraft that arise from aerodynamics and propulsion. Typically, thrust forces and moments are obtained from ground tests, so that aircraft system identification is applied to model the functional dependence of aerodynamic forces and moments on aircraft motion and control variables.

Modern computational methods and wind tunnel testing can provide, in many instances, comprehensive data about the aerodynamic characteristics of an aircraft. However, there are several motivations for identifying aircraft models from flight data, including:

1) Verifying, interpreting, and improving theoretical predictions and wind tunnel test results;

2) Obtaining more accurate and comprehensive mathematical models of aircraft dynamics, for use in designing stability augmentation and flight control systems;

3) Developing flight simulators, which require accurate representation of the aircraft in all flight regimes (many aircraft motions and flight conditions simply cannot be duplicated in the wind tunnel nor computed analytically with sufficient accuracy or computational efficiency);

4) Expanding the flight envelope for new aircraft, which can include quantifying stability and control impact of aircraft modifications, configuration changes, or special flight conditions;

5) Investigating aircraft accidents or anomalies;

6) Verifying aircraft specification compliance.

In the early days of powered flight, only the most basic information about aircraft aerodynamics was obtained from measurements in steady flight. One of the first approaches for obtaining static and dynamic parameters from flight data was given by Miliken [1.2] in 1947, using frequency response data and a simple semi-graphical method for the analysis. Four years later, Greenberg [1.3] and Shinbrot [1.4] established more general and rigorous ways to determine aerodynamic parameters from transient maneuvers. Parameter estimation methods introduced in these reports were based on ordinary and nonlinear least squares.

Dramatic improvement in aircraft aerodynamic modeling techniques came in the late 1960s and early 1970s, because of the availability of digital computers and progress in the new technical discipline known as system identification. There are many papers and books in the technical literature on this subject. Useful starting points would be the survey paper by Åström and Eykhoff [1.5], and the proceedings from the IFAC Symposia on Identification and System Parameter Estimation. The first of these was held in 1967, and others have followed in three-year intervals. The most relevant textbooks are those by Eykhoff [1.6], Goodwin and Payne [1.7], Ljung [1.8], Söderström and Stoica [1.9], Schweppe [1.10], Sage and Melsa [1.11], Hsia [1.12], and Norton [1.13]. However, none of these textbooks is aimed specifically at aircraft applications, as is this book.

Several authors made substantial contributions in the field of aircraft system identification during the late 1960s and early 1970s, e.g., Taylor et al. [1.14], Mehra [1.15], Stepner and Mehra [1.16], and Gerlach [1.17]. These contributions were mainly in the area of development and application of various estimation techniques. New challenges to aircraft system identification and parameter estimation were presented by the introduction of highly maneuverable and unstable aircraft. Some of these challenges are addressed by Klein [1.18] and Klein and Murphy [1.19]. An extensive bibliography for aircraft parameter estimation was compiled by Iliff and Maine [1.20] in 1986. Excellent theoretical and practical material on aircraft system identification is given by Maine and Iliff [1.21]-[1.22], with emphasis on the output-error method. Mulder [1.23] presented methods for experiment design, measured data compatibility, and parameter estimation. Broad overviews of aircraft system identification methods can be found in the works by Klein [1.24]-[1.25], Iliff [1.26], Hamel and Jategaonkar [1.27], and by the authors of two special issues of the *Journal of Aircraft* on applications of system identification to aircraft [1.28]-[1.29]. In 2006, the American Institute for Aeronautics and Astronautics (AIAA) published three books on aircraft system identification. One of them was the first edition of this book [1.30]. The other two were written by Jategaonkar [1.31] and Tischler and Remple [1.32], and both have been updated and expanded in second editions [1.33], [1.34]. All of these books are useful and informative. The present work contains a comprehensive explanation of aircraft system identification and widely-used methods, whereas the books by Jategaonkar [1.31], [1.33] and Tischler and Remple [1.32], [1.34] are more specialized treatments of time-domain and frequency-response approaches, respectively.

Apart from the contributions of individual authors, the following organizations must be mentioned as well: NASA Armstrong (formerly NASA Dryden) Flight Research Center, Langley Research Center, Ames Research Center, and Glenn Research Center, the Army Aeroflightdynamics Directorate at NASA Ames Research Center, the Deutsches Zentrum für Luft- und Raumfahrt (DLR) in Germany, the Delft University of Technology and the National Aerospace Laboratory (NLR) in the Netherlands, the Royal Aerospace Establishment (RAE) and later the Defense Evaluation and Research Agency (DERA) in the United Kingdom, the Aeronautical Research Laboratory (ARL) in Australia, and the National Research Council (NRC) in Canada. In these establishments, new techniques of system identification have been developed and applied to many different types of aircraft. The AIAA and the North Atlantic Treaty Organization (NATO) Advisory Group for Aerospace Research and Development (AGARD), and later the NATO Research and Technology Organization (RTO), have also played very constructive roles in the exchange of information and results, and in the education of aeronautical engineers and researchers in the area of aircraft system identification. References [1.35]-[1.39] are evidence of this, along with numerous papers presented at meetings organized by these agencies.

Based on the development of system identification methodology, it is now possible to determine the structure of aerodynamic model equations and estimate the model parameters involved, along with their confidence intervals, using data from a single flight test maneuver. Measurement noise in the output variables can be distinguished from external disturbances to the system caused by wind gusts or modeling errors. The analysis can also include prior knowledge of aircraft aerodynamic model parameters obtained from wind tunnel measurements and/or previous flight measurements. There are tools for estimating aircraft flying qualities parameters from measured pilot inputs and aircraft responses, and for obtaining more accurate measured data by reconstructing output variables and estimating systematic instrumentation errors, such as biases and scale factor errors. System identification techniques can also be used to design experiments that maximize information content in the measured data, which leads to more accurate models.

When formulating a system identification problem for aircraft (or any physical system), some general questions must be addressed:

1) What are the inputs and outputs?
2) How should the data be collected?
3) What is a reasonable form for the model to take, given the data and prior knowledge?
4) How can the unknown parameters in the model be accurately estimated based on the measured data?
5) How good is the identified model?
6) How will the results be used?

In the following chapters, theoretical and practical aspects of these questions and their answers for aircraft system identification will be explored in detail.

1.2 Outline of the Text

This book presents the requisite theory for the application of system identification methods to aircraft, including MATLAB® software to implement the methods, and some practical examples. An overview of the book is as follows.

In Chapter 2, two elements of system theory, modeling and system identification, are introduced. General mathematical model forms for dynamic systems are briefly reviewed. It is shown that in many practical situations, system identification can be reduced to parameter estimation. This is followed by an overview of aircraft system identification methodology.

Chapter 3 presents aircraft equations of motion and various forms of aircraft mathematical models, with emphasis on the aerodynamic model equations. The form of the aerodynamic model depends on the functions used to model the aerodynamic forces and moments. These functions are typically polynomials, polynomial splines, or indicial functions, computed using aircraft states and controls. The equations of motion are then simplified using analytical approximation methods or by substituting measured quantities into the nonlinear equations.

In Chapter 4, theory related to parameter estimation and state estimation is outlined. In addition to the linear and nonlinear behavior of a dynamic system, a distinction is made between models that are linear or nonlinear in their parameters. Three models for uncertainties in the parameters and the measurements are introduced. These models then lead to different estimation techniques, distinguished by their optimality criterion.

Chapter 5 deals with various forms of regression as a relatively simple technique for model structure determination and parameter estimation. Least-squares methods and various analytical and graphical diagnostic tools are introduced. The use of stepwise regression and orthogonal function theory for model structure determination is explained and demonstrated. Data partitioning is discussed as a data handling method, along with techniques for problems of near-linear dependence among model terms, known as data collinearity.

A second group of techniques often used in aircraft parameter estimation is based on maximum likelihood. Three methods are covered in Chapter 6. In the first method, the dynamic system is considered stochastic, due to the presence of noise in the dynamic equations, also called process noise. The process noise can be considered as an unmeasured input, e.g., turbulence. The second method considers the dynamic system as deterministic with measurable inputs. In both of these cases, the estimation problem is nonlinear in the parameters, regardless of whether the dynamic system model is linear or nonlinear in the states and controls. A method for accurately characterizing parameter estimate uncertainty in the context of aircraft problems is introduced and explained. The third method is an equation-error formulation of maximum likelihood parameter estimation, which has a simple and fast solution.

In Chapter 7, the previously mentioned methods for time-domain data, linear regression and maximum likelihood, are formulated in the frequency domain. As an example application, this chapter includes closed-loop dynamic modeling in the frequency domain, which is useful for flying qualities work. Another application example uses indicial function model forms to identify unsteady aerodynamic effects in the frequency domain.

Chapter 8 presents information on instrumentation, data collection, and experiment design. Flight test instrumentation requirements for aircraft system identification are given, with recommendations and rules of thumb. Input design for system identification flight test maneuvers is discussed and illustrated, along with recommendations for conducting an experimental flight test program. For highly augmented aircraft, short test times, or stringent experimental conditions, it is much more difficult to design inputs for system identification flight test maneuvers in a heuristic way. Chapter 8 explains various approaches for such situations, including optimal inputs and orthogonal optimized multisines. The chapter also includes a discussion of open-loop parameter estimation when the data are collected with the aircraft operating under closed-loop control.

In Chapter 9, measured data from aircraft sensors are used with the kinematic equations for rigid-body motion to estimate systematic instrumentation errors. After correcting the data for estimated systematic errors, the data should include only random errors. The result of this procedure, called data compatibility analysis, is more accurate data. Several methods for data compatibility analysis are explained. Chapter 9 also discusses corrections for sensor position and alignment that must be applied to the raw measurements.

The techniques covered through Chapter 7 describe methods based on having all measured data available. These are called batch methods or off-line methods. In Chapter 10, techniques are developed for updating model parameter estimates in real time, as the measured data become available. Previous estimates can be updated at each time step using recursive formulations and the current measured data, without reprocessing old data. Such formulations are called recursive or on-line. Alternatively, least squares parameter estimation in the time and frequency domains can be applied repeatedly to the most recent data. This approach is called sequential least squares. A variety of techniques for real-time modeling are discussed, including recursive least squares, extended Kalman filter, sequential least squares, modified sequential least squares, recursive orthogonalization, and sequential least squares in the frequency domain.

Important data analysis methods are explained in Chapter 11. Most of the techniques discussed could be classified as data processing. The chapter includes material on filtering, smoothing, numerical differentiation, practical computation methods for the finite Fourier transform and power spectrum estimates, signal comparisons, and data visualization.

Computer programs that implement the system identification techniques discussed in the book are described in Chapter 12. The programs are written as MATLAB® m-files [1.40] and are available by request from NASA Langley at https://software.nasa.gov/software/LAR-16100-1. Operation of the programs is explained, and demonstration examples are included.

The appendices contain background and supplemental information that is helpful for achieving a complete understanding of the material in the main body of the text. The appendices include mathematical background material (Appendix A); notes on probability, statistics, and random variables (Appendix B); reference information, including material on properties of the atmosphere, basic aerodynamics, aircraft geometry and mass properties (Appendix C); and a nonlinear F-16 simulation in MATLAB® that can be used to simulate flight testing (Appendix D).

Throughout the book, flight test data, wind tunnel data, and the F-16 nonlinear simulation are used to demonstrate various applications of system identification methodology. All of the examples are from work of the authors on aircraft system identification problems at NASA Langley Research Center.

References

[1.1] Zadeh, L.A., "From Circuit Theory to System Theory," *Proceedings of the IRE*, Vol. 50, May 1962, pp. 856-865.

[1.2] Milliken, W. F., Jr., "Progress in Stability and Control Research," *Journal of the Aeronautic Sciences*, Vol. 14, September 1947, pp. 494-519.

[1.3] Greenberg, H., "A Survey of Methods for Determining Stability Parameters of an Airplane from Dynamic Flight Measurements," NASA TN 2340, 1951.

[1.4] Shinbrot, M., "A Least Squares Curve Fitting Method with Application of the Calculation of Stability Coefficients from Transient-Response Data," NACA TN 2341, 1951.

[1.5] Åström, K.J., and Eykhoff, P., "System Identification – A Survey," *Automatica*, Vol. 7, March 1971, pp. 123-162.

[1.6] Eykhoff, P., *System Identification, Parameter and State Estimation*, John Wiley & Sons, New York, NY, 1974.

[1.7] Goodwin, G.C. and Payne, R.L., *Dynamic System Identification: Experiment Design and Data Analysis*, Academic Press, New York, NY, 1977.

[1.8] Ljung, L., *System Identification, Theory for the User*, 2nd Edition, Prentice-Hall, New York, NY, 1999.

[1.9] Söderström, T. and Stoica, P., *System Identification*, Prentice-Hall, New York, NY, 1989.

[1.10] Schweppe, F.C., *Uncertain Dynamic Systems*, Prentice-Hall, Englewood Cliffs, NJ, 1973.

[1.11] Sage, A.P. and Melsa, J.L., *System Identification*, Academic Press, New York, NY, 1971.

[1.12] Hsia, T.C., *System Identification*, Lexington Books, Lexington, MA, 1977.

[1.13] Norton, J.P., *An Introduction to Identification*, Academic Press, London, UK, 1986.

[1.14] Taylor, L.W., Iliff, K.W., and Powers, B.G., "A Comparison of Newton-Raphson and Other Methods for Determining Stability Derivatives from Flight Data," AIAA Paper 69-315, 1969.

[1.15] Mehra, R.K., "Maximum Likelihood Identification of Aircraft Parameters," *Proceedings of the Joint Automatic Control Conference*, Atlanta, GA, Paper 18-C, June 1970, pp. 442-444.

[1.16] Stepner, D.E. and Mehra, R.K., "Maximum Likelihood Identification and Optimal Input Design for Identifying Aircraft Stability and Control Derivatives," NASA CR-2200, 1973.

[1.17] Gerlach, O.H., "The Determination of Stability Derivatives and Performance Characteristics from Dynamic Maneuvers," Society of Automotive Engineers, Paper 700236, 1970.

[1.18] Klein, V., "Application of System Identification to High Performance Aircraft," *Proceedings of the 32nd IEEE Conference on Decision and Control*, San Antonio, TX, December 1993, pp. 2253-2259.

[1.19] Klein, V. and Murphy, P.C., "Aerodynamic Parameters of High Performance Aircraft Estimated from Wind Tunnel and Flight Test Data," *System Identification for Integrated Aircraft Development and Flight Testing*, RTO-MP-11, Paper 18, May 1999.

[1.20] Iliff, K.W. and Maine, R.E., "Bibliography for Aircraft Parameter Estimation," NASA TM 86804, 1986.

[1.21] Maine, R.E. and Iliff, K.W., "Identification of Dynamic Systems: Theory and Formulation," NASA RP 1138, 1985.

[1.22] Maine, R.E. and Iliff, K.W., "Application of Parameter Estimation to Aircraft Stability and Control: The Output Error Approach," NASA RP 1168, 1986.

[1.23] Mulder, J. A., "Design and Evaluation of Dynamic Flight Test Manoeuvres," Delft University of Technology, Department of Aerospace Engineering, Report LR-497, Delft, The Netherlands, 1986.

[1.24] Klein, V., "Estimation of Aircraft Aerodynamic Parameters from Flight Data," *Progress in Aerospace Sciences*, Vol. 26, No. 1, 1989, pp. 1-77.

[1.25] Klein, V., "Identification Evaluation Methods," *Parameter Identification*, AGARD-LS-104, Paper 2, 1972.

[1.26] Iliff, K.W., "Parameter Estimation for Flight Vehicles," *Journal of Guidance, Control, and Dynamics*, Vol. 12, No. 5, 1989, pp. 609-622.

[1.27] Hamel, P.G. and Jategaonkar, R., "Evolution of Flight Vehicle System Identification," *Journal of Aircraft*, Vol. 33, No. 1, January-February 1996, pp. 9-28.

[1.28] *Journal of Aircraft*, Vol. 41, No. 4, July-August 2004.

[1.29] *Journal of Aircraft*, Vol. 42, No. 1, January-February 2005.

[1.30] Klein, V. and Morelli, E.A., *Aircraft System Identification – Theory and Practice*, AIAA, Reston, VA, 2006.

[1.31] Jategaonkar, R.V., *Flight Vehicle System Identification – A Time Domain Methodology*, AIAA, Reston, VA, 2006.

[1.32] Tischler, M.B. and Remple, R.K., *Aircraft and Rotorcraft System Identification – Engineering Methods with Flight Test Examples*, AIAA, Reston, VA, 2006.

[1.33] Jategaonkar, R.V., *Flight Vehicle System Identification – A Time-Domain Methodology*, 2nd Edition, AIAA, Reston, VA, 2015.

[1.34] Tischler, M.B. and Remple, R.K., *Aircraft and Rotorcraft System Identification – Engineering Methods with Flight Test Examples*, 2nd Edition, AIAA, Reston, VA, 2012.

[1.35] *Methods for Aircraft State and Parameter Identification*, AGARD-CP-172, 1975.

[1.36] *Parameter Identification*, AGARD-LS-104, 1979.

[1.37] *Dynamic Stability Parameters*, AGARD-LS-114, 1981.

[1.38] *Rotorcraft System Identification*, AGARD-LS-178, 1991.

[1.39] *System Identification for Integrated Aircraft Development and Flight Testing*, NATO RTO-MP-11, 1999.

[1.40] *MATLAB® Primer, R2015b*, The MathWorks, Inc., Natick, MA, 2015.

Chapter 2

Elements of System Theory

The content of this book is closely related to system theory, a scientific discipline devoted to the study of mathematical properties of physical and nonphysical systems. System theory has become an extensive and rapidly-growing field which overlaps and interrelates with other fields of study, such as information theory, signal theory, stability and control theory, and others. There are many problems and areas covered by system theory, including mathematical modeling, system identification, dynamic system analysis, control system synthesis, and optimization, among others. In this chapter, two elements of system theory, mathematical modeling and system identification, will be discussed, with an emphasis on application to aircraft.

2.1 Mathematical Modeling

Mathematical modeling is the process of developing an adequate mathematical representation of some aspects of a physical system. Mathematical models can take various forms. One form that often results from direct application of physical laws is a set of differential equations relating input to output. An equivalent and more preferable form of the model is the state-space representation, which relates three variables: input, output, and state. The input $\boldsymbol{u}$, output $\boldsymbol{y}$, and state $\boldsymbol{x}$ can be vector or scalar quantities.

The input excites the system and can usually be specified by the experimenter. It is therefore an external disturbance that can be directly measured. The input must be distinguished from unmeasured disturbances, which are observed only through their influence on the system response. In some cases, the experimenter may not have the capability to specify the input, which means that inputs for normal operation of the system must be used.

The output is an observable signal indicating the system response to the input and disturbances. Typically, the output at any time t is a function of the current state and input. The state is a variable which completely specifies the status of the system at any given time. The state is not unique. For example, a change in coordinate system would give a different but equivalent state. For deterministic physical systems, knowledge of the state at time t_0, combined with knowledge of the system dynamics and inputs from time t_0 to time $t \geq t_0$, is sufficient to compute the state at time t. The state at time t reflects what has happened to the system up to time t, and can be used at time t in lieu of the system history. Mathematically, this can be expressed as:

$$x(t) = \phi\left\{x(t_0), u_{[t_0,t]}\right\} \tag{2.1}$$

$$y(t) = g\left[x(t), u(t)\right] \tag{2.2}$$

where $u_{[t_0,t]}$ denotes the input u over the time interval $[t_0, t]$. Equations (2.1) and (2.2) can represent the solution of model equations for the behavior of a dynamic system, as will be shown.

The principal categorizations of the various models for representing a physical system are: linear or nonlinear, time invariant or time varying, continuous time or discrete time, and deterministic or stochastic. These models belong to the class of parametric models. They are finite-dimensional, and can be implemented as state-space equations, differential equations, and transfer functions, among other forms. Nonparametric models, on the other hand, do not require explicit specification of the system dimension. They are inherently infinite-dimensional, and can be implemented in the form of impulse or step responses, frequency responses, correlation functions, or spectral densities, among other forms.

A short review of parametric and nonparametric models often used in the mathematical representation of physical systems is given next. The models are categorized as linear or nonlinear.

2.1.1 Linear Models

The state-space continuous-time representation of a linear, time-invariant deterministic system can be formulated as

$$\dot{x}(t) = Ax(t) + Bu(t) \qquad x(0) = x_0 \tag{2.3}$$

$$y(t) = Cx(t) + Du(t) \tag{2.4}$$

where A is the stability (or system) matrix, B is the control (or input) matrix, C and D are output transformation matrices, and x_0 is a vector of initial conditions for the state. The adjective "time-invariant" applies to the matrices A, B, C, and D, which contain constant elements. In general, the input $u \in \mathbb{R}^{n_i}$, output $y \in \mathbb{R}^{n_o}$, and state $x \in \mathbb{R}^{n_s}$ vary with time. The time scale is defined so that the initial time t_0 is zero. This will be done throughout the book.

The state equation for a scalar system with zero input (also called an autonomous or homogeneous system) reduces to:

$$\dot{x}(t) = ax(t) \qquad x(0) = x_0 \tag{2.5}$$

The solution can be found by separation of variables,

$$\int_{x_0}^{x(t)} \frac{dx}{x} = \int_0^t a\,dt \tag{2.6}$$

$$x(t) = e^{at} x_0 \tag{2.7}$$

The homogeneous form of vector equation (2.3) is

$$\dot{\boldsymbol{x}}(t) = \boldsymbol{A}\boldsymbol{x}(t) \qquad \boldsymbol{x}(0) = \boldsymbol{x}_0 \tag{2.8}$$

By analogy to the scalar case, the proposed solution is

$$\boldsymbol{x}(t) = e^{\boldsymbol{A}t} \boldsymbol{x}_0 \tag{2.9}$$

where the matrix exponential $e^{\boldsymbol{A}t}$ is defined by the infinite series

$$e^{\boldsymbol{A}t} = \boldsymbol{I} + \boldsymbol{A}t + \frac{\boldsymbol{A}^2 t^2}{2!} + \frac{\boldsymbol{A}^3 t^3}{3!} + \ldots \tag{2.10}$$

The time derivative of $e^{\boldsymbol{A}t}$ is

$$\frac{d}{dt}\left(e^{\boldsymbol{A}t}\right) = \boldsymbol{A} + \boldsymbol{A}^2 t + \frac{\boldsymbol{A}^3 t^2}{2!} + \ldots = \boldsymbol{A}e^{\boldsymbol{A}t} \tag{2.11}$$

It follows that the proposed solution in Eq. (2.9) satisfies Eq. (2.8). At the initial time $t = 0$, the state vector must equal $\boldsymbol{x}_0$. The solution in Eq. (2.9) also satisfies the initial condition, and therefore is the unique solution of the homogeneous Eq. (2.8).

Using Eq. (2.9) and assuming the state vector contains n_s elements, the solutions of Eq. (2.8) for the n_s initial condition vectors

$$\boldsymbol{x}(0) = \begin{bmatrix} 1 \\ 0 \\ 0 \\ \vdots \\ 0 \end{bmatrix}, \quad \boldsymbol{x}(0) = \begin{bmatrix} 0 \\ 1 \\ 0 \\ \vdots \\ 0 \end{bmatrix}, \quad \boldsymbol{x}(0) = \begin{bmatrix} 0 \\ 0 \\ 1 \\ \vdots \\ 0 \end{bmatrix}, \quad \ldots, \quad \boldsymbol{x}(0) = \begin{bmatrix} 0 \\ 0 \\ 0 \\ \vdots \\ 1 \end{bmatrix} \tag{2.12}$$

are the columns of the matrix $e^{\boldsymbol{A}t}$. The matrix $e^{\boldsymbol{A}t}$ is therefore analogous to the impulse response function for a scalar ordinary differential equation. Each solution for the initial conditions in Eq. (2.12) is a vector function of time, and the collection of these n_s vectors at time t make up the $n_s \times n_s$ matrix $e^{\boldsymbol{A}t}$.

The matrix e^{At} is one example of a general type of matrix called a state transition matrix, since multiplying the state at initial time $t=0$ by the state transition matrix $e^{At} = e^{A(t-0)}$ brings about a transition to the state at time t.

For the general linear state equation (2.3), the forcing function $\boldsymbol{Bu}(t)$ can be considered a sequence of impulses, the response to which can be computed using the impulse response matrix e^{At} inside a convolution integral to implement the superposition property of the linear system,

$$\boldsymbol{x}(t) = e^{At}\boldsymbol{x}_0 + \int_0^t e^{A(t-\tau)}\,\boldsymbol{B}\,\boldsymbol{u}(\tau)\,d\tau \tag{2.13}$$

Equations (2.13) and (2.4) are equivalent in form to Eqs. (2.1) and (2.2). The first term on the right side of Eq. (2.13) is the free response due to the initial conditions, and the convolution integral term is the forced response due to the input $\boldsymbol{u}(t)$. Note that the convolution integral uses $e^{A(t-\tau)}$ to compute the contribution to $\boldsymbol{x}(t)$ due to an impulse $\boldsymbol{B}\,\boldsymbol{u}(\tau)$ at time $\tau \le t$. Substituting Eq. (2.13) into Eq. (2.4), the output can be written as

$$\boldsymbol{y}(t) = \boldsymbol{C}\left[e^{At}\boldsymbol{x}_0 + \int_0^t e^{A(t-\tau)}\,\boldsymbol{B}\,\boldsymbol{u}(\tau)\,d\tau\right] + \boldsymbol{Du}(t) \tag{2.14}$$

Now define a matrix $\boldsymbol{H}(t)$ such that

$$\boldsymbol{H}(t) = \boldsymbol{C}e^{At}\boldsymbol{B} + \boldsymbol{D}\delta(t) \tag{2.15}$$

where $\delta(t)$ is the impulse function or Dirac delta function, with the properties

$$\delta(t) = \begin{cases} 0 & \text{for } t \ne 0 \\ \infty & \text{for } t = 0 \end{cases} \qquad \text{and} \qquad \int_{-\infty}^{\infty} \delta(t)\,dt = 1 \tag{2.16}$$

Then, the expression for the output in Eq. (2.14) can be written as

$$\boldsymbol{y}(t) = \boldsymbol{C}e^{At}\boldsymbol{x}_0 + \int_0^t \boldsymbol{H}(t-\tau)\,\boldsymbol{u}(\tau)\,d\tau \tag{2.17}$$

where $\boldsymbol{H}(t)$ is called a weighting function matrix.

If the convolution integral in Eq. (2.13) is taken over a small time step Δt for which the input vector $\boldsymbol{u}(t)$ can be considered constant, and assuming $\boldsymbol{A}$ is nonsingular,

$$\begin{aligned}\boldsymbol{x}(\Delta t) &= e^{A\Delta t}\boldsymbol{x}(0) + e^{A\Delta t}\int_0^{\Delta t} e^{-A\tau}\, d\tau\, \boldsymbol{B}\,\boldsymbol{u}(0) \\ &= e^{A\Delta t}\boldsymbol{x}(0) + e^{A\Delta t}\left[\boldsymbol{A}^{-1}\left(\boldsymbol{I} - e^{-A\Delta t}\right)\right]\boldsymbol{B}\,\boldsymbol{u}(0) \\ &= e^{A\Delta t}\boldsymbol{x}(0) + \boldsymbol{A}^{-1}\left(e^{A\Delta t} - \boldsymbol{I}\right)\boldsymbol{B}\,\boldsymbol{u}(0)\end{aligned} \tag{2.18}$$

where the last equality results from the fact that $e^{A\Delta t}\boldsymbol{A}^{-1} = \boldsymbol{A}^{-1}e^{A\Delta t}$, see Eq. (2.10). The same calculation for a single time step Δt can be based at any other starting time t, instead of at $t = 0$, as in Eq. (2.18). In discrete-time notation,

$$\boldsymbol{x}(i) = e^{A\Delta t}\boldsymbol{x}(i-1) + \boldsymbol{A}^{-1}\left(e^{A\Delta t} - \boldsymbol{I}\right)\boldsymbol{B}\,\boldsymbol{u}(i-1) \tag{2.19}$$

where $\boldsymbol{x}(i) \equiv \boldsymbol{x}(i\Delta t)$, etc., for nonnegative integers *i*. Equivalently,

$$\boldsymbol{x}(i) = \boldsymbol{\Phi}\,\boldsymbol{x}(i-1) + \boldsymbol{\Gamma}\boldsymbol{u}(i-1) \tag{2.20}$$

where

$$\boldsymbol{\Phi} \equiv e^{A\Delta t} \qquad \boldsymbol{\Gamma} \equiv \boldsymbol{A}^{-1}\left(e^{A\Delta t} - \boldsymbol{I}\right)\boldsymbol{B} \tag{2.21}$$

Equation (2.21) shows the connection between the continuous-time system and control matrices and the discrete-time form of these matrices. The $\boldsymbol{C}$ and $\boldsymbol{D}$ matrices in continuous time are unchanged in discrete time.

Returning to the homogeneous linear vector differential equation (2.8), consider the candidate solution

$$\boldsymbol{x}(t) = \boldsymbol{\xi}\, e^{\lambda t} \tag{2.22}$$

where λ is a scalar and $\boldsymbol{\xi}$ is a vector. Substituting into Eq. (2.8) gives

$$\lambda\boldsymbol{\xi} e^{\lambda t} = \boldsymbol{A}\boldsymbol{\xi}\, e^{\lambda t} \tag{2.23}$$

or

$$(\lambda\boldsymbol{I} - \boldsymbol{A})\,\boldsymbol{\xi} = \boldsymbol{0} \tag{2.24}$$

A nonzero solution to this set of algebraic equations exists if the determinant of the coefficient matrix equals zero,

$$\left|\lambda\boldsymbol{I} - \boldsymbol{A}\right| = 0 \tag{2.25}$$

The determinant in Eq. (2.25) is called the characteristic determinant, and the roots of Eq. (2.25), $\lambda_1, \lambda_2, \ldots, \lambda_{n_s}$ for an n_s-dimensional state vector $\boldsymbol{x}(t)$, are called the eigenvalues of the $n_s \times n_s$ matrix $\boldsymbol{A}$. If the eigenvalues are distinct, each eigenvalue λ_i, $i = 1,2,\ldots,n_s$, has a corresponding eigenvector $\boldsymbol{\xi}_i$, which is found by solving Eq. (2.24) with $\lambda = \lambda_i$. The solution corresponding to each real eigenvalue or to each complex pair of eigenvalues is called a mode. The solution of the homogeneous equation (2.8) is then a sum of the modal components,

$$\boldsymbol{x}(t) = c_1 \boldsymbol{\xi}_1 e^{\lambda_1 t} + c_2 \boldsymbol{\xi}_2 e^{\lambda_2 t} + \ldots + c_{n_s} \boldsymbol{\xi}_{n_s} e^{\lambda_{n_s} t} \tag{2.26}$$

with the values of the scalars c_i, $i = 1,2,\ldots,n_s$, determined by the requirement that the initial condition $\boldsymbol{x}(0) = \boldsymbol{x}_0$ be satisfied. This means that each c_i, $i = 1,2,\ldots,n_s$, is equal to the projection of the initial condition vector $\boldsymbol{x}_0$ along the direction of the corresponding eigenvector.

If the forcing function $\boldsymbol{B}\boldsymbol{u}(t)$ is again considered as a sequence of impulses, a similar analysis applies for the forcing function $\boldsymbol{B}\boldsymbol{u}(\tau)$ inside the convolution integral of Eq. (2.13), so that the participation of each mode in the forced solution depends on the projection of the forcing function along the eigenvectors.

It follows from the preceding discussion and Eq. (2.26) that the eigenvalues and eigenvectors are important properties of the linear dynamic system, providing information about response and stability. In general, the eigenvalues characterize the type of response associated with each mode, and the eigenvectors define the characteristic directions in state space for exciting particular modal responses.

Applying the Laplace transform (see Appendix A) to Eqs. (2.3) and (2.4) for $\boldsymbol{x}(0) = \boldsymbol{0}$ gives

$$s\,\tilde{\boldsymbol{x}}(s) = \boldsymbol{A}\tilde{\boldsymbol{x}}(s) + \boldsymbol{B}\tilde{\boldsymbol{u}}(s) \tag{2.27}$$

$$\tilde{\boldsymbol{y}}(s) = \boldsymbol{C}\tilde{\boldsymbol{x}}(s) + \boldsymbol{D}\tilde{\boldsymbol{u}}(s) \tag{2.28}$$

where s is a complex variable. The transformed state is

$$\tilde{\boldsymbol{x}}(s) = \int_0^\infty \boldsymbol{x}(t)\, e^{-st}\, dt \tag{2.29}$$

and similarly for $\tilde{\boldsymbol{y}}(s)$ and $\tilde{\boldsymbol{u}}(s)$. Solving Eqs. (2.27) and (2.28) for $\tilde{\boldsymbol{y}}(s)$,

$$\tilde{\boldsymbol{y}}(s) = \left[\boldsymbol{C}(s\boldsymbol{I} - \boldsymbol{A})^{-1}\boldsymbol{B} + \boldsymbol{D}\right]\tilde{\boldsymbol{u}}(s) = \boldsymbol{H}(s)\,\tilde{\boldsymbol{u}}(s) \tag{2.30}$$

This result is consistent with Eqs. (2.15) and (2.17) and the fact that convolution in the time domain is equivalent to multiplication in the Laplace domain (see Appendix A). Matrix $\boldsymbol{H}(s)$ is the transfer function matrix with elements

$$\left[H_{jk}\right] = \frac{num_{jk}(s)}{den(s)} \qquad \text{for} \quad \begin{cases} j = 1,2,\ldots,n_o \\ k = 1,2,\ldots,n_i \end{cases} \tag{2.31}$$

where n_o and n_i are the number of outputs and inputs, respectively. The quantities $num_{jk}(s)$ and $den(s)$ are both polynomials in s. The numerator polynomial $num_{jk}(s)$ corresponds to the transfer function of the jth output to the kth input, and the denominator polynomial $den(s)$ is the characteristic polynomial in s, which is the same for all elements of the $\boldsymbol{H}(s)$ matrix.

Setting $s = j\omega$, where $j = \sqrt{-1}$ and ω is the angular frequency, the Laplace transform changes to the Fourier transform,

$$\begin{aligned} \tilde{\boldsymbol{y}}(j\omega) &= \int_0^\infty \boldsymbol{y}(t)\, e^{-j\omega t}\, dt \\ \tilde{\boldsymbol{u}}(j\omega) &= \int_0^\infty \boldsymbol{u}(t)\, e^{-j\omega t}\, dt \end{aligned} \tag{2.32}$$

and the transfer function matrix becomes the frequency response matrix $\boldsymbol{H}(j\omega)$. This quantity can be determined experimentally, as a function of frequency. When this is done, the result is a nonparametric model in the frequency domain. This model type is discussed in Chapter 7.

Elements of the matrices $\boldsymbol{A}$, $\boldsymbol{B}$, $\boldsymbol{C}$, and $\boldsymbol{D}$ in Eqs. (2.3) and (2.4) are constant model parameters. These parameters do not depend on input $\boldsymbol{u}$, state $\boldsymbol{x}$, their derivatives, or time t. When the model parameters are time-varying, Eqs. (2.3) and (2.4) change to

$$\dot{\boldsymbol{x}}(t) = \boldsymbol{A}(t)\,\boldsymbol{x}(t) + \boldsymbol{B}(t)\,\boldsymbol{u}(t) \qquad \boldsymbol{x}(t_0) = \boldsymbol{x}_0 \tag{2.33}$$

$$\boldsymbol{y}(t) = \boldsymbol{C}(t)\,\boldsymbol{x}(t) + \boldsymbol{D}(t)\,\boldsymbol{u}(t) \tag{2.34}$$

where the time scale cannot be changed so that $t_0 = 0$, because the model depends explicitly on time. These equations represent a time-varying system. The solution is

$$\boldsymbol{y}(t) = \boldsymbol{C}(t)\boldsymbol{\Phi}(t,t_0)\boldsymbol{x}(t_0) + \int_{t_0}^{t} \boldsymbol{C}(t)\,\boldsymbol{\Phi}(t,\tau)\,\boldsymbol{B}(\tau)\,\boldsymbol{u}(\tau)\,d\tau + \boldsymbol{D}(t)\,\boldsymbol{u}(t) \tag{2.35}$$

In this case, the state transition matrix $\boldsymbol{\Phi}$ is a function of two variables: the time of application of the cause τ, and the time of observation of the effect t. The solution given by Eq. (2.35) involves a superposition integral and not a convolution integral as in Eq. (2.14). More detailed discussion of the state-space representation of a dynamic system can be found in Ref. [2.1].

Adding uncertain input disturbances to Eq. (2.33), the previously deterministic model is changed to a stochastic model,

$$\dot{\boldsymbol{x}}(t) = \boldsymbol{A}(t)\,\boldsymbol{x}(t) + \boldsymbol{B}(t)\,\boldsymbol{u}(t) + \boldsymbol{B}_w(t)\,\boldsymbol{w}(t) \tag{2.36}$$

$$\boldsymbol{y}(t) = \boldsymbol{C}(t)\,\boldsymbol{x}(t) + \boldsymbol{D}(t)\,\boldsymbol{u}(t) \tag{2.37}$$

where the vector $\boldsymbol{w}(t)$ is usually called process noise and $\boldsymbol{B}_w(t)$ is the control matrix associated with $\boldsymbol{w}(t)$. In many applications, it is assumed that $\boldsymbol{w}(t)$ is a white noise process specified by its mean and covariance matrix:

$$E\left[\boldsymbol{w}(t)\right] = \boldsymbol{0}$$

$$E\left[\boldsymbol{w}(t_i)\,\boldsymbol{w}^T(t_j)\right] = \boldsymbol{Q}(t_i)\,\delta(t_i - t_j)$$

where $E[\;]$ is the expectation operator (see Appendix B) and $\delta(t_i - t_j)$ is the Dirac delta function. To complete the modeling of stochastic system represented by Eqs. (2.36) and (2.37), it is necessary to specify the vector of initial conditions and define the interrelation between $\boldsymbol{x}(t_0)$ and $\boldsymbol{w}(t)$, usually as

$$E\left[\boldsymbol{x}(t_0)\right] = \bar{\boldsymbol{x}}_0$$

$$E\left\{\left[\boldsymbol{x}(t_0) - \bar{\boldsymbol{x}}_0\right]\left[\boldsymbol{x}(t_0) - \bar{\boldsymbol{x}}_0\right]^T\right\} = \boldsymbol{P}_0$$

$$E\left[\boldsymbol{x}(t_0)\,\boldsymbol{w}^T(t)\right] = \boldsymbol{0}$$

where $\boldsymbol{P}_0$ is a constant $n_s \times n_s$ error covariance matrix for the initial state vector. The symbol $\bar{\boldsymbol{x}}_0$ is used because the initial condition is the expected value of the stochastic quantity $\boldsymbol{x}(t_0)$.

Using the continuous-time formulation of a stochastic system creates some problems with the properties of $\boldsymbol{w}(t)$. For a fixed length of time, $\boldsymbol{w}(t)$ is a zero-mean random vector with infinite power, because the white noise power spectrum is theoretically a constant over an infinite frequency range. This means that $\boldsymbol{w}(t)$ does not exist in any physical sense.

To avoid this difficulty, it is preferable to formulate the model for a stochastic system in discrete-time form. The signals in this model are sampled at $t_0 + i\Delta t,\ i = 0,1,2,\ldots$, with a constant time increment Δt between samples. Then, using the shorthand notation

$$\boldsymbol{x}(i) \equiv \boldsymbol{x}(t_0 + i\Delta t) \qquad i = 0,1,2,\ldots \tag{2.38}$$

and similarly for $\boldsymbol{u}(i)$ and $\boldsymbol{w}(i)$, the discrete-time stochastic state-space model is given by

$$\boldsymbol{x}(i) = \boldsymbol{\Phi}(i-1)\,\boldsymbol{x}(i-1) + \boldsymbol{\Gamma}(i-1)\,\boldsymbol{u}(i-1) + \boldsymbol{\Gamma}_w(i-1)\,\boldsymbol{w}(i-1) \tag{2.39}$$

$$\boldsymbol{y}(i) = \boldsymbol{C}(i)\,\boldsymbol{x}(i) + \boldsymbol{D}(i)\,\boldsymbol{u}(i) \qquad i = 1,2,\ldots \tag{2.40}$$

with

$$E\left[\boldsymbol{x}(0)\right] = \bar{\boldsymbol{x}}_0$$

$$E\left[\boldsymbol{w}(i)\right] = \boldsymbol{0}$$

$$E\left[\boldsymbol{w}(i)\,\boldsymbol{w}^T(j)\right] \approx \boldsymbol{Q}(i)\,\delta_{ij}\,\Delta t$$

where δ_{ij} equals 1 for $i = j$, and zero for $i \neq j$, and the last expression is an approximation for small Δt.

The solution to Eqs. (2.39) and (2.40) is discussed in Chapter 4 in connection with state estimation.

2.1.2 Nonlinear Models

Most real-world systems are nonlinear. If these systems operate over a restricted range of conditions, then linear models can be used to approximate the nonlinear behavior. When such an approximation is not possible, a suitable nonlinear model must be postulated. For a stochastic, time-varying system, the model equations take the form

$$\dot{\boldsymbol{x}}(t) = \boldsymbol{f}\left[\boldsymbol{x}(t), \boldsymbol{u}(t), \boldsymbol{w}(t), t\right] \tag{2.41}$$

$$\boldsymbol{y}(t) = \boldsymbol{g}\left[\boldsymbol{x}(t), \boldsymbol{u}(t), t\right] \tag{2.42}$$

As for a linear system, these equations must be augmented with a model for $\boldsymbol{x}(t_0)$ and $\boldsymbol{w}(t)$. In the case of a deterministic, time-invariant system, Eqs. (2.41) and (2.42) are simplified to

$$\dot{\boldsymbol{x}}(t) = \boldsymbol{f}\left[\boldsymbol{x}(t), \boldsymbol{u}(t)\right] \qquad \boldsymbol{x}(0) = \boldsymbol{x}_0 \tag{2.43}$$

$$y(t) = g\left[x(t), u(t)\right] \tag{2.44}$$

In general, the solution of nonlinear differential equations must be computed using numerical methods such as fourth-order Runge-Kutta (e.g., see Appendix D).

For a wide class of nonlinear systems, a general nonparametric representation is the Volterra series, see Ref. [2.2]. The general form of the Volterra series for a scalar system can be expressed as

$$y(t) = \sum_{n=1}^{\infty} y_n(t) \tag{2.45}$$

where $y_n(t)$ is the nth functional defined as

$$\begin{aligned}
y_1(t) &= \int_0^t h_1(\tau_1) u(t-\tau_1)\, d\tau_1 \\
y_2(t) &= \int_0^t \int_0^t h_2(\tau_1, \tau_2) u(t-\tau_1) u(t-\tau_2)\, d\tau_1\, d\tau_2 \\
&\vdots \\
y_n(t) &= \int_0^t \int_0^t \cdots \int_0^t h_n(\tau_1, \tau_2, \ldots, \tau_n) u(t-\tau_1) \ldots u(t-\tau_n)\, d\tau_1 \ldots d\tau_n
\end{aligned} \tag{2.46}$$

The terms

$$h_1(\tau_1),\ h_2(\tau_1, \tau_2),\ \ldots,\ h_n(\tau_1, \tau_2, \ldots, \tau_n)$$

are called impulse responses or kernels. The description using Volterra series is therefore a direct generalization of the model for a linear system using a convolution integral. Practical use of the Volterra series for nonlinear modeling requires truncating the infinite series, usually after the first two or three terms.

More about mathematical models for dynamic systems can be found in Refs. [2.1]-[2.3].

2.2 System Identification and Parameter Estimation

System identification is the determination of a mathematical model from measured input-output data. The system identification problem is characterized by the selected inputs and outputs, the class of models from which the model will be chosen, and a criterion for equivalence of the model and the physical system. This is consistent with the definition of system identification quoted in Chapter 1.

System identification requires an experiment where the inputs typically are specified, taking into account prior knowledge about the system and the purpose of the identification. The class of models entering system identification procedures is also selected based on prior knowledge and the purpose of the identification, but also on data from the experiment.

In common practice, the equivalence of the observed physical system output z and the model output y is quantified in terms of a scalar cost function,

$$J = J(z, y) \tag{2.47}$$

Normally, the cost function J consists of a weighted sum of squared differences between z and y.

The relationship between the variables z and y is expressed by the measurement equation,

$$z = y + \boldsymbol{\nu} \tag{2.48}$$

where $\boldsymbol{\nu}$ is the measurement error, also called the measurement noise, which is assumed to be random. Measurement noise cannot be measured directly in the modeling context, so its properties are assumed. Measurement noise properties can also be estimated from measured data using filtering and smoothing techniques explained in Chapter 11.

Finding the best model based on the cost function in Eq. (2.47) could lead to the investigation of a large number of model candidates. To simplify the problem, system identification is changed to an optimization problem that requires finding a model $\mathfrak{M}$, selected from a class of models, such that $\mathfrak{M}$ minimizes the cost function. In many situations, a class of models can be limited to the models $\mathfrak{M}^*$ that have the same structure (mathematical form), but are distinguished by different values of the parameters $\boldsymbol{\theta}$ in the model,

$$\mathfrak{M}^* = \{\mathfrak{M}(\boldsymbol{\theta})\} \tag{2.49}$$

Data collected from the experiment is denoted by

$$\boldsymbol{Z}_N = \left[z(1), \boldsymbol{u}(1), z(2), \boldsymbol{u}(2), \ldots, z(N), \boldsymbol{u}(N)\right] \tag{2.50}$$

where N is the number of data points. Once the model structure is selected, system identification becomes the selection of a value for $\boldsymbol{\theta}$, based on the information in $\boldsymbol{Z}_N$, that minimizes a scalar cost function

$$J = J\left[\boldsymbol{Z}_N, \boldsymbol{Y}_N(\boldsymbol{\theta})\right] \tag{2.51}$$

where $\boldsymbol{Y}_N(\boldsymbol{\theta})$ represents the model outputs,

$$\boldsymbol{Y}_N(\boldsymbol{\theta}) = \left[y(1), y(2), \ldots, y(N)\right] \tag{2.52}$$

which depend on the parameter vector $\boldsymbol{\theta}$. Thus, system identification is reduced to model parameter estimation. Such a formulation makes it possible to exploit methods of statistical inference, mainly estimation theory. The selection of an estimation method will be influenced by assumptions made for measurement noise and whether the parameters are assumed to be random variables or unknown constants. Three principal estimation techniques will be explained in Chapter 4. Important references for the topic of parameter estimation are the books by Sorenson [2.3], Eykhoff [2.4], Ljung [2.5], and Schweppe [2.6].

Results from the parameter estimation process should include parameter estimates and their properties, including error bounds, and information for testing various statistical hypotheses. Some additional questions might arise during the identification process, such as the following:

1) Is the information content in the data sufficient to distinguish among different models?
2) Is the model formulated so that unique parameter values can be found?
3) Do the estimated parameters have physically realistic values and small error bounds?

These questions are related to the concept of identifiability, which is central to the identification procedure. Identifiability is discussed in a rigorous way in Ref. [2.5]. In this book, it will be mentioned mainly in connection with experiment design, model structure determination, and in examples.

2.3 Aircraft System Identification

When system identification is applied to an aircraft, the equations governing the aircraft dynamic motion are postulated and an experiment is designed to obtain measurements of input and output variables. The equations of motion for an aircraft come from the translational and rotational forms of Newton's second law of motion. Chapter 3 contains a detailed derivation of these equations, which describe the translational motion of the aircraft center of gravity and the rotational motion about the center of gravity. In vector form, the equations are:

$$m\dot{\boldsymbol{V}} + \boldsymbol{\omega} \times m\boldsymbol{V} = \boldsymbol{F}_G(\boldsymbol{\varsigma}) + \boldsymbol{F}_T + \boldsymbol{F}_A(\boldsymbol{V},\boldsymbol{\omega},\boldsymbol{u},\boldsymbol{\theta}) \tag{2.53}$$

$$\boldsymbol{I}\dot{\boldsymbol{\omega}} + \boldsymbol{\omega} \times \boldsymbol{I}\boldsymbol{\omega} = \boldsymbol{M}_T + \boldsymbol{M}_A(\boldsymbol{V},\boldsymbol{\omega},\boldsymbol{u},\boldsymbol{\theta}) \tag{2.54}$$

where m is the aircraft mass, $\boldsymbol{I}$ is the inertia tensor, $\boldsymbol{\varsigma}$ is a vector of Euler angles indicating the attitude of the aircraft relative to fixed earth axes, $\boldsymbol{V}$ and $\boldsymbol{\omega}$ are translational and angular velocity vectors for the aircraft motion, and $\boldsymbol{u}$ is the control vector. Applied forces in these equations come from gravity $(\boldsymbol{F}_G)$, thrust $(\boldsymbol{F}_T)$, and aerodynamics $(\boldsymbol{F}_A)$. Applied moments are generated by thrust $(\boldsymbol{M}_T)$ and aerodynamics $(\boldsymbol{M}_A)$. The gravity force is modeled by adding kinematic differential equations to describe the aircraft attitude relative to earth axes, assuming a constant gravitational acceleration vector. Usually, the applied forces and moments due to thrust are modeled using results from engine tests done on the ground, along with the geometry of the engine installation. Aircraft system identification then reduces to the determination of a model structure for the aerodynamic forces $(\boldsymbol{F}_A)$ and moments $(\boldsymbol{M}_A)$, and the estimation of unknown parameters in those model structures, based on measured data. The quantity $\boldsymbol{\theta}$ is a vector of parameters that in the present formulation specifies aerodynamic characteristics of the aircraft. The model structures identified for $\boldsymbol{F}_A$ and $\boldsymbol{M}_A$ are referred to as the aerodynamic model equations.

The dynamic equations (2.53) and (2.54) are augmented with output equations that specify the connection of aircraft states and controls to measured outputs, along with measurement equations describing the measurement process. The complete set of all these equations can be written as

$$\dot{\boldsymbol{x}}(t) = \boldsymbol{f}\left[\boldsymbol{x}(t), \boldsymbol{u}(t), \boldsymbol{\theta}\right] \qquad \boldsymbol{x}(0) = \boldsymbol{x}_0 \tag{2.55}$$

$$\boldsymbol{y}(t) = \boldsymbol{g}\left[\boldsymbol{x}(t), \boldsymbol{u}(t), \boldsymbol{\theta}\right] \tag{2.56}$$

$$\boldsymbol{z}(i) = \boldsymbol{y}(i) + \boldsymbol{v}(i) \qquad i = 1, 2, \ldots, N \tag{2.57}$$

where $\boldsymbol{x}$ is composed of $\boldsymbol{V}$, $\boldsymbol{\omega}$, and $\boldsymbol{\varsigma}$. The control vector $\boldsymbol{u}$ is generally composed of throttle position and control surface deflections. Elements of the output vector $\boldsymbol{y}$ are aircraft response variables, which usually include state variables. Discrete measured outputs $\boldsymbol{z}(i)$ are corrupted by measurement noise $\boldsymbol{v}(i)$.

Aircraft system identification can be defined as the determination, from input and output measurements, of model structures for $\boldsymbol{F}_A$ and $\boldsymbol{M}_A$, and estimation of the unknown parameters $\boldsymbol{\theta}$ contained in those model structures. In many practical applications, the structure of the models for the aerodynamic forces and moments is assumed to be known, and the system identification problem reduces to parameter estimation. The most common situation is that the aerodynamic forces and moments depend linearly on current values of the states and controls, leading to linear time-invariant aerodynamic model formulations. Chapter 3 details the various forms of the models for aerodynamic forces and moments.

In the literature, the process properly known as parameter estimation is often referred to as parameter identification, or an abbreviation of that term,

PID. It is also common to see the term "system identification" applied to a process that is really parameter estimation.

Aircraft system identification includes model postulation, experiment design, data compatibility analysis, model structure determination, parameter and state estimation, collinearity diagnostics, and model validation. These steps are necessary to identify a mathematical description of the functional dependence of the applied aerodynamic forces and moments on aircraft motion and control variables. A block diagram depicting the general approach to aircraft system identification is shown in Fig. 2.1. Each of the steps in the procedure will be described briefly.

Model Postulation

Model postulation is based on prior knowledge about the aircraft dynamics and aerodynamics. The postulated model influences the type of flight test maneuver used for system identification. It is common practice to express the aerodynamic forces and moments in terms of linear expansions, polynomials, or polynomial spline functions in the states and controls, with time-invariant parameters quantifying the contribution of each to the total aerodynamic force or moment. This formulation has been extended to cases with unsteady aerodynamic effects modeled by indicial functions or additional state equations. Formulations of the aerodynamic model and various forms of the equations of motion are covered in Chapter 3.

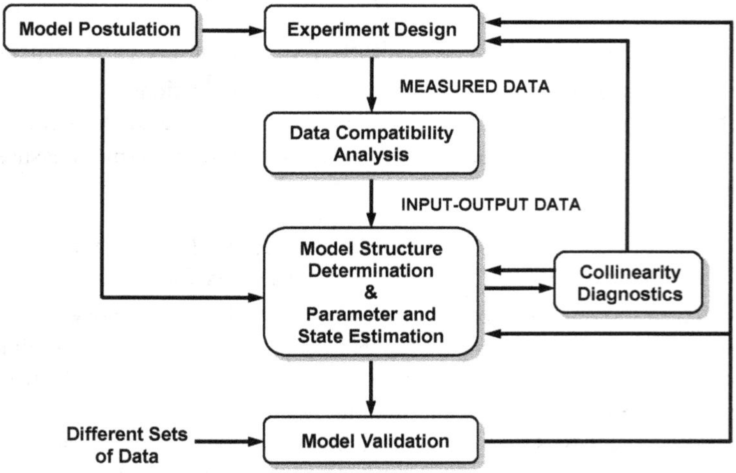

Figure 2.1 **Block diagram of aircraft system identification**

Experiment Design

Experiment design includes specification of the instrumentation system, aircraft configuration, flight conditions, and maneuvers for system identification. The instrumentation system is primarily required to measure input and output variables at regular sampling intervals during the maneuver. Input variables are throttle position and control surface deflections for open-loop or bare-airframe modeling. The output variables include quantities specifying the magnitude and direction of the air-relative velocity (airspeed, angle of attack, and sideslip angle), angular velocities, translational and angular accelerations, and Euler attitude angles. In addition to these variables, quantities defining flight conditions and aircraft configuration are recorded.

An important aspect of the experiment design is the selection of input forms for the flight test maneuvers. The input influences aircraft response, which in turn influences the accuracy of the system identification from flight measurements. Attempts to obtain parameter estimates with high accuracy in the most efficient manner has led researchers to the development of optimized inputs for aircraft parameter estimation. This and other experiment design issues are discussed in Chapter 8.

Data Compatibility Analysis

In practice, measured aircraft response data can contain systematic errors, even after careful instrumentation and experimental procedure. To verify data accuracy, data compatibility analysis can be applied to measured aircraft responses. Data compatibility analysis includes aircraft state estimation based on known rigid-body kinematics and available sensor measurements, estimation of systematic instrumentation errors, and a comparison of reconstructed responses with measured responses.

The state equations for the data compatibility analysis are kinematic relationships among the measured aircraft responses, and the model parameters typically are constant biases and scale factor errors for the sensors. The estimation techniques employed are similar to those used in estimation of aircraft aerodynamic parameters. Data compatibility analysis is described in Chapter 9.

Model Structure Determination

Model structure determination in aircraft system identification means selecting a specific form for the model from a class of models, based on measured data. For example, this might involve choosing an appropriate polynomial expansion in the aircraft motion and control variables to model a component of aerodynamic force acting on the aircraft, from the class of all possible polynomial models of order two or less. The model should be parsimonious to retain good prediction capability, while still adequately representing the physical phenomena. An adequate model will fit the data well, facilitate the successful estimation of unknown parameters associated with model terms whose existence can be substantiated, and exhibit good prediction capability.

Several techniques for selection of an aerodynamic model have been developed, and two of these are described in Chapter 5. In one of the techniques, known as stepwise regression, the determination of a model proceeds in three steps: postulation of terms which might enter the model, selection of an adequate model based on statistical metrics, and validation of the selected model. The other technique generates multivariate orthogonal modeling functions from the data to facilitate model structure determination. The orthogonality of the modeling functions makes it possible to automate the first two of the three steps listed above for model structure determination. Retained orthogonal functions can be decomposed without error into ordinary functions for the final model form.

Parameter and State Estimation

Four items are needed for implementation of aircraft system identification: an informative experiment, measured input-output data, a mathematical model of the aircraft being tested, and an estimation technique. Parameter and state estimation constitute a principal part of the aircraft system identification procedure.

Currently, two methods – equation error and output error – are used for most aircraft parameter estimation. The equation-error method is based on linear regression using the ordinary least squares principle. The unknown aerodynamic parameters are estimated by minimizing the sum of squared differences between measured and modeled aerodynamic forces and moments. Linear regression constitutes a linear estimation problem, meaning that the model output is linearly dependent on the model parameters. This simplifies the optimization required to find parameter estimates to the solution of an overdetermined set of linear equations, which can be found using well-known techniques from linear algebra.

In the output-error method, the unknown parameters are obtained by minimizing the sum of weighted square differences between the measured aircraft outputs and model outputs. The estimation problem is nonlinear because the unknown parameters appear in the equations of motion, which are integrated to compute the states. Outputs are computed from the states, controls, and parameters, using the output equations. Iterative nonlinear optimization techniques are required to solve this nonlinear estimation problem.

Theoretically, either the equation-error method or output-error method can be a maximum likelihood estimator, which means that the cost function optimization used for computing the unknown parameters is equivalent to maximizing the probability density function associated with the outcome from the experiment. In addition, both equation-error and output-error parameter estimation can be considered special cases of a more general approach based on Bayes's rule. These methods will be addressed in Chapters 4, 5, and 6.

Parameter estimation for linear dynamic systems based on maximum likelihood and the least squares principle can also be formulated in the frequency domain. In this approach, the measured data are first transformed from the time domain to the frequency domain using the Fourier transform. Parameter estimation methods can be applied to transformed input and output data, frequency response curves, or power spectral densities. The last two forms can also be used in nonparametric estimation methods. Both parameter estimation and nonparametric estimation in the frequency domain are covered in Chapter 7.

Sometimes during flight testing, the aircraft is subjected to random external disturbances, e.g., turbulence. In this case, the model becomes stochastic, and the states must be estimated, in addition to model parameters. The Kalman filter is predominantly used to estimate states in the case of a linear dynamic model. For a nonlinear dynamic model, the extended Kalman filter can be applied to estimate the states. The extended Kalman filter can also serve as a technique for obtaining simultaneous estimates of states and parameters, regardless of whether the dynamic model is linear or nonlinear. Chapters 4, 6, 7, and 10 contain material related to the use of the Kalman filter in aircraft system identification.

An overview of various estimation methods is given in Chapter 4. More detailed development of the methods and their practical application are discussed in Chapters 5, 6, and 7. Chapter 10 is concerned with real-time implementations of the estimation algorithms.

Collinearity Diagnostics

In almost all practical applications of linear regression, the model terms are correlated to some extent. Usually, the levels of correlation are low, and therefore not problematic. However, in some situations, the model terms are almost linearly related. When this happens, the problem of data collinearity exists, and inferences about the model based on the data can be misleading, or completely wrong.

The ability to diagnose data collinearity is important to users of linear regression or other parameter estimation techniques. Such a diagnostic consists of two basic steps: 1) detecting the presence of collinearity among the model terms, and 2) assessing the extent to which these relationships have adversely affected estimated parameters. Then, diagnostic information can aid in deciding what corrective actions are necessary and worthwhile. Data collinearity diagnostics, analysis, and remedies are discussed in Chapter 5.

Model Validation

Model validation is the last step in the identification process, and should be applied regardless of how the model was found. The identified model must have parameters with physically reasonable values and acceptable accuracy, and the model must exhibit good prediction capability on comparable maneuvers. Flight-determined parameter estimates should be compared with any available information about the aircraft aerodynamics, which can include theoretical predictions, wind tunnel measurements, or estimates from previous flight measurements using different maneuvers and/or different estimation techniques. During these comparisons, the limitations and accuracy of theoretical calculations, wind tunnel measurements, and the flight results must be taken into consideration.

Prediction capability of an identified model is checked on data not used in the identification process. The measured input for the prediction data is applied to the identified model to compute predicted responses, which are then compared with measured values. The differences between predicted values from the model and measured values should be random in nature, indicating that all deterministic components in the measured output have been characterized by the identified model. Examples of model validation are presented throughout the book.

2.4 Summary and Concluding Remarks

This chapter briefly outlined two elements of system theory that are relevant for this book – mathematical modeling and system identification.

An aircraft is a nonlinear dynamic system, so the mathematical model for the dynamic motion consists of systems of nonlinear differential equations. When the motion of the aircraft is restricted to small perturbations, the aircraft can be modeled as a linear dynamic system. In this chapter, state-space and transfer function mathematical models for continuous-time linear dynamic systems were developed, along with the relationship between these two model forms. The relationship between continuous-time and discrete-time forms of linear state-space dynamic models was also explained. In practical cases, the outputs from any of these models for given inputs and initial conditions are found numerically, because the inputs are measured time series that are not easily characterized analytically.

Once the decision has been made to model aircraft dynamic motion with a particular class of model, e.g., a continuous-time linear state-space model, the next tasks are to determine the structure of that model type, including deciding which inputs, outputs, and states will be included and determining the number and role of unknown parameters to be estimated, followed by estimating the unknown model parameters based on measured data. These steps, called model structure determination and parameter estimation, respectively, are important parts of system identification. System identification also includes other tasks that support and interact with model structure determination and parameter estimation. These tasks include experiment design, data compatibility analysis, collinearity diagnostics, and model validation. Brief descriptions of these aspects of system identification were included in the chapter.

The intent of this chapter was to give some background on the mathematical model forms used to characterize dynamic systems such as an aircraft, and to give a general overview of system identification applied to aircraft. Subsequent chapters will fill in details of the theory and application. The next chapter begins this process by explaining in detail the mathematical model forms used for aircraft system identification.

References

[2.1] DeRusso, P.M., Roy, R.J., and Close, C.M., *State Variables for Engineers*, John Wiley & Sons, New York, NY, 1965.

[2.2] Unbehauen, H. and Rao, G.P., *Identification of Continuous Systems*, North-Holland, New York, NY, 1987.

[2.3] Sorenson, H.W., *Parameter Estimation: Principles and Problems*, Marcel-Decker, New York, NY, 1980.

[2.4] Eykhoff, P., *System Identification, Parameter and State Estimation*, John Wiley & Sons, New York, NY, 1974.

[2.5] Ljung, L., *System Identification, Theory for the User*, 2nd Ed., Prentice-Hall, New York, NY, 1999.

[2.6] Schweppe, F.C., *Uncertain Dynamic Systems*, Prentice-Hall, Englewood Cliffs, NJ, 1973.

Problems

2.1 Answer the following, using words, graphs, equations, and/or diagrams:

a) Discuss two reasons that dynamic models based on flight data might be desired.

b) Does the transfer function model as formulated in this chapter include initial condition response? Why or why not?

c) Figure 2.1 includes several feedback and feedforward paths in the overall system identification process. Discuss some possible reasons or situations that might cause one or more of these paths to be exercised.

2.2 Trim and linearize the F-16 nonlinear simulation longitudinal dynamics at 5 degrees angle of attack and 10,000 feet altitude using gen_f16_model.m.

a) Find the eigenvalues of the system using MATLAB® function eig.m.

b) Convert the longitudinal state-space model to an equivalent transfer function model.

c) Convert the longitudinal state-space model to a discrete-time state-space model.

2.3 Repeat problem **2.2** for lateral dynamics at the same flight condition.

2.4 For the linear dynamic system in problem **2.2**, use lsims.m to demonstrate that a small bias error in the stabilator measurement at the trim condition will lead to a drift with time in the computed states and outputs.

Chapter 3

Mathematical Model of an Aircraft

Aircraft system identification is mainly concerned with providing a mathematical description for the aerodynamic forces and moments in terms of relevant quantities such as control surface deflections, angular velocities, airspeed or Mach number, and the orientation of the aircraft to the relative wind. Aerodynamic parameters quantify the functional dependence of the aerodynamic forces and moments on these quantities.

Estimation of aerodynamic parameters from flight test data requires that a mathematical model of the aircraft be postulated. The mathematical model includes both the aircraft equations of motion and the equations for aerodynamic forces and moments, known as the aerodynamic model equations. In this chapter, the equations of motion will be formulated as ordinary differential equations for the aircraft states, with algebraic equations for the measured outputs. The aerodynamic model equations will be developed first using linear terms, polynomials, and polynomial splines with time-invariant parameters, then generalized to include time-dependent terms representing unsteady aerodynamic effects.

Continuous-time differential equations are used almost exclusively for aircraft system identification. The main reasons are that the form of the aircraft equations of motion (which will be derived in this chapter) is known, and the parameters appearing in the continuous-time differential equations have physical significance for aircraft stability and control. It is therefore of interest to estimate the values of these parameters, and their associated uncertainties. Furthermore, results from wind tunnel tests and analytic computations are typically given as values of the physical parameters, and these values are often used as *a priori* information or for comparison with results from flight data analysis. Although it is possible to use discrete-time equations for flight data analysis, the parameters in discrete-time models are not the same as the physical parameters, and this introduces additional complexity.

The development given here is for conventional airplanes. Some fairly straightforward modifications can be made to model aircraft that vary from a conventional airplane in areas such as the controls available and the aircraft configuration. Throughout the book, the assumption that the aircraft is a rigid body is adopted. There is a vast literature concerning modeling the dynamics of flexible vehicles, including methods to address the important problems of flutter and structural divergence. However, in many practical applications of aircraft system identification, the aircraft can be approximated as a rigid body.

3.1 Reference Frames and Sign Conventions

Before developing the aircraft equations of motion, it is necessary to define reference frames and sign conventions. The reference frames required for studying aircraft dynamics and system identification are defined below. All reference frames are right-handed with mutually orthogonal axes.

3.1.1 Reference Frames

Inertial Axes

The origin of this reference frame is fixed or moving with a constant velocity relative to the distant stars, and the orientation is arbitrary and fixed. Newton's laws apply in an inertial reference frame defined this way.

Earth Axes $O x_E \, y_E \, z_E$

Origin is at an arbitrary point on the earth surface, with positive $O x_E$ axis pointing toward geographic north, positive $O y_E$ axis pointing east, and positive $O z_E$ axis pointing to the center of the earth. Earth axes are fixed with respect to the earth. For most aircraft system identification work, earth axes are assumed to be inertial axes, which is equivalent to ignoring the motion of the earth relative to the distant stars.

Vehicle-Carried Earth Axes $O x_V \, y_V \, z_V$

Origin is at the aircraft center of gravity (c.g.), orientation of the axes is parallel to earth axes. The center of gravity is the point about which the aircraft would balance if suspended by a cable. This reference frame is used to conveniently show the rotational orientation of the aircraft relative to earth axes.

Body Axes $O x y z$

Origin is at the aircraft c.g., with positive $O x$ axis pointing forward through the nose of the aircraft, positive $O y$ axis out the right wing, and positive $O z$ axis through the underside. The $O x z$ plane is usually a plane of symmetry for the aircraft. Body axes are fixed with respect to the aircraft body (see Fig. 3.1). It follows that the position vector between specified points on a rigid body is constant in body axes. Many of the variables associated with aircraft motion are referenced to body axes.

Stability Axes $O x_s \, y_s \, z_s$

Stability axes are a type of body axes, so they are fixed with respect to the aircraft. The orientation of stability axes is related to a reference flight condition, usually defined at the start of a maneuver. Origin is at the aircraft c.g., with positive Ox_s axis forward and aligned with the projection of the velocity vector of the aircraft c.g. through the air (also called the air-relative velocity) onto the Oxz plane in body axes. The positive Oy_s axis points out the right wing, and the positive Oz_s axis is directed through the underside.

Wind axes $Ox_w y_w z_w$

Origin is at the aircraft c.g., with positive Ox_w axis forward and aligned with the air-relative velocity vector, positive Oy_w axis out the right side of the aircraft, and positive Oz_w axis through the underside in the Oxz plane in body axes. The origin of the wind axes traces out the trajectory of the aircraft through the air. The Ox_w wind axis is always tangent to the air-relative trajectory, so wind axes are not fixed with respect to the aircraft body.

A typical jet fighter aircraft is shown in Fig. 3.1. Components of the translational velocity vector, angular velocity vector, applied aerodynamic force vector, and applied aerodynamic moment vector are expressed in components along the body axes. The notation shown is standard:

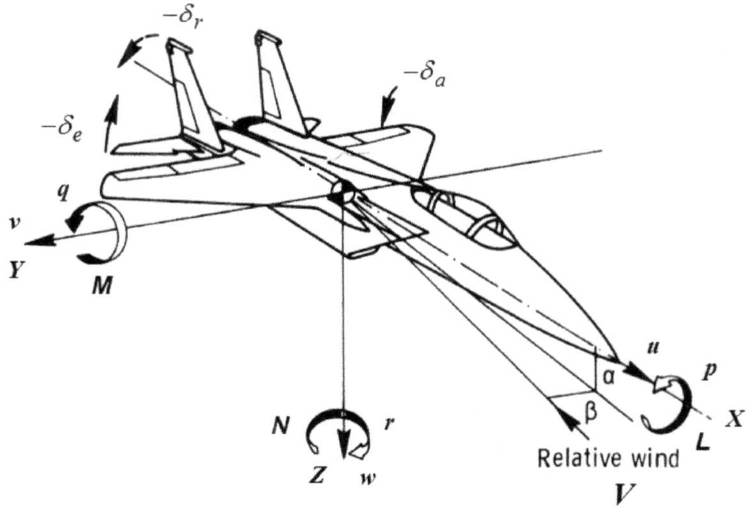

Figure 3.1 **Airplane notation and sign conventions**

u, v, w = body-axis components of aircraft velocity relative to earth axes
p, q, r = body-axis components of aircraft angular velocity
X, Y, Z = body-axis components of aerodynamic force acting on the aircraft
L, M, N = body-axis components of aerodynamic moment acting on the aircraft

3.1.2 Sign Conventions

For angular velocities or applied moments, the sign convention follows the right-hand rule. If the right hand thumb is pointed in the direction of a positive axis, the fingers curl in the direction of positive rotation. Angular velocities or applied moments about the x, y, and z body axes are described with the adjectives roll, pitch, and yaw, respectively.

Control surfaces are hinged surfaces that can be rotated about a hinge line to change the applied aerodynamic forces and moments on an aircraft. Conventional airplanes have elevator δ_e, aileron δ_a, and rudder δ_r control surfaces, see Fig. 3.2. These controls are intended primarily to produce moments about the body y (pitch) axis, x (roll) axis, and z (yaw) axis, respectively. A variety of other control surfaces may appear instead of or in addition to these three basic controls, such as flaps, strakes, canards, speed brakes, stabilators, and spoilers. Some aircraft also have the capability to change the line of action of the thrust, which is called thrust vectoring. Thrust vectoring controls are fundamentally different from aerodynamic control surfaces in that the change in the line of action of the thrust is the source of the applied force and moment change, rather than aerodynamics.

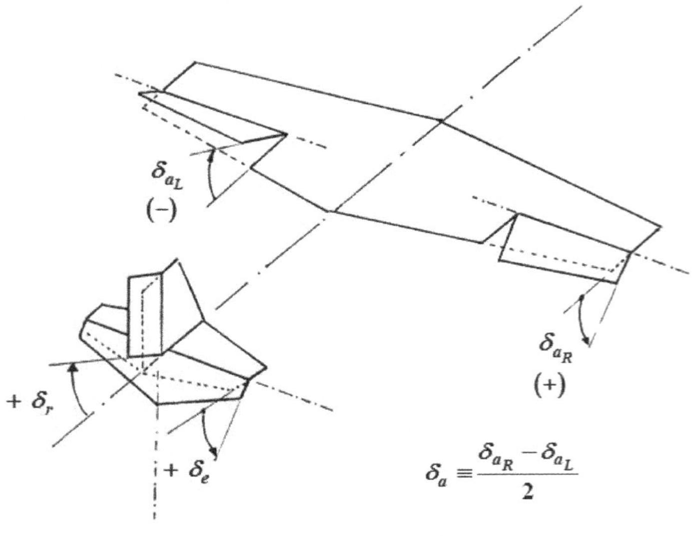

Figure 3.2 **Control surface sign conventions**

Individual control surface deflections also follow the right-hand rule. Some control surfaces, like the ailerons, are deflected simultaneously in an asymmetric manner, which means that the individual aileron control surfaces on each wing move in opposite directions. This requires another way to define a positive aileron deflection. Although there is no universal standard, in this text the aileron deflection is defined as one half the right aileron deflection minus the left aileron deflection,

$$\delta_a \equiv \frac{1}{2}\left(\delta_{a_R} - \delta_{a_L}\right) \tag{3.1}$$

Figure 3.2 illustrates asymmetric aileron deflection. Other asymmetric control surface deflections are defined similarly. Pairs of individual control surfaces can also be deflected symmetrically, which means the individual control surfaces are deflected together in the same direction with the same magnitude. The sign convention for control surface deflections is usually such that a positive control surface deflection produces a negative aerodynamic moment on the aircraft.

3.2 Rigid-Body Equations of Motion

A general formulation of aircraft flight dynamics would consider the dynamics of an elastic vehicle with varying mass density and moving component subsystems, subject to aerodynamic, propulsive, and gravitational forces, flying in nonstationary air. Common simplifying assumptions (used for most of this text) are:

1) The aircraft is a rigid body with fixed mass distribution and constant mass.
2) The air is at rest relative to the earth (no steady wind, gusts, or wind shears).
3) The earth is fixed in inertial space.
4) Flight in the earth atmosphere is close to the earth surface (on an astronomical scale), so the earth surface can be approximated as flat.
5) Gravity is uniform, so that the aircraft c.g. and the center of mass are coincident, and gravitational forces do not change with altitude.

The assumption that the aircraft is a rigid body means that dynamic effects due to fuel slosh, structural deformations, and relative motion of control surfaces are assumed negligible. The general motion of an aircraft can then be described by Newton's second law of motion in translational and rotational forms:

$$\boldsymbol{F} = \frac{d}{dt}(m\boldsymbol{V}) \tag{3.2}$$

$$\boldsymbol{M} = \frac{d}{dt}(\boldsymbol{I\omega}) \tag{3.3}$$

where $\boldsymbol{F}$ is the applied force, $m\boldsymbol{V}$ is the linear momentum, m is the mass, $\boldsymbol{V}$ is the translational velocity, $\boldsymbol{M}$ is the applied moment about the c.g., $\boldsymbol{I\omega}$ is the angular momentum about the c.g., $\boldsymbol{\omega}$ is the angular velocity, and $\boldsymbol{I}$ is the inertia matrix. Eqs. (3.2) and (3.3) are vector equations describing the translational motion of the c.g. and the rotational motion about the c.g., respectively. Each vector equation represents 3 scalar equations for the vector components, giving a total of 6 scalar equations for 6 degrees of freedom for the aircraft motion.

Equations (3.2) and (3.3) are valid in an inertial reference frame, but it is convenient to express the individual quantities in terms of their components in the body axes, which will generally be translating and rotating relative to inertial axes. This is because most measurements are made in the body-axis system, and the inertia matrix $\boldsymbol{I}$ is constant in body axes, but would be a function of time in inertial axes. Body-axis components of the quantities in Eqs. (3.2) and (3.3) are:

$$\boldsymbol{F} = \begin{bmatrix} F_x \\ F_y \\ F_z \end{bmatrix} \qquad \boldsymbol{V} = \begin{bmatrix} u \\ v \\ w \end{bmatrix} \tag{3.4}$$

$$\boldsymbol{M} = \begin{bmatrix} M_x \\ M_y \\ M_z \end{bmatrix} \qquad \boldsymbol{I} = \begin{bmatrix} I_x & -I_{xy} & -I_{xz} \\ -I_{yx} & I_y & -I_{yz} \\ -I_{zx} & -I_{zy} & I_z \end{bmatrix} \qquad \boldsymbol{\omega} = \begin{bmatrix} p \\ q \\ r \end{bmatrix} \tag{3.5}$$

where

$$I_x \equiv \int_{Volume} \left(y^2 + z^2\right) dm \qquad I_y \equiv \int_{Volume} \left(x^2 + z^2\right) dm \qquad I_z \equiv \int_{Volume} \left(x^2 + y^2\right) dm$$

$$I_{xy} \equiv \int_{Volume} xy\,dm = I_{yx} \qquad I_{yz} \equiv \int_{Volume} yz\,dm = I_{zy} \qquad I_{xz} \equiv \int_{Volume} xz\,dm = I_{zx}$$

$$\tag{3.6}$$

Sign conventions for the components of $\boldsymbol{F}$, $\boldsymbol{V}$, $\boldsymbol{M}$, and $\boldsymbol{\omega}$ are consistent with sign conventions for the body axes and the right-hand rule. The components of the angular velocity vector $\boldsymbol{\omega}$ are roll rate p (positive right wing down), pitch rate q (positive nose up), and yaw rate r (positive nose right).

The quantities x, y, and z inside the integrals in Eqs. (3.6) are body-axis coordinates of mass elements dm, which together compose the aircraft. Many texts contain derivations of the expressions in Eq. (3.6), including Etkin [3.1], Etkin and Reid [3.2], McGruer, Ashkenas, and Graham [3.3], Stevens and Lewis [3.4], and Roskam [3.5]. These references are also resources for the derivation of the rigid-body equations of motion given next.

From the definitions in Eqs. (3.6), it is clear that for a rigid body with symmetry relative to the Oxz plane in body axes, the inertia matrix $\boldsymbol{I}$ is symmetric, and $I_{xy} = I_{yx} = I_{yz} = I_{zy} = 0$. The inertia matrix then reduces to

$$\boldsymbol{I} = \begin{bmatrix} I_x & 0 & -I_{xz} \\ 0 & I_y & 0 \\ -I_{xz} & 0 & I_z \end{bmatrix} \tag{3.7}$$

so

$$\boldsymbol{I\omega} = \begin{bmatrix} I_x p - I_{xz} r \\ I_y q \\ -I_{xz} p + I_z r \end{bmatrix} \tag{3.8}$$

Note that translational velocity $\boldsymbol{V}$ and angular velocity $\boldsymbol{\omega}$ represent the aircraft motion relative to inertial axes, but expressed in body-axis components. All aircraft motion relative to body axes is zero by definition of the body axes.

For rotating axis systems like the body axes, the derivative operator applied to vectors has two parts – one that accounts for the rate of change of the vector components expressed in the rotating system, and one that accounts for the axis system rotation (see Ref. [3.2]),

$$\frac{d}{dt}(\cdot) = \frac{\delta}{\delta t}(\cdot) + \boldsymbol{\omega} \times (\cdot) \tag{3.9}$$

Combining Eqs. (3.2), (3.3), and (3.9) with the rigid body and constant mass assumptions, and using a dot superscript notation for $\delta/\delta t$,

$$\boldsymbol{F} = m\dot{\boldsymbol{V}} + \boldsymbol{\omega} \times m\boldsymbol{V} \tag{3.10}$$

$$\boldsymbol{M} = \boldsymbol{I}\dot{\boldsymbol{\omega}} + \boldsymbol{\omega} \times \boldsymbol{I\omega} \tag{3.11}$$

Equations (3.10) and (3.11) are the vector forms of the equations of motion written in body axes. Substituting Eqs. (3.4), (3.5), (3.7), and (3.8) into Eqs. (3.10) and (3.11) gives the body-axis component form of the equations:

Force Equations

$$F_x = m(\dot{u} + qw - rv) \tag{3.12a}$$

$$F_y = m(\dot{v} + ru - pw) \tag{3.12b}$$

$$F_z = m(\dot{w} + pv - qu) \tag{3.12c}$$

Moment Equations

$$M_x = \dot{p}I_x - \dot{r}I_{xz} + qr(I_z - I_y) - qpI_{xz} \tag{3.13a}$$

$$M_y = \dot{q}I_y + pr(I_x - I_z) + (p^2 - r^2)I_{xz} \tag{3.13b}$$

$$M_z = \dot{r}I_z - \dot{p}I_{xz} + pq(I_y - I_x) + qrI_{xz} \tag{3.13c}$$

For airplanes, the applied forces and moments on the left sides of the above equations arise from aerodynamics, gravity, and propulsion. Since gravity acts through the c.g., and the gravity field is assumed uniform, there is no gravity moment acting on the airplane. Equations (3.10) and (3.11) can therefore be written as:

$$\boldsymbol{F}_A + \boldsymbol{F}_T + \boldsymbol{F}_G = m\dot{\boldsymbol{V}} + \boldsymbol{\omega} \times m\boldsymbol{V} \tag{3.14}$$

$$\boldsymbol{M}_A + \boldsymbol{M}_T = \boldsymbol{I}\dot{\boldsymbol{\omega}} + \boldsymbol{\omega} \times \boldsymbol{I}\boldsymbol{\omega} \tag{3.15}$$

Aerodynamics

Aerodynamic forces and moments acting on the aircraft result from the relative motion of the air and the aircraft. Components of the aerodynamic forces and moments can be expressed in terms of nondimensional coefficients:

$$\boldsymbol{F}_A = \bar{q}S \begin{bmatrix} C_X \\ C_Y \\ C_Z \end{bmatrix} \tag{3.16}$$

$$\boldsymbol{M}_A = \bar{q}S \begin{bmatrix} bC_l \\ \bar{c}C_m \\ bC_n \end{bmatrix} \tag{3.17}$$

where $\bar{q} = (1/2)\rho V^2$ is the dynamic pressure, V is the magnitude of the air-relative velocity (also called the airspeed), ρ is the air density, S is the wing reference area, b is the wing span, and $\bar{c}$ is the mean aerodynamic chord of the wing (see Appendix C).

In general, the nondimensional aerodynamic force and moment coefficients depend nonlinearly on the aircraft translational and angular velocity vector components and the control surface deflections, plus possibly their time derivatives, and/or other nondimensional quantities, such as Mach number and Reynolds number. A discussion of how this dependence can be characterized mathematically is given later in the chapter.

<u>Gravity</u>

Aircraft weight is assumed constant in both magnitude and direction relative to earth axes, acting along the z earth axis. The components of the aircraft weight along the body axes change with orientation of the aircraft relative to earth axes. Gravity components in body axes therefore depend on the aircraft orientation relative to earth axes, and this dependence can be described based on the relative orientation of the body axes to vehicle-carried earth axes. Aircraft orientation with respect to vehicle-carried earth axes can be described in many ways, but the most common method is using Euler angles.

Figure 3.3 shows how the orientation of one right-handed coordinate system can be defined relative to another. The sequence for rotating vehicle-carried earth axes into alignment with body axes is a yaw angle rotation ψ about the Oz_V axis, followed by a pitch angle rotation θ about an intermediate y axis, completed by a roll angle rotation ϕ about the Ox body axis.

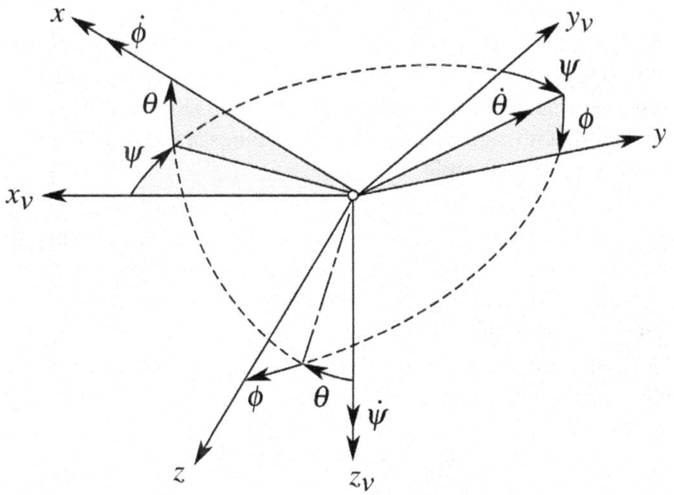

Figure 3.3 **Rotation from vehicle-carried earth axes to body axes**

Components of the gravity vector in body axes can be found by means of a product of three rotation matrices (see Appendix A):

$$\boldsymbol{F}_G = \begin{bmatrix} g_x \\ g_y \\ g_z \end{bmatrix}_B = \begin{bmatrix} 1 & 0 & 0 \\ 0 & \cos\phi & \sin\phi \\ 0 & -\sin\phi & \cos\phi \end{bmatrix} \begin{bmatrix} \cos\theta & 0 & -\sin\theta \\ 0 & 1 & 0 \\ \sin\theta & 0 & \cos\theta \end{bmatrix} \begin{bmatrix} \cos\psi & \sin\psi & 0 \\ -\sin\psi & \cos\psi & 0 \\ 0 & 0 & 1 \end{bmatrix} \begin{bmatrix} 0 \\ 0 \\ g \end{bmatrix}_V$$

$$\begin{bmatrix} g_x \\ g_y \\ g_z \end{bmatrix}_B = \begin{bmatrix} -g\sin\theta \\ g\sin\phi\cos\theta \\ g\cos\phi\cos\theta \end{bmatrix}$$

so,

$$\boldsymbol{F}_G = m\begin{bmatrix} g_x \\ g_y \\ g_z \end{bmatrix}_B = \begin{bmatrix} -mg\sin\theta \\ mg\sin\phi\cos\theta \\ mg\cos\phi\cos\theta \end{bmatrix} \tag{3.18}$$

Propulsion

Assuming the thrust from the propulsion system acts along the x body axis and through the c.g., the thrust appears only as an applied force along the x body axis,

$$\boldsymbol{F}_T = \begin{bmatrix} T \\ 0 \\ 0 \end{bmatrix} \tag{3.19a}$$

When the line of action of the thrust is not directed along the aircraft longitudinal body axis and through the c.g., thrust terms appear in other applied force and moment body-axis components. This would be the case for aircraft equipped with thrust vectoring or engines mounted below the wings and below the c.g., for example.

Sometimes it is necessary to account for the effect of rotating mass in the propulsion system, e.g., propellers or rotors of jet engines. The gyroscopic terms associated with the rotating mass must be considered an applied moment, because the equations of motion have been formulated assuming the aircraft is a rigid body with no internal moving parts.

Assuming the thrust acts along the x body axis of the aircraft, the angular momentum of the rotating mass in body axes is

$$\boldsymbol{h}_p = \begin{bmatrix} I_p\Omega_p & 0 & 0 \end{bmatrix}^T \tag{3.19b}$$

where I_p is the inertia of the rotating mass and Ω_p is the angular velocity. If the angular velocity of the rotating mass is constant, then $I_p\dot{\Omega}_p = 0$, and the gyroscopic moment from the rotating mass in the propulsion system is given by [cf. Eq. (3.9)],

$$\boldsymbol{M}_T = \frac{d}{dt}(\boldsymbol{h}_p) = \boldsymbol{\omega} \times \boldsymbol{h}_p = \begin{bmatrix} 0 & -r & q \\ r & 0 & -p \\ -q & p & 0 \end{bmatrix} \begin{bmatrix} I_p\Omega_p \\ 0 \\ 0 \end{bmatrix} = \begin{bmatrix} 0 \\ I_p\Omega_p r \\ -I_p\Omega_p q \end{bmatrix} \quad (3.19c)$$

The components in Eq. (3.19c) are added on the left side of the moment equations (3.13).

<u>Applied Forces and Moments</u>

Combining the expressions for the aerodynamic forces and moments with Eqs. (3.18) and (3.19), the body-axis components of the applied forces and moments are:

$$F_x = \overline{q}SC_X - mg\sin\theta + T \quad (3.20a)$$

$$F_y = \overline{q}SC_Y + mg\cos\theta\sin\phi \quad (3.20b)$$

$$F_z = \overline{q}SC_Z + mg\cos\theta\cos\phi \quad (3.20c)$$

$$M_x = \overline{q}SbC_l \quad (3.21a)$$

$$M_y = \overline{q}S\overline{c}C_m + I_p\Omega_p r \quad (3.21b)$$

$$M_z = \overline{q}SbC_n - I_p\Omega_p q \quad (3.21c)$$

Substituting the preceding expressions into the dynamic equations (3.12)-(3.13) gives

<u>Force Equations</u>

$$m\dot{u} = m(rv - qw) + \overline{q}SC_X - mg\sin\theta + T \quad (3.22a)$$

$$m\dot{v} = m(pw - ru) + \overline{q}SC_Y + mg\cos\theta\sin\phi \quad (3.22b)$$

$$m\dot{w} = m(qu - pv) + \overline{q}SC_Z + mg\cos\theta\cos\phi \quad (3.22c)$$

Moment Equations

$$\dot{p}I_x - \dot{r}I_{xz} = \bar{q}SbC_l - qr\left(I_z - I_y\right) + qpI_{xz} \tag{3.23a}$$

$$\dot{q}I_y = \bar{q}S\bar{c}C_m - pr\left(I_x - I_z\right) - \left(p^2 - r^2\right)I_{xz} + I_p\Omega_p r \tag{3.23b}$$

$$\dot{r}I_z - \dot{p}I_{xz} = \bar{q}SbC_n - pq\left(I_y - I_x\right) - qrI_{xz} - I_p\Omega_p q \tag{3.23c}$$

3.3 Rotational Kinematic Equations

The rotational kinematic equations relate the rate of change of the Euler angles to the body-axis components of angular velocity. The relationships can be found by applying the technique in Appendix A for expressing vector components in a rotated axis system,

$$\begin{bmatrix} p \\ q \\ r \end{bmatrix} = \begin{bmatrix} \dot{\phi} \\ 0 \\ 0 \end{bmatrix} + \begin{bmatrix} 1 & 0 & 0 \\ 0 & \cos\phi & \sin\phi \\ 0 & -\sin\phi & \cos\phi \end{bmatrix} \begin{bmatrix} 0 \\ \dot{\theta} \\ 0 \end{bmatrix}$$
$$+ \begin{bmatrix} 1 & 0 & 0 \\ 0 & \cos\phi & \sin\phi \\ 0 & -\sin\phi & \cos\phi \end{bmatrix} \begin{bmatrix} \cos\theta & 0 & -\sin\theta \\ 0 & 1 & 0 \\ \sin\theta & 0 & \cos\theta \end{bmatrix} \begin{bmatrix} 0 \\ 0 \\ \dot{\psi} \end{bmatrix} \tag{3.24}$$

or

$$\begin{bmatrix} p \\ q \\ r \end{bmatrix} = \begin{bmatrix} 1 & 0 & -\sin\theta \\ 0 & \cos\phi & \sin\phi\cos\theta \\ 0 & -\sin\phi & \cos\phi\cos\theta \end{bmatrix} \begin{bmatrix} \dot{\phi} \\ \dot{\theta} \\ \dot{\psi} \end{bmatrix} \tag{3.25}$$

Inverting the last relationship gives differential equations for the Euler angles, which describe the rotational kinematics:

$$\dot{\phi} = p + \tan\theta\left(q\sin\phi + r\cos\phi\right) \tag{3.26a}$$

$$\dot{\theta} = q\cos\phi - r\sin\phi \tag{3.26b}$$

$$\dot{\psi} = \frac{q\sin\phi + r\cos\phi}{\cos\theta} \tag{3.26c}$$

Equations (3.22), (3.23), and (3.26) are coupled nonlinear first-order differential equations for nine aircraft states: three translational velocity components *u*, *v*, and *w*; three angular velocity components *p*, *q*, and *r*; and three Euler angles ϕ, θ, and ψ.

3.4 Navigation Equations

The navigation equations are written by expressing the aircraft velocity vector in earth axes, starting with body-axis components,

$$\begin{bmatrix} \dot{x}_E \\ \dot{y}_E \\ \dot{z}_E \end{bmatrix} = \begin{bmatrix} cos\psi & -sin\psi & 0 \\ sin\psi & cos\psi & 0 \\ 0 & 0 & 1 \end{bmatrix} \begin{bmatrix} cos\theta & 0 & sin\theta \\ 0 & 1 & 0 \\ -sin\theta & 0 & cos\theta \end{bmatrix} \begin{bmatrix} 1 & 0 & 0 \\ 0 & cos\phi & -sin\phi \\ 0 & sin\phi & cos\phi \end{bmatrix} \begin{bmatrix} u \\ v \\ w \end{bmatrix} \quad (3.27)$$

Introducing altitude (height above the ground) $h = -z_E$,

$$\begin{aligned} \dot{x}_E &= u \, cos\psi \, cos\theta + v \left(cos\psi \, sin\theta \, sin\phi - sin\psi \, cos\phi \right) \\ &\quad + w \left(cos\psi \, sin\theta \, cos\phi + sin\psi \, sin\phi \right) \end{aligned} \quad (3.28a)$$

$$\begin{aligned} \dot{y}_E &= u \, sin\psi \, cos\theta + v \left(sin\psi \, sin\theta \, sin\phi + cos\psi \, cos\phi \right) \\ &\quad + w \left(sin\psi \, sin\theta \, cos\phi - cos\psi \, sin\phi \right) \end{aligned} \quad (3.28b)$$

$$\dot{h} = u \, sin\theta - v \, cos\theta \, sin\phi - w \, cos\theta \, cos\phi \quad (3.28c)$$

3.5 Force Equations in Wind Axes

Aircraft sensors measure airspeed $V \equiv |V|$, angle of attack α, and sideslip angle β or flank angle β_f, rather than body-axis velocities *u*, *v*, and *w*. In addition, the nondimensional aerodynamic force and moment coefficients are generally characterized as functions of α, β, and Mach number V/a, where a is the speed of sound in air. Therefore, it is often useful to write the force equations in terms of V, α, and β instead of *u*, *v*, and *w*. To do this, first note the definition of α and β shown in Figs. 3.1 and 3.4. The sequence of rotating the wind axes into alignment with body axes is a negative β rotation about the Oz_w wind axis, followed by a positive α rotation about the Oy body axis. These two aerodynamic angles are defined as

$$\alpha = tan^{-1}\left(\frac{w}{u}\right) \quad (3.29)$$

$$\beta = sin^{-1}\left(\frac{v}{V}\right) \quad (3.30)$$

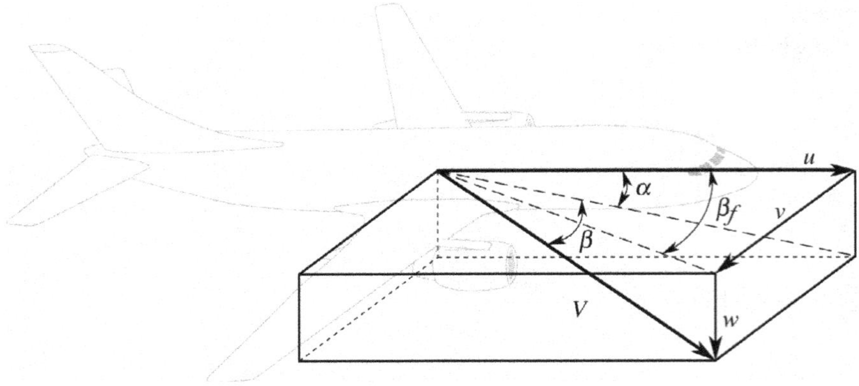

Figure 3.4 **Aerodynamic angle definitions**

and the airspeed V is given by

$$V \equiv |\mathbf{V}| = \sqrt{u^2 + v^2 + w^2} \tag{3.31}$$

Body-axis velocity components are related to V, α, and β by

$$u = V \cos\alpha \cos\beta \tag{3.32a}$$

$$v = V \sin\beta \tag{3.32b}$$

$$w = V \sin\alpha \cos\beta \tag{3.32c}$$

Differentiating Eqs. (3.29) to (3.31) with respect to time gives

$$\dot{V} = \frac{1}{V}\left(u\dot{u} + v\dot{v} + w\dot{w}\right) \tag{3.33a}$$

$$\dot{\alpha} = \left(\frac{u\dot{w} - w\dot{u}}{u^2 + w^2}\right) \tag{3.33b}$$

$$\dot{\beta} = \left(\frac{V\dot{v} - v\dot{V}}{V^2}\right)\left[\frac{1}{\sqrt{1-(v/V)^2}}\right] = \left[\frac{V\dot{v} - v\dot{V}}{V\sqrt{u^2 + w^2}}\right]$$

$$= \frac{\left(u^2 + v^2 + w^2\right)\dot{v} - v\left(u\dot{u} + v\dot{v} + w\dot{w}\right)}{V^2\sqrt{u^2 + w^2}} \tag{3.33c}$$

$$= \frac{\left(u^2 + w^2\right)\dot{v} - v\left(u\dot{u} + w\dot{w}\right)}{V^2\sqrt{u^2 + w^2}}$$

Substituting in Eqs. (3.33) for $\dot{u}$, $\dot{v}$, and $\dot{w}$ from Eqs. (3.22), and for u, v, and w from Eqs. (3.32), gives the force equations written in terms of V, α, and β:

$$\begin{aligned}\dot{V} &= -\frac{\bar{q}S}{m}C_{D_W} + \frac{T}{m}\cos\alpha\cos\beta \\ &\quad + g\left(\cos\phi\cos\theta\sin\alpha\cos\beta + \sin\phi\cos\theta\sin\beta - \sin\theta\cos\alpha\cos\beta\right)\end{aligned} \tag{3.34a}$$

$$\begin{aligned}\dot{\alpha} &= -\frac{\bar{q}S}{mV\cos\beta}C_L + q - \tan\beta\left(p\cos\alpha + r\sin\alpha\right) \\ &\quad + \frac{g}{V\cos\beta}\left(\cos\phi\cos\theta\cos\alpha + \sin\theta\sin\alpha\right) - \frac{T\sin\alpha}{mV\cos\beta}\end{aligned} \tag{3.34b}$$

$$\begin{aligned}\dot{\beta} &= \frac{\bar{q}S}{mV}C_{Y_W} + p\sin\alpha - r\cos\alpha + \frac{g}{V}\cos\beta\sin\phi\cos\theta \\ &\quad + \frac{\sin\beta}{V}\left(g\cos\alpha\sin\theta - g\sin\alpha\cos\phi\cos\theta - \frac{T\cos\alpha}{m}\right)\end{aligned} \tag{3.34c}$$

where

$$C_L = -C_Z\cos\alpha + C_X\sin\alpha \tag{3.35a}$$

$$C_D = -C_X\cos\alpha - C_Z\sin\alpha \tag{3.35b}$$

$$\begin{aligned}C_{D_W} &= -C_X\cos\alpha\cos\beta - C_Y\sin\beta - C_Z\sin\alpha\cos\beta \\ &= C_D\cos\beta - C_Y\sin\beta\end{aligned} \tag{3.35c}$$

$$\begin{aligned}C_{Y_W} &= -C_X\cos\alpha\sin\beta + C_Y\cos\beta - C_Z\sin\alpha\sin\beta \\ &= C_Y\cos\beta + C_D\sin\beta\end{aligned} \tag{3.35d}$$

The nondimensional coefficients on the left sides of Eqs. (3.35) are obtained from body-axes components by rotation through α and β. Positive lift coefficient C_L and drag coefficient C_D are directed along the $-z_S$ and $-x_S$ stability axes, respectively, while positive wind-axes drag coefficient C_{D_W} and side force coefficient C_{Y_W} are directed along the $-x_W$ and $+y_W$ wind axes, respectively.

3.6 Collected Equations of Motion

The equations of motion in body axes, containing force and moment dynamic equations, and the rotational kinematic equations, are:

Force Equations

$$\dot{u} = rv - qw + \frac{\overline{q}S}{m}C_X - g\sin\theta + \frac{T}{m} \tag{3.36a}$$

$$\dot{v} = pw - ru + \frac{\overline{q}S}{m}C_Y + g\cos\theta\sin\phi \tag{3.36b}$$

$$\dot{w} = qu - pv + \frac{\overline{q}S}{m}C_Z + g\cos\theta\cos\phi \tag{3.36c}$$

Moment Equations

$$\dot{p} - \frac{I_{xz}}{I_x}\dot{r} = \frac{\overline{q}Sb}{I_x}C_l - \frac{\left(I_z - I_y\right)}{I_x}qr + \frac{I_{xz}}{I_x}qp \tag{3.37a}$$

$$\dot{q} = \frac{\overline{q}S\overline{c}}{I_y}C_m - \frac{\left(I_x - I_z\right)}{I_y}pr - \frac{I_{xz}}{I_y}\left(p^2 - r^2\right) + \frac{I_p}{I_y}\Omega_p r \tag{3.37b}$$

$$\dot{r} - \frac{I_{xz}}{I_z}\dot{p} = \frac{\overline{q}Sb}{I_z}C_n - \frac{\left(I_y - I_x\right)}{I_z}pq - \frac{I_{xz}}{I_z}qr - \frac{I_p}{I_z}\Omega_p q \tag{3.37c}$$

Kinematic Equations

$$\dot{\phi} = p + \tan\theta\left(q\sin\phi + r\cos\phi\right) \tag{3.38a}$$

$$\dot{\theta} = q\cos\phi - r\sin\phi \tag{3.38b}$$

$$\dot{\psi} = \frac{q\sin\phi + r\cos\phi}{\cos\theta} \tag{3.38c}$$

Wind Axes Force Equations

$$\dot{V} = -\frac{\overline{q}S}{m}C_{D_W} + \frac{T}{m}\cos\alpha\cos\beta + g\left(\cos\phi\cos\theta\sin\alpha\cos\beta + \sin\phi\cos\theta\sin\beta - \sin\theta\cos\alpha\cos\beta\right) \quad (3.39a)$$

$$\dot{\alpha} = -\frac{\overline{q}S}{mV\cos\beta}C_L + q - \tan\beta\left(p\cos\alpha + r\sin\alpha\right) - \frac{T\sin\alpha}{mV\cos\beta} + \frac{g}{V\cos\beta}\left(\cos\phi\cos\theta\cos\alpha + \sin\theta\sin\alpha\right) \quad (3.39b)$$

$$\dot{\beta} = \frac{\overline{q}S}{mV}C_{Y_W} + p\sin\alpha - r\cos\alpha + \frac{g}{V}\cos\beta\sin\phi\cos\theta + \frac{\sin\beta}{V}\left(g\cos\alpha\sin\theta - g\sin\alpha\cos\phi\cos\theta - \frac{T\cos\alpha}{m}\right) \quad (3.39c)$$

where

$$C_D = -C_X\cos\alpha - C_Z\sin\alpha \quad (3.40a)$$

$$C_L = -C_Z\cos\alpha + C_X\sin\alpha \quad (3.40b)$$

$$C_{D_W} = C_D\cos\beta - C_Y\sin\beta \quad (3.40c)$$

$$C_{Y_W} = C_Y\cos\beta + C_D\sin\beta \quad (3.40d)$$

Conversely,

$$C_X = -C_D\cos\alpha + C_L\sin\alpha \quad (3.40e)$$

$$C_Z = -C_L\cos\alpha - C_D\sin\alpha \quad (3.40f)$$

$$C_D = C_{D_W}\cos\beta + C_{Y_w}\sin\beta \quad (3.40g)$$

$$C_Y = C_{Y_W}\cos\beta - C_{D_w}\sin\beta \quad (3.40h)$$

State-Space Moment Equations

$$\dot{p} = \left(c_1 r + c_2 p - c_4 I_p \Omega_p\right)q + \overline{q}Sb\left(c_3 C_l + c_4 C_n\right) \quad (3.41a)$$

$$\dot{q} = \left(c_5 p + c_7 I_p \Omega_p\right)r - c_6\left(p^2 - r^2\right) + c_7\,\overline{q}S\overline{c}C_m \quad (3.41b)$$

$$\dot{r} = \left(c_8 p - c_2 r - c_9 I_p \Omega_p\right)q + \overline{q}Sb\left(c_9 C_n + c_4 C_l\right) \quad (3.41c)$$

where $c_1, c_2, \ldots, c_9$ = inertia constants dependent on body-axis moments of inertia,

$$c_1 = \left[\left(I_y - I_z\right)I_z - I_{xz}^2\right]/\Gamma \qquad \Gamma = I_x I_z - I_{xz}^2$$

$$c_2 = \left[\left(I_x - I_y + I_z\right)I_{xz}\right]/\Gamma \qquad c_3 = I_z/\Gamma$$

$$c_4 = I_{xz}/\Gamma \qquad c_5 = \left(I_z - I_x\right)/I_y \tag{3.42}$$

$$c_6 = I_{xz}/I_y \qquad c_7 = 1/I_y$$

$$c_8 = \left[\left(I_x - I_y\right)I_x + I_{xz}^2\right]/\Gamma \qquad c_9 = I_x/\Gamma$$

Navigation Equations

The navigation equations can be written in terms of V, α, and β, by substituting Eqs. (3.32) into Eqs. (3.28),

$$\begin{aligned}\dot{x}_E &= V\cos\alpha\cos\beta\cos\psi\cos\theta + V\sin\beta\left(\cos\psi\sin\theta\sin\phi - \sin\psi\cos\phi\right)\\ &\quad + V\sin\alpha\cos\beta\left(\cos\psi\sin\theta\cos\phi + \sin\psi\sin\phi\right)\end{aligned} \tag{3.43a}$$

$$\begin{aligned}\dot{y}_E &= V\cos\alpha\cos\beta\sin\psi\cos\theta + V\sin\beta\left(\sin\psi\sin\theta\sin\phi + \cos\psi\cos\phi\right)\\ &\quad + V\sin\alpha\cos\beta\left(\sin\psi\sin\theta\cos\phi - \cos\psi\sin\phi\right)\end{aligned} \tag{3.43b}$$

$$\dot{h} = V\cos\alpha\cos\beta\sin\theta - V\sin\beta\cos\theta\sin\phi - V\sin\alpha\cos\beta\cos\theta\cos\phi \tag{3.43c}$$

The equations in this section were developed under these assumptions:

1) The earth is fixed in inertial space (i.e., earth axes are an inertial reference frame).
2) The aircraft is a rigid body.
3) Aircraft mass and mass distribution are constant.
4) The aircraft is symmetric about the Oxz plane in body axes.
5) The atmosphere is fixed relative to earth axes.
6) The earth has negligible curvature ("flat earth").
7) Gravitational acceleration is constant in magnitude and direction.
8) Thrust is directed along the x body axis and through the aircraft c.g.

The equations of motion can be expressed in the general form of a nonlinear first-order vector differential equation for the aircraft state,

$$\dot{\boldsymbol{x}} = \boldsymbol{f}(\boldsymbol{x},\boldsymbol{u}) \tag{3.44}$$

where $\boldsymbol{x}$ is a vector of state variables $u, v, w, p, q, r, \phi, \theta, \psi, x_E, y_E, h$ or $V, \beta, \alpha, p, q, r, \phi, \theta, \psi, x_E, y_E, h$, and $\boldsymbol{u}$ is a vector of input variables that usually is composed of throttle position and control surface deflections. The input variables are not explicitly shown in the preceding collected force and moment equations, but are included implicitly, because they influence the thrust and the aerodynamic forces and moments acting upon the aircraft.

3.7 Output Equations

Output variables for the mathematical model of an aircraft are the measured aircraft responses $V,\alpha,\beta,p,q,r,\phi,\theta,\psi,h,a_x,a_y,a_z,\dot{p},\dot{q}$, and $\dot{r}$. The quantities x_E and y_E, which define aircraft x and y position relative to earth axes, are not included, because these position coordinates are not relevant to aircraft dynamics. Altitude has an indirect effect on aircraft dynamics, which is manifested through changes in the air density, and consequently the dynamic pressure. Neither altitude h nor heading angle ψ affects aircraft dynamics directly, but these quantities are important piloting parameters that are usually measured. Additionally, these measurements can be used in data compatibility analysis to check other measurements, as discussed in Chapter 9. For these reasons, the altitude h and heading angle ψ are retained as aircraft states in the development given here.

The output equations specify the analytical connection between the output variables and the aircraft states, state derivatives, and controls. For specifying the output equations, the following assumptions are made:

1) Sensor calibrations are known and have been applied to the raw measurements.

2) Measurements are corrected for misalignment with the body axes, and are corrected to the aircraft c.g.

3) Sensor dynamics are negligible.

Chapter 9 contains a detailed discussion of the steps that need to be taken so that the preceding assumptions are applicable.

In general, the output equations take the following form:

$$\boldsymbol{y} = \boldsymbol{g}(\boldsymbol{x},\dot{\boldsymbol{x}},\boldsymbol{u}) \tag{3.45}$$

The precise form of the output equations in (3.45) depends on which aircraft responses are included as outputs, the aircraft instrumentation, and how the equations of motion are formulated. For example, if the force equations are written in body axes [cf. Eqs. (3.36)], and the output variables are $V, \alpha,$ and β , then the output equations are:

$$V = \sqrt{u^2 + v^2 + w^2} \tag{3.46a}$$

$$\alpha = tan^{-1}\left(\frac{w}{u}\right) \tag{3.46b}$$

$$\beta = sin^{-1}\left(\frac{v}{V}\right) \tag{3.46c}$$

These relationships, which were introduced earlier as Eqs. (3.29)-(3.31), show a nonlinear dependence of the output variables on the states. On the other hand, if the force equations are written in wind axes [cf. Eqs. (3.39)], the outputs $V, \alpha,$ and β are also states, so the output equations are simply:

$$V = V \tag{3.47a}$$

$$\alpha = \alpha \tag{3.47b}$$

$$\beta = \beta \tag{3.47c}$$

which indicate a linear relationship between the outputs and the states. Similar output equations could be written for $p, q, r, \phi, \theta, \psi, \dot{p}, \dot{q},$ and $\dot{r}$, because all of these quantities are either states or state derivatives, which are available directly from the equations of motion.

Airspeed V is usually obtained from dynamic pressure measurements. The aerodynamic angles α and β are normally measured using wind vanes mounted on the aircraft. These vanes directly measure angle of attack α in the Oxz plane, and flank angle β_f in the Oxy plane, see Fig. 3.4. The flank angle β_f can be expressed as:

$$\beta_f = tan^{-1}\left(\frac{v}{u}\right) \tag{3.48}$$

The sideslip angle β is defined as the angle between the velocity vector and its projection onto the Oxz plane. Sideslip angle is related to flank angle by

$$tan\,\beta_f = \frac{v}{u} = \frac{V\,sin\,\beta}{V\,cos\,\alpha\,cos\,\beta} = \frac{tan\,\beta}{cos\,\alpha}$$

$$\beta = tan^{-1}\left(tan\,\beta_f \; cos\,\alpha\right) \tag{3.49}$$

For small β and β_f in radians,

$$\beta \approx \beta_f \; cos\,\alpha \tag{3.50a}$$

If angle of attack α is also small,

$$\beta \approx \beta_f \tag{3.50b}$$

Accelerometers measure the translational acceleration due to applied forces, excluding gravity. In vector notation, the output equation for the translational acceleration is

$$\boldsymbol{a} = \dot{\boldsymbol{V}} + \boldsymbol{\omega} \times \boldsymbol{V} - \frac{\boldsymbol{F}_G}{m} = \frac{1}{m}\left(\boldsymbol{F}_A + \boldsymbol{F}_T\right) \tag{3.51}$$

which follows from Eq. (3.14). In scalar form,

$$a_x = \dot{u} - rv + qw + g\,sin\theta \tag{3.52a}$$

$$a_y = \dot{v} - pw + ru - g\,cos\,\theta\,sin\phi \tag{3.52b}$$

$$a_z = \dot{w} - qu + pv - g\,cos\,\theta\,cos\,\phi \tag{3.52c}$$

The accelerometer output equations are nonlinear in the states and linear in the state derivatives. From Eq. (3.51), the accelerometer outputs in g units can be related to the applied forces,

$$\boldsymbol{a} = \frac{\left(\boldsymbol{F}_A + \boldsymbol{F}_T\right)}{mg} \tag{3.53}$$

or in scalar form [cf. Eqs. (3.16) and (3.19a)],

$$a_x = \frac{1}{mg}\left(\bar{q}SC_X + T\right) \tag{3.54a}$$

$$a_y = \frac{1}{mg}\left(\bar{q}SC_Y\right) \tag{3.54b}$$

$$a_z = \frac{1}{mg}\left(\bar{q}SC_Z\right) \tag{3.54c}$$

3.8 Aerodynamic Model Equations

Modeling the aircraft aerodynamics raises the fundamental question of what the mathematical structure of the model should be. Although a complicated model structure can be justified for accurate description of the aerodynamic forces and moments, it is not always clear what the relationship between model complexity and information in the measured data should be. If too many model parameters are sought for a limited amount of data, reduced accuracy of estimated parameters can be expected, or the attempts to estimate all the parameters in the model might fail.

In general, the nondimensional aerodynamic force and moment coefficients depend nonlinearly on present and past values of airspeed, angles of incidence of the air-relative velocity with respect to the aircraft body, aircraft rigid-body rotation rates, control surface deflections, and other nondimensional quantities. The functional dependencies can be quite complicated, so a variety of experiments are used to determine an adequate characterization.

There are also analytic methods to find the aerodynamic functional dependencies, such as: computational fluid dynamics (CFD), which involves solving the partial differential equations governing the motion of the air about the aircraft; panel methods and strip theory, which involve dicing the aircraft into sections that approximate two-dimensional airfoil sections, then summing up lift, drag, and pitching moment from each section to get resultant forces and moments on the airplane; and U.S. Air Force DATCOM, which is an extensive set of rules of thumb for aerodynamic dependencies, based on experience and airplane geometry. Although these methods work well in some cases, typically for low angles of attack and low rotational rates, the best aerodynamic predictions are obtained using experimental methods. The experimental methods include wind tunnel tests (static tests, forced oscillation tests, rotary balance tests, and spin tests), as well as flight test measurements in steady and maneuvering flight.

Based on dimensional analysis, the nondimensional aerodynamic force and moment coefficients for rigid aircraft can be characterized as a function of nondimensional quantities as follows:

$$C_i = C_i\left(\alpha,\beta,\delta,\frac{\Omega l}{V},\frac{\dot{\Omega} l^2}{V^2},\frac{\dot{V} l}{V^2},\frac{\omega l}{V},\frac{\rho V l}{\mu},\frac{V^2}{lg},\frac{V}{a},\frac{m}{\rho l^3},\frac{I}{\rho l^5},\frac{tV}{l}\right)$$

for $i = D, Y, L, l, m, n,$ where

$\delta =$ control surface deflections, deg or rad

$\Omega =$ stability axis rotation rate, rad/sec

$l =$ characteristic length, ft or m

$\omega =$ oscillation frequency, rad/sec

$\omega l/V \equiv Str =$ Strouhal number (unsteady oscillatory flow effects)

$\rho Vl/\mu = Vl/\nu \equiv Re =$ Reynolds number (fluid inertial forces / viscous forces)

$\nu \equiv \mu/\rho =$ kinematic viscosity, ft^2/sec or m^2/sec

$V^2/gl \equiv Fr =$ Froude number (inertial forces / gravitational forces)

$V/a \equiv M =$ Mach number (fluid compressibility effects)

$a =$ speed of sound, ft/sec or m/sec

$t =$ time, sec

Common simplifications are that the airplane mass and inertia are significantly larger than the surrounding air mass and inertia, fluid properties change slowly, and Froude number effects are small. In addition, the flow is often assumed to be quasi-steady, which means that the flow field adjusts instantaneously to changes. One exception to this is the retention of Strouhal number effects, also called reduced frequency effects. These assumptions reduce the relationship above to

$$C_i = C_i\left(\alpha, \beta, \delta, \frac{\Omega l}{V}, \frac{\omega l}{V}, \frac{\rho Vl}{\mu}, \frac{V}{a}\right)$$

for $i = D, Y, L, l, m, n$.

In wind tunnel testing, experimental characterization of the functional dependencies in the above equation is usually broken down as follows:

$$C_i = f_1\left(\alpha, \beta, \delta, \frac{V}{a}\right) \quad \text{(static wind tunnel tests)}$$

$$+ \quad f_2\left(\frac{\rho Vl}{\mu}\right) \quad \text{(Reynolds number corrections)}$$

$$+ \quad f_3\left(\frac{\Omega l}{V}\right) \quad \text{(rotary balance tests)}$$

$$+ \quad f_4\left(\frac{\omega l}{V}\right) \quad \text{(forced oscillation tests)}$$

for $i = D, Y, L, l, m, n$.

This decomposition assumes that the effects are separable and that superposition can be used – assumptions normally associated with linear and uncoupled functional dependencies. These assumptions are not always valid.

For full-scale aircraft flight testing, Reynolds number effects are not relevant, since this nondimensional quantity changes only slightly for the flight experiment. The effects of rotary motion and forced oscillation are usually modeled as a function of the body-axis angular rates, air incidence angles, and their first time derivatives. For a full-scale conventional airplane in quasi-steady flow at low Mach number, the functional form for the nondimensional force and moment coefficients becomes

$$C_i = C_i\left(\frac{V}{V_o}, \alpha, \beta, \frac{pb}{2V_o}, \frac{q\bar{c}}{2V_o}, \frac{rb}{2V_o}, \frac{\dot{\alpha}\bar{c}}{2V_o}, \frac{\dot{\beta}b}{2V_o}, \delta\right)$$

$$\text{for } i = D, Y, L, l, m, n$$

for $i = D, Y, L, l, m, n,$ where V_o is the airspeed at a reference condition, and δ represents all the aircraft controls. Note that the angular rates are nondimensionalized using V_o, which is a constant. This avoids comingling airspeed with other explanatory variables. However this practice is not universal, and sometimes the angular rates are nondimensionalized using the airspeed variable V.

The variables inside the parentheses of the above equation are sometimes called independent variables, although in flight these variables cannot be changed independently. They are more accurately called explanatory variables. In general, the functional dependencies of the nondimensional force and moment coefficients on the explanatory variables are nonlinear. The airspeed and angular rates must be nondimensionalized as shown above, for dimensional consistency. The nondimensional quantities are also denoted by

$$\hat{V} \equiv \frac{V}{V_o} \qquad \hat{p} \equiv \frac{pb}{2V_o} \qquad \hat{q} \equiv \frac{q\bar{c}}{2V_o} \qquad \hat{r} \equiv \frac{rb}{2V_o} \qquad \hat{\dot{\alpha}} \equiv \frac{\dot{\alpha}\bar{c}}{2V_o} \qquad \hat{\dot{\beta}} \equiv \frac{\dot{\beta}b}{2V_o}$$

In general, aerodynamic forces and moments are functionals of the state variables u, v, w, p, q, r or $V, \beta, \alpha, p, q, r,$ and the control variables, denoted collectively by δ. For a conventional, tailed airplane, δ includes elevator, aileron, and rudder deflections that can change during the maneuver, as well as flap deflections and power settings, which usually are constant throughout a maneuver.

Considering only the angle of attack dependence of the aerodynamic force $\boldsymbol{F}_A$,

$$\boldsymbol{F}_A(t) = \boldsymbol{F}_A\left[\alpha(\tau)\right] \qquad -\infty < \tau \le t$$

The aerodynamic force depends on the instantaneous value of α, but also on its entire history. In the majority of practical applications, the dependence on past values of α can be neglected by considering the flow to be quasi-steady. This assumption presumes that the flow reaches a steady state instantaneously, so that dependence on the history of the explanatory variable is neglected. The aerodynamic model is then converted from a functional that depends on the entire history of α into a function that depends only on the current value $\alpha(t)$. In practice, the aerodynamic forces and moments depend on multiple states and controls, not just α, which makes the modeling much more complex.

In the following, a linear aerodynamic model for quasi-steady flow will be considered first. This approach leads to the concept of stability and control derivatives, also called aerodynamic derivatives. This concept will be extended to nonlinear aerodynamic models, formulated as polynomials or polynomial splines. Finally, the problem of modeling aerodynamics in unsteady flow conditions will be briefly addressed. In this case, the dependence of the aerodynamic forces and moments on past values of the explanatory variables must be restored.

3.8.1 Quasi-Steady Flow

The next subsections describe linear, nonlinear, and spline model forms commonly used with the quasi-steady flow assumption.

Linear Model

The form of aerodynamic model equations based on quasi-steady flow can be given by a linear Taylor series expansion of the aerodynamic forces and moments about a reference condition in terms of aircraft states and controls,

$$X = X_o + X_u \Delta u + X_v \Delta v + \ldots + X_\delta \Delta \delta \tag{3.55}$$

and similarly for $Y, Z, L, M,$ and N. The expansion can also be done in terms of V, α, β instead of u, v, w. The expansion will include a term of the form $X_\delta \Delta \delta$ for each control that affects X. The notation represents all of these analogous terms with a single $X_\delta \Delta \delta$ term, to simplify the expression. In Eq. (3.55), X_o represents the force component at the reference condition, which is usually specified as steady symmetric flight with $v = p = q = r = 0$. The quantities $X_u = \left.\frac{\partial X}{\partial u}\right|_o, X_v = \left.\frac{\partial X}{\partial v}\right|_o, \ldots$ are the partial derivatives of X with respect to $u, v, \ldots, \delta$, evaluated at the reference condition. Eq. (3.55) is a Taylor series approximation, truncated to retain only the linear terms. The approximation is only valid for small perturbations in the explanatory states and controls, relative to their values at the reference condition. For these small perturbations, the omitted higher order terms involving multiplication of small perturbation quantities are negligible compared to the first order terms. The model parameters $X_u, X_v, \ldots, X_\delta$, etc., are constants associated with the reference condition. Strictly speaking, the reference flight condition should be specified by reference values of the all the explanatory variables shown in the lengthy functional dependence derived from dimensional analysis, shown earlier. In practice, a reference flight condition of straight and level trimmed flight is specified by reference angle of attack α_o, Mach number M_o or airspeed V_o, altitude h_o, mass m_o, and inertia properties I_o. For a steady turn, add bank angle ϕ_o; for steady climbing or descending flight, add path angle γ_o.

Based on vehicle symmetry and experience, the model equations like (3.55) can be simplified by neglecting:

1) Dependence of symmetric (longitudinal) forces and moment $X, Z, M,$ on asymmetric (lateral) variables v, p, r;

2) Dependence of asymmetric (lateral) force and moments $Y, L, N,$ on symmetric (longitudinal) variables u, w, q.

The simplified equations are then augmented by adding two terms $Z_{\dot{w}} \Delta \dot{w}$ and $M_{\dot{w}} \Delta \dot{w}$, or equivalently $Z_{\dot{\alpha}} \Delta \dot{\alpha}$ and $M_{\dot{\alpha}} \Delta \dot{\alpha}$. These terms are necessary for obtaining closer correlation between predicted and observed aircraft longitudinal motion, as first mentioned in Ref. [3.6]. For aircraft with a conventional wing and tail configuration, these terms can be approximated by the change in the vertical aerodynamic force and pitching moment of the tail, resulting from the lag in the downwash from the wing, see Ref. [3.1]. In general, the parameters $Z_{\dot{w}}$ and $M_{\dot{w}}$, or $Z_{\dot{\alpha}}$ and $M_{\dot{\alpha}}$, can be determined on the basis of unsteady aerodynamic modeling, as will be explained later.

Simplified aerodynamic models can be further augmented by adding three terms, $Y_{\dot{v}} \Delta \dot{v}$, $L_{\dot{v}} \Delta \dot{v}$, and $N_{\dot{v}} \Delta \dot{v}$, or equivalently $Y_{\dot{\beta}} \Delta \dot{\beta}$, $L_{\dot{\beta}} \Delta \dot{\beta}$, and $N_{\dot{\beta}} \Delta \dot{\beta}$. In this case, the coefficients associated with each term can be determined only on the basis of unsteady aerodynamics. The augmentation by $\dot{v}$ or $\dot{\beta}$ terms will not be considered in equations describing quasi-steady flow.

To maintain consistency in the aerodynamic model equations, $Z_{\dot{w}}$ and $M_{\dot{w}}$ are interpreted as

$$Z_{\dot{w}} = \left. \frac{\partial Z}{\partial \dot{w}} \right|_o \quad \text{and} \quad M_{\dot{w}} = \left. \frac{\partial M}{\partial \dot{w}} \right|_o \tag{3.56}$$

and similarly for $Z_{\dot{\alpha}}$ and $M_{\dot{\alpha}}$. This interpretation is mathematically incorrect, since neither $\dot{w}$ nor $\dot{\alpha}$ can be varied independently of w or α. However, use of the term "derivative" for the parameters $Z_{\dot{w}}$ and $M_{\dot{w}}$, or $Z_{\dot{\alpha}}$ and $M_{\dot{\alpha}}$, is fully embedded in the flight dynamics literature, and will also appear in this book.

Considering the simplifications and term additions mentioned above, the linear aerodynamic model equations can be written as follows:

<u>*Longitudinal*</u>

$$X = X_o + X_u \Delta u + X_w \Delta w + X_q q + X_\delta \Delta\delta \quad (3.57a)$$

$$Z = Z_o + Z_u \Delta u + Z_w \Delta w + Z_{\dot{w}} \Delta \dot{w} + Z_q q + Z_\delta \Delta\delta \quad (3.57b)$$

$$M = M_o + M_u \Delta u + M_w \Delta w + M_{\dot{w}} \Delta \dot{w} + M_q q + M_\delta \Delta\delta \quad (3.57c)$$

<u>*Lateral*</u>

$$Y = Y_o + Y_v \Delta v + Y_p p + Y_r r + Y_\delta \Delta\delta \quad (3.58a)$$

$$L = L_o + L_v \Delta v + L_p p + L_r r + L_\delta \Delta\delta \quad (3.58b)$$

$$N = N_o + N_v \Delta v + N_p p + N_r r + N_\delta \Delta\delta \quad (3.58c)$$

As before, a control term of the form shown would be included for each control affecting the aerodynamic force or moment. The angular rate perturbations do not have the Δ notation, because the angular rates at the assumed reference condition are zero, so the angular rates and their perturbation values are identical. For state variables $V, \alpha, \beta, p, q, r$ and the aerodynamic force components in terms of lift L and drag D, the equations change to:

<u>*Longitudinal*</u>

$$D = D_o + D_V \Delta V + D_\alpha \Delta\alpha + D_q q + D_\delta \Delta\delta \quad (3.59a)$$

$$L = L_o + L_V \Delta V + L_\alpha \Delta\alpha + L_{\dot{\alpha}} \Delta\dot{\alpha} + L_q q + L_\delta \Delta\delta \quad (3.59b)$$

$$M = M_o + M_V \Delta V + M_\alpha \Delta\alpha + M_{\dot{\alpha}} \Delta\dot{\alpha} + M_q q + M_\delta \Delta\delta \quad (3.59c)$$

Lateral

$$Y = Y_o + Y_\beta \Delta\beta + Y_p p + Y_r r + Y_\delta \Delta\delta \tag{3.60a}$$

$$L = L_o + L_\beta \Delta\beta + L_p p + L_r r + L_\delta \Delta\delta \tag{3.60b}$$

$$N = N_o + N_\beta \Delta\beta + N_p p + N_r r + N_\delta \Delta\delta \tag{3.60c}$$

The model parameters representing partial derivatives in Eqs. (3.57)-(3.60) are called dimensional derivatives. These can be expressed in terms of partial derivatives of the nondimensional aerodynamic coefficients. For example, in Eq. (3.59b),

$$L_V = \frac{\partial}{\partial V}\left(\frac{1}{2}\rho V^2 S C_L\right)\bigg|_o = \rho_o V_o S C_{L_o} + \frac{1}{2}\rho_o V_o^2 S \frac{\partial C_L}{\partial V}\bigg|_o$$

$$L_\alpha = \frac{1}{2}\rho_o V_o^2 S \frac{\partial C_L}{\partial \alpha}\bigg|_o \qquad L_{\dot\alpha} = \frac{1}{2}\rho_o V_o^2 S \frac{\partial C_L}{\partial \dot\alpha}\bigg|_o \tag{3.61}$$

$$L_q = \frac{1}{2}\rho_o V_o^2 S \frac{\partial C_L}{\partial q}\bigg|_o \qquad L_\delta = \frac{1}{2}\rho_o V_o^2 S \frac{\partial C_L}{\partial \delta}\bigg|_o$$

and similarly for the other derivatives in Eqs. (3.57)-(3.60). From (3.61) it is apparent that the dimensional derivatives depend on the reference airspeed V_o, and the air density ρ_o, which changes with atmospheric conditions and altitude.

For aircraft system identification, it is more convenient to use nondimensional derivatives of the nondimensional aerodynamic force and moment coefficients C_D or C_X, C_Y, C_L or C_Z, C_l, C_m, and C_n. This removes the known dependence on the airspeed and air density (dynamic pressure), and normalizes the partial derivatives. These derivatives are obtained from the following relationships:

$$\begin{aligned} &C_X = C_X(u, w, q, \delta) \\ &C_a = C_a(u, w, \dot w, q, \delta) \qquad \text{for } a = Z \text{ or } m \end{aligned} \tag{3.62}$$

or

$$\begin{aligned} &C_D = C_D(V, \alpha, q, \delta) \\ &C_a = C_a(V, \alpha, \dot\alpha, q, \delta) \qquad \text{for } a = L \text{ or } m \end{aligned} \tag{3.63}$$

and

$$C_a = C_a(\beta, p, r, \delta) \qquad \text{for } a = Y, l, \text{ or } n \tag{3.64}$$

Then

$$C_X = C_{X_o} + C_{X_u}\frac{\Delta u}{u_o} + C_{X_w}\frac{\Delta w}{u_o} + C_{X_q}\frac{q\bar{c}}{2u_o} + C_{X_\delta}\Delta\delta$$

$$C_a = C_{a_o} + C_{a_u}\frac{\Delta u}{u_o} + C_{a_w}\frac{\Delta w}{u_o} + C_{a_{\dot{w}}}\left[\frac{\dot{w}}{u_o}\right]\frac{\bar{c}}{2u_o} + C_{a_q}\frac{q\bar{c}}{2u_o} + C_{a_\delta}\Delta\delta \quad (3.65)$$

for $a = Z$ or m

or

$$C_D = C_{D_o} + C_{D_V}\frac{\Delta V}{V_o} + C_{D_\alpha}\Delta\alpha + C_{D_q}\frac{q\bar{c}}{2V_o} + C_{D_\delta}\Delta\delta$$

$$C_a = C_{a_o} + C_{a_V}\frac{\Delta V}{V_o} + C_{a_\alpha}\Delta\alpha + C_{a_{\dot{\alpha}}}\frac{\dot{\alpha}\bar{c}}{2V_o} + C_{a_q}\frac{q\bar{c}}{2V_o} + C_{a_\delta}\Delta\delta \quad (3.66)$$

for $a = L$ or m

and

$$C_a = C_{a_o} + C_{a_\beta}\Delta\beta + C_{a_p}\frac{pb}{2V_o} + C_{a_r}\frac{rb}{2V_o} + C_{a_\delta}\Delta\delta \qquad \text{for } a = Y, l, \text{ or } n$$

(3.67)

In Eqs. (3.65) and (3.66),

$$C_{a_u} = u_o\left.\frac{\partial C_a}{\partial u}\right|_o \qquad C_{a_w} = u_o\left.\frac{\partial C_a}{\partial w}\right|_o \qquad C_{a_{\dot{w}}} = \frac{2u_o^2}{\bar{c}}\left.\frac{\partial C_a}{\partial \dot{w}}\right|_o$$

(3.68)

$$C_{a_q} = \frac{2u_o}{\bar{c}}\left.\frac{\partial C_a}{\partial q}\right|_o \qquad C_{a_\delta} = \left.\frac{\partial C_a}{\partial \delta}\right|_o \qquad \text{for } a = X, Z, \text{ or } m$$

$$C_{a_V} = V_o\left.\frac{\partial C_a}{\partial V}\right|_o \qquad C_{a_\alpha} = \left.\frac{\partial C_a}{\partial \alpha}\right|_o \qquad C_{a_{\dot{\alpha}}} = \frac{2V_o}{\bar{c}}\left.\frac{\partial C_a}{\partial \dot{\alpha}}\right|_o$$

(3.69)

$$C_{a_q} = \frac{2V_o}{\bar{c}}\left.\frac{\partial C_a}{\partial q}\right|_o \qquad C_{a_\delta} = \left.\frac{\partial C_a}{\partial \delta}\right|_o \qquad \text{for } a = D, L, \text{ or } m$$

Similarly, in Eq. (3.67)

$$C_{a_\beta} = \left.\frac{\partial C_a}{\partial \beta}\right|_o \quad C_{a_p} = \left.\frac{2V_o}{b}\frac{\partial C_a}{\partial p}\right|_o \quad C_{a_r} = \left.\frac{2V_o}{b}\frac{\partial C_a}{\partial r}\right|_o \quad C_{a_\delta} = \left.\frac{\partial C_a}{\partial \delta}\right|_o \tag{3.70}$$

for $a = Y, l,$ or n

The quantities defined in Eqs. (3.65)-(3.70) are called nondimensional stability and control derivatives. Stability derivatives involve partial derivatives with respect to states; control derivatives involve partial derivatives with respect to controls. The stability derivatives are further divided into static stability derivatives for derivatives associated with air-relative velocity quantities $(u, v, w, V, \alpha, \beta)$, dynamic stability derivatives for derivatives associated with angular rates (p, q, r), and derivatives associated with unsteady aerodynamics $(\dot{w}, \dot{\alpha})$.

Note that the angular rates are nondimensionalized using u_o in Eq. (3.65) and V_o in Eq. (3.66). For maneuvers at low values of trim angle of attack, the difference is negligible, but at high angles of attack, the result is a slightly different definition of the parameters associated with $\dot{\alpha}$ and q [cf. Eq. (3.32a)].

At very high speeds, e.g., hypersonic flight, a problem occurs with the nondimensionalization of the angular rates. In that case, V_o and u_o are very large, making the nondimensionalized rates very small. This can cause numerical problems in the parameter estimation, because of low sensitivity for these parameters (see Chapter 6). One solution is to identify dimensional parameters such as $\left.(\partial C_m/\partial q)\right|_o$, then nondimensionalize the results using Eqs. (3.68)-(3.70) after the parameter estimation is finished.

If Eq. (3.66) is used for the model equations, an identifiability problem arises with the estimation of the $\dot{\alpha}$ and q derivatives for many typical aircraft maneuvers, see Refs. [3.6] and [3.7]. The problem is that the time series for $\dot{\alpha}$ and q are very similar, which leads to an indeterminacy in the modeling, as discussed in Chapter 5. To avoid this problem, the $\dot{\alpha}$ and q terms can be lumped together, and a single equivalent derivative is then identified. The model equations would be

$$\begin{aligned} C_D &= C_{D_o} + C_{D_V}\frac{\Delta V}{V_o} + C_{D_\alpha}\Delta\alpha + C_{D_q}\frac{q\bar{c}}{2V_o} + C_{D_\delta}\Delta\delta \\ C_a &= C_{a_o} + C_{a_V}\frac{\Delta V}{V_o} + C_{a_\alpha}\Delta\alpha + \bar{C}_{a_q}\frac{q\bar{c}}{2V_o} + C_{a_\delta}\Delta\delta \quad \text{for } a = L \text{ or } m \end{aligned} \tag{3.71}$$

where

$$\bar{C}_{a_q} = C_{a_{\dot{\alpha}}} + C_{a_q} \qquad \text{for } a = L \text{ or } m \tag{3.72}$$

Nonlinear Model

In many practical situations, the aerodynamic models in Eqs. (3.66) and (3.67) are good representations of the aerodynamic forces and moments. However, for large amplitudes or rapid excursions from the reference flight condition, it is necessary to extend the linear models by adding nonlinear terms. Taking the lift coefficient dependence on angle of attack and pitch rate as an example, $C_L = C_L(\alpha, q)$, and the Taylor series expansion can be written as

$$\begin{aligned} C_L = C_{L_o} &+ \frac{\partial C_L}{\partial \alpha}\Delta\alpha + \frac{\partial C_L}{\partial q} q \\ &+ \frac{1}{2}\left[\frac{\partial^2 C_L}{\partial \alpha^2}(\Delta\alpha)^2 + 2\frac{\partial^2 C_L}{\partial \alpha\,\partial q}(\Delta\alpha\, q) + \frac{\partial^2 C_L}{\partial q^2} q^2\right] + \ldots \end{aligned} \tag{3.73}$$

Introducing nonlinear aerodynamic derivatives,

$$\begin{aligned} C_L = C_{L_o} &+ C_{L_\alpha}\Delta\alpha + C_{L_q}\frac{q\bar{c}}{2V_o} \\ &+ \frac{1}{2}\left[C_{L_{\alpha^2}}(\Delta\alpha)^2 + 2C_{L_{\alpha q}}\left(\Delta\alpha\frac{q\bar{c}}{2V_o}\right) + C_{L_{q^2}}\left(\frac{q\bar{c}}{2V_o}\right)^2\right] + \ldots \end{aligned} \tag{3.74}$$

where coefficients of the nonlinear terms are defined in a manner analogous to Eqs. (3.69).

Another way to include nonlinear effects is to combine the static terms and treat the dynamic stability derivatives and control derivatives as functions of important explanatory variables, e.g., angle of attack. In the previous example, the nonlinear model would be

$$C_L = C_{L_o}(\alpha) + C_{L_q}(\alpha)\frac{q\bar{c}}{2V_o} \tag{3.75}$$

This approach is sometimes preferable in nonlinear modeling, because the nonlinear dependencies are partitioned differently and simplified. The idea can be generalized to all the aerodynamic coefficients. Common assumptions made are:

1) For subsonic flight, airspeed changes do not affect the aerodynamic coefficients.
2) The $\dot{\alpha}$ contributions to C_L and C_m are included in the q terms.

3) The dependence of longitudinal and lateral coefficients on the states and controls are:

$$
\begin{aligned}
C_a &= C_a(\alpha,\beta,q,\delta) \qquad &\text{for } a = D, L, \text{ or } m \\
C_a &= C_a(\alpha,\beta,p,r,\delta) \qquad &\text{for } a = Y, l, \text{ or } n
\end{aligned}
\tag{3.76}
$$

4) The aerodynamic coefficients are modeled as the sum of a static term that includes nonlinear angle of attack and sideslip angle dependencies, and dynamic and control terms that are linear in p, q, r, and δ. The second group of terms usually involves derivatives that depend nonlinearly on angle of attack, and sometimes also sideslip angle or Mach number.

Under the preceding assumptions, the aerodynamic model equations can be written as

$$
C_a = C_{a_o}(\alpha,\beta)_{q=\delta=0} + \bar{C}_{a_q}(\alpha)\frac{q\bar{c}}{2V_o} + C_{a_\delta}(\alpha)\delta \qquad \text{for } a = D, L, \text{ or } m \tag{3.77}
$$

$$
C_a = C_{a_o}(\alpha,\beta)_{p=r=\delta=0} + C_{a_p}(\alpha)\frac{pb}{2V_o} + C_{a_r}(\alpha)\frac{rb}{2V_o} + C_{a_\delta}(\alpha)\delta \qquad \text{for } a = Y, l, \text{ or } n \tag{3.78}
$$

These expressions for the aerodynamic coefficients are similar to those used in wind tunnel testing. In that case, the functions shown would be implemented as tables of measured values. Note that the explanatory variables shown are typically not perturbations, because the nonlinear model formulation can be used to model the aerodynamic coefficients over the entire physical range of the explanatory variables. When this is not true, then perturbation quantities must again be used, and the model validity is localized.

The first terms on the right side of Eqs. (3.77) and (3.78) represent the static part, with controls fixed at zero deflection. The remaining terms represent contributions of dynamic stability derivatives, control derivatives, and their dependence on angle of attack.

Equations (3.77) and (3.78) are fairly general formulations of the aerodynamic forces and moments. The functional dependencies shown are based on wind tunnel and flight testing experience. In each particular case, however, the aerodynamic model equations should reflect any available prior knowledge based on wind tunnel experiments and/or theoretical aerodynamic calculations.

Each of the stability and control derivative functions in Eqs. (3.77) and (3.78) can be approximated by either polynomials or polynomial splines in the explanatory variables. For example,

$$C_{m_o}(\alpha,\beta)_{q=\delta=0} = \theta_{m_{o0}} + \theta_{m_{o1}}\alpha + \theta_{m_{o2}}\beta + \theta_{m_{o3}}\alpha\beta + \theta_{m_{o4}}\alpha^2$$
$$C_{l_p}(\alpha) = \theta_{l_{p0}} + \theta_{l_{p1}}\alpha \tag{3.79}$$

Spline Model

For large amplitude maneuvers and flight at high angles of attack, the behavior of aerodynamic coefficients over different ranges of angle of attack may be different and totally unrelated. In these cases, the polynomial approximation for some aerodynamic nonlinearities can be inadequate. The choices then are to add more terms to the model or identify separate models for partitions of the explanatory variable space. Practically, the second option usually works better, because it is easier to solve several smaller and simpler sub-problems compared to one large problem with a complicated model structure.

Polynomials can follow a curve in one interval but depart from that curve or oscillate elsewhere. Even if a high order polynomial approximates the aerodynamic function sufficiently, the increase in the number of terms needed can lead to inaccurate parameter estimates from measured data.

To avoid the disadvantages of the polynomial representation, spline functions can be used. Splines remove some difficulties of polynomials because they are defined only on selected intervals, and low-order terms defined on limited intervals can approximate nonlinearities quite well.

Spline functions are defined as piecewise polynomials of a given degree, see Appendix A, and Refs. [3.8]-[3.9]. When continuity restrictions are included, the function values and derivatives agree at the points where the piecewise polynomials join. These points are called knots, and are defined by their location in the explanatory variable space.

As examples of using splines in formulating aerodynamic models, the lift coefficient C_L and the yawing moment coefficient C_n are considered. In the first case,

$$C_L = C_{L_o}(\alpha)_{q=\delta=0} + \bar{C}_{L_q}(\alpha)\frac{q\bar{c}}{2V_o} + C_{L_\delta}(\alpha)\delta \tag{3.80}$$

and the terms on the right side of Eq. (3.80) are approximated by

$$C_{L_o}(\alpha) = C_{L_o}(0) + C_{L_\alpha}\alpha + \sum_{i=1}^{k} D_{\alpha_i}(\alpha - \alpha_i)_+^1 \tag{3.81a}$$

$$\bar{C}_{L_q}(\alpha) = \bar{C}_{L_q}(0) + \bar{C}_{L_{q_\alpha}}\alpha + \bar{C}_{L_{q_{\alpha^2}}}\alpha^2 + \sum_{i=1}^{k} D_{q_i}(\alpha - \alpha_i)_+^2 \tag{3.81b}$$

$$C_{L_\delta}(\alpha) = C_{L_\delta}(0) + \sum_{i=1}^{k} D_{\delta_i}(\alpha - \alpha_i)_+^0 \tag{3.81c}$$

where

$$(\alpha - \alpha_i)_+^m = \begin{cases} (\alpha - \alpha_i)^m & \alpha \geq \alpha_i \\ 0 & \alpha < \alpha_i \end{cases} \tag{3.82}$$

and α_i are constant values of α, which are the knots, and D_{α_i}, D_{q_i}, and D_{δ_i} are constant parameters that quantify each spline contribution, in a manner similar to stability and control derivatives.

Equations (3.81) indicate that $C_{L_o}(\alpha)$ is approximated by piecewise linear polynomials (first-degree splines), $\bar{C}_{L_q}(\alpha)$ by piecewise quadratic polynomials (second-degree splines), and $C_{L_\delta}(\alpha)$ by piecewise constants (zero-degree splines). These spline types are sketched in Fig. 3.5 for the same knots.

In the second example, the yawing moment coefficient dependence on α and β only is considered. Using a two-dimensional spline, $C_n(\alpha,\beta)$ can be approximated as

$$C_n(\alpha,\beta) = C_{n_o} + C_{n_\beta}\beta + \sum_{i=1}^{k}\left(A_{0_i} + A_{1_i}\beta\right)(\alpha - \alpha_i)_+^0$$
$$+ \sum_{j=1}^{l} B_{0_j}\beta\left(1 - \frac{\beta_j}{|\beta|}\right)_+ + \sum_{i=1}^{k}\sum_{j=1}^{l} D_{ij}\beta\left(1 - \frac{\beta_j}{|\beta|}\right)_+ (\alpha - \alpha_i)_+^0 \tag{3.83}$$

where

$$\beta\left(1 - \frac{\beta_j}{|\beta|}\right)_+ = \begin{cases} 0 & |\beta| \leq \beta_j \\ \beta - \beta_j & \beta > \beta_j \\ \beta + \beta_j & \beta < -\beta_j \end{cases} \tag{3.84}$$

and A_0, A_{1_i}, B_{0_j}, and D_{ij} are constant parameters. The knots β_j are always positive values, and C_n is an odd function of β. The spline function defined in Eq. (3.84) could be called a two-sided spline. Equation (3.83) is of the same general form as Eq. (3.79), where polynomials have been replaced by splines.

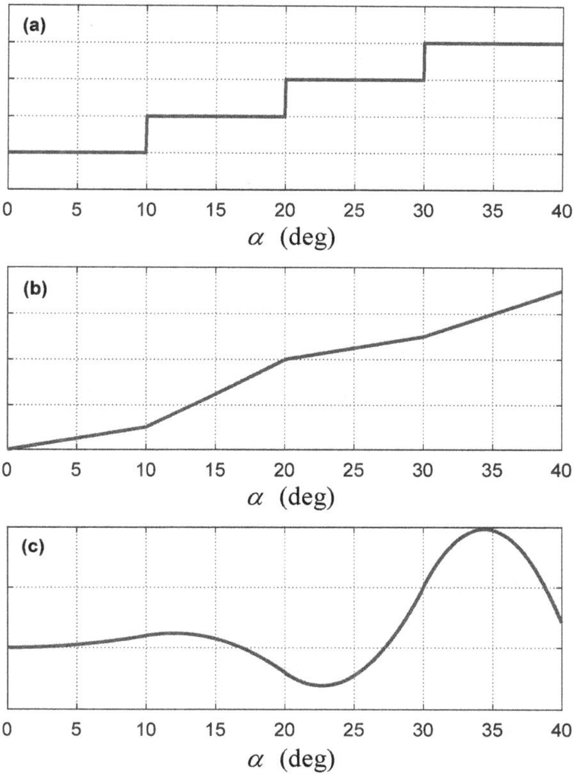

Figure 3.5 **Polynomial splines: (a) zero degree, (b) first degree, (c) second degree**

3.8.2 Unsteady Flow

In the preceding section, it was shown how the aerodynamic forces and moments acting on an aircraft in arbitrary motion can be approximated by linear terms, polynomials, or polynomial splines. It was assumed that the parameters appearing in these approximations were time-invariant, and only the current values of the explanatory variables were required. These assumptions, however, have been questioned many times, based on studies of unsteady aerodynamics that go back to the 1920s, e.g., Ref. [3.10]. A formulation of linear unsteady aerodynamics in the aircraft longitudinal equations in terms of indicial functions was introduced by Tobak [3.11]. Later, Tobak and Schiff [3.12] expressed the aerodynamic coefficients as functionals of the state and input variables. This very general approach includes linear unsteady aerodynamics as a special case.

Using results from Ref. [3.12], aircraft aerodynamic characteristics can be formulated as

$$C_a(t) = C_a(\infty) + \int_0^t C_{a_{\xi_1}}\left[t-\tau;\xi(\tau)\right]^T \frac{d}{d\tau}\xi_1(\tau)\,d\tau \\ + \frac{l}{V}\int_0^t C_{a_{\xi_2}}\left[t-\tau;\xi(\tau)\right]^T \frac{d}{d\tau}\xi_2(\tau)\,d\tau \tag{3.85}$$

where

$a = D, L, m, Y, l,$ or n

$C_a(t)$ = aerodynamic force or moment coefficient

$C_a(\infty)$ = steady-state value of the aerodynamic force or moment coefficient

$C_{a_\xi}(t)$ = vector of indicial functions whose elements are the responses to a unit step in ξ

$\xi_1 = \left[\alpha \;\; \beta\right]^T$, $\xi_2 = \left[p \;\; q \;\; r\right]^T$

$\xi = \left[\xi_1^T \;\; \xi_2^T\right]^T = \left[\alpha \;\; \beta \;\; p \;\; q \;\; r\right]^T$

l = characteristic length, $l = \bar{c}/2$ or $l = b/2$

The indicial functions approach steady-state values as the argument $(t-\tau)$ increases. To indicate this property, each indicial function can be expressed as

$$C_{a_{\xi_j}}\left[t-\tau;\xi(\tau)\right] = C_{a_{\xi_j}}\left[\infty;\xi(\tau)\right] - F_{a_{\xi_j}}\left[t-\tau;\xi(\tau)\right] \tag{3.86}$$

where

$C_{a_{\xi_j}}\left[\infty;\xi(\tau)\right]$ = steady-state rate of change of the coefficient C_a with respect to ξ_j, with the remaining variables in ξ fixed at the instantaneous values $\xi(\tau)$

$F_{a_{\xi_j}}\left[t-\tau;\xi(\tau)\right]$ = the deficiency function, which approaches zero as $(t-\tau)\rightarrow\infty$

When Eq. (3.86) is substituted into Eq. (3.85), the terms involving the steady-state parameters can be integrated, and Eq. (3.85) becomes

$$C_a(t)=C_a\left[\infty;\xi(t)\right]-\int_0^t F_{a_{\xi_1}}\left[t-\tau;\xi(\tau)\right]^T\frac{d}{d\tau}\xi_1(\tau)d\tau$$
$$-\frac{l}{V}\int_0^t F_{a_{\xi_2}}\left[t-\tau;\xi(\tau)\right]^T\frac{d}{d\tau}\xi_2(\tau)d\tau \qquad (3.87)$$

where

$C_a\left[\infty;\xi(t)\right]$ = the total aerodynamic coefficient for steady flow with ξ fixed at the instantaneous values $\xi(t)$

$F_{a_{\xi_1}}\left[t-\tau;\xi(\tau)\right]$ and $F_{a_{\xi_2}}\left[t-\tau;\xi(\tau)\right]$ = vector deficiency functions

Further simplification of Eq. (3.87) is achieved by assuming that the coefficients C_a are linearly dependent on the motion rates ξ_2. Applying a Taylor series expansion about $\xi_2=\mathbf{0}$ to the terms in Eq. (3.87), and keeping only the linear terms, results in

$$C_a(t)=C_a\left[\infty;\xi_1(t),\mathbf{0}\right]-\int_0^t F_{a_{\xi_1}}\left[t-\tau;\xi_1(\tau),\mathbf{0}\right]^T\frac{d}{d\tau}\xi_1(\tau)d\tau$$
$$-\frac{l}{V}C_{a_{\xi_2}}\left[\infty;\xi_1(\tau),\mathbf{0}\right]^T\xi_2 \qquad (3.88)$$

or, using different notation,

$$C_a(t)=C_a\left[\infty;\alpha(t),\beta(t)\right]+\frac{l}{V}C_{a_p}\left[\infty;\alpha(t),\beta(t)\right]p(t)$$
$$+\frac{l}{V}C_{a_q}\left[\infty;\alpha(t),\beta(t)\right]q(t)+\frac{l}{V}C_{a_r}\left[\infty;\alpha(t),\beta(t)\right]r(t)$$
$$-\int_0^t F_{a_\alpha}\left[t-\tau;\alpha(\tau),\beta(\tau)\right]\dot{\alpha}(\tau)\,d\tau \qquad (3.89)$$
$$-\int_0^t F_{a_\beta}\left[t-\tau;\alpha(\tau),\beta(\tau)\right]\dot{\beta}(\tau)\,d\tau$$

If the indicial responses are only functions of elapsed time, and the steady flow coefficients are assumed to be independent of $\alpha(t)$ and $\beta(t)$, Eq. (3.89) is simplified as

$$
\begin{aligned}
C_a(t) = {} & C_{a_o}(\infty) + C_{a_\alpha}(\infty)\alpha(t) + C_{a_\beta}(\infty)\beta(t) \\
& + \frac{l}{V}C_{a_p}(\infty)p(t) + \frac{l}{V}C_{a_q}(\infty)q(t) + \frac{l}{V}C_{a_r}(\infty)r(t) \\
& - \int_0^t F_{a_\alpha}(t-\tau)\,\dot{\alpha}(\tau)d\tau - \int_0^t F_{a_\beta}(t-\tau)\dot{\beta}(\tau)d\tau
\end{aligned}
\tag{3.90}
$$

where $C_{a_\alpha}(\infty), C_{a_\beta}(\infty), \ldots, C_{a_r}(\infty)$ are constants representing the classical stability derivatives.

As an example, the equation for the pitching moment coefficient in terms of indicial functions is considered. Assuming linear aerodynamics with $C_{m_o}(\infty) = 0$, and neglecting the effects of $\dot{q}(t)$ and lateral variables, the resulting equation follows from Eq. (3.90) as

$$
C_m(t) = C_{m_\alpha}(\infty)\alpha(t) + \frac{\overline{c}}{2V}C_{m_q}(\infty)q(t) - \int_0^t F_{m_\alpha}(t-\tau)\,\dot{\alpha}(\tau)d\tau \tag{3.91}
$$

For fixed controls, the linear model for pitching moment is formulated as

$$
C_m(t) = C_{m_\alpha}\alpha(t) + \frac{\overline{c}}{2V}C_{m_q}q(t) + \frac{\overline{c}}{2V}C_{m_{\dot{\alpha}}}\dot{\alpha}(t) \tag{3.92}
$$

It is therefore expected that the integral in Eq. (3.91) should be a counterpart of the term $(\overline{c}/2V)C_{m_{\dot{\alpha}}}\dot{\alpha}(t)$. If the unsteady effect on $C_m(t)$ is neglected, the integral in Eq. (3.91) can be reduced to a constant $\dot{\alpha}$ term,

$$
\int_0^t F_{m_\alpha}(t-\tau)\,\dot{\alpha}(\tau)d\tau = \dot{\alpha}(t)\int_0^t F_{m_\alpha}(\tau)d\tau \tag{3.93}
$$

as demonstrated in Ref. [3.11] or [3.12]. It follows that the counterpart of $C_{m_{\dot{\alpha}}}$ is proportional to the area under the deficiency function curve, i.e.,

$$
C_{m_{\dot{\alpha}}} = -\frac{2V}{\overline{c}}\int_0^\infty F_{m_\alpha}(\tau)\,d\tau \tag{3.94}
$$

When indicial functions are used in aerodynamic model equations, it is not clear, either from theory or experiment, what the analytical form of the indicial functions should be. In general, the indicial function represents a sum of two contributions. The first is a noncirculatory part which decays rapidly, and the second is a circulatory part which approaches the steady state value of the indicial function with increased time (see Ref. [3.12]). Jones [3.13] developed approximation formulae for the lift indicial function for angle of attack on an elliptical wing with finite aspect ratio as

$$
C_{L_\alpha}(t) = a_1\left(1 - c_1 e^{-b_1 t}\right) \tag{3.95}
$$

which leads to a deficiency function of the form

$$F_{L_\alpha}(t) = ae^{-b_1 t} \tag{3.96}$$

For nonlinear aerodynamics, the deficiency function in Eq. (3.96) was generalized in Ref. [3.14] as

$$F_{L_\alpha}(t;\alpha) = h(t;\alpha)\, a(\alpha) \tag{3.97}$$

where

$$h(t;\alpha) = \sum_{j=0}^{m} c_j e^{-b_j(\alpha)t} \tag{3.98}$$

and $a(\alpha)$ and $b(\alpha)$ are polynomials in α.

The modeling of unsteady aerodynamics becomes more complicated for a wing/tail configuration, as discussed in Refs. [3.15]-[3.17].

Goman et al. [3.18] proposed a different approach to modeling unsteady aerodynamics, using a concept of internal state variables. This approach retains the state-space formulation of aircraft dynamics,

$$\dot{\boldsymbol{x}} = \boldsymbol{f}\left[\boldsymbol{x}(t), \boldsymbol{u}(t), \boldsymbol{\theta}\right] \qquad \boldsymbol{x}(0) = \boldsymbol{x}_o \tag{3.99}$$

by augmenting the aircraft states with the additional state variable $\eta(t)$. Then, the aerodynamic coefficients are formulated as

$$C_a(t) = C_a\left[\xi(t), \eta(t)\right] \tag{3.100}$$

where $\dot{\eta}$ is a function of $\eta(t)$, $\xi(t)$, and $\dot{\xi}(t)$,

$$\dot{\eta} = \phi\left[\eta(t), \xi(t), \dot{\xi}(t)\right] \tag{3.101}$$

and

$$\xi(t) = \begin{bmatrix} \boldsymbol{x}(t) \\ \boldsymbol{u}(t) \end{bmatrix} \tag{3.102}$$

An example of Eq. (3.101) for longitudinal aircraft dynamics was given in Ref. [3.19]. In that work, the internal state variable represents the vortex burst point location along the chord of a triangular wing, modeled by

$$T_1\dot{\eta} + \eta = \eta_o\left(\alpha - T_\alpha \dot{\alpha}\right) \qquad |\eta| \le 1 \tag{3.103}$$

where

η_o	=	vortex burst point location under steady conditions
T_1	=	time constant in the vortical flow development

T_α = time lag in the vortical flow development caused by the angle of attack rate of change

Klein and Noderer [3.20] showed that in certain cases of linear aerodynamics, the formulations using either indicial functions or internal state variables lead to identical models.

The first example of aerodynamic model equations with unsteady terms used in aircraft parameter estimation was given by Goman et al. [3.18], and later by Goman and Khrabrov [3.19]. Examples using wind tunnel data can be found in Refs. [3.14], [3.21], and [3.22]. Some results obtained from flight data were reported in Ref. [3.23].

3.9 Simplifying the Equations of Motion

The aircraft equations of motion in general form are a set of coupled nonlinear differential equations. However, for many applications, these equations can be simplified. The most commonly used simplification is obtained by linearizing the equations about a reference condition. When the reference condition is selected as steady, wings-level flight with no sideslip, the linearized equations decouple into two independent sets: one describing the longitudinal motion in the plane of symmetry (V,α,q,θ), and the other describing the lateral motion out of the plane of symmetry (β,p,r,ϕ).

Linearized equations have been extensively and successfully used in stability and control analysis and also in system identification. There are three main reasons for the wide use of linearized models:

1) Many common flight motions can be described by small changes in linear and angular velocities from a reference condition.

2) The main aerodynamic effects are very well described by linear functions of state and control variables in many cases.

3) Practical stability analysis and control system design are based on linear dynamic models.

For some flight motions, such as spins, flight near the stall, and maneuvers involving large changes in amplitudes and/or high angular rates, linear models are usually inadequate. In these cases, nonlinear models must be used.

3.9.1 Linearization

The nonlinear equations of motion can be linearized by applying small-disturbance theory. Each variable is assumed to be comprised of two parts – a constant component associated with the steady reference condition, and a perturbation associated with the linear model. In the following development, the reference condition is chosen as steady, wings-level flight with no sideslip. Other reference flight conditions, such as a steady turn, can also be used, but the steady, wings-level flight condition is the most common. Two axis systems are used for the model equations:

1) Wind axes for the force equations
2) Body axes for the moment equations

Steady values for each variable are denoted by subscript o, and perturbations are denoted by the prefix Δ,

$$\begin{array}{lll} V = V_o + \Delta V & \alpha = \alpha_o + \Delta\alpha & \beta = \beta_o + \Delta\beta \\ p = p_o + \Delta p & q = q_o + \Delta q & r = r_o + \Delta r \\ \phi = \phi_o + \Delta\phi & \theta = \theta_o + \Delta\theta & \psi = \psi_o + \Delta\psi \\ C_{D_w} = C_{D_{w_o}} + \Delta C_{D_w} & C_{Y_w} = C_{Y_{w_o}} + \Delta C_{Y_w} & C_L = C_{L_o} + \Delta C_L \\ C_l = C_{l_o} + \Delta C_l & C_m = C_{m_o} + \Delta C_m & C_n = C_{n_o} + \Delta C_n \\ \delta = \delta_o + \Delta\delta & & \end{array} \tag{3.104}$$

In cases where the steady value is zero, the perturbation and the value of the total variable are the same, so the Δ prefix will be dropped. Similarly, time derivatives do not have the Δ prefix, because the time derivative of the total variable is the same as the time derivative of its perturbation component.

For steady, symmetric, wings-level flight at a reference condition,

$$\beta_o = p_o = q_o = r_o = \phi_o = 0 \tag{3.105}$$

and all the perturbations are zero. The equations of motion (3.39) and (3.37) then reduce to:

$$0 = -\frac{\bar{q}S}{m} C_{D_o} - g \sin\gamma_o + \frac{T_o}{m} \cos\alpha_o \tag{3.106a}$$

$$0 = -\frac{\bar{q}S}{mV_o} C_{L_o} + \frac{g}{V_o} \cos\gamma_o - \frac{T_o \sin\alpha_o}{mV_o} \tag{3.106b}$$

$$0 = C_{Y_o} \tag{3.106c}$$

$$0 = C_{l_o} = C_{m_o} = C_{n_o} \tag{3.106d}$$

where γ_o is the steady flight path angle

$$\gamma_o = \theta_o - \alpha_o \tag{3.107}$$

and the kinematic equations (3.38) reduce to zero on both sides. Next, small perturbation theory is applied, by replacing all the variables in the nonlinear equations (3.39), (3.37), and (3.38) with expansions from Eq. (3.104), then subtracting steady-state equations (3.106), applying the small angle approximations $\sin x \approx x$, $\cos x \approx 1$, and $\tan x \approx x$, for x in radians, and dropping terms that are second order and higher in the perturbation quantities. In this case, all the angular perturbations are small angles. The following relationships, which come from the small-angle approximations and trigonometric identities, are useful in deriving the linearized equations:

$$\cos(\phi_o + \Delta\phi) \approx \cos\phi_o - \sin\phi_o \Delta\phi \tag{3.108a}$$

$$\sin(\phi_o + \Delta\phi) \approx \sin\phi_o + \cos\phi_o \Delta\phi \tag{3.108b}$$

$$\tan(\theta_o + \Delta\theta) \approx \tan\theta_o + \Delta\theta \tag{3.108c}$$

The airspeed dependence is also simplified by assuming that $V \approx V_o$ and $\bar{q} \approx \bar{q}_o$, which are good assumptions for airspeeds associated with fixed-wing aircraft, where $V_o \gg \Delta V$. The gyroscopic terms in the moment equations from the rotating mass of the propulsion system are usually dropped, because these terms are small for small angular rate perturbations. Similarly, the terms multiplied by $\sin\beta/V$ in Eq. (3.39c) are also dropped, because their contribution is usually small. The resulting linearized equations are:

$$\dot{V} = -\frac{\bar{q}_o S}{m}\Delta C_D - g\cos\gamma_o(\Delta\theta - \Delta\alpha) - \frac{T_o \sin\alpha_o}{m}\Delta\alpha + \frac{\cos\alpha_o}{m}\Delta T \tag{3.109a}$$

$$\dot{\alpha} = -\frac{\bar{q}_o S}{mV_o}\Delta C_L + q - \frac{g\sin\gamma_o}{V_o}(\Delta\theta - \Delta\alpha) - \frac{T_o\cos\alpha_o}{mV_o}\Delta\alpha - \frac{\sin\alpha_o}{mV_o}\Delta T \tag{3.109b}$$

$$\dot{\beta} = \frac{\bar{q}_o S}{mV_o}\Delta C_Y + p\sin\alpha_o - r\cos\alpha_o + \frac{g\cos\theta_o}{V_o}\phi \tag{3.109c}$$

$$\dot{p} - \frac{I_{xz}}{I_x}\dot{r} = \frac{\bar{q}_o S b}{I_x}\Delta C_l \tag{3.109d}$$

$$\dot{q} = \frac{\bar{q}_o S \bar{c}}{I_y}\Delta C_m \tag{3.109e}$$

$$\dot{r} - \frac{I_{xz}}{I_z}\dot{p} = \frac{\bar{q}_o S b}{I_z}\Delta C_n \tag{3.109f}$$

$$\dot{\phi} = p + \tan\theta_o r \tag{3.109g}$$

$$\dot{\theta} = q \tag{3.109h}$$

$$\dot{\psi} = sec\,\theta_o\, r \tag{3.109i}$$

The perturbations in the nondimensional aerodynamic force and moment coefficients $\Delta C_D, \Delta C_L, \Delta C_Y, \Delta C_l, \Delta C_m,$ and ΔC_n, are replaced by the linear expansions in Eqs. (3.66) and (3.67), excluding the constant terms with subscript o, which were subtracted with the trim equations (3.106).

Linear expansions in terms of propulsion derivatives for ΔT can be found in Ref. [3.1] for various propulsion types. However, flight testing for aerodynamic parameter estimation is usually done at constant power setting for each maneuver, which means the ΔT term is zero. To estimate the effects of thrust or power level on aerodynamic parameters, several maneuvers can be run at different power settings to collect the required flight data. In this sense, the power setting is treated like a flight condition parameter, similar to Mach number. The power setting can affect the aerodynamic force and moment coefficients, due to changes in the flow field caused by flow entrainment from jet exhaust plumes or from propeller slipstreams washing over the aircraft.

Assuming constant power setting, $\Delta T = 0$, and substituting linear expansions for the force and moment coefficient perturbations, Eqs. (3.109) can be divided into two decoupled subsets describing the longitudinal and lateral motion:

Longitudinal Equations

$$\begin{aligned}\dot{V} &= -\frac{\bar{q}_o S}{m}\left(C_{D_V}\frac{\Delta V}{V_o} + C_{D_\alpha}\Delta\alpha + C_{D_q}\frac{q\bar{c}}{2V_o} + C_{D_\delta}\Delta\delta \right)\\ &\quad - g\cos\gamma_o\left(\Delta\theta - \Delta\alpha \right) - \frac{T_o \sin\alpha_o}{m}\Delta\alpha\end{aligned} \tag{3.110a}$$

$$\begin{aligned}\dot{\alpha} &= -\frac{\bar{q}_o S}{mV_o}\left(C_{L_V}\frac{\Delta V}{V_o} + C_{L_\alpha}\Delta\alpha + C_{L_{\dot{\alpha}}}\frac{\dot{\alpha}\bar{c}}{2V_o} + C_{L_q}\frac{q\bar{c}}{2V_o} + C_{L_\delta}\Delta\delta \right)\\ &\quad + q - \frac{g\sin\gamma_o}{V_o}\left(\Delta\theta - \Delta\alpha \right) - \frac{T_o\cos\alpha_o}{mV_o}\Delta\alpha\end{aligned} \tag{3.110b}$$

$$\dot{q} = \frac{\bar{q}_o S\bar{c}}{I_y}\left(C_{m_V}\frac{\Delta V}{V_o} + C_{m_\alpha}\Delta\alpha + C_{m_{\dot{\alpha}}}\frac{\dot{\alpha}\bar{c}}{2V_o} + C_{m_q}\frac{q\bar{c}}{2V_o} + C_{m_\delta}\Delta\delta \right) \tag{3.110c}$$

$$\dot{\theta} = q \tag{3.110d}$$

<u>*Lateral Equations*</u>

$$\dot{\beta} = \frac{\bar{q}_o S}{m V_o}\left(C_{Y_\beta}\beta + C_{Y_p}\frac{pb}{2V_o} + C_{Y_r}\frac{rb}{2V_o} + C_{Y_\delta}\delta \right)$$
$$+ p\, sin\alpha_o - r\, cos\alpha_o + \frac{g\, cos\theta_o}{V_o}\phi \tag{3.111a}$$

$$\dot{p} - \frac{I_{xz}}{I_x}\dot{r} = \frac{\bar{q}_o S b}{I_x}\left(C_{l_\beta}\beta + C_{l_p}\frac{pb}{2V_o} + C_{l_r}\frac{rb}{2V_o} + C_{l_\delta}\delta \right) \tag{3.111b}$$

$$\dot{r} - \frac{I_{xz}}{I_z}\dot{p} = \frac{\bar{q}_o S b}{I_z}\left(C_{n_\beta}\beta + C_{n_p}\frac{pb}{2V_o} + C_{n_r}\frac{rb}{2V_o} + C_{n_\delta}\delta \right) \tag{3.111c}$$

$$\dot{\phi} = p + tan\theta_o r \tag{3.111d}$$

$$\dot{\psi} = sec\,\theta_o\, r \tag{3.111e}$$

Lateral variables with zero values at the reference condition of steady, wings-level flight with no sideslip $(\beta,\ \phi,$ and lateral controls $\delta)$ are shown without the Δ notation, similar to the notation for the body-axis angular rates. The longitudinal variables $\Delta V,\ \Delta\alpha,\ q,$ and $\Delta\theta$ appear only in the longitudinal equations, and the lateral variables $\beta, p, r, \phi,$ and $\Delta\psi$ appear only in the lateral equations. In addition, the longitudinal controls are assumed to affect only longitudinal forces and moments, and similarly for the lateral controls. Therefore, these two sets of equations are decoupled, and can be solved and analyzed separately. This simplification is important for aircraft system identification problems, as well as aircraft dynamic analysis and control system design.

The linear model states $\Delta V, \Delta\alpha, \beta, p, q, r, \phi, \Delta\theta,$ and $\Delta\psi$ can also be outputs, so the linear output equations are simply

$$\begin{aligned} \Delta V &= \Delta V \\ \Delta\alpha &= \Delta\alpha \\ &\vdots \\ \Delta\psi &= \Delta\psi \end{aligned} \tag{3.112}$$

For translational accelerometer outputs in g units, the linear output equations follow from Eqs. (3.54), with constant power setting $(\Delta T = 0)$,

$$\Delta a_x = \frac{\overline{q}_o S}{mg}\left(2C_{X_o}\frac{\Delta V}{V_o} + \Delta C_X\right) \tag{3.113a}$$

$$\Delta a_y = \frac{\overline{q}_o S}{mg}\Delta C_Y \tag{3.113b}$$

$$\Delta a_z = \frac{\overline{q}_o S}{mg}\left(2C_{Z_o}\frac{\Delta V}{V_o} + \Delta C_Z\right) \tag{3.113c}$$

The terms involving C_{X_o} and C_{Z_o} come from the dependence of the translational accelerations on $\overline{q}$. No analogous term appears in the Δa_y equation because $C_{Y_o} = 0$.

Using Eqs. (3.40e) and (3.40f), which can be can be determined by trigonometry from Fig. 3.6,

$$C_X = -C_D \cos\alpha + C_L \sin\alpha \tag{3.114a}$$

$$C_Z = -C_L \cos\alpha - C_D \sin\alpha \tag{3.114b}$$

The perturbations ΔC_X and ΔC_Z are obtained from (3.114) as

$$\Delta C_X = \left(C_{D_o}\sin\alpha_o + C_{L_o}\cos\alpha_o\right)\Delta\alpha - \cos\alpha_o\,\Delta C_D + \sin\alpha_o\,\Delta C_L \tag{3.115a}$$

$$\Delta C_Z = \left(C_{L_o}\sin\alpha_o - C_{D_o}\cos\alpha_o\right)\Delta\alpha - \cos\alpha_o\,\Delta C_L - \sin\alpha_o\,\Delta C_D \tag{3.115b}$$

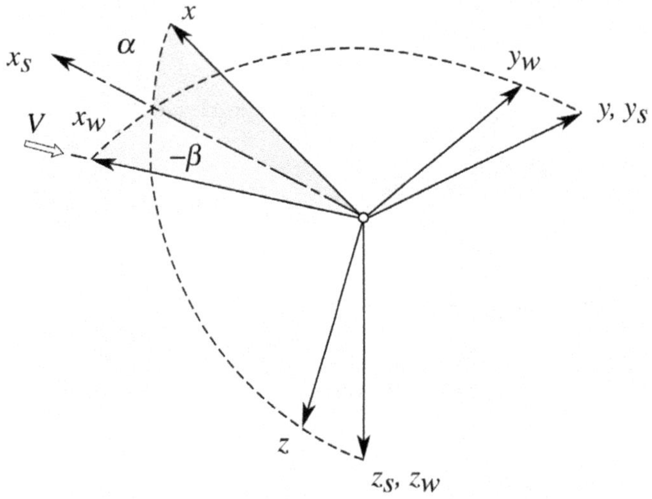

Figure 3.6 **Rotation from wind axes to stability axes and body axes**

When output equations (3.113a) and (3.113c) are used with the longitudinal equations (3.110), the expressions for Δa_x and Δa_z in Eqs. (3.113a) and (3.113c) must be modified by substituting for ΔC_X and ΔC_Z from Eqs. (3.115), and replacing ΔC_D and ΔC_L with their linear expansions, specified in Eq. (3.66).

The linearized output equation for lateral acceleration measured in g units comes from combining Eq. (3.113b) and Eq. (3.67),

$$\Delta a_y = \frac{\bar{q}S}{mg}\left(C_{Y_\beta}\beta + C_{Y_p}\frac{pb}{2V_o} + C_{Y_r}\frac{rb}{2V_o} + C_{Y_\delta}\delta \right) \tag{3.116}$$

Using dimensional stability and control derivatives, the longitudinal linearized equations (3.110) and (3.115) are:

$$\begin{aligned} \dot{V} = &-D_V \Delta V + \left(g\cos\gamma_o - \frac{T_o \sin\alpha_o}{m} - D_\alpha \right)\Delta\alpha - D_q q \\ &- D_\delta \Delta\delta - g\cos\gamma_o \Delta\theta \end{aligned} \tag{3.117a}$$

$$\begin{aligned} \dot{\alpha} = &-L_V \Delta V + \left(\frac{g\sin\gamma_o}{V_o} - \frac{T_o\cos\alpha_o}{mV_o} - L_\alpha \right)\Delta\alpha \\ &- L_{\dot{\alpha}}\dot{\alpha} + \left(1 - L_q\right)q - L_\delta \Delta\delta - \frac{g\sin\gamma_o}{V_o}\Delta\theta \end{aligned} \tag{3.117b}$$

$$\dot{q} = M_V \Delta V + M_\alpha \Delta\alpha + M_{\dot{\alpha}}\dot{\alpha} + M_q q + M_\delta \Delta\delta \tag{3.117c}$$

$$\dot{\theta} = q \tag{3.117d}$$

$$\begin{aligned} \Delta a_x = &\frac{1}{g}\left(-\cos\alpha_o \Delta D + V_o \sin\alpha_o \Delta L\right) \\ &+ \frac{1}{mg}\left(D_o \sin\alpha_o + L_o\cos\alpha_o\right)\Delta\alpha \end{aligned} \tag{3.117e}$$

$$\begin{aligned} \Delta a_z = &\frac{1}{g}\left(-V_o\cos\alpha_o \Delta L - \sin\alpha_o \Delta D\right) \\ &+ \frac{1}{mg}\left(L_o\sin\alpha_o - D_o\cos\alpha_o\right)\Delta\alpha \end{aligned} \tag{3.117f}$$

where

$$D_V = \frac{\overline{q}_o S}{mV_o} C_{D_V} \quad D_\alpha = \frac{\overline{q}_o S}{m} C_{D_\alpha} \quad D_q = \frac{\overline{q}_o S\overline{c}}{2mV_o} C_{D_q} \quad D_\delta = \frac{\overline{q}_o S}{m} C_{D_\delta} \tag{3.118a}$$

$$\begin{aligned} L_V &= \frac{\overline{q}_o S}{mV_o^2} C_{L_V} \quad & L_\alpha &= \frac{\overline{q}_o S}{mV_o} C_{L_\alpha} \quad & L_{\dot\alpha} &= \frac{\overline{q}_o S\overline{c}}{2mV_o^2} C_{L_{\dot\alpha}} \\ L_q &= \frac{\overline{q}_o S\overline{c}}{2mV_o^2} C_{L_q} \quad & L_\delta &= \frac{\overline{q}_o S}{mV_o} C_{L_\delta} \end{aligned} \tag{3.118b}$$

$$\begin{aligned} M_V &= \frac{\overline{q}_o S\overline{c}}{V_o I_y} C_{m_V} \quad & M_\alpha &= \frac{\overline{q}_o S\overline{c}}{I_y} C_{m_\alpha} \quad & M_{\dot\alpha} &= \frac{\overline{q}_o S\overline{c}^2}{2V_o I_y} C_{m_{\dot\alpha}} \\ M_q &= \frac{\overline{q}_o S\overline{c}^2}{2V_o I_y} C_{m_q} \quad & M_\delta &= \frac{\overline{q}_o S\overline{c}}{I_y} C_{m_\delta} \end{aligned} \tag{3.118c}$$

$$\Delta D = D_V \Delta V + \left(D_\alpha + \frac{T_o \sin\alpha_o}{m} \right) \Delta\alpha + D_q q + D_\delta \Delta\delta \tag{3.118d}$$

$$\Delta L = L_V \Delta V + \left(\frac{T_o \cos\alpha_o}{mV_o} + L_\alpha \right) \Delta\alpha + L_\alpha \dot\alpha + L_q q + L_\delta \Delta\delta \tag{3.118e}$$

The lateral linearized equations (3.111) and (3.116) can be written using dimensional stability and control derivatives as:

$$\dot\beta = Y_\beta \beta + \left(Y_p + \sin\alpha_o \right) p + \left(Y_r - \cos\alpha_o \right) r + \frac{g\cos\theta_o}{V_o} \phi + Y_\delta \delta \tag{3.119a}$$

$$\dot p - \frac{I_{xz}}{I_x} \dot r = L_\beta \beta + L_p p + L_r r + L_\delta \delta \tag{3.119b}$$

$$\dot r - \frac{I_{xz}}{I_z} \dot p = N_\beta \beta + N_p p + N_r r + N_\delta \delta \tag{3.119c}$$

$$\dot\phi = p + \tan\theta_o r \tag{3.119d}$$

$$\dot\psi = \sec\theta_o \, r \tag{3.119e}$$

$$\Delta a_y = \frac{V_o}{g} \left(Y_\beta \beta + Y_p p + Y_r r + Y_\delta \delta \right) \tag{3.119f}$$

where

$$Y_\beta = \frac{\bar{q}_o S}{m V_o} C_{Y_\beta} \quad Y_p = \frac{\bar{q}_o S b}{2 m V_o^2} C_{Y_p} \quad Y_r = \frac{\bar{q}_o S b}{2 m V_o^2} C_{Y_r} \quad Y_\delta = \frac{\bar{q}_o S}{m V_o} C_{Y_\delta} \tag{3.120a}$$

$$L_\beta = \frac{\bar{q}_o S b}{I_x} C_{l_\beta} \quad L_p = \frac{\bar{q}_o S b^2}{2 V_o I_x} C_{l_p} \quad L_r = \frac{\bar{q}_o S b^2}{2 V_o I_x} C_{l_r} \quad L_\delta = \frac{\bar{q}_o S b}{I_x} C_{l_\delta} \tag{3.120b}$$

$$N_\beta = \frac{\bar{q}_o S b}{I_z} C_{n_\beta} \quad N_p = \frac{\bar{q}_o S b^2}{2 V_o I_z} C_{n_p} \quad N_r = \frac{\bar{q}_o S b^2}{2 V_o I_z} C_{n_r} \quad N_\delta = \frac{\bar{q}_o S b}{I_z} C_{n_\delta} \tag{3.120c}$$

Note that the dimensional derivatives defined in Eqs. (3.118) and (3.120) are not the same as the dimensional partial derivatives of the aerodynamic forces and moments defined earlier in Eqs. (3.57)-(3.60). This notational ambiguity is unfortunately common practice.

Further simplification of the linearized longitudinal equations can be achieved using the short-period approximation, which is a low-order linear dynamic model that accurately describes the short-period response of the aircraft. Most longitudinal flight maneuvers of interest for stability and control purposes occur over relatively short time periods, and therefore involve the short-period response.

The short-period approximation is obtained by introducing the following assumptions:

1) $\Delta V = 0$, and the airspeed dynamic equation is dropped
2) $\gamma_o = 0$ for level flight at the reference condition
3) $\frac{T_o \cos \alpha_o}{m V_o} \Delta\alpha$ is relatively small, and can be neglected
4) Variations in pitch rate q and angle of attack rate $\dot{\alpha}$ are similar
5) $C_L = C_L(\alpha, q, \delta)$
6) $C_m = C_m(\alpha, q, \delta)$

The short-period linear model equations then have the form

$$\dot{\alpha} = q - \frac{\bar{q}_o S}{m V_o}\left(C_{L_\alpha} \Delta\alpha + \bar{C}_{L_q} \frac{q \bar{c}}{2 V_o} + C_{L_\delta} \Delta\delta \right) \tag{3.121a}$$

$$\dot{q} = \frac{\bar{q}_o S \bar{c}}{I_y}\left(C_{m_\alpha} \Delta\alpha + \bar{C}_{m_q} \frac{q \bar{c}}{2 V_o} + C_{m_\delta} \Delta\delta \right) \tag{3.121b}$$

The derivatives $\bar{C}_{L_q}$ and $\bar{C}_{m_q}$ in Eqs. (3.121) indicate the combined effects of pitch rate and angle of attack rate. In practice, these effects cannot be separated using data from typical aircraft short period motion, because the variations in pitch rate and angle of attack rate are similar. Consequently, a single combined derivative is used for the combined pitch rate and angle of attack rate effects.

The linearized output equations for the states of the short-period model have the form [cf. Eq. (3.112)],

$$\begin{aligned} \Delta\alpha &= \Delta\alpha \\ q &= q \end{aligned} \tag{3.122}$$

Since the airspeed dynamic equation has been dropped, the Δa_x output equation is also omitted. For the vertical acceleration output, the linearized equation comes from Eq. (3.113c) with $\Delta V = 0$,

$$\Delta a_z = \frac{\bar{q}_o S}{mg} \Delta C_Z \tag{3.123}$$

Dropping the ΔC_D term from Eq. (3.115b) and substituting into Eq. (3.123) gives the linearized output equation for the vertical acceleration in the short-period approximation,

$$\Delta a_z = \frac{\bar{q}_o S}{mg}\left[\left(C_{L_o} \sin\alpha_o - C_{D_o} \cos\alpha_o\right)\Delta\alpha - \cos\alpha_o\, \Delta C_L\right] \tag{3.124}$$

Substituting the expression for ΔC_L from inside the parentheses on the right side of Eq. (3.121a),

$$\begin{aligned} \Delta a_z = \frac{\bar{q}_o S}{mg}\Bigg[&\left(C_{L_o} \sin\alpha_o - C_{D_o} \cos\alpha_o\right)\Delta\alpha \\ &- \cos\alpha_o\left(C_{L_\alpha}\Delta\alpha + \bar{C}_{L_q}\frac{q\bar{c}}{2V_o} + C_{L_\delta}\Delta\delta\right)\Bigg] \end{aligned} \tag{3.125}$$

For low values of trim angle of attack α_o, the first $\Delta\alpha$ term is small and $\cos\alpha_o \approx 1$, so that

$$\Delta a_z \approx -\frac{\bar{q}_o S}{mg}\left(C_{L_\alpha}\Delta\alpha + \bar{C}_{L_q}\frac{q\bar{c}}{2V_o} + C_{L_\delta}\Delta\delta\right) \tag{3.126}$$

Equations (3.121) and (3.126) are the linear equations for the short-period approximation,

$$\dot{\alpha} = q - \frac{\bar{q}_o S}{mV_o}\left(C_{L_\alpha} \Delta\alpha + \bar{C}_{L_q} \frac{q\bar{c}}{2V_o} + C_{L_\delta} \Delta\delta \right) \tag{3.127a}$$

$$\dot{q} = \frac{\bar{q}_o S\bar{c}}{I_y}\left(C_{m_\alpha} \Delta\alpha + \bar{C}_{m_q} \frac{q\bar{c}}{2V_o} + C_{m_\delta} \Delta\delta \right) \tag{3.127b}$$

$$\Delta a_z = -\frac{\bar{q}_o S}{mg}\left(C_{L_\alpha} \Delta\alpha + \bar{C}_{L_q} \frac{q\bar{c}}{2V_o} + C_{L_\delta} \Delta\delta \right) \tag{3.127c}$$

These equations are sometimes written in terms of dimensional stability and control derivatives,

$$\dot{\alpha} = -L_\alpha \Delta\alpha + \left(1 - \bar{L}_q\right) q - L_\delta \Delta\delta \tag{3.128a}$$

$$\dot{q} = M_\alpha \Delta\alpha + \bar{M}_q q + M_\delta \Delta\delta \tag{3.128b}$$

$$\Delta a_z = -\frac{V_o}{g}\left(L_\alpha \Delta\alpha + \bar{L}_q q + L_\delta \Delta\delta \right) \tag{3.128c}$$

where

$$L_\alpha = \frac{\bar{q}_o S}{mV_o} C_{L_\alpha} \qquad \bar{L}_q = \frac{\bar{q}_o S\bar{c}}{2mV_o^2} \bar{C}_{L_q} \qquad L_\delta = \frac{\bar{q}_o S}{mV_o} C_{L_\delta} \tag{3.129a}$$

$$M_\alpha = \frac{\bar{q}_o S\bar{c}}{I_y} C_{m_\alpha} \qquad \bar{M}_q = \frac{\bar{q}_o S\bar{c}^2}{2V_o I_y} \bar{C}_{m_q} \qquad M_\delta = \frac{\bar{q}_o S\bar{c}}{I_y} C_{m_\delta} \tag{3.129b}$$

At low angles of attack, $-L \approx Z \cos\alpha \approx Z$, so that Eqs. (3.129) can also be written as

$$\dot{\alpha} = Z_\alpha \Delta\alpha + \left(1 + \bar{Z}_q\right) q + Z_\delta \Delta\delta \tag{3.130a}$$

$$\dot{q} = M_\alpha \Delta\alpha + \bar{M}_q q + M_\delta \Delta\delta \tag{3.130b}$$

$$\Delta a_z = \frac{V_o}{g}\left(Z_\alpha \Delta\alpha + \bar{Z}_q q + Z_\delta \Delta\delta \right) \tag{3.130c}$$

where

$$Z_\alpha = \frac{\bar{q}_o S}{mV_o} C_{Z_\alpha} \qquad \bar{Z}_q = \frac{\bar{q}_o S\bar{c}}{2mV_o^2} \bar{C}_{Z_q} \qquad Z_\delta = \frac{\bar{q}_o S}{mV_o} C_{Z_\delta} \tag{3.131a}$$

$$M_\alpha = \frac{\bar{q}_o S \bar{c}}{I_y} C_{m_\alpha} \quad \bar{M}_q = \frac{\bar{q}_o S \bar{c}^2}{2V_o I_y} \bar{C}_{m_q} \quad M_\delta = \frac{\bar{q}_o S \bar{c}}{I_y} C_{m_\delta} \tag{3.131b}$$

Linearized dynamic equations for translational motion of the aircraft can also be developed in body axes, using the same procedure, starting from the nonlinear equations (3.36). The result is

$$\dot{u} = -w_o q + \frac{\bar{q}S}{m} \Delta C_X - g\cos\theta_o \Delta\theta + \frac{\Delta T}{m} \tag{3.132a}$$

$$\dot{v} = w_o p - u_o r + \frac{\bar{q}S}{m} \Delta C_Y + g\cos\theta_o\, \phi \tag{3.132b}$$

$$\dot{w} = u_o q + \frac{\bar{q}S}{m} \Delta C_Z - g\sin\theta_o \Delta\theta \tag{3.132c}$$

The linear output equations for air-data quantities normally measured on aircraft $(\text{i.e., } V, \alpha, \text{ and } \beta)$ follow from Eqs. (3.46) and (3.48) with $u_o \approx V_o$,

$$\Delta V = \Delta u + w_o \frac{\Delta w}{u_o} \tag{3.133a}$$

$$\Delta\alpha = \Delta w / u_o \tag{3.133b}$$

$$\beta = v / u_o \tag{3.133c}$$

The linear output equations for the accelerometers for $\Delta T = 0$ are similar to those given in Eqs. (3.113),

$$\Delta a_x = \frac{\bar{q}_o S}{mg} \left\{ 2C_{X_o} \left[\frac{\Delta u}{u_o} + \left(\frac{w_o}{u_o} \right) \frac{\Delta w}{u_o} \right] + \Delta C_X \right\} \tag{3.134a}$$

$$\Delta a_y = \frac{\bar{q}_o S}{mg} \Delta C_Y \tag{3.134b}$$

$$\Delta a_z = \frac{\bar{q}_o S}{mg} \left\{ 2C_{Z_o} \left[\frac{\Delta u}{u_o} + \left(\frac{w_o}{u_o} \right) \frac{\Delta w}{u_o} \right] + \Delta C_Z \right\} \tag{3.134c}$$

Note that if the body axes used to derive Eqs. (3.132) are rotated into stability axes, then $w_o = 0$, $\alpha_o = 0$, $\Delta\gamma = \Delta\theta$, $u = V$, $C_Z = -C_L$ and $C_X = -C_D$. In addition, the thrust will have components along the x and z stability axes, $T\cos\alpha$ and $-T\sin\alpha$, respectively, with corresponding derivatives that are more complicated than the simple ΔT appearing in the x body axis equation. Making these substitutions in Eqs. (3.132) and using the approximations in Eqs. (3.133), results in the translational linear dynamic equations in stability axes, Eqs. (3.109a)-(3.109c).

The linearized lateral equations can be simplified to isolate the roll mode, which is the dominant aircraft dynamic response to conventional aileron inputs. Using only the linearized rolling moment equation, and dropping all states and controls except roll rate p and aileron δ_a ,

$$\dot{p} = \frac{\bar{q}_o S b}{I_x}\left(C_{l_p} \frac{pb}{2V_o} + C_{l_{\delta_a}} \delta_a \right) \tag{3.135a}$$

Using dimensional derivatives,

$$\dot{p} = L_p p + L_{\delta_a} \delta_a \tag{3.135b}$$

Roll rate p is the only output. Eqs. (3.135a) and (3.135b) are called the roll mode approximation using nondimensional and dimensional derivatives, respectively.

3.9.2 Substituting Measured Values

Data used for aircraft system identification come mostly from maneuvers that excite the longitudinal short period dynamics or the lateral modes associated with body-axis roll, spiral motion, and lateral oscillations. For these maneuvers, the airspeed is expected to be constant and the longitudinal and lateral motions are decoupled. In practice, the airspeed exhibits some changes from the reference value, and there are variations in the lateral quantities during a longitudinal maneuver, and vice versa. To keep the equations relevant to the maneuvers in a simple form, the dynamic pressure $\bar{q}$, airspeed V, and the variables that would cause coupling between the longitudinal and lateral motion can be replaced by their measured values. For longitudinal motion, dropping the airspeed and kinematic equations, the longitudinal equations (3.39b), (3.37b), and (3.38b) are simplified by using measured values from the experiment, $\bar{q}_E, T_E, V_E, \alpha_E, \theta_E, \beta_E, p_E, r_E,$ and ϕ_E , where α_E and θ_E are used only in the nonlinear coupling terms. The result is

$$\begin{aligned}\dot{\alpha} = &-\frac{\bar{q}_E S}{m V_E \cos\beta_E} C_L + q - \tan\beta_E \left(p_E \cos\alpha_E + r_E \sin\alpha_E \right) - \frac{T_E \sin\alpha_E}{m V_E \cos\beta_E} \\ &+ \frac{g}{V_E \cos\beta_E}\left(\cos\phi_E \cos\theta_E \cos\alpha_E + \sin\theta_E \sin\alpha_E \right)\end{aligned} \tag{3.136a}$$

$$\dot{q} = \frac{\bar{q}_E S\bar{c}}{I_y} C_m - \frac{(I_x - I_z)}{I_y} p_E r_E - \frac{I_{xz}}{I_y}\left(p_E^2 - r_E^2 \right) + \frac{I_p}{I_y} \Omega_p r_E \tag{3.136b}$$

$$\dot{\theta} = q \cos\phi_E - r_E \sin\phi_E \tag{3.136c}$$

Assuming measured sideslip angle β_E is small, the simplified longitudinal equations have the form

$$\begin{aligned}\dot{\alpha} = &-\frac{\overline{q}_E S}{mV_E} C_L + q - \beta_E \left(p_E \cos\alpha_E + r_E \sin\alpha_E \right) - \frac{T_E \sin\alpha_E}{mV_E} \\ &+ \frac{g}{V_E} \left(\cos\phi_E \cos\theta_E \cos\alpha_E + \sin\theta_E \sin\alpha_E \right)\end{aligned} \tag{3.137a}$$

$$\dot{q} = \frac{\overline{q}_E S\overline{c}}{I_y} C_m - \frac{(I_x - I_z)}{I_y} p_E r_E - \frac{I_{xz}}{I_y} \left(p_E^2 - r_E^2 \right) + \frac{I_p}{I_y} \Omega_p r_E \tag{3.137b}$$

$$\dot{\theta} = q \cos\phi_E - r_E \sin\phi_E \tag{3.137c}$$

Using measured values removes nonlinearities and also decouples the longitudinal equations from the lateral equations. Linear expansions from Eq. (3.71) are substituted for C_L and C_m. Note that this expansion must retain the constant terms C_{L_o} and C_{m_o}, because the measured quantities include steady trim values. Further simplifications to the equations can be made by dropping individual terms with small relative magnitudes for the specific maneuver being studied.

Using measured values, the linearized lateral equations are

$$\begin{aligned}\dot{\beta} = &\frac{\overline{q}_E S}{mV_E} C_{Y_W} + p \sin\alpha_E - r \cos\alpha_E + \frac{g}{V_E} \cos\beta_E \sin\phi_E \cos\theta_E \\ &+ \frac{\sin\beta_E}{V_E} \left(g \cos\alpha_E \sin\theta_E - g \sin\alpha_E \cos\phi_E \cos\theta_E - \frac{T_E \cos\alpha_E}{m} \right)\end{aligned} \tag{3.138a}$$

$$\dot{p} - \frac{I_{xz}}{I_x} \dot{r} = \frac{\overline{q}_E Sb}{I_x} C_l - \frac{\left(I_z - I_y\right)}{I_x} q_E r + \frac{I_{xz}}{I_x} q_E p \tag{3.138b}$$

$$\dot{r} - \frac{I_{xz}}{I_z} \dot{p} = \frac{\overline{q}_E Sb}{I_z} C_n - \frac{\left(I_y - I_x\right)}{I_z} q_E p - \frac{I_{xz}}{I_z} q_E r - \frac{I_p}{I_z} \Omega_p q_E \tag{3.138c}$$

$$\dot{\phi} = p + \tan\theta_E \left(q_E \sin\phi_E + r \cos\phi_E \right) \tag{3.138d}$$

$$\dot{\psi} = \frac{q_E \sin\phi_E + r \cos\phi_E}{\cos\theta_E} \tag{3.138e}$$

Assuming measured sideslip angle β_E is small, so that $cos\,\beta_E \approx 1$, $C_{Y_W} \approx C_Y$, and $sin\,\beta_E / V_E \approx 0$, and dropping the heading angle equation, which has no influence on the aircraft dynamics, the lateral equations are simplified as

$$\dot{\beta} = \frac{\overline{q}_E S}{mV_E} C_Y + p\,sin\alpha_E - r\,cos\,\alpha_E + \frac{g}{V_E}\,sin\phi_E\,cos\,\theta_E \tag{3.139a}$$

$$\dot{p} - \frac{I_{xz}}{I_x}\dot{r} = \frac{\overline{q}_E Sb}{I_x} C_l - \frac{(I_z - I_y)}{I_x} q_E r + \frac{I_{xz}}{I_x} q_E p \tag{3.139b}$$

$$\dot{r} - \frac{I_{xz}}{I_z}\dot{p} = \frac{\overline{q}_E Sb}{I_z} C_n - \frac{(I_y - I_x)}{I_z} q_E p - \frac{I_{xz}}{I_z} q_E r - \frac{I_p}{I_z}\Omega_p q_E \tag{3.139c}$$

$$\dot{\phi} = p + tan\theta_E\,(q_E\,sin\phi_E + r\,cos\,\phi_E) \tag{3.139d}$$

The equations linearized in this way can be used for larger-amplitude maneuvers or significantly coupled maneuvers, because none of the nonlinear terms were dropped due to multiplications of small perturbation quantities. However, when measured values are substituted into the equations of motion, the accuracy of the solution will be affected by systematic and random errors in the measured quantities.

3.10 Summary and Concluding Remarks

In this chapter, the rigid-body equations of motion for an airplane were derived. The resulting set of coupled nonlinear ordinary differential equations include terms representing the aerodynamic forces and moments acting on the vehicle. Since the rigid-body equations of motion, also called the Euler equations, are well known, the focus of aircraft system identification is typically on obtaining good models for the dependencies of the aerodynamic forces and moments on aircraft states and controls. The equations developed for this purpose are called aerodynamic model equations. Several mathematical constructs were introduced for the aerodynamic model equations, including linear expansions, multivariate polynomials, polynomial splines, and indicial functions. Parameters in these mathematical models are the unknowns to be estimated based on measured data. Each model form has implications in terms of the relationships that can be described and the experimentation and measurements required for successful modeling. These issues will be explored further in the following chapters.

The full nonlinear equations of motion were then simplified using two methods: linearization for small motions about a reference condition, and substituting measured values in the nonlinear equations to linearized them. The resulting sets of longitudinal and lateral equations are linear and decoupled, with fewer equations and fewer model parameters, which simplifies the analysis.

The next chapter covers the theory required for estimating the unknown parameters in postulated models of the forms developed in this chapter. Chapters 5-7 explain the practical application of this theory to aircraft system identification problems.

References

[3.1] Etkin, B., *Dynamics of Atmospheric Flight – Stability and Control*, 2nd Ed., John Wiley & Sons, New York, NY, 1982.

[3.2] Etkin, B. and Reid, L.D., *Dynamics of Flight – Stability and Control*, 3rd Ed., John Wiley & Sons, New York, NY, 1996.

[3.3] McRuer, D., Ashkenas, I. and Graham, D., *Aircraft Dynamics and Automatic Control*, Princeton University Press, Princeton, NJ, 1973.

[3.4] Stevens, B.L. and Lewis, F.L., *Aircraft Control and Simulation*, John Wiley & Sons, New York, NY, 1992.

[3.5] Roskam, J., *Airplane Flight Dynamics and Automatic Flight Controls*, Parts I and II, Roskam Aviation and Engineering, Lawrence, KS, 1979.

[3.6] Greenberg, H., "Determination of Stability Derivatives From Flight Data," *Journal of the Aeronautical Sciences*, Vol. 16, No. 1, 1949, p. 62.

[3.7] Greenberg, H., "A Survey of Methods for Determining Stability Parameters of an Airplane from Dynamic Flight Measurements," NASA TN 2340, 1951.

[3.8] Schumacher, L.L., *Spline Function Basic Theory*, John Wiley & Sons, New York, NY, 1981.

[3.9] Klein, V. and Batterson, J.G., "Determination of Airplane Model Structure From Flight Data Using Splines and Stepwise Regression," NASA TP-2126, 1983.

[3.10] Cowley, W.L. and Glauert, H., "The Effect of the Lag of the Downwash on the Longitudinal Stability of an Airplane and on the Rotary Derivative," Reports and Memoranda, No. 718, 1921.

[3.11] Tobak, M., "On the Use of the Indicial Function Concept in the Analysis of Unsteady Motion of Wing and Wing-Tail Combinations," NACA Report 1188, 1954.

[3.12] Tobak, M. and Schiff, L.B., "On the Formulation of the Aerodynamic Characteristics in Aircraft Dynamics," NASA TR R-456, 1976.

[3.13] Jones, R.T., "The Unsteady Lift of a Wing of Finite Aspect Ratio," NACA Report No. 681, 1939.

[3.14] Klein, V. and Murphy, P.C., "Estimation of Aircraft Nonlinear Unsteady Parameters From Wind Tunnel Data," NASA TM-1998-208969, 1998.

[3.15] Jones, R.T. and Fehlner, L.F., "Transient Effects of the Wing Wake on the Horizontal Tail," NASA TN No. 771, 1940.

[3.16] Klein, V., "Modeling of Longitudinal Unsteady Aerodynamics of a Wing-Tail Combination," NASA CR-1999-209547, 1999.

[3.17] Khrabrov, A., Vinogradov, Y., and Abramov, N., "Mathematical Modelling of Aircraft Unsteady Aerodynamics at High Incidence with Account of Wing-Tail Interaction," AIAA Paper 2004-5278, *AIAA Atmospheric Flight Mechanics Conference*, Providence, RI, 2004.

[3.18] Goman, M.G., Stolyarov, G.J., Tartyshnikov, S.L., Usokev, S.P, and Khrabrov, A.N., "Mathematical Description of Aerodynamic Forces and Moments at Nonstationary Flow Regimes with a Nonunique Structure," *Proceedings of TsAGI*, Issue 2195, Moscow, Russia, 1983 (in Russian).

[3.19] Goman, M.G. and Khrabrov, A.N., "State-Space Representation of Aerodynamic Characteristics of an Aircraft at High Angles of Attack," AIAA Paper 92-4651, *AIAA Atmospheric Flight Mechanics Conference*, Hilton Head Island, SC, 1992.

[3.20] Klein, V. and Noderer, K.D., "Modeling of Aircraft Unsteady Aerodynamic Characteristics, Part 1 - Postulated Models," NASA TM 109120, 1994.

[3.21] Abramov, N., Goman, M., and Khrabrov, A., "Aircraft Dynamics at High Incidence Flight with Account of Unsteady Aerodynamic Effects," AIAA Paper 2004-5274, *AIAA Atmospheric Flight Mechanics Conference*, Providence, RI, 2004.

[3.22] Murphy, P.C. and Klein, V., "Estimation of Aircraft Unsteady Aerodynamic Parameters From Dynamic Wind Tunnel Testing," AIAA Paper 2001-4016, *AIAA Atmospheric Flight Mechanics Conference*, Montreal, Canada, 2001.

[3.23] Fishenberg, D., "Identification of an Unsteady Aerodynamic Stall Model From Flight Test Data," AIAA Paper 95-3438, 1995.

Problems

3.1 Answer the following, using words, graphs, equations, and/or diagrams:

a) Explain why it is important to know the precise location of the force and moment measuring device installed on an aircraft model in a wind tunnel experiment.

b) Explain why nonlinear aircraft simulations use numerical integration routines.

c) A fighter aircraft engaged in air combat takes a hit in the left wing, putting a hole in the wing, and causing a significant steady fuel leak from the left wing fuel tank. The right wing tank is separate and undamaged. Discuss the changes in the dynamics and control of the damaged aircraft compared to the undamaged aircraft.

d) Explain why differential equations for the Euler angles are needed in the equations of motion for an aircraft.

e) Compare and contrast wind tunnel testing and flight testing for aerodynamic modeling of fixed-wing airplanes.

f) What are nondimensional stability and control derivatives, and why are they important?

g) Explain the quasi-steady flow assumption used for aerodynamic modeling.

h) Explain why rotorcraft system identification is done using equations of motion written in body axes rather than wind axes.

3.2 Fly a longitudinal doublet maneuver on the F-16 nonlinear simulation using fly_f16.m. Make the maneuver length about 10 seconds, and use approximately a 1 second pulse width for the doublet. The data from your maneuver is assembled into the standard `fdata` matrix by fly_f16.m.

a) Cut the data for your maneuver using cutftd.m.

b) Plot angle of attack, pitch rate, body-axis z acceleration, and stabilator deflection in MATLAB®, using time `t` and the `fdata` array.

c) The script fly_f16.m trims and linearizes the nonlinear simulation at the start. Use `damps.m` and input the A matrix to find the characteristic roots for the longitudinal dynamic modes.

3.3 Using gen_f16_model.m, find the longitudinal linear model for the F-16 at 5 deg angle of attack, $x_{cg} = 0.25\bar{c}$, and 10,000 ft altitude.

a) Identify the natural frequency, damping, and/or time constants of the modes.

b) Is the aircraft stable in pitch at this flight condition?

c) What are the values of M_α and M_{δ_s} ?

3.4 Using gen_f16_model.m, find the lateral linear model for the F-16 at 5 deg angle of attack, $x_{cg} = 0.25\bar{c}$, and 10,000 ft altitude.

a) Identify the natural frequency, damping, and/or time constants of the modes.

b) How do you expect the F-16 to fly at this flight condition? Test fly the F-16 and discuss your findings.

3.5 Investigate the effects of changing the longitudinal c.g. position by $\pm 0.1\bar{c}$ on the longitudinal dynamics of the F-16, for flight at 5 deg angle of attack and 10,000 ft altitude. Make time history comparisons for linear and nonlinear simulations using a doublet input on the stabilator, generated with mksqw.m. Discuss the comparison and the airplane dynamic response.

3.6 Find the values of the longitudinal and lateral nondimensional stability and control derivatives for the F-16 in level flight at 500 ft/s with $x_{cg} = 0.25\bar{c}$. Use the short period approximation for the longitudinal model, and the full fourth-order dynamics for the lateral model. Write an m-file that will do this for any given linear system matrices.

3.7 For the F-16 flying straight and level at 15 degrees trim angle of attack, $x_{cg} = 0.25\bar{c}$, and 20,000 feet altitude, answer the following questions:

a) What are the trim conditions for angle of attack, stabilator deflection, and throttle position?

b) Find the $\tilde{q}/\tilde{\delta}_s$ transfer function for the short period approximation.

c) Find the natural frequency and damping ratio for the short period mode. How does this compare to the values found for the full linear model?

d) What is the value of L_q in the short period approximation?

3.8 Write an m-file that will find the nondimensional stability and control derivatives from any given linear system matrices, using the short period approximation for the longitudinal model and the fourth-order lateral dynamics for the lateral model. Demonstrate the use of this m-file to find the nondimensional longitudinal and lateral stability and control derivatives for the F-16 in straight and level flight at 500 ft/s airspeed and 10,000 ft altitude, with $x_{cg} = 0.25\bar{c}$. (*Hint:* Fix the airspeed at 500 ft/s in gen_f16_model.m, and let the angle of attack be free and determined by the trim routine).

3.9 The linearized lateral equations of motion can be simplified to a Dutch roll approximation, as follows:

$$\dot{\beta} = Y_\beta \beta + (Y_r - 1) r + Y_{\delta_r} \delta_r$$

$$\dot{r} = N'_\beta \beta + N'_r r + N'_{\delta_r} \delta_r$$

a) Find the $\tilde{r}/\tilde{\delta}_r$ transfer function for the Dutch roll approximation.

b) Find the natural frequency and damping ratio for the Dutch roll mode.

c) Write an expression for N'_β in terms of non-dimensional derivatives.

c) Assume there is yaw rate feedback to the rudder, so that $\delta_r = Kr$, where K is selectable. Show how using this type of feedback can change the Dutch roll damping.

3.10 Starting with Eqs. (3.2) and (3.3), derive the equations of motion for time-varying mass and inertia.

3.11 Derive the linearized equations of motion when the reference condition is a steady, level, coordinated turn, where

$$\beta_o = 0\,,\; p = p_o + \Delta p\,,\; q = q_o + \Delta q\,,\; r = r_o + \Delta r\,,\; \phi = \phi_o + \Delta\phi$$

Are the longitudinal and lateral linear equations still decoupled in this case?

3.12 Modify the linear short period mode equations (3.128) for the case when there is a steady wind that changes the angle of attack by a known constant value $\Delta\alpha_g$. How would the equations change if the wind were not steady, so that $\Delta\alpha_g$ is not constant?

3.13 For the Twin Otter flight data in the file totter_demo_data.mat, show that $T_o \cos\alpha_o \Delta\alpha / mV_o$ is negligible, compared to the terms retained in Eq. (3.117a).

3.14 Make a table of values for the full expressions compared to the approximations for $\sin x \approx x$, $\cos x \approx 1$ and $\tan x \approx x$ using various values of x in radians. How large can the angle x be before the error in using the approximation exceeds 10 percent of the exact value? Do a similar analysis for the approximations in Eqs. (3.108).

3.15 A fighter flying straight and level at 625 ft/s and 30,000 ft has the following longitudinal linear equations of motion:

$$\dot{q} = -0.85q - \alpha - 1.2\delta_e$$

$$\dot{V} = 225\delta_{th} + 0.035\alpha - 9.81\theta - 0.18V$$

$$\dot{\alpha} = q - 0.2V - 0.6\alpha - 0.035\delta_e$$

$$\dot{\theta} = q$$

a) Write the equations for the short period approximation.

b) Find the $\tilde{q}/\tilde{\delta}_e$ transfer function for the short period approximation.

c) Find the natural frequency and damping ratio for the short period mode.

d) What is the value of L_q ? Explain your answer.

3.16 Compare the F-16 outputs for a longitudinal doublet input with pulse width of 1 s (see mksqw.m) using the full nonlinear F-16 simulation (see f16.m) and the short-period approximation (see gen_f16_model.m and lsims.m). Discuss quantitatively the quality of the short-period approximation.

Chapter 4

Outline of Estimation Theory

The parameter estimation process consists of finding values of unknown model parameters $\boldsymbol{\theta}$ in an assumed model structure, based on noisy measurements z. An estimator is a function of the random variable z that produces an estimate $\hat{\boldsymbol{\theta}}$ of the unknown parameters $\boldsymbol{\theta}$. Since the estimator computes $\hat{\boldsymbol{\theta}}$ based on noisy measurements z, $\hat{\boldsymbol{\theta}}$ is a random variable.

Parameter estimation requires specification of the following:

1) A model structure with unknown parameters $\boldsymbol{\theta}$ to be estimated;
2) Observations or measurements z;
3) A mathematical model for the measurement process;
4) Assumptions about the uncertainty in the model parameters $\boldsymbol{\theta}$ and the measurement noise $\boldsymbol{v}$.

In Chapter 2, a distinction was made between linear and nonlinear dynamic systems, based on the relation between state time derivatives and the state and control variables. For parameter estimation, however, the relation between the measured outputs and model parameters is of much greater importance.

A model is called linear in the parameters if the output y is given by

$$y = \boldsymbol{X\theta} \tag{4.1}$$

where the matrix $\boldsymbol{X}$ is assumed to be known. Then the measurement equation can be expressed as

$$z = \boldsymbol{X\theta} + \boldsymbol{v} \tag{4.2}$$

A model that is nonlinear in the parameters has a measurement equation of the form

$$z = \boldsymbol{\varphi}(\boldsymbol{\theta}) + \boldsymbol{v} \tag{4.3}$$

where the form of the function $\boldsymbol{\varphi}(\boldsymbol{\theta})$ is assumed to be known.

In general, there are n_o measured outputs, and a vector of measurements is taken at each sample *i*, where $i = 1, 2, \ldots, N$, and N is the number of sampled data points.

In this chapter, a single measured output is considered, so that $n_o = 1$, and z is a vector composed of N scalar measurements. The extension to multiple measured outputs is discussed in Chapters 6 and 7. measurement vectors.

The notion of a linear or nonlinear dynamic system has no connection to the model being linear or nonlinear in the parameters, as can be seen from the following example.

Example 4.1

Consider a linear, first-order, scalar system described as

$$\dot{x} + \theta x = u \tag{4.4}$$

with output equation

$$y = \dot{x} = -\theta x + u \tag{4.5}$$

and measurement equation

$$z = y + v = -\theta x + u + v \tag{4.6}$$

The output is linear in the parameter θ. If instead the output equation is

$$y = x = \int_0^t e^{-\theta(t-\tau)} u(\tau) d\tau \tag{4.7}$$

so that

$$z = \int_0^t e^{-\theta(t-\tau)} u(\tau) d\tau + v \tag{4.8}$$

then the output is nonlinear in the parameter θ. In this simple example, the same dynamic system exhibits output equations that are linear or nonlinear in the parameter, depending on how the model output is defined. ■

The notion of linearity and nonlinearity in the parameters will be further clarified when the various parameter estimation methods are introduced. Models for uncertainty in $\boldsymbol{\theta}$ and $\boldsymbol{v}$ will be specified by probability density functions $p(\boldsymbol{\theta})$ and $p(\boldsymbol{v})$, respectively.

Knowledge about the estimated parameters can be expressed in terms of the probability density function $p(\hat{\boldsymbol{\theta}} \mid z)$, which represents the probability density of the parameter estimate $\hat{\boldsymbol{\theta}}$, given the measurements z. This probability density function would be the most complete information that could be derived by applying statistical techniques. In practice, however, it can be quite difficult to find the solution to the estimation problem in the form of a probability density function, because the expected value and higher-order moments for a random variable vector would have to be estimated. For that reason, the solution is reduced from the probability density function $p(\hat{\boldsymbol{\theta}} \mid z)$ to its most significant properties, which are:

$$\text{expected value: } E(\hat{\boldsymbol{\theta}} \mid z)$$

$$\text{covariance: } E\left\{\left[\hat{\boldsymbol{\theta}} - E(\hat{\boldsymbol{\theta}} \mid z)\right]\left[\hat{\boldsymbol{\theta}} - E(\hat{\boldsymbol{\theta}} \mid z)\right]^T\right\} \tag{4.9}$$

$$\text{bias: } E(\hat{\boldsymbol{\theta}} \mid z) - E(\boldsymbol{\theta} \mid z)$$

During the estimation process, an attempt is made to obtain a good estimate of $\boldsymbol{\theta}$. To achieve this, it is necessary to define what is meant by "good". The main properties that are used to characterize the quality of an estimator are defined next. This is followed by the development of estimators for a system described by Eq. (4.2) or (4.3), and by three different models of the uncertainty in $\boldsymbol{\theta}$ and $\boldsymbol{v}$. Then, state estimation procedures for dynamic systems will be introduced and described.

4.1 Properties of Estimators

The following definitions of estimator properties are presented without development or proof. Rigorous treatment of these definitions can be found in Refs. [4.1] and [4.2].

Definition 4.1

An estimator is linear if $\hat{\boldsymbol{\theta}}$ is obtained as a linear function of measurements. If $\hat{\boldsymbol{\theta}}$ is obtained as a nonlinear function of measurements, the estimator is nonlinear.

Definition 4.2

An estimator is unbiased if the expected value of $\hat{\boldsymbol{\theta}}$ is equal to the expected value of $\boldsymbol{\theta}$ for different sample sizes, i.e.,

$$E(\hat{\boldsymbol{\theta}}) = E(\boldsymbol{\theta}) \quad \text{for each } N \text{ and all } \boldsymbol{\theta} \tag{4.10}$$

Definition 4.3

An estimator is called a minimum mean squared error estimator if it minimizes the mean squared error (MSE):

$$\text{MSE} = E\left[(\hat{\boldsymbol{\theta}} - \boldsymbol{\theta})^T (\hat{\boldsymbol{\theta}} - \boldsymbol{\theta})\right] \tag{4.11}$$

Note that the MSE for an estimate $\hat{\boldsymbol{\theta}}$ is equal to the trace of the corresponding error covariance matrix,

$$\text{MSE} = E\left[\left(\hat{\boldsymbol{\theta}}-\boldsymbol{\theta}\right)^T\left(\hat{\boldsymbol{\theta}}-\boldsymbol{\theta}\right)\right] = Tr\left\{E\left[\left(\hat{\boldsymbol{\theta}}-\boldsymbol{\theta}\right)\left(\hat{\boldsymbol{\theta}}-\boldsymbol{\theta}\right)^T\right]\right\} \tag{4.12}$$

In general, the MSE includes both the variance (random) error and the squared bias (systematic) error,

$$\text{MSE} = \text{ variance } + \left(\text{bias}\right)^2$$

For an unbiased estimate $\hat{\boldsymbol{\theta}}$, a minimum mean squared error estimator is a minimum variance estimator.

<u>*Definition 4.4*</u>

An estimator is called a best linear unbiased estimator of $\boldsymbol{\theta}$ if it has minimum MSE among the class of unbiased estimators that are linear functions of the measurements.

<u>*Definition 4.5*</u>

The Fisher information matrix $\boldsymbol{M}$ is defined as

$$\boldsymbol{M} \equiv E\left[\left(\frac{\partial \ln \mathbb{L}}{\partial \boldsymbol{\theta}}\right)\left(\frac{\partial \ln \mathbb{L}}{\partial \boldsymbol{\theta}}\right)^T\right] = -E\left(\frac{\partial^2 \ln \mathbb{L}}{\partial \boldsymbol{\theta}\,\partial \boldsymbol{\theta}^T}\right) \tag{4.13}$$

where $\mathbb{L}$ is the likelihood function, which is equal to the probability density function of z given $\boldsymbol{\theta}$,

$$\mathbb{L}\left(z;\boldsymbol{\theta}\right) \equiv p\left(z \mid \boldsymbol{\theta}\right) \tag{4.14}$$

In Eq. (4.13), the first equality is a definition; the second equality is derived in Appendix B.

The likelihood function is regarded as a function of the unknown parameter vector $\boldsymbol{\theta}$, with z denoting the measurements. Then, an unbiased estimator is called efficient if the covariance matrix equals the inverse of the Fisher information matrix,

$$Cov\left(\hat{\boldsymbol{\theta}}\right) = E\left[\left(\hat{\boldsymbol{\theta}}-\boldsymbol{\theta}\right)\left(\hat{\boldsymbol{\theta}}-\boldsymbol{\theta}\right)^T\right] = \boldsymbol{M}^{-1} \tag{4.15}$$

The matrix $\boldsymbol{M}^{-1}$ is known as the Cramér-Rao lower bound, and the expression

$$Cov\left(\hat{\boldsymbol{\theta}}\right) \geq \boldsymbol{M}^{-1} \tag{4.16}$$

is the Cramér-Rao inequality for an unbiased estimator $\hat{\boldsymbol{\theta}}$, see Appendix B. This inequality indicates that any unbiased estimator can have a covariance matrix no smaller than $\boldsymbol{M}^{-1}$. If the Cramér-Rao inequality becomes an equality as $N \to \infty$, the estimator is called asymptotically efficient.

The Cramér-Rao inequality for a constant unknown parameter vector $\boldsymbol{\theta}$ and an unbiased estimator will be discussed further in Chapter 6. The Cramér-Rao inequality for a random parameter vector $\boldsymbol{\theta}$ and/or a biased estimator is covered in Ref. [4.2].

Definition 4.6

Let $\hat{\boldsymbol{\theta}}(N)$ be an estimate based on N samples. Then an estimator is called consistent if $\hat{\boldsymbol{\theta}}(N)$ converges to the true value $\boldsymbol{\theta}$ as N increases,

$$\lim_{N\to\infty} \hat{\boldsymbol{\theta}}(N) = \boldsymbol{\theta} \tag{4.17}$$

4.2 Parameter Estimation

In this section, the cost functions to be optimized for various estimators are introduced. Linear and nonlinear models for the observations as functions of the model parameters will be considered, together with three different models for the uncertainties in $\boldsymbol{\theta}$ and $\boldsymbol{\nu}$. Applications of the estimation theories presented here will be described in later chapters.

Measurement equations that are linear and nonlinear in the parameters have already been introduced as

$$z = \boldsymbol{X\theta} + \boldsymbol{\nu} \tag{4.18}$$

and

$$z = \boldsymbol{\varphi}(\boldsymbol{\theta}) + \boldsymbol{\nu} \tag{4.19}$$

Three models for the uncertainties in the parameters and the measurements will be considered. They are designated according to Schweppe [4.3] as the Bayesian model, the Fisher model, and the least-squares model, formed as follows.

Bayesian Model

1) $\boldsymbol{\theta}$ is a vector of random variables with probability density $p(\boldsymbol{\theta})$.
2) $\boldsymbol{\nu}$ is a random vector with probability density $p(\boldsymbol{\nu})$.

Fisher Model

1) $\boldsymbol{\theta}$ is a vector of unknown constant parameters.
2) $\boldsymbol{v}$ is a random vector with probability density $p(\boldsymbol{v})$.

Least-Squares Model:

1) $\boldsymbol{\theta}$ is a vector of unknown constant parameters.
2) $\boldsymbol{v}$ is a random vector of measurement noise.

For measurement equations that are nonlinear in the parameters, Eq. (4.18) in all three models is replaced by Eq. (4.19).

4.2.1 Estimator for the Bayesian Model

The development of an estimator for the Bayesian model follows from the Bayesian estimation theory explained, for example, in Refs. [4.3] and [4.4]. The probability densities $p(\boldsymbol{\theta})$ and $p(\boldsymbol{v})$ are assumed to be known *a priori*. The conditional density of parameter vector $\boldsymbol{\theta}$, given the observations $\boldsymbol{z}$, designated by $p(\boldsymbol{\theta}\,|\,\boldsymbol{z})$, is sometimes called the *a posteriori* probability density. This probability density is related to the *a priori* probability densities by Bayes's rule (see Appendix B):

$$p(\boldsymbol{\theta}\,|\,\boldsymbol{z})=\frac{p(\boldsymbol{z}\,|\,\boldsymbol{\theta})\,p(\boldsymbol{\theta})}{p(\boldsymbol{z})} \tag{4.20}$$

There are several ways to form an estimator, as reviewed in Refs. [4.2] and [4.3]. In this section, only the estimator that selects $\hat{\boldsymbol{\theta}}$ as a value which maximizes the conditional probability density $p(\boldsymbol{\theta}\,|\,\boldsymbol{z})$ is considered. If the vectors $\boldsymbol{\theta}$ and $\boldsymbol{v}$ have Gaussian distributions and are independent, their mean values and variances are

$$E(\boldsymbol{\theta})=\boldsymbol{\theta}_p \quad , \quad Cov(\boldsymbol{\theta})=\boldsymbol{\Sigma}_p \tag{4.21}$$

$$E(\boldsymbol{v})=\boldsymbol{0} \quad , \quad Cov(\boldsymbol{v})=\boldsymbol{R} \tag{4.22}$$

which can also be stated as (see Appendix B)

$$\begin{gathered}\boldsymbol{\theta} \text{ is } \mathbb{N}(\boldsymbol{\theta}_p,\boldsymbol{\Sigma}_p)\\ \boldsymbol{v} \text{ is } \mathbb{N}(\boldsymbol{0},\boldsymbol{R})\\ \boldsymbol{\theta} \text{ and } \boldsymbol{v} \text{ are independent}\end{gathered} \tag{4.23}$$

Under these assumptions,

$$p(\boldsymbol{\theta}) = \left[(2\pi)^{n_p} \left|\boldsymbol{\Sigma}_p\right|\right]^{-\frac{1}{2}} exp\left[-\frac{1}{2}\left(\boldsymbol{\theta}-\boldsymbol{\theta}_p\right)^T \boldsymbol{\Sigma}_p^{-1}\left(\boldsymbol{\theta}-\boldsymbol{\theta}_p\right)\right] \tag{4.24}$$

where n_p is the number of unknown parameters. For the model of Eq. (4.18),

$$p(\boldsymbol{z} \mid \boldsymbol{\theta}) = \left[(2\pi)^N |\boldsymbol{R}|\right]^{-\frac{1}{2}} exp\left[-\frac{1}{2}(\boldsymbol{z}-\boldsymbol{X\theta})^T \boldsymbol{R}^{-1}(\boldsymbol{z}-\boldsymbol{X\theta})\right] \tag{4.25}$$

The probability density function $p(\boldsymbol{\theta} \mid \boldsymbol{z})$ is obtained from Eq. (4.20) as

$$p(\boldsymbol{\theta} \mid \boldsymbol{z}) = \left[\frac{1}{p(\boldsymbol{z})}\right]\left[(2\pi)^{N+n_p} |\boldsymbol{R}|\left|\boldsymbol{\Sigma}_p\right|\right]^{-\frac{1}{2}}$$

$$\cdot exp\left[-\frac{1}{2}(\boldsymbol{z}-\boldsymbol{X\theta})^T \boldsymbol{R}^{-1}(\boldsymbol{z}-\boldsymbol{X\theta})-\frac{1}{2}\left(\boldsymbol{\theta}-\boldsymbol{\theta}_p\right)^T \boldsymbol{\Sigma}_p^{-1}\left(\boldsymbol{\theta}-\boldsymbol{\theta}_p\right)\right] \tag{4.26}$$

Thus, the most probable estimate

$$\hat{\boldsymbol{\theta}} = \max_{\boldsymbol{\theta}} p(\boldsymbol{\theta} \mid \boldsymbol{z}) \tag{4.27}$$

minimizes

$$J(\boldsymbol{\theta}) = \frac{1}{2}(\boldsymbol{z}-\boldsymbol{X\theta})^T \boldsymbol{R}^{-1}(\boldsymbol{z}-\boldsymbol{X\theta})+\frac{1}{2}\left(\boldsymbol{\theta}-\boldsymbol{\theta}_p\right)^T \boldsymbol{\Sigma}_p^{-1}\left(\boldsymbol{\theta}-\boldsymbol{\theta}_p\right) \tag{4.28}$$

The quantity $J(\boldsymbol{\theta})$ is usually called the cost function. The probability density $p(\boldsymbol{z})$ does not depend on $\boldsymbol{\theta}$, and therefore has no influence on the cost function for parameter estimation, as follows from Eqs. (4.26) and (4.27).

4.2.2 Estimator for the Fisher Model

An estimator for the Fisher model is based on the Fisher estimation theory [4.5], using the concept of a likelihood function,

$$\mathbb{L}(\boldsymbol{z};\boldsymbol{\theta}) = p(\boldsymbol{z} \mid \boldsymbol{\theta}) \tag{4.29}$$

Because $\boldsymbol{\theta}$ is now assumed to be a vector of unknown constants, and not a random variable, the probability density function $p(\boldsymbol{\theta})$ is not defined, and Bayes's rule does not hold.

The most common estimator for the Fisher model is the maximum likelihood (ML) estimator, which is equal to the value of $\boldsymbol{\theta}$ that maximizes

$\mathbb{L}(z;\boldsymbol{\theta})$ for given z. In the case of a Gaussian $p(z)$, where $\boldsymbol{v}$ is $\mathbb{N}(\mathbf{0},\boldsymbol{R})$, the likelihood function takes the form

$$\mathbb{L}(z;\boldsymbol{\theta})=\left[(2\pi)^{N}|\boldsymbol{R}|\right]^{-\frac{1}{2}} exp\left[-\frac{1}{2}(z-\boldsymbol{X\theta})^{T}\boldsymbol{R}^{-1}(z-\boldsymbol{X\theta})\right] \tag{4.30}$$

Then the maximum likelihood estimate

$$\hat{\boldsymbol{\theta}}=\max_{\boldsymbol{\theta}} \mathbb{L}(z;\boldsymbol{\theta}) \tag{4.31}$$

minimizes

$$J(\boldsymbol{\theta})=\frac{1}{2}(z-\boldsymbol{X\theta})^{T}\boldsymbol{R}^{-1}(z-\boldsymbol{X\theta}) \tag{4.32}$$

or, in the case of nonlinear observations,

$$J(\boldsymbol{\theta})=\frac{1}{2}\left[z-\boldsymbol{\varphi}(\boldsymbol{\theta})\right]^{T}\boldsymbol{R}^{-1}\left[z-\boldsymbol{\varphi}(\boldsymbol{\theta})\right] \tag{4.33}$$

The development of ML parameter estimators for a dynamic system, with direct application to aircraft parameter estimation, is covered in Chapter 6.

4.2.3 Estimator for the Least-Squares Model

In specifying the form of the least-squares model, no uncertainty models for $\boldsymbol{\theta}$ or $\boldsymbol{v}$ are used, i.e., there are no probability statements concerning $\boldsymbol{\theta}$ or $\boldsymbol{v}$. An estimate for the least-squares model can be obtained by the reasoning that, given z, the "best" estimate of $\boldsymbol{\theta}$ comes from minimizing the weighted sum of squared differences between the measured outputs and the model outputs,

$$J(\boldsymbol{\theta})=\frac{1}{2}(z-\boldsymbol{X\theta})^{T}\boldsymbol{R}^{-1}(z-\boldsymbol{X\theta}) \tag{4.34}$$

where $\boldsymbol{R}^{-1}$ is now a positive definite weighting matrix, chosen by judgment. Optimization of $J(\boldsymbol{\theta})$ leads to the well-known weighted least-squares (WLS) estimator. In the special case where $\boldsymbol{R}^{-1}=\boldsymbol{I}$, the ordinary least-squares (OLS) estimator is obtained, with cost function

$$J(\boldsymbol{\theta})=\frac{1}{2}(z-\boldsymbol{X\theta})^{T}(z-\boldsymbol{X\theta}) \tag{4.35}$$

For a nonlinear observation model, the minimization of

$$J(\boldsymbol{\theta})=\frac{1}{2}\left[z-\boldsymbol{\varphi}(\boldsymbol{\theta})\right]^{T}\left[z-\boldsymbol{\varphi}(\boldsymbol{\theta})\right] \tag{4.36}$$

leads to the nonlinear least-squares estimator. The application of various least-squares estimators to aircraft system identification problems is covered in Chapter 5.

4.3 State Estimation

For a deterministic dynamic system, the time histories of state variables are obtained by integrating the state equations for given input and initial conditions. For a stochastic system, however, there are random variables in the dynamic equations, so the time histories of the state variables must be estimated using a statistical method. There are many possible formulations of the state estimation problem. In this section, state estimation algorithms are outlined using probability density functions describing the state and measured variables. This probabilistic formulation is usually called the Bayesian approach to state estimation, and has many similarities to the development of parameter estimation using the Bayesian model for the uncertainties, discussed in the previous section.

Because of technical difficulties associated with continuous-time white noise, mentioned in Chapter 2, the development of state estimation algorithms begins with a discrete-time system. The discrete-time results are then extended to the case of a continuous-time dynamic system model with discrete-time measurements, which is a more realistic model of the practical aircraft problem. One of the resulting linear estimation formulas is known as the Kalman filter. This filter is a widely-used algorithm for state estimation.

4.3.1 Linear State Estimator

Recalling Eqs. (2.39) and (2.40), the discrete-time linear stochastic system can be modeled as

$$\boldsymbol{x}(i)=\boldsymbol{\Phi}(i-1)\boldsymbol{x}(i-1)+\boldsymbol{\Gamma}(i-1)\boldsymbol{u}(i-1)+\boldsymbol{\Gamma}_w(i-1)\boldsymbol{w}(i-1) \tag{4.37}$$

$$\boldsymbol{y}(i)=\boldsymbol{C}(i)\boldsymbol{x}(i)+\boldsymbol{D}(i)\boldsymbol{u}(i) \tag{4.38}$$

$$\boldsymbol{z}(i)=\boldsymbol{y}(i)+\boldsymbol{v}(i) \qquad i=1,2,\ldots,N \tag{4.39}$$

The matrices $\boldsymbol{\Phi}(i)$, $\boldsymbol{\Gamma}(i)$, $\boldsymbol{\Gamma}_w(i)$, $\boldsymbol{C}(i)$, and $\boldsymbol{D}(i)$ are assumed to be known functions of time $t_i=i\Delta t$, and $\boldsymbol{w}(i)$ and $\boldsymbol{v}(i)$ are white noise sequences. The Bayesian model for uncertainties in $\boldsymbol{x}(0), \boldsymbol{w}(i)$, and $\boldsymbol{v}(i)$ is given as

$$
\begin{aligned}
&E\left[\boldsymbol{x}(0)\right]=\bar{\boldsymbol{x}}_0 && E\left\{\left[\boldsymbol{x}(0)-\bar{\boldsymbol{x}}_0\right]\left[\boldsymbol{x}(0)-\bar{\boldsymbol{x}}_0\right]^T\right\}=\boldsymbol{P}_0 \\
&E\left[\boldsymbol{w}(i)\right]=\boldsymbol{0} && E\left[\boldsymbol{w}(i)\,\boldsymbol{w}^T(j)\right]=\boldsymbol{Q}(i)\delta_{ij} \\
&E\left[\boldsymbol{v}(i)\right]=\boldsymbol{0} && E\left[\boldsymbol{v}(i)\,\boldsymbol{v}^T(j)\right]=\boldsymbol{R}(i)\delta_{ij}
\end{aligned}
\tag{4.40}
$$

$\boldsymbol{x}(0)$, $\boldsymbol{w}(i)$, and $\boldsymbol{v}(i)$ are uncorrelated

In addition, the following notation is used:

1. $\boldsymbol{Z}_i=\left[z(1)\;\; z(2)\;\; \dots\;\; z(i)\right]^T$ contains all measurements up to and including time $t_i=i\Delta t$
2. $\hat{\boldsymbol{x}}(i_1 \mid i_2)$ is the "best" estimate of the state $\boldsymbol{x}(i_1)$ using measurements $\boldsymbol{Z}_{i_2}=\left[z(1)\;\; z(2)\;\; \dots\;\; z(i_2)\right]^T$
3. $\boldsymbol{e}_x(i_1 \mid i_2)$ is the error in the estimate $\hat{\boldsymbol{x}}(i_1 \mid i_2)$, i.e.,

$$
\boldsymbol{e}_x(i_1 \mid i_2)=\boldsymbol{x}(i_1)-\hat{\boldsymbol{x}}(i_1 \mid i_2)
\tag{4.41}
$$

The state error covariance matrix, sometimes referred to as the state covariance matrix, is

$$
\boldsymbol{P}(i_1 \mid i_2)=E\left[\boldsymbol{e}_x(i_1 \mid i_2)\boldsymbol{e}_x^T(i_1 \mid i_2)\right]
\tag{4.42}
$$

assuming that $\hat{\boldsymbol{x}}(i_1 \mid i_2)$ is an unbiased estimate.

For the system specified by Eqs. (4.37)-(4.40), with measurements $\boldsymbol{Z}_{i_2}=\left[z(1)\;\; z(2)\;\; \dots\;\; z(i_2)\right]^T$, three types of state estimation can be considered:

1. Prediction: $i_1 > i_2$
2. Filtering: $i_1 = i_2$
3. Smoothing: $1 \le i_1 < i_2$

Of particular interest are unbiased and consistent estimators which minimize the state estimation error in a well-defined statistical sense.

Prediction

A one-step prediction is formulated as the calculation of an estimate of $\boldsymbol{x}(i)$, given the measured data $\boldsymbol{Z}_{i-1}$. From Eq. (4.37),

$$\boldsymbol{x}(i)=\boldsymbol{\Phi}(i-1)\,\boldsymbol{x}(i-1)+\boldsymbol{\Gamma}(i-1)\,\boldsymbol{u}(i-1)+\boldsymbol{\Gamma}_w(i-1)\,\boldsymbol{w}(i-1) \tag{4.43}$$

Assuming that $\hat{\boldsymbol{x}}(i-1|i-1)$ is available, the best estimate of $\boldsymbol{\Phi}(i-1)\,\boldsymbol{x}(i-1)$ in Eq. (4.43) is $\boldsymbol{\Phi}(i-1)\,\hat{\boldsymbol{x}}(i-1|i-1)$. Because $\boldsymbol{w}(i-1)$ represents a random white sequence, the best prediction of $\boldsymbol{w}(i-1)$ that can be made from the measurements $\boldsymbol{Z}_{i-1}$ is the expected value $E\left[\boldsymbol{w}(i-1)\right]=\boldsymbol{0}$, and $\boldsymbol{\Gamma}(i-1)\,\boldsymbol{u}(i-1)$ is deterministic. Therefore,

$$\hat{\boldsymbol{x}}(i|i-1)=\boldsymbol{\Phi}(i-1)\,\hat{\boldsymbol{x}}(i-1|i-1)+\boldsymbol{\Gamma}(i-1)\,\boldsymbol{u}(i-1) \tag{4.44}$$

The one-step prediction of the estimation error at time step i follows from Eqs. (4.41), (4.43), and (4.44) as

$$\boldsymbol{e}_x(i|i-1)=\boldsymbol{\Phi}(i-1)\,\boldsymbol{e}_x(i-1|i-1)+\boldsymbol{\Gamma}_w(i-1)\,\boldsymbol{w}(i-1) \tag{4.45}$$

The state covariance matrix of the one-step prediction is given by

$$\begin{aligned}\boldsymbol{P}(i|i-1)&=E\left[\boldsymbol{e}_x(i|i-1)\boldsymbol{e}_x^T(i|i-1)\right]\\&=\boldsymbol{\Phi}(i-1)\boldsymbol{P}(i-1|i-1)\boldsymbol{\Phi}^T(i-1)\\&\quad+\boldsymbol{\Gamma}_w(i-1)E\left[\boldsymbol{w}(i-1)\boldsymbol{e}_x^T(i-1|i-1)\right]\boldsymbol{\Phi}^T(i-1)\\&\quad+\boldsymbol{\Phi}(i-1)E\left[\boldsymbol{e}_x(i-1|i-1)\boldsymbol{w}^T(i-1)\right]\boldsymbol{\Gamma}_w^T(i-1)\\&\quad+\boldsymbol{\Gamma}_w(i-1)\boldsymbol{Q}(i-1)\boldsymbol{\Gamma}_w^T(i-1)\end{aligned} \tag{4.46}$$

The quantities $E\left[\boldsymbol{w}(i-1)\boldsymbol{e}_x^T(i-1|i-1)\right]$ and $E\left[\boldsymbol{e}_x(i-1|i-1)\boldsymbol{w}^T(i-1)\right]$ equal zero, because $\boldsymbol{w}$ is a white sequence, and $\boldsymbol{w}(i-1)$ affects $\boldsymbol{e}_x(i|i-1)$ by Eq. (4.45), but not $\boldsymbol{e}_x(i-1|i-1)$. Therefore,

$$\boldsymbol{P}(i|i-1)=\boldsymbol{\Phi}(i-1)\boldsymbol{P}(i-1|i-1)\boldsymbol{\Phi}^T(i-1)+\boldsymbol{\Gamma}_w(i-1)\boldsymbol{Q}(i-1)\boldsymbol{\Gamma}_w^T(i-1) \tag{4.47}$$

Filtering

The Bayesian approach to filtering requires an update of the knowledge about the state of the dynamic system from the probability density $p(\boldsymbol{x})$ to the conditional probability density $p(\boldsymbol{x} \mid \boldsymbol{z})$, using the measurements $\boldsymbol{z}$. The updating procedure can be repeated every time a new measurement vector is available.

It is therefore necessary to find a recursive relationship which changes $p\left[\boldsymbol{x}(i-1) \mid \boldsymbol{Z}_{i-1}\right]$ into $p\left[\boldsymbol{x}(i) \mid \boldsymbol{Z}_i\right]$. The development of this relationship, which is shown, for example, in Refs. [4.6] and [4.7], results in the *a posteriori* density function

$$p\left[\boldsymbol{x}(i) \mid \boldsymbol{Z}_i\right] = \frac{p\left[\boldsymbol{z}(i) \mid \boldsymbol{x}(i)\right] \, p\left[\boldsymbol{x}(i) \mid \boldsymbol{Z}_{i-1}\right]}{p\left[\boldsymbol{z}(i) \mid \boldsymbol{Z}_{i-1}\right]} \tag{4.48}$$

Expressions for filtering can be obtained from Eq. (4.48) in different ways. Refs. [4.4] and [4.7] use a Gaussian assumption in the model for uncertainties. This assumption also means that all $\hat{\boldsymbol{x}}(i)$ and $\boldsymbol{z}(i)$ will be Gaussian, as will be all conditional probability densities. In this case, all probability densities involved can be fully described by their mean value and covariance matrix.

Using the Gaussian assumption, the conditional probability density $p\left[\boldsymbol{x}(i) \mid \boldsymbol{Z}_i\right]$ takes the form

$$\begin{aligned} p\left[\boldsymbol{x}(i) \mid \boldsymbol{Z}_i\right] = & \left[(2\pi)^{n_s} \left|\boldsymbol{P}(i)\right|\right]^{-\frac{1}{2}} \\ & \cdot exp\left\{-\frac{1}{2}\left[\boldsymbol{x}(i) - \hat{\boldsymbol{x}}(i)\right]^T \boldsymbol{P}^{-1}(i)\left[\boldsymbol{x}(i) - \hat{\boldsymbol{x}}(i)\right]\right\} \end{aligned} \tag{4.49}$$

with mean value given by

$$\hat{\boldsymbol{x}}(i \mid i) = \hat{\boldsymbol{x}}(i \mid i-1) + \boldsymbol{K}(i)\left[\boldsymbol{z}(i) - \boldsymbol{C}(i)\,\hat{\boldsymbol{x}}(i \mid i-1) - \boldsymbol{D}(i)\,\boldsymbol{u}(i)\right] \tag{4.50a}$$

and state covariance matrix

$$\boldsymbol{P}(i \mid i) = \left[\boldsymbol{I} - \boldsymbol{K}(i)\boldsymbol{C}(i)\right]\boldsymbol{P}(i \mid i-1) \tag{4.50b}$$

where

$$\boldsymbol{K}(i) = \boldsymbol{P}(i \mid i-1)\,\boldsymbol{C}^T(i)\left[\boldsymbol{C}(i)\,\boldsymbol{P}(i \mid i-1)\,\boldsymbol{C}^T(i) + \boldsymbol{R}(i)\right]^{-1} \tag{4.50c}$$

The algorithm for sequential estimation is completed by Eqs. (4.44) and (4.47) for propagating the state estimate from one sample time to the next, along with the initial conditions

$$\hat{x}(0|0) = \bar{x}_0 = E[x(0)]$$
$$P(0|0) = P_0 = E\left\{[x(0) - \bar{x}_0][x(0) - \bar{x}_0]^T\right\} \tag{4.50d}$$

The development of Eqs. (4.50) can be found in Ref. [4.4], [4.7], or [4.8].

Smoothing

The smoothing process produces an optimal state estimate $\hat{x}(i_1 | i_2)$ for $1 \le i_1 < i_2$, using measurements $Z_{i_2} = [z(1) \;\; z(2) \;\; \dots \;\; z(i_2)]^T$. Three types of smoothing can be defined, as shown in Fig. 4.1. They are called fixed-interval, fixed-point, and fixed-lag smoothing. In fixed-interval smoothing, i_2 is fixed and i_1 varies from 1 to i_2. For fixed-point smoothing, i_1 is fixed while i_2 increases. Finally, in fixed-lag smoothing, both i_1 and i_2 vary, but the interval between i_1 and i_2 remains fixed.

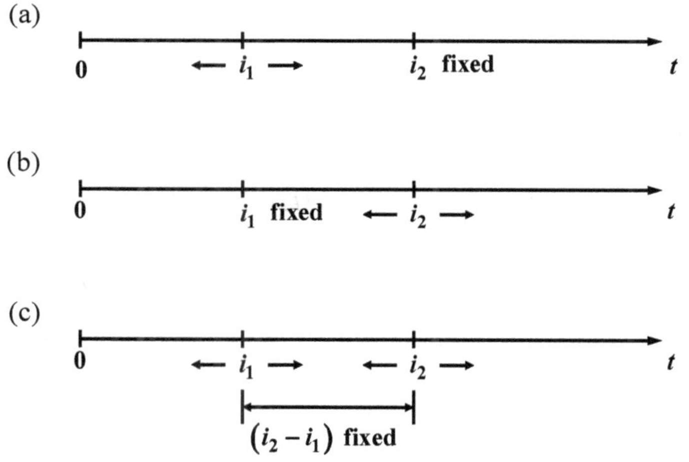

Figure 4.1 **Three types of discrete-time smoothing: (a) fixed-interval, (b) fixed-point, (c) fixed-lag**

There are many algorithms for optimal smoothing, see Refs. [4.8] and [4.9]. A fixed-interval smoother will be introduced in Chapter 10 in connection with estimation of systematic instrumentation errors and flight path reconstruction.

4.3.2 Kalman Filter

Equations (4.50) together with Eqs. (4.44) and (4.47) are complete recursive relationships for computing the optimal estimate of the state $\hat{\boldsymbol{x}}$ and its covariance matrix $\boldsymbol{P}$, at each time step. The equations are summarized as follows:

Initial Conditions

$$\begin{aligned}\hat{\boldsymbol{x}}(0\,|\,0) &= \bar{\boldsymbol{x}}_0 \\ \boldsymbol{P}(0\,|\,0) &= \boldsymbol{P}_0\end{aligned} \tag{4.51a}$$

Prediction

$$\hat{\boldsymbol{x}}(i\,|\,i-1) = \boldsymbol{\Phi}(i-1)\,\hat{\boldsymbol{x}}(i-1\,|\,i-1) + \boldsymbol{\Gamma}(i-1)\,\boldsymbol{u}(i-1) \tag{4.51b}$$

$$\boldsymbol{P}(i\,|\,i-1) = \boldsymbol{\Phi}(i-1)\,\boldsymbol{P}(i-1\,|\,i-1)\boldsymbol{\Phi}^T(i-1) + \boldsymbol{\Gamma}_w(i-1)\,\boldsymbol{Q}(i-1)\boldsymbol{\Gamma}_w^T(i-1) \tag{4.51c}$$

Measurement Update

$$\hat{\boldsymbol{x}}(i\,|\,i) = \hat{\boldsymbol{x}}(i\,|\,i-1) + \boldsymbol{K}(i)\left[\boldsymbol{z}(i) - \boldsymbol{C}(i)\,\hat{\boldsymbol{x}}(i\,|\,i-1) - \boldsymbol{D}(i)\,\boldsymbol{u}(i)\right] \tag{4.51d}$$

$$\boldsymbol{P}(i\,|\,i) = \left[\boldsymbol{I} - \boldsymbol{K}(i)\boldsymbol{C}(i)\right]\boldsymbol{P}(i\,|\,i-1) \tag{4.51e}$$

$$\boldsymbol{K}(i) = \boldsymbol{P}(i\,|\,i-1)\,\boldsymbol{C}^T(i)\left[\boldsymbol{C}(i)\,\boldsymbol{P}(i\,|\,i-1)\,\boldsymbol{C}^T(i) + \boldsymbol{R}(i)\right]^{-1} \tag{4.51f}$$

Equations (4.51) are identical to the Kalman filter algorithm originally developed in Ref. [4.10], using a different approach. In Eq. (4.51d), the second term on the right represents a correction to the propagated estimate $\hat{\boldsymbol{x}}(i\,|\,i-1)$, based on the innovations, defined as

$$\boldsymbol{\upsilon}(i) = \boldsymbol{z}(i) - \boldsymbol{C}(i)\,\hat{\boldsymbol{x}}(i\,|\,i-1) - \boldsymbol{D}(i)\,\boldsymbol{u}(i) \tag{4.52}$$

The estimator (4.51d) can therefore be viewed as a linear feedback system with time-varying gain $\boldsymbol{K}(i)$. The block diagram of the system described by Eqs. (4.51) is shown in Fig. 4.2.

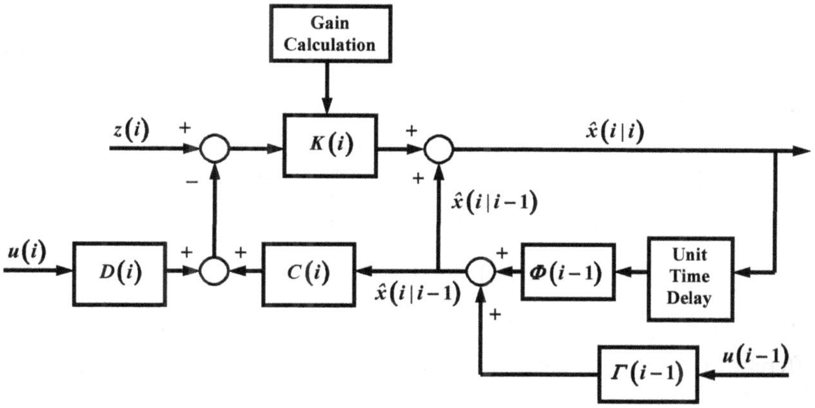

Figure 4.2 **Block diagram of the discrete-time Kalman filter**

The matrix difference equation (4.51e) for $\boldsymbol{P}(i|i)$ is called a Riccati equation. Because $\boldsymbol{P}(i|i)$ represents the error covariance matrix for the state estimate, the Kalman filter provides an error analysis that indicates how well the state estimation is being performed. In Eq. (4.51e), the second term on the right shows a change in the uncertainty of the state estimate, due to process noise and the noisy measurement [cf. Eq. (4.51f)]. Because Eqs. (4.51e) and (4.51f) do not depend on the state estimates, they can be computed before the measurement is taken.

The two steps in the state estimation, i.e., the prediction and measurement update, are visualized for a simple scalar system in Fig. 4.3. The various quantities involved in the discrete optimal filter equations (4.51) are shown for a single time step.

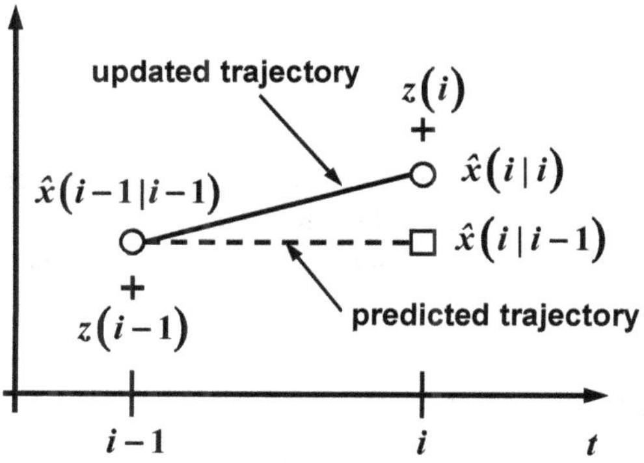

Figure 4.3 **Two steps in the discrete-time Kalman filter**

Steady-State Kalman Filter

So far, the Kalman filter equations have been presented for a time-varying discrete model specified by Eqs. (4.37) and (4.38). The filter equations can be simplified when the dynamic system is time-invariant, i.e., the matrices $\boldsymbol{\Phi}, \boldsymbol{\Gamma}, \boldsymbol{\Gamma}_w, \boldsymbol{C}, \boldsymbol{D}, \boldsymbol{Q},$ and $\boldsymbol{R}$ are constant. Further simplification is possible when the estimator becomes time-invariant as well. This can happen if the state covariance matrix $\boldsymbol{P}(i\,|\,i)$ converges to a constant matrix for every i,

$$\boldsymbol{P}(i\,|\,i) = \boldsymbol{P}(i-1\,|\,i-1) \equiv \boldsymbol{P}_\infty \tag{4.53}$$

Then $\boldsymbol{P}(i\,|\,i-1) \equiv \boldsymbol{P}$ will also be a constant matrix which is, in general, different from $\boldsymbol{P}_\infty$, and the filter equations for the steady-state estimator are

$$\hat{\boldsymbol{x}}(i\,|\,i-1) = \boldsymbol{\Phi}\,\hat{\boldsymbol{x}}(i-1\,|\,i-1) + \boldsymbol{\Gamma}\,\boldsymbol{u}(i-1) \tag{4.54a}$$

$$\boldsymbol{P} = \boldsymbol{\Phi}\,\boldsymbol{P}_\infty\,\boldsymbol{\Phi}^T + \boldsymbol{\Gamma}_w\,\boldsymbol{Q}\,\boldsymbol{\Gamma}_w^T \tag{4.54b}$$

$$\hat{\boldsymbol{x}}(i\,|\,i) = \hat{\boldsymbol{x}}(i\,|\,i-1) + \boldsymbol{K}\left[\boldsymbol{z}(i) - \boldsymbol{C}\,\hat{\boldsymbol{x}}(i\,|\,i-1) - \boldsymbol{D}\,\boldsymbol{u}(i)\right] \tag{4.54c}$$

$$\boldsymbol{P}_\infty = \boldsymbol{P} - \boldsymbol{KCP} \tag{4.54d}$$

$$\boldsymbol{K} = \boldsymbol{P}\,\boldsymbol{C}^T\left[\boldsymbol{C}\,\boldsymbol{P}\,\boldsymbol{C}^T + \boldsymbol{R}\right]^{-1} \tag{4.54e}$$

and the initial conditions are given by Eq. (4.51a). The filter represented by Eqs. (4.54) is known as a steady-state filter. It can be easily implemented once $\boldsymbol{P}_\infty$ has been determined. The matrix $\boldsymbol{P}_\infty$ can be obtained by solving the matrix difference equation (4.51e) repeatedly until it converges to a steady-state value. The other possibility is to find a positive definite solution from the three algebraic equations (4.54b), (4.54d), and (4.54e), or equivalently from the combined equation

$$\boldsymbol{P} = \boldsymbol{\Phi}\left[\boldsymbol{P} - \boldsymbol{P}\boldsymbol{C}^T\left(\boldsymbol{C}\boldsymbol{P}\boldsymbol{C}^T + \boldsymbol{R}\right)^{-1}\boldsymbol{C}\boldsymbol{P}\right]\boldsymbol{\Phi}^T + \boldsymbol{\Gamma}_w\,\boldsymbol{Q}\,\boldsymbol{\Gamma}_w^T \tag{4.55}$$

Conditions for the existence of a solution to the preceding equation are discussed in Ref. [4.3], and a method for the solution is discussed in Refs. [4.3] and [4.11]. More about the steady-state Kalman filter can be found in Ref. [4.9].

4.3.3 Continuous-Discrete Kalman Filter

For some state estimation problems, it is preferable to use the state equation in continuous-time form. As shown in Chapter 3, this is the form taken by the aircraft linearized equations of motion. When combined with a discrete measurement equation, the combined continuous-discrete model equations corresponding to Eqs. (4.37) and (4.39) are

$$\dot{\boldsymbol{x}}(t) = \boldsymbol{A}(t)\,\boldsymbol{x}(t) + \boldsymbol{B}(t)\,\boldsymbol{u}(t) + \boldsymbol{B}_w(t)\,\boldsymbol{w}(t) \tag{4.56}$$

$$\boldsymbol{z}(i) = \boldsymbol{C}(i)\,\boldsymbol{x}(i) + \boldsymbol{D}(i)\,\boldsymbol{u}(i) + \boldsymbol{v}(i) \qquad i = 1, 2, \ldots, N \tag{4.57}$$

where

$$\begin{aligned}
&E\left[\boldsymbol{x}(0)\right] = \bar{\boldsymbol{x}}_0 && E\left\{\left[\boldsymbol{x}(0) - \bar{\boldsymbol{x}}_0\right]\left[\boldsymbol{x}(0) - \bar{\boldsymbol{x}}_0\right]^T\right\} = \boldsymbol{P}_0 \\
&E\left[\boldsymbol{w}(t)\right] = \boldsymbol{0} && E\left[\boldsymbol{w}(t_i)\,\boldsymbol{w}^T(t_j)\right] = \boldsymbol{Q}(t_i)\,\delta(t_i - t_j) \\
&E\left[\boldsymbol{v}(i)\right] = \boldsymbol{0} && E\left[\boldsymbol{v}(i)\,\boldsymbol{v}^T(j)\right] = \boldsymbol{R}(i)\,\delta_{ij}
\end{aligned} \tag{4.58}$$

and $\boldsymbol{x}(0)$, $\boldsymbol{w}(t)$, and $\boldsymbol{v}(i)$ are uncorrelated.

To develop the prediction equations, it is assumed that at time $t_i = i\Delta t$, the Kalman filter has a state estimate $\hat{\boldsymbol{x}}(i-1\,|\,i-1)$ with covariance $\boldsymbol{P}(i-1\,|\,i-1)$. As in the previous discrete case, a one-step ahead prediction is formulated as the calculation of an estimate $\hat{\boldsymbol{x}}(t\,|\,i-1)$, given the measured data $\boldsymbol{Z}_{i-1}$. The equation for the predicted state estimates can be obtained by taking conditional expectations on both sides of Eq. (4.56), which gives

$$\frac{d}{dt}\left[\hat{\boldsymbol{x}}(t\,|\,i-1)\right] = \boldsymbol{A}(t)\,\hat{\boldsymbol{x}}(t\,|\,i-1) + \boldsymbol{B}(t)\,\boldsymbol{u}(t) \tag{4.59}$$

for $(i-1)\Delta t \le t \le i\Delta t$.

The state covariance matrix of the one-step ahead prediction can be obtained by limiting arguments applied to Eq. (4.51c) (cf. Ref. [4.8]), or by using covariance equations from Ref. [4.6]. The resulting differential equation is

$$\frac{d}{dt}\left[\boldsymbol{P}(t\,|\,i-1)\right] = \boldsymbol{A}(t)\,\boldsymbol{P}(t\,|\,i-1) + \boldsymbol{P}(t\,|\,i-1)\,\boldsymbol{A}^T(t) + \boldsymbol{B}_w(t)\,\boldsymbol{Q}(t)\,\boldsymbol{B}_w^T(t) \tag{4.60}$$

The measurement update equations in the continuous-discrete Kalman filter are identical to those given in Eqs. (4.51d)-(4.51f). The Kalman filter equations for the continuous-discrete system specified by Eqs. (4.56)-(4.58) are summarized as follows.

Initial Conditions

$$\hat{\boldsymbol{x}}(0|0) = \bar{\boldsymbol{x}}_0$$
$$\boldsymbol{P}(0|0) = \boldsymbol{P}_0 \tag{4.61a}$$

Prediction

$$\frac{d}{dt}\left[\hat{\boldsymbol{x}}(t|i-1)\right] = \boldsymbol{A}(t)\,\hat{\boldsymbol{x}}(t|i-1) + \boldsymbol{B}(t)\,\boldsymbol{u}(t) \tag{4.61b}$$

$$\frac{d}{dt}\left[\boldsymbol{P}(t|i-1)\right] = \boldsymbol{A}(t)\,\boldsymbol{P}(t|i-1) + \boldsymbol{P}(t|i-1)\,\boldsymbol{A}^T(t) + \boldsymbol{B}_w(t)\,\boldsymbol{Q}(t)\,\boldsymbol{B}_w^T(t) \tag{4.61c}$$

Measurement Update

$$\hat{\boldsymbol{x}}(i|i) = \hat{\boldsymbol{x}}(i|i-1) + \boldsymbol{K}(i)\left[\boldsymbol{z}(i) - \boldsymbol{C}(i)\,\hat{\boldsymbol{x}}(i|i-1) - \boldsymbol{D}(i)\,\boldsymbol{u}(i)\right] \tag{4.61d}$$

$$\boldsymbol{P}(i|i) = \left[\boldsymbol{I} - \boldsymbol{K}(i)\boldsymbol{C}(i)\right]\boldsymbol{P}(i|i-1) \tag{4.61e}$$

$$\boldsymbol{K}(i) = \boldsymbol{P}(i|i-1)\,\boldsymbol{C}^T(i)\left[\boldsymbol{C}(i)\,\boldsymbol{P}(i|i-1)\,\boldsymbol{C}^T(i) + \boldsymbol{R}(i)\right]^{-1} \tag{4.61f}$$

4.3.4 Nonlinear State Estimator

There are several model forms for nonlinear dynamic systems, expressing various degrees of complexity. In the following discussion, a discrete-time nonlinear model is considered in the form

$$\boldsymbol{x}(i) = \boldsymbol{f}\left[\boldsymbol{x}(i-1), \boldsymbol{u}(i-1), i-1\right] + \boldsymbol{\Gamma}_w(i-1)\,\boldsymbol{w}(i-1) \tag{4.62a}$$

$$\boldsymbol{y}(i) = \boldsymbol{g}\left[\boldsymbol{x}(i), \boldsymbol{u}(i), i\right] \tag{4.62b}$$

$$\boldsymbol{z}(i) = \boldsymbol{g}\left[\boldsymbol{x}(i), \boldsymbol{u}(i), i\right] + \boldsymbol{v}(i) \qquad i = 1, 2, \ldots, N \tag{4.62c}$$

$$E\left[\boldsymbol{x}(0)\right] = \bar{\boldsymbol{x}}_0 \qquad E\left\{\left[\boldsymbol{x}(0) - \bar{\boldsymbol{x}}_0\right]\left[\boldsymbol{x}(0) - \bar{\boldsymbol{x}}_0\right]^T\right\} = \boldsymbol{P}_0$$
$$E\left[\boldsymbol{w}(i)\right] = \boldsymbol{0} \qquad E\left[\boldsymbol{w}(i)\,\boldsymbol{w}^T(j)\right] = \boldsymbol{Q}(i)\,\delta_{ij} \tag{4.62d}$$
$$E\left[\boldsymbol{v}(i)\right] = \boldsymbol{0} \qquad E\left[\boldsymbol{v}(i)\,\boldsymbol{v}^T(j)\right] = \boldsymbol{R}(i)\,\delta_{ij}$$

where $\boldsymbol{x}(0)$, $\boldsymbol{w}(i)$, and $\boldsymbol{v}(i)$ are uncorrelated, and $\boldsymbol{f}$ and $\boldsymbol{g}$ are vector functions of the state vector $\boldsymbol{x}(i)$, control vector $\boldsymbol{u}(i)$, and time index i. The same Bayesian model for the uncertainties that was stated earlier for a linear system is considered.

As in the linear case, the state estimator uses measurements to update the knowledge of the state from the *a priori* probability density $p(\boldsymbol{x})$, to the *a posteriori* probability density $p(\boldsymbol{x} \mid z)$. The recursive relation between these two probability densities has already been presented as

$$p\left[\boldsymbol{x}(i) \mid \boldsymbol{Z}_i\right] = \frac{p\left[z(i) \mid \boldsymbol{x}(i)\right] \, p\left[\boldsymbol{x}(i) \mid \boldsymbol{Z}_{i-1}\right]}{p\left[z(i) \mid \boldsymbol{Z}_{i-1}\right]} \tag{4.63}$$

Computing probability densities when the state and output equations are nonlinear is difficult in practice, because nonlinear functions of Gaussian probability densities are not Gaussian, so expected values and higher order moments are needed to accurately characterize the probability densities.

To avoid this difficulty, several approximate solutions to the nonlinear filtering problem have been developed, e.g., in Refs. [4.3], [4.8], and [4.12]. Most of these algorithms are based on linearization of the nonlinear model equations. They differ, however, by the selection of the point about which the linearization is done.

As an example, the filter equations will be presented for the model specified by Eqs. (4.62). The nonlinear terms $\boldsymbol{f}\left[\boldsymbol{x}(i-1), \boldsymbol{u}(i-1), i-1\right]$ and $\boldsymbol{g}\left[\boldsymbol{x}(i), \boldsymbol{u}(i), i\right]$ will be expanded about the most recent available estimates of the state. The linearization yields

$$\begin{aligned} \boldsymbol{f}\left[\boldsymbol{x}(i-1), \boldsymbol{u}(i-1), i-1\right] = \boldsymbol{f}\left[\hat{\boldsymbol{x}}(i-1 \mid i-1), \boldsymbol{u}(i-1), i-1\right] \\ + \boldsymbol{\Phi}(i-1)\left[\boldsymbol{x}(i-1) - \hat{\boldsymbol{x}}(i-1 \mid i-1)\right] \end{aligned} \tag{4.64}$$

$$\boldsymbol{g}\left[\boldsymbol{x}(i), \boldsymbol{u}(i), i\right] = \boldsymbol{g}\left[\hat{\boldsymbol{x}}(i \mid i-1), \boldsymbol{u}(i), i\right] + \boldsymbol{C}(i)\left[\boldsymbol{x}(i) - \hat{\boldsymbol{x}}(i \mid i-1)\right] \tag{4.65}$$

where

$$\boldsymbol{\Phi}(i-1) = \left. \frac{\partial \boldsymbol{f}\left[\boldsymbol{x}, \boldsymbol{u}(i-1), i-1\right]}{\partial \boldsymbol{x}} \right|_{\boldsymbol{x}=\hat{\boldsymbol{x}}(i-1 \mid i-1)} \tag{4.66a}$$

$$\boldsymbol{C}(i) = \left. \frac{\partial \boldsymbol{g}\left[\boldsymbol{x}, \boldsymbol{u}(i), i\right]}{\partial \boldsymbol{x}} \right|_{\boldsymbol{x}=\hat{\boldsymbol{x}}(i \mid i-1)} \tag{4.66b}$$

Substituting Eq. (4.64) and (4.65) into (4.62), the linearized model takes the form

$$x(i)=\boldsymbol{\Phi}(i-1)x(i-1)+\boldsymbol{v}(i-1)+\boldsymbol{\Gamma}_w(i-1)w(i-1) \tag{4.67}$$

$$z(i)=C(i)x(i)+b(i)+\nu(i) \tag{4.68}$$

where $\boldsymbol{v}(i-1)$ and $b(i)$ are known at the ith time step,

$$\boldsymbol{v}(i-1)=f\left[\hat{x}(i-1|i-1),u(i-1),i-1\right]-\boldsymbol{\Phi}(i-1)\hat{x}(i-1|i-1) \tag{4.69}$$

$$b(i)=g\left[\hat{x}(i|i-1),u(i),i\right]-C(i)\hat{x}(i|i-1) \tag{4.70}$$

For the filter equations, assuming that the best estimate of $x(i-1)$ for Z_{i-1}, $\hat{x}(i-1|i-1)$, is known, then the best estimate of $\boldsymbol{\Phi}(i-1)x(i-1)$ is $\boldsymbol{\Phi}(i-1)\hat{x}(i-1|i-1)$. Therefore, the prediction equation for the state is

$$\begin{aligned}\hat{x}(i|i-1)&=\boldsymbol{\Phi}(i-1)\hat{x}(i-1|i-1)+\boldsymbol{v}(i-1)\\&=f\left[\hat{x}(i-1|i-1),u(i-1),i-1\right]\end{aligned} \tag{4.71a}$$

Similarly, for the output prediction,

$$\hat{y}(i|i-1)=g\left[\hat{x}(i|i-1),u(i),i\right] \tag{4.71b}$$

The remaining prediction equation for the state covariance matrix is given by Eq. (4.51c) as

$$P(i|i-1)=\boldsymbol{\Phi}(i-1)P(i-1|i-1)\boldsymbol{\Phi}^T(i-1)+\boldsymbol{\Gamma}_w(i-1)Q(i-1)\boldsymbol{\Gamma}_w^T(i-1) \tag{4.71c}$$

The measurement update equations are obtained from Eqs. (4.51d)-(4.51f) after only small modification,

$$\hat{x}(i|i)=\hat{x}(i|i-1)+K(i)\left[z(i)-\hat{y}(i|i-1)\right] \tag{4.71d}$$

$$P(i|i)=\left[I-K(i)C(i)\right]P(i|i-1) \tag{4.71e}$$

$$K(i)=P(i|i-1)C^T(i)\left[C(i)P(i|i-1)C^T(i)+R(i)\right]^{-1} \tag{4.71f}$$

From Eqs. (4.71f) and (4.66b), the matrix $K(i)$ is dependent on the estimator $\hat{x}(i|i-1)$. This means that the gain matrix cannot be pre-computed, so the complete set of filter equations (4.71) must be solved simultaneously. Eqs. (4.71) are referred to as the extended Kalman filter. Application of this filter to state and parameter estimation is discussed in Chapters 9 and 10. State estimation algorithms for a nonlinear system using the Fisher model or the least-squares model for the uncertainties can also be developed. More on these possibilities is available in Ref. [4.3].

4.4 Summary and Concluding Remarks

This chapter contains a brief introduction to parameter and state estimation theory for dynamic systems. Parameter estimation was recognized as a process for finding values of unknown system parameters $\boldsymbol{\theta}$, based on noisy measurements z. Because the parameter estimates depend on the random variable z, estimated parameters are expressed in terms of the most important properties of the probability density function $p(\theta \mid z)$, i.e., the mean and covariance. To describe the quality of the estimates of $\boldsymbol{\theta}$, estimator properties were defined.

Two measurement equations relating the measurements z, measurement errors $\boldsymbol{v}$, and parameters $\boldsymbol{\theta}$, were considered:

$$z = \boldsymbol{X\theta} + \boldsymbol{v}$$

$$z = \boldsymbol{\varphi}(\boldsymbol{\theta}) + \boldsymbol{v}$$

where the matrix $\boldsymbol{X}$ and the form of the function $\boldsymbol{\varphi}(\boldsymbol{\theta})$ are assumed to be completely known. The first equation is linear in the parameters, and the second is nonlinear in the parameters, leading to linear and nonlinear estimation problems, respectively.

The other distinction between measurement equations is based on the mathematical models for the uncertainties in the parameters and measurements. The models considered were the Bayesian model, Fisher model, and least-squares model.

In the first model, the unknown parameters are assumed to be random variables. These parameters are estimated using Bayes's rule, which assumes the existence of an *a priori* probability density for the parameters and measurements. From this information, Bayes's rule provides the *a posteriori* probability density for the parameters, which is equal to the conditional probability density for the parameters, given the measurements. Despite the generality of the Bayes estimator, the method has not found wide application in aircraft parameter estimation. The main reason is the difficulty in making an explicit statement about the form of the *a priori* probability density for the parameters. There are, however, Bayes-like methods which combine *a priori* information about the parameters with measured data. These methods will be described in Chapters 5 and 6 in connection with the linear regression and maximum likelihood techniques.

In the Fisher model, the parameters are assumed to be unknown constants, and the probability density function of the measurement noise is specified.

This estimator is based on maximization of a likelihood function, which is equal to the conditional probability density of the measurements, given the parameters. Detailed treatment of the maximum likelihood estimator and its application to aircraft parameter estimation is given in Chapter 6.

In the least-squares model, the parameters are assumed to be unknown constants, and the measurement noise is assumed to be random. An estimator is obtained by application of the least-squares principle. The least-squares model is a basis for the regression analysis to be discussed in Chapter 5, along with its practical application to aircraft parameter estimation problems.

The three uncertainty models have corresponding formulations in the frequency domain (see Chapter 7) and as real-time estimators (see Chapter 10). Broader explanations of the parameter estimation problem can be found in many textbooks, e.g., Sorensen [4.2] and Mendel [4.13].

The discussion of state estimation was based on the Bayesian approach to the problem. Three estimation techniques – prediction, filtering, and smoothing – were covered, with special emphasis on the Kalman filter for a linear discrete-time system and a continuous-time dynamic system with discrete-time measurements. The steady-state version of the Kalman filter and an extension to a nonlinear discrete-time dynamic system were also introduced. Application of the Kalman filter in the maximum likelihood parameter estimation algorithm for a stochastic system is covered in Chapter 6. The nonlinear filtering problem appears in Chapters 9 and 10, in the context of simultaneous state and parameter estimation and recursive time-varying parameter estimation.

There have been numerous books and papers published on optimal filtering, optimal smoothing, and specifically on the Kalman filter. Recommended books are those by Schweppe [4.3], Gelb [4.8], Minkler and Minkler [4.9], Jazwinski [4.14], and the paper by Sorensen [4.7]. The presentation here was intended to provide background information and theory for the next four chapters, which are concerned with application of the theory to practical aircraft system identification problems.

References

[4.1] Goodwin, G.C. and Payne, R.L., *Dynamic System Identification: Experiment Design and Data Analysis*, Academic Press, New York, NY, 1977.

[4.2] Sorenson, H.W., *Parameter Estimation: Principles and Problems*, Marcel-Decker, New York, NY, 1980.

[4.3] Schweppe, F.C., *Uncertain Dynamic Systems*, Prentice-Hall, Englewood Cliffs, NJ, 1973.

[4.4] Smith, Gerald L., "On the Theory and Methods of Statistical Inference," NASA TR R-251, 1967.

[4.5] Fisher, R.A., "On an Absolute Criterion for Fitting Frequency Curves," *Messenger of Mathematics*, Vol. 41, 1912, pp. 155-160.

[4.6] Bryson, A.E. and Ho, Y.C., *Applied Optimal Control*, Hemisphere Publishing, Washington, DC, 1975.

[4.7] Sorenson, H.W., "Comparison of Kalman, Bayesian, and Maximum Likelihood Estimation Techniques," *Theory and Application of Kalman Filtering*, AGARD AG-139, 1970, pp. 121-142.

[4.8] Gelb, A. (editor), *Applied Optimal Estimation*, MIT Press, Cambridge, MA, 1974.

[4.9] Minkler, G. and Minkler, J., *Theory and Application of Kalman Filtering*, Magellan Book Co., Baltimore, MD, 1993.

[4.10] Kalman, R.E., "A New Approach to Linear Filtering and Prediction Problems," *Transactions of the ASME,* Series D, Vol. 82, 1960, pp. 35-45.

[4.11] Vaugham, D.R., "A Nonrecursive Algebraic Solution for the Discrete Riccati Equation," *IEEE Transactions on Automatic Control*, Vol. AC-15, 1970, pp. 597-599.

[4.12] *Advances in the Techniques and Technology of the Application of Nonlinear Filters and Kalman Filter*, AGARD-AG-256, 1982.

[4.13] Mendel, J.M., *Discrete Techniques of Parameter Estimation, The Equation Error Formulation*, Marcel Dekker, Inc., New York, NY, 1973.

[4.14] Jazwinski, A.H., *Stochastic Processes and Filtering Theory*, Academic Press, New York, NY, 1970.

[4.15] Bach, R.E., Jr., "State Estimation Applications in Aircraft Flight Data Analysis," NASA RP 1252, 1991.

[4.16] Sorenson, H.W., "Least-squares estimation: from Gauss to Kalman," *IEEE Spectrum*, July 1970, pp. 63-68.

Problems

4.1 Answer the following, using words, graphs, equations, and/or diagrams:

a) Explain the difference between a linear and nonlinear parameter estimation problem.

b) Compare and contrast Bayes estimation, Fisher estimation, and least squares estimation.

c) What is the central limit theorem, and how does it relate to aircraft system identification?

d) For most dynamic models, the model parameters to be found are assumed to be unknown constants. However, the modeling results are estimated mean values of the unknown model parameters, with standard errors. Explain this.

e) Explain why statistical analysis of experimental data is based on the assumption that the noise processes are Gaussian.

f) Describe the components of the mean squared error (MSE) for a parameter estimate.

g) Discuss the difference between a filter and a smoother. If there is a difference, is it important? Why or why not?

Chapter 5

Regression Methods

In the context of aircraft system identification, regression refers to a statistical technique for modeling and investigating relationships among measured variables. An example of a regression is the following model relating the nondimensional pitching moment coefficient C_m, to angle of attack α and Mach number M, for data collected during a wind tunnel test:

$$C_m = C_{m_o} + C_{m_\alpha}\alpha + C_{m_M}M + C_{m_{\alpha M}}\alpha M + v_m \tag{5.1}$$

In this example, α and M are variables that are set to selected values for each experimental test point, and measurements of α, M, and C_m are made. The pitching moment coefficient C_m is assumed to depend on α and M in the manner postulated by the form of the model shown in Eq. (5.1), where C_{m_o}, C_{m_α}, C_{m_M}, and $C_{m_{\alpha M}}$ are constant model parameters to be determined. Accordingly, α and M are called explanatory variables, and C_m is called the dependent variable or the response variable. There may be other variables that could affect C_m, but these are held fixed during the test, insofar as it is known what the influential variables might be.

Since the explanatory variables α and M are set by the experimenter for each data point, it is assumed that they are known without error. The dependent variable C_m is subject to random measurement errors, and is therefore a random variable. The random error term v_m includes random effects of unknown influences and random measurement errors in the dependent variable.

The model in Eq. (5.1) is linear in the parameters, although the quantities that the parameters multiply are both linear and nonlinear functions of the explanatory variables. The term "linear regression" refers to the linearity of the equation for the model output [e.g., Eq. (5.1)] with respect to the model parameters, not with respect to the explanatory variables. As shown in Eq. (5.1), a linear regression problem can have modeling functions that are linear or nonlinear functions of the explanatory variables.

In the preceding wind tunnel example, measured values of C_m can be obtained directly by nondimensionalizing the pitching moment measured by a strain-gage balance installed between the model and the mounting system in the wind tunnel. For flight test data, the measured values of nondimensional force and moment coefficients cannot be obtained in this way, but must instead be computed from measurements of the aircraft translational and rotational motion, along with geometric and mass/inertia properties of the aircraft, and the equations of motion. Values of C_m are therefore computed from other measurements, rather than measured directly, as in the wind tunnel experiment. For the case of pitching moment coefficient C_m, the measured values are computed from Eq. (3.37b),

$$C_m = \frac{1}{\bar{q}S\bar{c}}\left[I_y\dot{q} + \left(I_x - I_z\right)pr + I_{xz}\left(p^2 - r^2\right) - I_p\Omega_p r\right] \tag{5.2}$$

where the pitch rate derivative $\dot{q}$ is normally not measured directly, but rather is obtained using a smoothed numerical derivative of measured pitch rate q (see Chapter 11).

Data are collected for explanatory variables and response variables at each sample time, and the modeling problem based on flight test data is then similar to that for the wind tunnel experiment. In the case of flight test data analysis, control surface deflections and motion variables such as nondimensional pitch rate can also enter into the model for C_m. Wind tunnel tests can also include these explanatory variables, but the essential difference is that in flight testing, more than one of the explanatory variables change at the same time, in a manner dictated by the aircraft dynamics. As a result, these variables cannot be varied independently in flight, as is done in wind tunnel experiments.

For example, changing the angle of attack on an aircraft flying in still air requires a control surface deflection, whereas in a wind tunnel experiment the angle of attack can be changed at will with the control surface deflections held constant. It follows that the explanatory variables in a flight test are not really independent, although the terminology "independent variables" is often used in practice instead of "explanatory variables". This issue complicates experiment design for flight testing, but the fundamental modeling problem of finding a mathematical relationship among measured variables is the same.

For flight test data, identifying a model that matches the C_m values obtained from the pitching moment equation in a least-squares sense will minimize the squared error in the pitching moment equation when the identified model is used for C_m [cf. Eq. (5.2)]. For this reason, the method is also called equation error.

The modeling example in Eq. (5.1) can be generalized to the following model form for relating explanatory variables to a dependent variable,

$$y = \theta_o + \sum_{j=1}^{n} \theta_j \xi_j \tag{5.3}$$

where y is the dependent variable, ξ_j are linear or nonlinear functions of m explanatory variables $x_1, x_2, \ldots, x_m$, and the model parameters $\theta_o, \theta_1, \theta_2, \ldots, \theta_n$ are constants that quantify the influence of each term on the dependent variable y. The notation has been simplified by omitting the explicit dependence of the modeling functions ξ_j on the explanatory variables, i.e., $\xi_j \equiv \xi_j\left(x_1, x_2, \ldots x_m\right)$. The parameter θ_o can be assumed to multiply the constant 1, and models the bias in the dependent variable. The measured values of the dependent variable are corrupted by random measurement noise, so that

$$z(i) = \theta_o + \sum_{j=1}^{n} \theta_j \xi_j(i) + v(i) \qquad i = 1,2,\ldots,N \tag{5.4}$$

where $z(i)$ are the dependent variable measurements, N is the number of data points, and the $\xi_j(i)$ depend on the m explanatory variables $x_1, x_2, \ldots, x_m$ at the ith data point. Eq. (5.4) is called the regression equation, and ξ_j, $j = 1,2,\ldots,n$ are called the regressors. It is assumed that the explanatory variables $x_1, x_2, \ldots, x_m$ are measured without error, and the mathematical forms of the dependence of regressor functions ξ_j on the explanatory variables are postulated and therefore known. As a result, the regressors are assumed to be known without error. The quantity v is the measurement error for the dependent variable. Sometimes the dependent variable y is also called the output variable, and the explanatory variables x_k, $k = 1,2,\ldots,m$ are called input variables.

Following the material in Chapter 4, the least-squares model will be assumed initially, wherein the model parameters are assumed to be unknown constants, and the output measurements are corrupted by a vector of random noise. The next task then is to find an estimate $\hat{\boldsymbol{\theta}}$ of the parameter vector $\boldsymbol{\theta}^T = \left[\theta_0 \;\; \theta_1 \;\; \ldots \;\; \theta_n\right]$, based on the least-squares principle. The solution to this problem will be discussed in detail in this chapter, along with methods for characterizing the accuracy of the estimated parameters.

There is also practical interest choosing an appropriate mathematical model structure for the regression equation, i.e., deciding what terms should appear in model equations like Eq. (5.1), or its more general form, Eq. (5.4). In the foregoing discussion, it was assumed that the postulated model structure was adequate to characterize the measured data. In practice, an adequate model structure is not always known *a priori*, and therefore must be identified from the measured data using a model structure determination procedure.

Approaches to diagnosing model structure deficiencies and solving the model structure determination problem are discussed in this chapter.

Additional problems related to accurate parameter estimation can be caused by near-linear dependencies among the regressors. Methods for detecting and assessing these dependencies must be used, along with new methods for obtaining accurate parameter estimates. Later parts of the chapter are concerned with this problem.

This chapter contains essential information for using linear regression in aircraft system identification. Comprehensive treatments of linear regression can be found in Refs. [5.1]-[5.3].

5.1 Ordinary Least Squares

The general form of the model equation (5.3) and the regression equation (5.4) can be written using vector and matrix notation as

$$y = X\theta \tag{5.5}$$

and

$$z = X\theta + \nu \tag{5.6}$$

where

$z = \left[z(1) \;\; z(2) \;\; \ldots \;\; z(N) \right]^T = N \times 1$ response vector

$\boldsymbol{\theta} = [\theta_0 \;\; \theta_1 \;\; \ldots \;\; \theta_n]^T = n_p \times 1$ vector of unknown parameters, $n_p = n + 1$

$\boldsymbol{X} = [\mathbf{1} \;\; \xi_1 \;\; \ldots \;\; \xi_n] = N \times n_p$ matrix of vectors of ones and regressors

$\boldsymbol{\nu} = \left[\nu(1) \;\; \nu(2) \;\; \ldots \;\; \nu(N) \right]^T = N \times 1$ measurement error vector

The regressor vectors ξ_j, $j = 1, 2, \ldots, n$, are known postulated functions of the vectors of explanatory variables. The regressors ξ_j, $j = 1, 2, \ldots, n$, can be arbitrary functions of the explanatory variables. Usually, at least some of the regressors are equal to the explanatory variables themselves.

Regression equation (5.6) is equivalent to measurement equation (4.2), except that the measured output at each sample time is now a scalar, and the z vector has been redefined as a vector of the entire set of N scalar measurements. For the least-squares model, there are no probability statements regarding $\boldsymbol{\theta}$ or $\boldsymbol{\nu}$, but $\boldsymbol{\nu}$ is assumed to be zero mean and uncorrelated, with constant variance,

$$E(\boldsymbol{\nu}) = \mathbf{0} \qquad E\left(\boldsymbol{\nu}\boldsymbol{\nu}^T\right) = \sigma^2 \boldsymbol{I} \tag{5.7}$$

As discussed in Chapter 4, the best estimator of $\boldsymbol{\theta}$ in a least-squares sense comes from minimizing the sum of squared differences between the measurements and the model,

$$J(\boldsymbol{\theta}) = \frac{1}{2}(\boldsymbol{z} - \boldsymbol{X\theta})^T (\boldsymbol{z} - \boldsymbol{X\theta}) \tag{5.8}$$

Because the measured output is a scalar, the $\boldsymbol{R}^{-1}$ matrix included in Eq. (4.34) for weighting multiple outputs is irrelevant, and therefore omitted. Note that Eq. (5.8) includes data for the entire set of N data points.

The parameter estimate $\hat{\boldsymbol{\theta}}$ that minimizes the cost function $J(\boldsymbol{\theta})$ must satisfy

$$\frac{\partial J}{\partial \boldsymbol{\theta}} = -\boldsymbol{X}^T \boldsymbol{z} + \boldsymbol{X}^T \boldsymbol{X} \hat{\boldsymbol{\theta}} = \boldsymbol{0} \tag{5.9a}$$

which can be rearranged as

$$\boldsymbol{X}^T \boldsymbol{X} \hat{\boldsymbol{\theta}} = \boldsymbol{X}^T \boldsymbol{z} \tag{5.9b}$$

or

$$\boldsymbol{X}^T \left(\boldsymbol{z} - \boldsymbol{X}\hat{\boldsymbol{\theta}}\right) = \boldsymbol{0} \tag{5.9c}$$

The $n_p = n + 1$ equations represented in Eqs. (5.9) are called the normal equations. The solution of these equations for the unknown parameter vector $\boldsymbol{\theta}$ gives the formula for the least-squares estimator, also called the ordinary least-squares estimator,

$$\hat{\boldsymbol{\theta}} = \left(\boldsymbol{X}^T \boldsymbol{X}\right)^{-1} \boldsymbol{X}^T \boldsymbol{z} \tag{5.10}$$

The $n_p \times n_p$ matrix $\boldsymbol{X}^T \boldsymbol{X}$ is always symmetric. If the regressor vectors that make up the columns of $\boldsymbol{X}$ are linearly independent, then $\boldsymbol{X}^T \boldsymbol{X}$ is positive definite, the eigenvalues of $\boldsymbol{X}^T \boldsymbol{X}$ are positive real numbers, the associated eigenvectors are mutually orthogonal, and $\left(\boldsymbol{X}^T \boldsymbol{X}\right)^{-1}$ exists (see Appendix A). Note also from Eq. (5.9a) that the second gradient of the cost function with respect to the parameter vector is $\boldsymbol{X}^T \boldsymbol{X}$, which is positive definite, indicating a minimum, rather than a maximum.

The covariance matrix of the parameter estimate $\hat{\boldsymbol{\theta}}$, also known as the covariance matrix of the estimation error $\hat{\boldsymbol{\theta}} - \boldsymbol{\theta}$, is

$$Cov\left(\hat{\boldsymbol{\theta}}\right) \equiv E\left[\left(\hat{\boldsymbol{\theta}}-\boldsymbol{\theta}\right)\left(\hat{\boldsymbol{\theta}}-\boldsymbol{\theta}\right)^T\right]$$
$$= E\left\{\left(\boldsymbol{X}^T\boldsymbol{X}\right)^{-1}\boldsymbol{X}^T(z-y)(z-y)^T\boldsymbol{X}\left(\boldsymbol{X}^T\boldsymbol{X}\right)^{-1}\right\} \tag{5.11}$$
$$= \left(\boldsymbol{X}^T\boldsymbol{X}\right)^{-1}\boldsymbol{X}^T E\left(\boldsymbol{v}\boldsymbol{v}^T\right)\boldsymbol{X}\left(\boldsymbol{X}^T\boldsymbol{X}\right)^{-1}$$

where the true parameter vector $\boldsymbol{\theta}$ is related to the true output $\boldsymbol{y}$ by $\boldsymbol{\theta} = \left(\boldsymbol{X}^T\boldsymbol{X}\right)^{-1}\boldsymbol{X}^T\boldsymbol{y}$. Assuming the measurement errors are uncorrelated and have constant variance σ^2, $E\left(\boldsymbol{v}\boldsymbol{v}^T\right) = \sigma^2\boldsymbol{I}$ [cf. Eq. (5.7)], the expression for the covariance matrix of $\hat{\boldsymbol{\theta}}$ is then simplified to

$$Cov\left(\hat{\boldsymbol{\theta}}\right) = E\left[\left(\hat{\boldsymbol{\theta}}-\boldsymbol{\theta}\right)\left(\hat{\boldsymbol{\theta}}-\boldsymbol{\theta}\right)^T\right] = \sigma^2\left(\boldsymbol{X}^T\boldsymbol{X}\right)^{-1} \tag{5.12}$$

Note that the matrix $\left(\boldsymbol{X}^T\boldsymbol{X}\right)^{-1}$ was also required to compute the least-squares parameter estimate in Eq. (5.10). Defining the matrix $\boldsymbol{\mathcal{D}}$ as

$$\boldsymbol{\mathcal{D}} \equiv \left(\boldsymbol{X}^T\boldsymbol{X}\right)^{-1} = \left[d_{jk}\right] \qquad j,k = 1,2,\ldots,n_p \tag{5.13}$$

the variance of the *j*th estimated parameter in the parameter vector $\hat{\boldsymbol{\theta}}$ is the *j*th diagonal element of the covariance matrix, or

$$Var\left(\hat{\theta}_j\right) = \sigma^2 d_{jj} \equiv s^2\left(\hat{\theta}_j\right) \qquad j = 1,2,\ldots,n_p \tag{5.14}$$

and the covariance between two estimated parameters $\hat{\theta}_j$ and $\hat{\theta}_k$ is

$$Cov\left(\hat{\theta}_j,\hat{\theta}_k\right) = \sigma^2 d_{jk} \qquad j,k = 1,2,\ldots,n_p \tag{5.15}$$

The correlation coefficient r_{jk} is defined as

$$r_{jk} \equiv \frac{d_{jk}}{\sqrt{d_{jj}d_{kk}}} = \frac{Cov\left(\hat{\theta}_j,\hat{\theta}_k\right)}{\sqrt{Var\left(\hat{\theta}_j\right)Var\left(\hat{\theta}_k\right)}} \qquad j,k = 1,2,\ldots,n_p \tag{5.16a}$$

$$-1 \le r_{jk} \le 1 \qquad j,k = 1,2,\ldots,n_p \tag{5.16b}$$

The correlation coefficient r_{jk} is a measure of the pairwise correlation between parameter estimates $\hat{\theta}_j$ and $\hat{\theta}_k$. A value of $r_{jk} = 1$ means that estimated parameters $\hat{\theta}_j$ and $\hat{\theta}_k$ are linearly related, or equivalently, their corresponding regressors are linearly dependent. When $r_{jk} = -1$, the same statements apply, except that the linear relationship includes a negative sign. Arranging all the values of r_{jk} in a $n_p \times n_p$ matrix forms the parameter correlation matrix $Corr(\hat{\boldsymbol{\theta}})$, which is symmetric with ones on the main diagonal,

$$Corr(\hat{\boldsymbol{\theta}}) = \left[r_{jk} \right] \qquad j,k = 1,2,\ldots,n_p \tag{5.17a}$$

$$r_{jj} = 1 \qquad j = 1,2,\ldots,n_p \tag{5.17b}$$

The parameter correlation matrix can also be computed by the following matrix multiplication,

$$Corr(\hat{\boldsymbol{\theta}}) = \begin{bmatrix} \frac{1}{s(\hat{\theta}_1)} & 0 & \cdots & 0 \\ 0 & \frac{1}{s(\hat{\theta}_2)} & & 0 \\ \vdots & & \ddots & \vdots \\ 0 & 0 & \cdots & \frac{1}{s(\hat{\theta}_{n_p})} \end{bmatrix} Cov(\hat{\boldsymbol{\theta}}) \begin{bmatrix} \frac{1}{s(\hat{\theta}_1)} & 0 & \cdots & 0 \\ 0 & \frac{1}{s(\hat{\theta}_2)} & & 0 \\ \vdots & & \ddots & \vdots \\ 0 & 0 & \cdots & \frac{1}{s(\hat{\theta}_{n_p})} \end{bmatrix} \tag{5.18}$$

Using Eqs. (5.5) and (5.10), the estimated dependent variable vector $\hat{\boldsymbol{y}}$, based on the vector of measured outputs $\boldsymbol{z}$ and the regressor matrix $\boldsymbol{X}$, is given by

$$\hat{\boldsymbol{y}} = \boldsymbol{X}\hat{\boldsymbol{\theta}} = \boldsymbol{X}\left(\boldsymbol{X}^T\boldsymbol{X}\right)^{-1}\boldsymbol{X}^T\boldsymbol{z} = \boldsymbol{K}\boldsymbol{z} \tag{5.19}$$

where

$$\boldsymbol{K} = \boldsymbol{X}\left(\boldsymbol{X}^T\boldsymbol{X}\right)^{-1}\boldsymbol{X}^T \tag{5.20}$$

is the $N \times N$ prediction matrix that maps the measured outputs to the estimated outputs. The differences between the measured values $\boldsymbol{z}$ and the estimated values $\hat{\boldsymbol{y}}$ are the residuals, which form the vector $\boldsymbol{v}$,

$$\begin{aligned} \boldsymbol{v} &= \boldsymbol{z} - \hat{\boldsymbol{y}} = \boldsymbol{z} - \boldsymbol{X}\hat{\boldsymbol{\theta}} \\ &= \boldsymbol{z} - \boldsymbol{X}\left(\boldsymbol{X}^T\boldsymbol{X}\right)^{-1}\boldsymbol{X}^T\boldsymbol{z} \\ &= \left(\boldsymbol{I} - \boldsymbol{K}\right)\boldsymbol{z} \end{aligned} \tag{5.21}$$

The prediction matrix and the residuals will be discussed further in Section 5.1.4.

The calculation of the parameter covariance matrix from Eq. (5.12) requires the measurement error variance σ^2, which was assumed constant for all the data points when deriving Eq. (5.12). In practice, σ^2 is usually not known *a priori* and therefore must be estimated from the measured data. An unbiased estimate of σ^2 can be computed from repeated measurements at the same explanatory variable settings using

$$\hat{\sigma}^2 = \frac{1}{\left(n_r - 1\right)} \sum_{i=1}^{n_r} \left[z_r(i) - \bar{z}_r\right]^2 \equiv s^2 \tag{5.22}$$

where n_r is the number of repeated measurements z_r, and $\bar{z}_r$ is the mean value of the n_r repeated measurements,

$$\bar{z}_r = \frac{1}{n_r} \sum_{i=1}^{n_r} z_r(i) \tag{5.23}$$

For dynamic flight data, where it is unlikely that the same explanatory variable settings will be exactly repeated, σ^2 can be estimated independently using smoothing methods (see Chapter 11). Assuming the model structure is adequate, an unbiased estimate for σ^2 can also be obtained based on the residuals,

$$\hat{\sigma}^2 = \frac{\boldsymbol{v}^T\boldsymbol{v}}{\left(N - n_p\right)} = \frac{\sum_{i=1}^{N}\left[z(i) - \hat{y}(i)\right]^2}{\left(N - n_p\right)} \equiv s^2 \tag{5.24}$$

The square root of s^2 is sometimes called the fit error, which indicates how close the estimates $\hat{y}(i)$ are to the measured values $z(i)$, using the same units as the measured and estimated values, rather than their squares. Note that the estimator in Eq. (5.24) depends on the model, because of the appearance of $\hat{y}(i)$ in the calculation. Consequently, when Eq. (5.24) is used, the estimate of σ^2 depends on the postulated model structure and associated parameter estimates, instead of being based solely on the measured data.

Another metric that quantifies the closeness of $\hat{y}(i)$ to $z(i)$ is the coefficient of determination, R^2. The definition of R^2 follows from partitioning the total sum of squared variations of the measured output z about its mean value into the sum of squared variations of the estimate $\hat{y}$ about the same mean value, plus sum of squared variations of the measurement z about the estimate $\hat{y}$. These quantities are called the total sum of squares SS_T, the regression sum of squares SS_R, and the residual sum of squares SS_E, respectively. The three sums are defined by

$$SS_T \equiv \sum_{i=1}^{N}\left[z(i)-\bar{z}\right]^2 = \boldsymbol{z}^T\boldsymbol{z} - N\bar{z}^2 \tag{5.25}$$

where

$$\bar{z} = \frac{1}{N}\sum_{i=1}^{N} z(i) \tag{5.26}$$

Then

$$SS_R \equiv \sum_{i=1}^{N}\left[\hat{y}(i)-\bar{z}\right]^2 \tag{5.27}$$

$$\begin{aligned} SS_E \equiv \sum_{i=1}^{N}\left[z(i)-\hat{y}(i)\right]^2 &= \left(\boldsymbol{z}-\boldsymbol{X}\hat{\boldsymbol{\theta}}\right)^T\left(\boldsymbol{z}-\boldsymbol{X}\hat{\boldsymbol{\theta}}\right) \\ &= \boldsymbol{z}^T\boldsymbol{z} - 2\hat{\boldsymbol{\theta}}^T\boldsymbol{X}^T\boldsymbol{z} + \hat{\boldsymbol{\theta}}^T\boldsymbol{X}^T\boldsymbol{X}\hat{\boldsymbol{\theta}} \\ &= \boldsymbol{z}^T\boldsymbol{z} - 2\hat{\boldsymbol{\theta}}^T\boldsymbol{X}^T\boldsymbol{z} + \hat{\boldsymbol{\theta}}^T\boldsymbol{X}^T\boldsymbol{z} \\ &= \boldsymbol{z}^T\boldsymbol{z} - \hat{\boldsymbol{\theta}}^T\boldsymbol{X}^T\boldsymbol{z} \end{aligned} \tag{5.28}$$

where the third line in Eq. (5.28) follows from the normal equations (5.9b).

The relationship among the three sums SS_T, SS_R, and SS_E is

$$SS_T = SS_R + SS_E \tag{5.29}$$

which is derived in Appendix B. From Eqs. (5.29), (5.25), and (5.28),

$$SS_R = SS_T - SS_E = \hat{\boldsymbol{\theta}}^T\boldsymbol{X}^T\boldsymbol{z} - N\bar{z}^2 \tag{5.30}$$

Equation (5.29) shows that the total sum of squares of the measurements about the mean can be partitioned into the sum of squares of the model about the same mean, plus the sum of squares of the measurements about the model.

For good models, SS_E will include only noise, and SS_R will be large relative to SS_E.

The coefficient of determination R^2 represents the proportion of the variation in the measured output that is explained by the model,

$$R^2 = \frac{SS_R}{SS_T} = \frac{SS_T - SS_E}{SS_T} = \frac{\hat{\theta}^T X^T z - N\bar{z}^2}{z^T z - N\bar{z}^2} \tag{5.31}$$

Values of R^2 vary from 0 to 1, where 1 represents a perfect fit to the data; however, R^2 is usually expressed as a percentage. Further information on the coefficient of determination will be given in Section 5.4.

5.1.1 Properties of the Least-Squares Estimator

The least-squares estimator was developed under the assumptions of a postulated model that is linear in the parameters, deterministic regressors, and white measurement noise with constant variance. Under these assumptions, and using the definitions in Chapter 4, the properties of the least-squares estimates are as follows:

1) The least-squares estimator is unbiased:

$$\begin{aligned} E(\hat{\theta}) &= E\left[\left(X^T X\right)^{-1} X^T z\right] = E\left[\left(X^T X\right)^{-1} X^T (X\theta + \nu)\right] \\ &= E(\theta) + \left(X^T X\right)^{-1} X^T E(\nu) = \theta \end{aligned} \tag{5.32}$$

2) The least-squares estimator is a minimum variance and efficient estimator. This means $\hat{\theta}$ is the best unbiased estimate that is a linear function of the measurements, and the parameter accuracy is given by the parameter variance from Eq. (5.14).

3) The least-squares estimator is consistent. The practical meaning of this property is that as the number of data points increases, the parameter estimates will converge to their true values.

4) If the measurement noise is assumed Gaussian, where ν is $\mathbb{N}\left(0, \sigma^2 I\right)$, then the least-squares model for the measurement process becomes the Fisher model. It follows from Chapter 4 that in this situation the least-squares estimator is a maximum likelihood estimator.

In practical aircraft problems, the assumptions stated here are often violated, so that these theoretical properties do not hold. Section 5.2 and the following material in this chapter describe approaches used in practice to account for violation of the assumptions in the theory.

5.1.2 Confidence Intervals

When estimating the model parameters in a linear regression problem, it is of interest to know the quality of the parameter estimates obtained. The quality of the parameter estimates can be evaluated in terms of confidence intervals for the model parameter estimates, the estimated output, and the prediction of outputs for new data that were not used to identify the model. Procedures used to quantify these confidence intervals require an additional assumption that the measurement errors are normally distributed, i.e.,

$$\boldsymbol{v} \text{ is } \mathbb{N}\left(\mathbf{0}, \sigma^2 \boldsymbol{I}\right)$$

Then, from the linearity of the least-squares estimator with respect to the measurements [cf. Eq. (5.10)], along with the assumption that the regressors are deterministic, it follows that

$$\begin{aligned} \hat{\boldsymbol{\theta}} &\text{ is } \mathbb{N}\left[\boldsymbol{\theta}, \sigma^2 \left(\boldsymbol{X}^T \boldsymbol{X}\right)^{-1}\right] \\ \hat{\boldsymbol{y}} &\text{ is } \mathbb{N}\left(\boldsymbol{X\theta}, \sigma^2 \boldsymbol{K}\right) \end{aligned} \tag{5.33}$$

because any linear function of a Gaussian random variable is also Gaussian (see Appendix B). The parameter estimates from repeated experiments should therefore have a Gaussian distribution about the true value, and similarly for the estimated outputs.

<u>*Confidence Interval on Estimated Parameters*</u>

The normal distribution of the estimated parameter vector $\hat{\boldsymbol{\theta}}$ implies that each element $\hat{\theta}_j$ is also normally distributed, with mean value θ_j and variance $\sigma^2 d_{jj}$, where d_{jj} is the jth diagonal element of the $\left(\boldsymbol{X}^T \boldsymbol{X}\right)^{-1}$ matrix. Consequently, each of the statistics

$$t = \frac{\hat{\theta}_j - \theta_j}{\sqrt{\hat{\sigma}^2 d_{jj}}} = \frac{\hat{\theta}_j - \theta_j}{s\left(\hat{\theta}_j\right)} \qquad j = 1, 2, \ldots, n_p \tag{5.34}$$

has the t-distribution with $N - n_p$ degrees of freedom (see Appendix B). A common value selected for the confidence level is 95 percent, so that the fraction of the two-sided t-distribution excluded is $\alpha = 0.05$. Accordingly, a $100(1-\alpha)$ percent confidence interval for the parameter $\hat{\theta}_j$ is

$$\hat{\theta}_j - t\left(\alpha/2, N-n_p\right)s\left(\hat{\theta}_j\right) \le \theta_j \le \hat{\theta}_j + t\left(\alpha/2, N-n_p\right)s\left(\hat{\theta}_j\right)$$

$$j = 1,2,\ldots,n_p \qquad (5.35)$$

where $\alpha = 0.05$ for a 95 percent confidence level. From the table of t-distribution values, e.g., in Ref. [5.3], $t\left(\alpha/2, N-n_p\right) \approx 1.96$ for $N - n_p > 100$, and $\alpha = 0.05$. Then the confidence interval can be written as

$$\theta_j = \hat{\theta}_j \pm 1.96\, s\left(\hat{\theta}_j\right) \qquad j = 1,2,\ldots,n_p \qquad (5.36)$$

In practical flight test applications, $N - n_p \gg 100$, and the t-distribution approaches the normal distribution, so that a confidence interval corresponding to two standard deviations is often used:

$$\theta_j = \hat{\theta}_j \pm 2\, s\left(\hat{\theta}_j\right) \qquad j = 1,2,\ldots,n_p \qquad (5.37)$$

For normally distributed measurement errors, this can be interpreted as a 95 percent probability that the interval $\left[\hat{\theta}_j - 2\, s\left(\hat{\theta}_j\right), \hat{\theta}_j + 2\, s\left(\hat{\theta}_j\right)\right]$ will contain the true value of the parameter θ_j.

Confidence Interval on Estimated Output

The confidence interval for the estimated output can be constructed at each data point. At the ith data point, the vector of regressors is

$$\boldsymbol{x}^T(i) \equiv \begin{bmatrix} 1 & \xi_1(i) & \xi_2(i) & \ldots & \xi_n(i) \end{bmatrix} \qquad (5.38)$$

The quantity $\boldsymbol{x}^T(i)$ is the ith row of the regressor matrix $\boldsymbol{X}$. The regressors $\xi_1(i), \xi_2(i), \ldots, \xi_n(i)$ are functions of the measured explanatory variables (typically aircraft states and controls) at the ith data point, although this is not shown in the notation.

The estimated output at that point is

$$\hat{y}(i) = \boldsymbol{x}^T(i)\hat{\boldsymbol{\theta}} \qquad (5.39)$$

and the variance of $\hat{y}(i)$ is

$$\begin{aligned}
Var\left[\hat{y}(i)\right] &= E\left\{\left[\hat{y}(i)-y(i)\right]\left[\hat{y}(i)-y(i)\right]^T\right\} \\
&= E\left\{\left[\boldsymbol{x}^T(i)\hat{\boldsymbol{\theta}}-\boldsymbol{x}^T(i)\boldsymbol{\theta}\right]\left[\boldsymbol{x}^T(i)\hat{\boldsymbol{\theta}}-\boldsymbol{x}^T(i)\boldsymbol{\theta}\right]^T\right\} \\
&= \boldsymbol{x}^T(i)E\left[\left(\hat{\boldsymbol{\theta}}-\boldsymbol{\theta}\right)\left(\hat{\boldsymbol{\theta}}-\boldsymbol{\theta}\right)^T\right]\boldsymbol{x}(i) \\
&= \sigma^2\boldsymbol{x}^T(i)\left(\boldsymbol{X}^T\boldsymbol{X}\right)^{-1}\boldsymbol{x}(i) \\
&\equiv s^2\left[\hat{y}(i)\right]
\end{aligned} \tag{5.40}$$

The $100(1-\alpha)$ percent confidence interval for the output estimate at the *i*th data point is

$$y(i) = \hat{y}(i) \pm t\left(\alpha/2, N-n_p\right) s\left[\hat{y}(i)\right] \tag{5.41}$$

where $s\left[\hat{y}(i)\right]$ is computed from Eq. (5.40), with σ^2 replaced by its estimate $\hat{\sigma}^2$ from Eq. (5.24) or from an independent estimate, as described earlier.

Confidence Interval on Predicted Output

The quantity $\hat{y}(i)$ can be interpreted in two ways. First, it is the estimate of the output at the *i*th data point. In this context, $\hat{y}(i)$ is sometimes called the fitted value to the measurement $z(i)$. The variance of $\hat{y}(i)$ given by Eq. (5.40) reflects the variation of $\hat{y}$ at $\boldsymbol{x}(i)$ for repeated regressions with the same sample size and the same regressors.

In the second interpretation, $\hat{y}(i)$ is the predicted value of y at $\boldsymbol{x}(i)$. For assessment of $\hat{y}(i)$ as a predictor, consider $z(i)$ to be a new single measurement at $\boldsymbol{x}(i)$. This measurement is assumed to be independent of $\hat{y}(i)$. The prediction error variance is

$$\begin{aligned}
Var\left[z(i)-\hat{y}(i)\right] &= E\left\{\left[z(i)-\hat{y}(i)\right]\left[z(i)-\hat{y}(i)\right]\right\} \\
&= \sigma^2 E\left(\left\{v(i)-\left[\hat{y}(i)-y(i)\right]\right\}\left\{v(i)-\left[\hat{y}(i)-y(i)\right]\right\}\right)
\end{aligned}$$

$$
\begin{aligned}
Var\left[z(i)-\hat{y}(i)\right] &= E\left[v(i)v(i)\right]+E\left\{\left[\hat{y}(i)-y(i)\right]\left[\hat{y}(i)-y(i)\right]\right\} \\
&= \sigma^2\left[1+\boldsymbol{x}^T(i)\left(\boldsymbol{X}^T\boldsymbol{X}\right)^{-1}\boldsymbol{x}(i)\right] \\
&\equiv s^2\left[z(i)-\hat{y}(i)\right]
\end{aligned}
\tag{5.42}
$$

Note that the variance of $z(i)-\hat{y}(i)$ is the sum of the measurement error variance σ^2 and the variance of the estimated output $\hat{y}(i)$ from Eq. (5.40). Using Eq. (5.42) with an estimated measurement error variance $\hat{\sigma}^2$, the prediction interval is given by

$$
y(i)=\hat{y}(i)\pm t\left(\alpha/2,N-n_p\right)s\left[z(i)-\hat{y}(i)\right] \tag{5.43}
$$

5.1.3 Hypothesis Testing

In the analysis of the linear regression problem up to this point, it has been assumed that the regressors necessary for an adequate model were known. In practice, this is often not true, so there must be some tests for adding or deleting terms from a proposed model structure, based on statistics that can be computed from the measured data. Three of these tests will be outlined, and later used in the selection of an adequate model structure from measured data. In the following development, the Fisher model

$$
\begin{gathered}
z = \boldsymbol{X}\boldsymbol{\theta}+\boldsymbol{v} \\
\boldsymbol{v} \text{ is } \mathbb{N}\left(\boldsymbol{0},\sigma^2\boldsymbol{I}\right)
\end{gathered}
\tag{5.44}
$$

with n regressors and $n_p = n+1$ parameters will be considered. This is identical to a least-squares model with weighting matrix chosen to be the same as for a Fisher model with white Gaussian measurement noise of constant variance.

Test for Significance of the Regression

The objective of the test for the significance of the regression is to determine if there is a linear relationship between the measured dependent variable vector z and any of a set of candidate regressor vectors $\xi_j,\ j=1,2,\ldots,n$. The bias term in the model, which has a regressor vector consisting of a vector of ones, is included in the model by default, because an estimate of the bias term is always useful, even if it turns out to be zero.

The related hypotheses, called the null hypothesis H_0, and the alternative hypothesis H_1, are stated as

$$\begin{aligned} &H_0 : \theta_1 = \theta_2 = \ldots = \theta_n = 0 \\ &H_1 : \theta_j \neq 0 \ \text{ for at least one } j \end{aligned} \tag{5.45}$$

Since the constant term is always assumed to be included in the model, it is not included in H_0 or H_1. If the null hypothesis is rejected, then at least one of the regressors ξ_j, $j = 1,2,\ldots,n$, significantly contributes to modeling the variation in the dependent variable. If the null hypothesis H_0 is true, then the sum of squares due to the regression, SS_R, divided by its degrees of freedom will be an estimate of the fit error variance, using only the bias term in the model. The residual sum of squares SS_E divided by its degrees of freedom is also an estimate of the fit error variance. The relevant test statistic for H_0 is therefore the F-statistic for the ratio of squared random variables (see e.g. Ref. [5.3]),

$$\begin{aligned} F_0 &= \frac{SS_R/n}{SS_E/\left(N - n_p\right)} \\ &= \frac{\hat{\boldsymbol{\theta}}^T \boldsymbol{X}^T \boldsymbol{z} - N\bar{z}^2}{n s^2} \end{aligned} \tag{5.46}$$

The statistic F_0 is a random variable that follows the F-distribution (see Appendix B), with the number of degrees of freedom n for the numerator and $N - n_p$ for the denominator. The null hypothesis is rejected if $F_0 > F\left(\alpha; n, N - n_p\right)$, where $F\left(\alpha; n, N - n_p\right)$ can be found from tabulated values of the F-distribution for a selected significance level α, $0 \le \alpha \le 1$. If the null hypothesis is rejected at the selected significance level α, that is equivalent to the statement that not all of the model parameters are zero, with $100\left(1 - \alpha\right)$ percent confidence. The situation is also sometimes described by saying that F_0 is statistically significant at the 100α percent level. If the null hypothesis is rejected, then one or more of the model terms is explaining some significant part of the variation in the dependent variable, which causes the numerator of F_0 to be larger than it would be if none of the model terms were significant. Note that the selected confidence level $\left(1 - \alpha\right)$ and the statistical inference is necessary because of the uncertainty associated with chance variations introduced by the noise. For aircraft system identification, a typical choice is $\alpha = 0.05$, which corresponds to a 95 percent confidence level.

Test for Significance of a Subset of Model Parameters

The statistical significance testing just described can also be used to investigate the contribution of a subset of one or more regressors to modeling the variation in the dependent variable. For this investigation, the parameter vector is partitioned as follows:

$$\boldsymbol{\theta} = \begin{bmatrix} \boldsymbol{\theta}_1 \\ \boldsymbol{\theta}_2 \end{bmatrix} \tag{5.47}$$

where $\boldsymbol{\theta}_1$ is a $p \times 1$ vector of model parameters that includes the bias term, and $\boldsymbol{\theta}_2$ is a $q \times 1$ vector of unknown parameters associated with terms being considered for inclusion in the model, where $q < n$ and $p + q = n_p$. For the partitioned parameter vector, the regression equation changes to

$$\boldsymbol{z} = \boldsymbol{X}_1\boldsymbol{\theta}_1 + \boldsymbol{X}_2\boldsymbol{\theta}_2 + \boldsymbol{\nu} \tag{5.48}$$

To determine if the subset of regressors in the columns of $\boldsymbol{X}_2$ contribute significantly to the model, the hypotheses are:

$$\begin{aligned} H_0 &: \boldsymbol{\theta}_2 = \boldsymbol{0} \\ H_1 &: \boldsymbol{\theta}_2 \neq \boldsymbol{0} \end{aligned} \tag{5.49}$$

For the full model, it is known that

$$\begin{aligned} \hat{\boldsymbol{\theta}} &= \left(\boldsymbol{X}^T\boldsymbol{X}\right)^{-1}\boldsymbol{X}^T\boldsymbol{z} \\ SS_R\left(\hat{\boldsymbol{\theta}}\right) &= \hat{\boldsymbol{\theta}}^T\boldsymbol{X}^T\boldsymbol{z} - N\bar{z}^2 \\ SS_E\left(\hat{\boldsymbol{\theta}}\right) &= \boldsymbol{z}^T\boldsymbol{z} - \hat{\boldsymbol{\theta}}^T\boldsymbol{X}^T\boldsymbol{z} \end{aligned} \tag{5.50}$$

In order to find the contribution of the $\boldsymbol{X}_2\boldsymbol{\theta}_2$ terms to the model, it is necessary to estimate the model parameters assuming that the null hypothesis is true. The reduced model is

$$\boldsymbol{z} = \boldsymbol{X}_1\boldsymbol{\theta}_1 + \boldsymbol{\nu} \tag{5.51}$$

with the solution

$$\begin{aligned} \hat{\boldsymbol{\theta}}_1 &= \left(\boldsymbol{X}_1^T\boldsymbol{X}_1\right)^{-1}\boldsymbol{X}_1^T\boldsymbol{z} \\ SS_R\left(\hat{\boldsymbol{\theta}}_1\right) &= \hat{\boldsymbol{\theta}}_1^T\boldsymbol{X}_1^T\boldsymbol{z} - N\bar{z}^2 \\ SS_E\left(\hat{\boldsymbol{\theta}}_1\right) &= \boldsymbol{z}^T\boldsymbol{z} - \hat{\boldsymbol{\theta}}_1^T\boldsymbol{X}_1^T\boldsymbol{z} \end{aligned} \tag{5.52}$$

Then the regression sum of squares due to $\boldsymbol{X}_2\boldsymbol{\theta}_2$ given that $\boldsymbol{X}_1\boldsymbol{\theta}_1$ is already in the model, denoted by $SS_R\left(\hat{\boldsymbol{\theta}}_2 \mid \hat{\boldsymbol{\theta}}_1\right)$, follows from the relation

$$SS_R\left(\hat{\boldsymbol{\theta}}_2 \mid \hat{\boldsymbol{\theta}}_1\right) = SS_R\left(\hat{\boldsymbol{\theta}}\right) - SS_R\left(\hat{\boldsymbol{\theta}}_1\right) \tag{5.53}$$

Using Eqs. (5.50) and (5.52),

$$SS_R\left(\hat{\boldsymbol{\theta}}_2 \mid \hat{\boldsymbol{\theta}}_1\right) = \left[\hat{\boldsymbol{\theta}}^T \boldsymbol{X}^T \boldsymbol{z} - \hat{\boldsymbol{\theta}}_1^T \boldsymbol{X}_1^T \boldsymbol{z}\right] = \left[\hat{\boldsymbol{\theta}}^T \boldsymbol{X}^T - \hat{\boldsymbol{\theta}}_1^T \boldsymbol{X}_1^T\right]\boldsymbol{z}$$

$$= \boldsymbol{z}^T\left[\boldsymbol{X}\left(\boldsymbol{X}^T\boldsymbol{X}\right)^{-1}\boldsymbol{X}^T - \boldsymbol{X}_1\left(\boldsymbol{X}_1^T\boldsymbol{X}_1\right)^{-1}\boldsymbol{X}_1^T\right]\boldsymbol{z} \tag{5.54}$$

The associated F statistic for testing the null hypothesis (5.49) is

$$F_0 = \frac{SS_R\left(\hat{\boldsymbol{\theta}}_2 \mid \hat{\boldsymbol{\theta}}_1\right)}{q s^2} \tag{5.55}$$

If $F_0 > F\left(\alpha; q, N - n_p\right)$, the null hypothesis is rejected, so at least one of the parameters in $\boldsymbol{\theta}_2$ is not zero. Therefore, at least one of the regressors in $\boldsymbol{X}_2$ contributes significantly to the regression model. Significance testing using the statistic (5.55) is sometimes called a partial F-test, because it is concerned with the contribution of the regressors in $\boldsymbol{X}_2$, given that the other regressors in $\boldsymbol{X}_1$ are already in the model.

Test for Significance of a Single Model Parameter

If the parameter vector $\boldsymbol{\theta}_2$ in the preceding development is reduced to a scalar, then the significance of an individual regressor to the regression model can be tested by the hypotheses

$$\begin{aligned} H_0 &: \theta_2 = 0 \\ H_1 &: \theta_2 \neq 0 \end{aligned} \tag{5.56}$$

where $\boldsymbol{X}_2$ is now a single vector regressor and θ_2 is a scalar. Then the F statistic has one degree of freedom in the numerator, and $N - n_p$ degrees of freedom in the denominator. Defining a new dependent variable z_1 as the original dependent variable adjusted for the terms $\boldsymbol{X}_1\hat{\boldsymbol{\theta}}_1$,

$$z_1 \equiv \boldsymbol{z} - \boldsymbol{X}_1\hat{\boldsymbol{\theta}}_1 \tag{5.57}$$

and the regression equation is

$$z_1 = \boldsymbol{X}_2\theta_2 + \boldsymbol{\nu} \tag{5.58}$$

The mean value of z_1 is zero because the regressor matrix X_1 includes the bias term. The partial F statistic is

$$F_0 = \frac{\hat{\theta}_2 X_2^T z_1}{s^2} = \frac{\left(X_2^T X_2\right)\hat{\theta}_2^2}{s^2} = \frac{\hat{\theta}_2^2}{s^2\left(\hat{\theta}_2\right)} \tag{5.59}$$

where the second equality follows from the normal equation (5.9b), and the last equality comes from Eq. (5.14), because for the single regressor X_2, $s^2\left(\hat{\theta}_2\right) = s^2/\left(X_2^T X_2\right)$. Equation (5.59) shows that the partial F statistic in this case is identical to the square of the t statistic for parameter θ_2 with the true value of θ_2 equal to zero (in accordance with the null hypothesis), and $N - n_p$ degrees of freedom [cf. Eq. (5.34)]. The distribution of the squared t statistic is the same as the F distribution, so that equivalent significance testing can be done using the t statistic

$$t_0 = \frac{\hat{\theta}_2}{s\left(\hat{\theta}_2\right)} \tag{5.60}$$

The null hypothesis is rejected if $|t_0| > t\left(\alpha/2, N - n_p\right)$. This can be interpreted as a test for $\hat{\theta}_2$ being statistically different from zero. From Eqs. (5.59) and (5.60), $t_0^2 = F_0$.

5.1.4 Analysis of Residuals

The residuals $v(i)$, $i = 1,2,\ldots,N$, are defined as the difference between the measured output $z(i)$ and the estimated output $\hat{y}(i)$

$$v(i) = z(i) - \hat{y}(i) \qquad i = 1,2,\ldots,N \tag{5.61}$$

The residuals can be interpreted as samples of the measurement errors $v(i)$. Any departure from the underlying assumptions on the errors should be seen in the residuals. Analysis of the residuals is an effective method for discovering various types of model deficiencies. Unfortunately, it is not simple to do this, as can be seen from the following development.

Returning to Eq. (5.21), the vector of residuals can be expressed as

$$v = (I - K)z \tag{5.62a}$$

$$K = X\left(X^T X\right)^{-1} X^T \tag{5.62b}$$

where the prediction matrix $\boldsymbol{K}$ is symmetric and idempotent, which means that $\boldsymbol{K} = \boldsymbol{K}^T$ and $\boldsymbol{K}\boldsymbol{K} = \boldsymbol{K}$. From Eqs. (5.62) and (5.21), the expected value of the residuals is zero,

$$E(\boldsymbol{v}) = E(\boldsymbol{z}) - \boldsymbol{K}E(\boldsymbol{z}) = \boldsymbol{X\theta} - \boldsymbol{X}\left(\boldsymbol{X}^T\boldsymbol{X}\right)^{-1}\boldsymbol{X}^T\boldsymbol{X\theta} = \boldsymbol{0} \tag{5.63}$$

and the covariance matrix is

$$\begin{aligned} Cov(\boldsymbol{v}) &= E\left(\boldsymbol{v}\boldsymbol{v}^T\right) = E\left[(\boldsymbol{I}-\boldsymbol{K})\boldsymbol{z}\boldsymbol{z}^T(\boldsymbol{I}-\boldsymbol{K})^T\right] \\ &= E\left[(\boldsymbol{I}-\boldsymbol{K})(\boldsymbol{X\theta}+\boldsymbol{v})(\boldsymbol{X\theta}+\boldsymbol{v})^T(\boldsymbol{I}-\boldsymbol{K})^T\right] \\ &= E\left[(\boldsymbol{I}-\boldsymbol{K})\boldsymbol{v}\boldsymbol{v}^T(\boldsymbol{I}-\boldsymbol{K})^T\right] = (\boldsymbol{I}-\boldsymbol{K})E\left(\boldsymbol{v}\boldsymbol{v}^T\right)(\boldsymbol{I}-\boldsymbol{K})^T \\ &= \sigma^2\left(\boldsymbol{I}-2\boldsymbol{K}+\boldsymbol{K}^2\right) \\ &= \sigma^2(\boldsymbol{I}-\boldsymbol{K}) \end{aligned} \tag{5.64}$$

For the ith residual $v(i)$,

$$Var\left[v(i)\right] = \sigma^2(1-\kappa_{ii}) \tag{5.65}$$

$$Cov\left[v(i),v(j)\right] = -\sigma^2\left(\kappa_{ij}\right) \tag{5.66}$$

where $\boldsymbol{K} = \left[\kappa_{ij}\right]$ for $i,j = 1,2,\ldots,N$. Equations (5.65) and (5.66) indicate that the residuals have different variances and they are correlated. However, it is shown in Ref. [5.4] that for large N, the average values of the elements of the $\boldsymbol{K}$ matrix approach zero, so that $Cov(\boldsymbol{v})$ from Eq. (5.64) approaches $\sigma^2\boldsymbol{I}$.

In practice, the analysis of residuals involves the use of various types of plots. The residuals are plotted against time order, because in many cases, including aircraft system identification, the time sequence in which the data were collected is known. The ideal appearance of this type of residual plot is sketched in Fig. 5.1(a). The residuals in this figure form a random pattern around zero. On the other hand, the plot in Fig. 5.1(b) shows some deterministic components in the residuals, which might be the result of a deficiency in the model structure, also called deterministic modeling error.

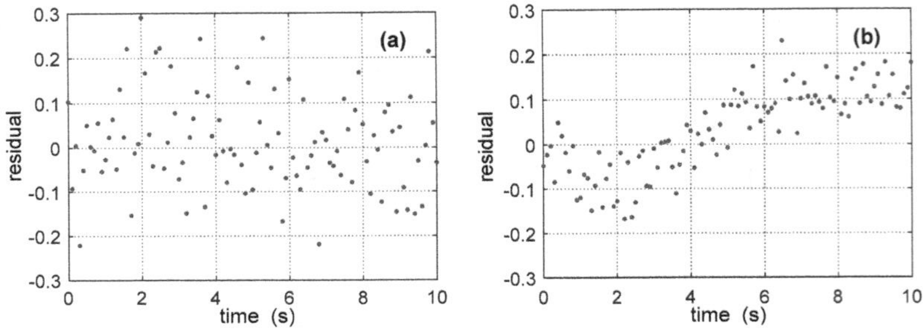

Figure 5.1 **Residual plots: (a) random (satisfactory); (b) deterministic component remaining (unsatisfactory)**

The whiteness of the residuals can be checked by estimating the autocorrelation function of the residuals,

$$\hat{r}_{vv}(k) = \frac{1}{(N-k)} \sum_{i=1}^{N-k} v(i)v(i+k) \qquad k = 0,1,2,\ldots,N-1 \tag{5.67}$$

This calculation of $\hat{r}_{vv}(k)$ gets very inaccurate for large lag values k, which make $(N-k)$ small. Instead, the following expression can be used

$$\hat{r}_{vv}(k) = \frac{1}{N} \sum_{i=1}^{N-k} v(i)v(i+k) \qquad k = 0,1,2,\ldots,N-1 \tag{5.68}$$

Theoretically, the $\hat{r}_{vv}(k)$ values from Eq. (5.68) are biased (see Ref. [5.5]). However, in the important practical cases where N is large relative to k, the last two expressions give nearly the same result. For large values of k relative to N, Eq. (5.68) gives more accurate estimates, in spite of the fact that these estimates are biased. Consequently, Eq. (5.68) is used to compute $\hat{r}_{vv}(k)$ for residuals.

If the residuals are completely uncorrelated, then it should be that $\hat{r}_{vv}(k) = 0$, $k \neq 0$. In practice, even for an adequate model structure, the values of $\hat{r}_{vv}(k)$, $k \neq 0$, are never exactly zero, but instead vary slightly around zero. However, zero should be within ± 2 standard errors of the residual autocorrelation estimate. The standard error for the residual autocorrelation estimate can be approximated by (see Ref. [5.6])

$$s\left[\hat{r}_{vv}(k \neq 0)\right] \approx \frac{\hat{r}_{vv}(0)}{\sqrt{N}} \tag{5.69}$$

A plot of residuals $v(i)$ against the corresponding estimated values $\hat{y}(i)$ can be useful for detecting several common types of model inadequacy. Two of them are demonstrated in Fig. 5.2. The pattern in Fig. 5.2(a) indicates that the variance of the residual is not constant. Figure 5.2(b) shows some nonlinearity not captured by the model.

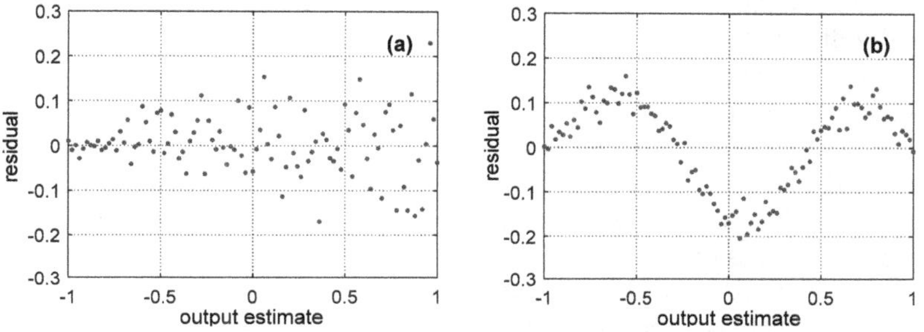

Figure 5.2 **Residual plots: (a) variance of residuals not constant; (b) unmodeled nonlinearity**

Another type of residual plot can be used for detecting the deviation of the probability distribution of the residuals from a Gaussian distribution. Small differences between the assumed normal distribution and the existing distribution do not substantially affect the model. However, large differences can be serious for confidence and prediction intervals, and for hypothesis testing.

A simple method for checking the normality assumption on the residuals is to plot the residuals on the abscissa in ascending order, accounting for the signs, against the following quantity on the ordinate:

$$\Phi^{-1}\left[P(i)\right] \qquad i = 1, 2, \ldots, N \tag{5.70}$$

where

$$\begin{aligned} P(i) &= \left(i - \frac{1}{2}\right)\Big/N \\ \Phi(z) &= \frac{1}{\sqrt{2\pi}} \int_{-\infty}^{z} e^{-u^2/2}\, du \end{aligned} \tag{5.71}$$

The quantity $\Phi(z)$ gives the cumulative probability distribution of a variable z that is normally distributed with zero mean and unit variance, i.e., z is $\mathbb{N}(0,1)$. The symbol Φ^{-1} represents the inverse of Φ, so that the expression (5.70) returns the z value corresponding to a given value of cumulative probability $P(i)$. The cumulative probability must lie in the interval $[0, 1]$.

The plot of sorted residuals versus $\Phi^{-1}\left[P(i)\right]$ should form approximately a straight line, if the residuals are in fact normally distributed. The function in Eq. (5.70) straightens the cumulative probability density curve for ordered normally distributed residuals from an "S" shape into a straight line, using the inverse of the ideal normal cumulative probability distribution (see Fig. B.2, Appendix B). Deviations from a normal distribution will make the plot deviate from a straight line. The degree to which the line is straight is usually determined visually. Points near the middle of the plot are more important than the points at either edge. The mean value of the residuals can be determined by picking the abscissa value corresponding to an ordinate of 0 (the mean value of the standardized z variable), and the standard error can be obtained from the difference between the abscissa values at ordinate values of 1 and 0, which is the slope of the line.

Examples of various residual plots resulting from flight data analysis will be given in the next section.

5.1.5 Application to Aircraft

In applying linear regression to aircraft parameter estimation, the regressors ξ_j, $j = 1,2,\ldots,n$, in Eq. (5.4) or (5.6) are computed from direct measurements of the aircraft states and control variables. As noted earlier, often at least some of the regressors are simply the aircraft states and controls themselves. These regressors are linear functions of the explanatory variables. Other regressors are postulated based on experience, or selected from a pool of candidate regressors using statistical model structure determination techniques, which will be described later.

The vector of ones for the bias term in the model is always included as a regressor. If the modeling is to be referenced to values of the explanatory variables that are different from zero, then the explanatory variables should have their constant part removed before calculating the regressors. There are two reasons for this. First, the bias term in the model is already being used to account for the constant part of the dependent variable, so any other regressor with a constant part will be correlated with the bias term. Regressor correlation causes problems with the parameter estimation because of indeterminacy in assigning model dependencies. This will be studied in detail later in the chapter. Second, removing the constant part from the explanatory variables will make any associated multivariate polynomial regressors consistent with the assumption of using terms from a multivariable Taylor series expansion [cf. Eq. (3.73)] to model the dependent variable. If the biases are not removed from the measured explanatory variables, that is equivalent to assuming the associated multivariate Taylor series expansion is being made about a reference value of zero for every explanatory variable. The reference values for any multivariate Taylor series expansion must be carefully noted, because the parameter values depend on what these reference values are, and the reference values are important when comparing results with information from other sources.

For aerodynamic parameter estimation, nondimensional aerodynamic force and moment coefficients are used as the dependent variable in linear regressions. A separate linear regression problem is solved for each force or moment coefficient, corresponding to minimizing the equation error in each individual equation of motion for the six degrees of freedom of the aircraft. This approach divides the overall modeling problem into several simpler problems, and avoids the need to choose a weighting matrix $\boldsymbol{R}^{-1}$ [cf. Eq. (4.34)], because only one equation is analyzed at a time.

As explained earlier, values for the aerodynamic force and moment coefficients cannot be measured directly in flight, but instead must be computed from other measurements and the equations of motion. The necessary equations follow from Eqs. (3.54), (3.40), and (3.37):

$$C_X \equiv -C_A = \frac{1}{\bar{q}S}\left(m\,a_x - T\right) \tag{5.72a}$$

$$C_Y = \frac{m\,a_y}{\bar{q}S} \tag{5.72b}$$

$$C_Z = -C_N = \frac{m\,a_z}{\bar{q}S} \tag{5.72c}$$

$$C_L = -C_Z \cos\alpha + C_X \sin\alpha \tag{5.72d}$$

$$C_D = -C_X \cos\alpha - C_Z \sin\alpha \tag{5.72e}$$

$$C_l = \frac{1}{\bar{q}Sb}\left[I_x\dot{p} - I_{xz}\left(pq + \dot{r}\right) + \left(I_z - I_y\right)qr\right] \tag{5.73a}$$

$$C_m = \frac{1}{\bar{q}S\bar{c}}\left[I_y\dot{q} + \left(I_x - I_z\right)pr + I_{xz}\left(p^2 - r^2\right) - I_p\Omega_p r\right] \tag{5.73b}$$

$$C_n = \frac{1}{\bar{q}Sb}\left[I_z\dot{r} - I_{xz}\left(\dot{p} - qr\right) + \left(I_y - I_x\right)pq + I_p\Omega_p q\right] \tag{5.73c}$$

Eqs. (5.72) and (5.73) show that the quantities required to calculate the nondimensional force and moment coefficients are: translational and angular accelerations, angular velocities, thrust, angular momentum of rotating mass in the propulsion system, mass/inertia properties, dynamic pressure, and reference geometry. Errors in any of these measured quantities show up as errors in the dependent variables of the regression problems. This makes it important to have good instrumentation with small systematic and random errors when using the linear regression approach. Data compatibility analysis described in Chapter 10 can be used to obtain a consistent data set with small systematic instrumentation errors. Note that measurements of the aircraft attitude angles (Euler angles) are not required to compute nondimensional force and moment coefficients.

Reference geometry and mass properties also appear in Eqs. (5.72) and (5.73), so these must be determined accurately as well. Relevant information is included in Appendix C. The effects of inaccuracy in these quantities on aircraft parameter estimation results were studied in Refs. [5.7] and [5.8].

Thrust is usually not measured directly in flight. Often, the thrust is obtained by interpolating ground test data or using a model identified from ground test data. Alternatively, it is possible to identify stability and control derivatives that quantify the combined effect of aerodynamics and thrust. In this case, the subtraction of the thrust term in Eq. (5.72a) is omitted, and the identified model characterizes the combined effect of aerodynamics and thrust. This approach might be used when the interest is more in accurate simulation as opposed to accurately characterizing the aerodynamics. Obviously, the same method can be used for other force and moment coefficients when the thrust is not directed along the x body axis and through the c.g. However, if the effects of aerodynamics and thrust are to be accurately separated, a good measurement or estimate of the thrust magnitude and direction is required.

This issue is not to be confused with thrust-induced aerodynamic effects, where local airflow over the aircraft is affected by the operation of the engine, resulting in changes in the apparent aerodynamic characteristics of the aircraft. The aerodynamic modeling then typically includes propulsion variables such as throttle position or power level, but the same principles discussed earlier regarding separating aerodynamic and thrust effects apply.

For the lift coefficient C_L and drag coefficient C_D, the measured angle of attack is also required [cf. Eqs. (5.72d) and (5.72e)]. These coefficients might be used instead of C_X and C_Z when there is a desire to compare parameter estimation results with wind tunnel values, for example. However, each calculation for C_L and C_D involves both translational accelerations a_x and a_z, as well as measured angle of attack and thrust. Using all these measurements to compute C_L and C_D can lead to multiple error sources and significantly increased noise levels. Because of this, C_X (or axial force coefficient $C_A = -C_X$) and C_Z (or normal force coefficient $C_N = -C_Z$) are usually preferred for aerodynamic parameter estimation, although this is not a universal practice.

In Chapter 3, it was seen that the applied forces and moments appearing in the equations of motion act at the aircraft c.g. However, it is sometimes necessary to compute nondimensional aerodynamic coefficients at a location different from the aircraft c.g. This happens when comparing flight results to wind tunnel data, where the forces and moments were measured in the wind tunnel at a reference location different from the c.g. of the aircraft in flight. The force coefficients are unaltered for a change in reference point; the moment coefficients at the reference point can be computed from

$$\begin{bmatrix} C_l \\ C_m \\ C_n \end{bmatrix}_{ref} = \begin{bmatrix} C_l \\ C_m \\ C_n \end{bmatrix}_{c.g.} + \begin{bmatrix} 1/b & 0 & 0 \\ 0 & 1/\bar{c} & 0 \\ 0 & 0 & 1/b \end{bmatrix} \left\{ \begin{bmatrix} (x_{cg} - x_{ref}) \\ (y_{cg} - y_{ref}) \\ (z_{cg} - z_{ref}) \end{bmatrix} \times \begin{bmatrix} C_X \\ C_Y \\ C_Z \end{bmatrix} \right\} \tag{5.74}$$

where $\begin{bmatrix} x_{cg} & y_{cg} & z_{cg} \end{bmatrix}$ and $\begin{bmatrix} x_{ref} & y_{ref} & z_{ref} \end{bmatrix}$ are the coordinates of the aircraft c.g. and the reference point, respectively. If this conversion is done before the modeling begins, then the estimated aerodynamic model parameters will be associated with the reference point, rather than the aircraft c.g., which can simplify comparisons with wind tunnel data.

Angular accelerations $\dot{p}, \dot{q},$ and $\dot{r}$ are usually not measured directly. Instead, they are obtained by a smoothed numerical differentiation of the angular rates. Effective algorithms for obtaining accurate smoothed derivatives of measured data are presented in Chapter 11.

When applying linear regression using flight test data, the regressors are assembled from measured data, which are noisy. This violates the assumption made in the linear regression analysis that the regressors are deterministic. The result is that the estimated parameters are biased and inefficient, as discussed in Refs. [5.2], [5.3], and [5.9]. The extent to which this occurs increases with increasing noise levels on the measurements used to assemble the regressors, as will be shown next.

Assume that the matrix of regressors $\boldsymbol{X}$ is comprised of noise-free regressors $\boldsymbol{X}_t$, plus a zero-mean random noise matrix $\boldsymbol{X}_\varepsilon$,

$$\boldsymbol{X} = \boldsymbol{X}_t + \boldsymbol{X}_\varepsilon \tag{5.75}$$

From Eqs. (5.6) and (5.10), the expected value of the estimated parameter vector $\hat{\boldsymbol{\theta}}$ is computed as

$$E(\hat{\boldsymbol{\theta}}) = E\left[\left(\boldsymbol{X}^T\boldsymbol{X}\right)^{-1}\boldsymbol{X}^T\left(\boldsymbol{X}_t\boldsymbol{\theta} + \boldsymbol{v}\right)\right] \tag{5.76}$$

Because $\boldsymbol{X}$ and $\boldsymbol{v}$ are uncorrelated [cf. Eq. (5.9c)],

$$E(\hat{\boldsymbol{\theta}}) = E\left[\left(\boldsymbol{X}^T\boldsymbol{X}\right)^{-1}\boldsymbol{X}^T\boldsymbol{X}_t\boldsymbol{\theta}\right]$$

$$E(\hat{\boldsymbol{\theta}}) = E\left[\left(\boldsymbol{X}^T\boldsymbol{X}\right)^{-1}\boldsymbol{X}^T(\boldsymbol{X}-\boldsymbol{X}_\varepsilon)\boldsymbol{\theta}\right]$$

$$E(\hat{\boldsymbol{\theta}}) = \boldsymbol{\theta} - E\left[\left(\boldsymbol{X}^T\boldsymbol{X}\right)^{-1}\boldsymbol{X}^T\boldsymbol{X}_\varepsilon\,\boldsymbol{\theta}\right] \tag{5.77}$$

The parameter estimate $\hat{\boldsymbol{\theta}}$ computed using noisy regressors is biased by the quantity $-E\left[\left(\boldsymbol{X}^T\boldsymbol{X}\right)^{-1}\boldsymbol{X}^T\boldsymbol{X}_\varepsilon\,\boldsymbol{\theta}\right]$. This bias error approaches zero as the regressor noise $\boldsymbol{X}_\varepsilon$ gets smaller. Reference [5.9] shows that the bias error from noisy regressors can be removed by applying data smoothing to the regressor data in the time domain (see Chapter 11) or by transforming the regressor data into the frequency domain using a limited bandwidth (see Chapter 7), which is the functional equivalent of smoothing the regressor data.

Linear regression can also be applied to the linearized state-space aircraft equations of motion, such as Eqs. (3.128a)-(3.128b) and (3.119a)-(3.119c). In this case, the state derivative terms on the left sides of the equations are considered the dependent variable, and the perturbation states and controls are the regressors. The estimated parameters are the dimensional stability and control derivatives. A similar approach can be used with the linearized output equations (3.128c) and (3.119f). Example 6.3 in the next chapter demonstrates this approach using flight data.

The same technique can also be used with transfer function models and measured data transformed into the frequency domain (see Chapter 7). In both the state-space and transfer function models, the dimensional model parameters combine the nondimensional aerodynamic stability and control derivatives with dynamic pressure, aircraft reference geometry, and mass/inertia properties [cf. Eqs. (3.127)-(3.129)]. Consequently, the dimensional parameters can vary throughout the maneuver as the dynamic pressure and mass/inertia properties change. This introduces some inaccuracy in the estimates of these parameters when the parameter estimation algorithms assume that the model parameters are unknown constants throughout the maneuver. The problem is avoided by using nondimensional aerodynamic coefficients as the dependent variable, as described earlier, or by applying real-time methods, where the unknown parameters are assumed to vary with time (see Chapter 10).

Example 5.1

In this example, linear regression is applied to aircraft flight test data to estimate nondimensional stability and control derivatives. The test aircraft was the NASA Twin Otter aircraft, which is a twin-engine turboprop commuter aircraft, shown in Fig. 5.3.

Flight test data were collected for two lateral maneuvers initiated from the same steady trim condition and executed using rudder and aileron deflections. The flight control system was unaugmented, so the pilot commands were implemented directly at the control surfaces through the control linkage. Measured flight data from run 1 were intended for aerodynamic parameter estimation; data from run 2 were for model validation.

Figure 5.3 **NASA Twin Otter aircraft**

The aircraft characteristics and flight condition are specified as follows:

$\bar{c} = 6.5$ ft	$I_x = 20{,}900$ slug-ft^2	$V_o = 238$ ft/s
$b = 65$ ft	$I_y = 24{,}261$ slug-ft^2	$\alpha_o = 0.2$ deg
$S = 422.5$ ft^2	$I_z = 38{,}469$ slug-ft^2	$\bar{q}_o = 56.6$ lbf/ft^2
$m = 340$ slugs	$I_{xz} = 1{,}128$ slug-ft^2	$g = 32.17$ ft/s^2

The input and output variables were sampled at intervals of 0.02 s, corresponding to a 50 Hz sampling rate. Figure 5.4 shows measured data for run 1, which is a lateral maneuver implemented by a series of rudder pulses, followed by an aileron doublet.

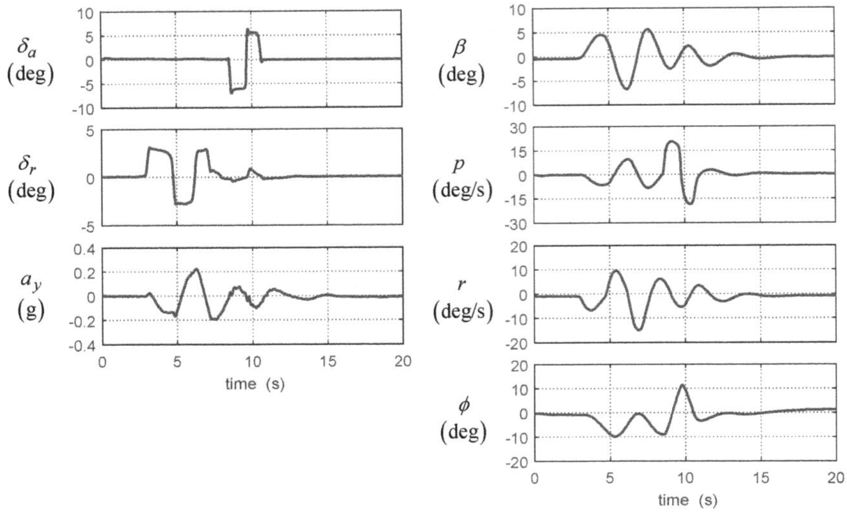

Figure 5.4 **Measured input and output data for lateral maneuver, run 1**

The regression equations for lateral aerodynamic force and moment coefficients were:

$$C_Y(i) = C_{Y_o} + C_{Y_\beta}\beta(i) + C_{Y_r}\frac{b}{2V_o}r(i) + C_{Y_{\delta_r}}\delta_r(i) + \nu_Y(i) \tag{5.78a}$$

$$\begin{aligned} C_l(i) &= C_{l_o} + C_{l_\beta}\beta(i) + C_{l_p}\frac{b}{2V_o}p(i) + C_{l_r}\frac{b}{2V_o}r(i) \\ &\quad + C_{l_{\delta_a}}\delta_a(i) + C_{l_{\delta_r}}\delta_r(i) + \nu_l(i) \end{aligned} \tag{5.78b}$$

$$\begin{aligned} C_n(i) &= C_{n_o} + C_{n_\beta}\beta(i) + C_{n_p}\frac{b}{2V_o}p(i) + C_{n_r}\frac{b}{2V_o}r(i) \\ &\quad + C_{n_{\delta_a}}\delta_a(i) + C_{n_{\delta_r}}\delta_r(i) + \nu_n(i) \end{aligned} \tag{5.78c}$$

for $i = 1, 2, \ldots, N$. The error terms are assumed to be zero mean with constant variance, i.e., $E\left[\nu_Y(i)\right] = 0$ and $Var\left[\nu_Y(i)\right] = E\left[\nu_Y^2(i)\right] = \sigma_Y^2$, etc. The dependent variable values on the left sides of the above equations were computed from Eqs. (5.72b), (5.73a), and (5.73c), respectively. The angular accelerations $\dot{p}$ and $\dot{r}$ in Eqs. (5.73a) and (5.73c) were obtained by smoothed local numerical differentiation of the measured angular velocities p and r, as described in Chapter 11.

The least-squares estimate of the aerodynamic parameters in the above equations is given by Eq. (5.10),

$$\hat{\boldsymbol{\theta}} = \left(\boldsymbol{X}^T \boldsymbol{X}\right)^{-1} \boldsymbol{X}^T \boldsymbol{z}$$

For the yawing moment coefficient C_n,

$$\boldsymbol{X} = \begin{bmatrix} 1 & \beta(1) & \dfrac{b}{2V_o} p(1) & \dfrac{b}{2V_o} r(1) & \delta_a(1) & \delta_r(1) \\ 1 & \beta(2) & \dfrac{b}{2V_o} p(2) & \dfrac{b}{2V_o} r(2) & \delta_a(2) & \delta_r(2) \\ \vdots & \vdots & \vdots & \vdots & \vdots & \vdots \\ 1 & \beta(N) & \dfrac{b}{2V_o} p(N) & \dfrac{b}{2V_o} r(N) & \delta_a(N) & \delta_r(N) \end{bmatrix}$$

$$\boldsymbol{\theta} = \begin{bmatrix} C_{n_o} & C_{n_\beta} & C_{n_p} & C_{n_r} & C_{n_{\delta_a}} & C_{n_{\delta_r}} \end{bmatrix}^T$$

$$\boldsymbol{z} = \left[C_n(1) \;\; C_n(2) \;\; \ldots \;\; C_n(N)\right]^T$$

and similarly for C_Y and C_l.

The results for the yawing moment coefficient are summarized in Table 5.1, including parameter estimates, standard errors, t statistics, fit error, and coefficient of determination. The parameter estimates were computed from Eq. (5.10). Standard errors for the parameter estimates came from the square root of the diagonal elements of the covariance matrix computed using Eq. (5.12), with the fit error estimated by Eq. (5.24). The t statistic for the addition of each single model term was computed from Eq. (5.60). The coefficient of determination came from Eq. (5.31), with Eq. (5.26). Pairwise correlations for the estimated parameters were obtained from Eq. (5.18), and shown in Table 5.2. Results for C_Y and C_l can be computed in the same way.

In Table 5.1, note that the $|t_0|$ values are very high for all parameters except $C_{n_{\delta_a}}$. The estimate of $C_{n_{\delta_a}}$ is also close to zero. This parameter quantifies the effect of aileron on the yawing moment. Airplanes are designed so that the ailerons affect primarily rolling moment, and produce as little yawing moment as possible. Because of this, $C_{n_{\delta_a}}$ is a normally a weak parameter, i.e., a parameter with relatively small magnitude. Based on this information, it might be that the $C_{n_{\delta_a}} \delta_a$ term is not necessary in the model. Section 5.4 gives more detail on methods that can be used to address this issue of model structure determination, using statistical metrics based on the measured data.

Table 5.1 **Least-squares parameter estimation results, aerodynamic yawing moment coefficient, run 1**

Parameter	$\hat{\theta}$	$s(\hat{\theta})$	$\lvert t_0 \rvert$	$100\left[s(\hat{\theta})/\lvert\hat{\theta}\rvert\right]$
C_{n_β}	8.54 x 10^{-2}	3.58 x 10^{-4}	238.9	0.4
C_{n_p}	–5.15 x 10^{-2}	1.43 x 10^{-3}	35.9	2.8
C_{n_r}	–1.98 x 10^{-1}	1.30 x 10^{-3}	151.8	0.7
$C_{n_{\delta_a}}$	2.34 x 10^{-3}	5.00 x 10^{-4}	4.7	21.4
$C_{n_{\delta_r}}$	–1.31 x 10^{-1}	5.97 x 10^{-4}	218.5	0.5
C_{n_o}	–4.60 x 10^{-4}	7.42 x 10^{-6}	62.0	1.6
$s = \hat{\sigma}$	2.25 x 10^{-4}			
R^2 (%)	99.6			

The last column of Table 5.1 gives the standard error as a percentage of the parameter estimate. The row associated with $C_{n_{\delta_a}}$ demonstrates that this metric must be interpreted carefully, because a small parameter estimate will lead to a large percentage error, even for a very small standard error. This is the case for $C_{n_{\delta_a}}$ in Table 5.1.

Table 5.2 shows pairwise correlations for terms in the linear model structure of Eq. (5.78c). Although there are some moderately high correlations, none has absolute value that exceeds 0.90, which is an empirically-determined cutoff value for acceptable regressor pairwise correlation. The relatively high correlation between $pb/2V_o$ and δ_a may be partially responsible for the low significance of $C_{n_{\delta_a}}$. Generally, C_{n_p} and $C_{n_{\delta_a}}$ are both weak parameters, meaning that only a relatively small part of the variation in the response variable (C_n in this case) can be explained by the associated model terms. When $pb/2V_o$ and δ_a have high correlation, one of the associated parameters can exhibit low significance, because the other has already accounted for the relatively small part of the variation that can be explained mostly by either.

Table 5.2 **Parameter correlation matrix, aerodynamic yawing moment coefficient, run 1**

	C_{n_β}	C_{n_p}	C_{n_r}	$C_{n_{\delta_a}}$	$C_{n_{\delta_r}}$	C_{n_o}
C_{n_β}	1	0.82	0.15	0.67	–0.02	0.03
C_{n_p}		1	0.30	0.89	–0.18	–0.02
C_{n_r}			1	0.29	0.79	–0.21
$C_{n_{\delta_a}}$				1	0.16	–0.01
$C_{n_{\delta_r}}$					1	–0.28
C_{n_o}						1

A comparison between the measured yawing moment coefficient and the identified model is shown in Fig. 5.5. The average 95% confidence interval on the mean response $\hat{C}_n$, computed from Eqs. (5.40) and (5.41), is approximately $\pm 3.0\times 10^{-5}$. These interval limits about the estimated output are not plotted, because the interval is so small that the associated lines are not distinguishable.

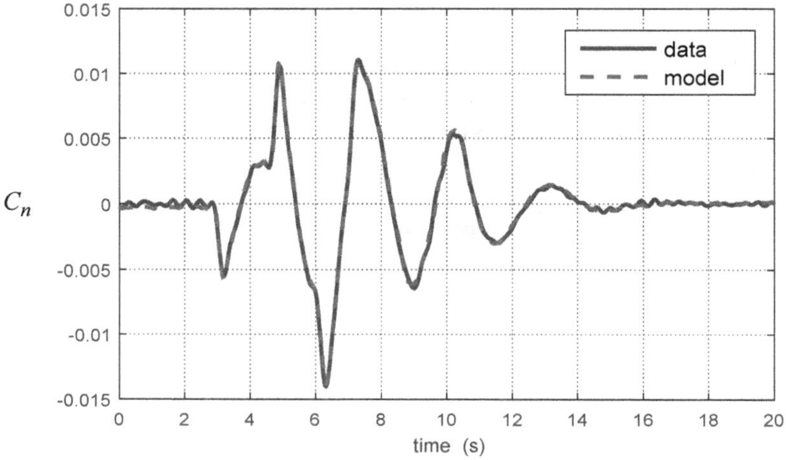

Figure 5.5 **Equation-error model fit to yawing moment coefficient for lateral maneuver, run 1**

Residuals are plotted against time and $\hat{C}_n$ in Fig. 5.6. The dashed lines in Fig. 5.6 represent the average 95% confidence interval for prediction, which was $\pm 4.5\times 10^{-4}$, computed from Eqs. (5.42) and (5.43). The residuals have a random character and lie mostly within the 95 percent confidence bounds, which are characteristics indicative of a good model.

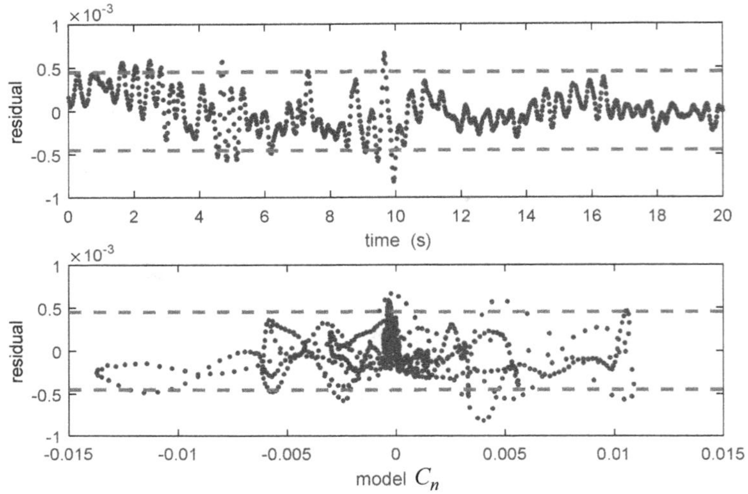

Figure 5.6 **Yawing moment coefficient residuals for lateral maneuver, run 1**

Figure 5.7 shows the cumulative probability plot computed using the ordered residuals and Eqs. (5.70)-(5.71). A straight line indicates that the residuals are normally distributed. Only small deviations from a straight line are evident.

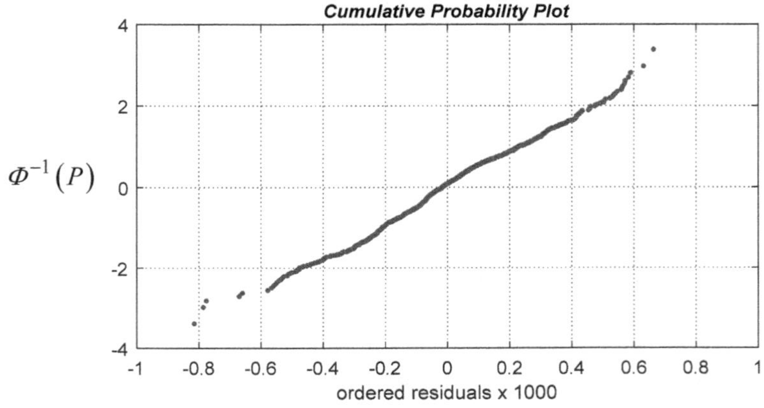

Figure 5.7 **Diagnostic plot for yawing moment coefficient residuals**

In Fig. 5.8, autocorrelation estimates computed from Eq. (5.68) are plotted along with the estimated 95 percent confidence interval, which is $\pm 2\hat{r}_{vv}(0)/\sqrt{N}$, from Eq. (5.69).

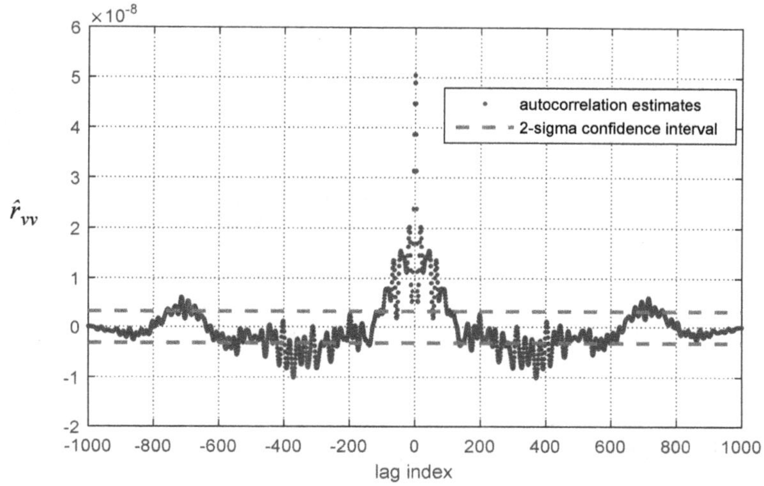

Figure 5.8 **Autocorrelation plot for yawing moment coefficient residuals for lateral maneuver, run 1**

If the residual sequence were uncorrelated, then the autocorrelation estimates for nonzero lags would lie almost completely inside the 95 percent confidence interval shown, with a single large value at zero lag. Since this is not true, the residual sequence is colored. This has a significant effect on the parameter standard errors, as will be shown in Example 5.2.

The statistical metrics and parameter estimation results shown indicate that the linear model was adequate to characterize the measured data and allowed accurate calculation of the model parameters. The coefficient of determination was very high $\left(R^2 = 99.6\,\%,\ \text{from Table 5.1}\right)$, and the model fit to the data was excellent. Residuals showed little deterministic content, with magnitudes that fell mostly within the 95% confidence interval for response prediction. All of these characteristics are indicative of a good model.

The model identified for C_n from run 1 was validated using data from run 2. Measured data for run 2 appear in Fig. 5.9. Note that the polarities of the initial rudder input and the aileron doublet are reversed compared with run 1. This maneuver is a simple doublet sequence at the same flight condition as specified earlier for run 1. Parameter estimates from run 1, listed in Table 5.1, were used with regressor data from run 2 to predict yawing moment coefficient for run 2.

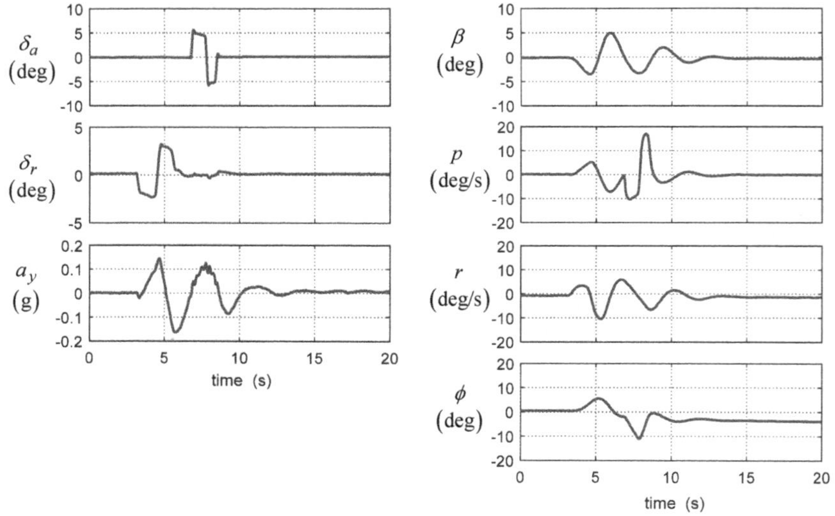

Figure 5.9 **Measured input and output data for lateral maneuver, run 2**

Figure 5.10 shows the comparison of measured C_n from run 2 with the predicted values. The residuals for this prediction case are shown in Fig. 5.11, where the dashed lines mark the average 95% confidence interval for the prediction, computed using the modeling data of run 1. These are the same confidence intervals plotted in Fig. 5.6. The residuals for the prediction case are mostly within the 95% confidence bounds for the prediction, as expected.

The results shown here indicate that the linear model structure given above is adequate for characterizing the aerodynamic yawing moment coefficient, and that the identified model predicts well for a different maneuver at the same flight condition. Good prediction capability makes the model useful, and gives confidence that the identified model is a good characterization of the aircraft aerodynamics.

Chapter 11 includes a description of a global smoothing method that uses Fourier analysis to separate deterministic signal from random noise. This method can be used to isolate the random component of the dependent variable, which can then be compared to residuals from the modeling. If the model structure used in the analysis was adequate, then the deterministic part of the dependent variable would be described by the identified model, and the residuals should be the same as the random component found with the global Fourier smoother. The global Fourier smoother is a nonparametric method that does not depend on any assumed model, only on the measured data. Therefore, the global Fourier smoother produces a model-independent estimate of the random component of the dependent variable measurement. This random component cannot be modeled with deterministic regressors.

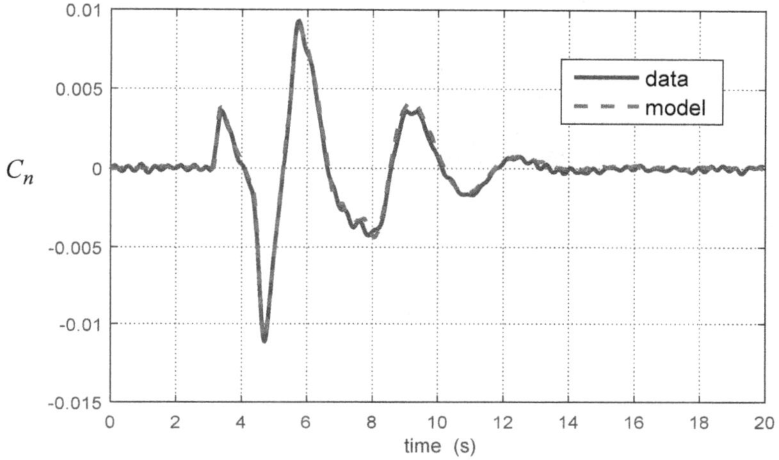

Figure 5.10 **Model prediction of yawing moment coefficient for lateral maneuver, run 2**

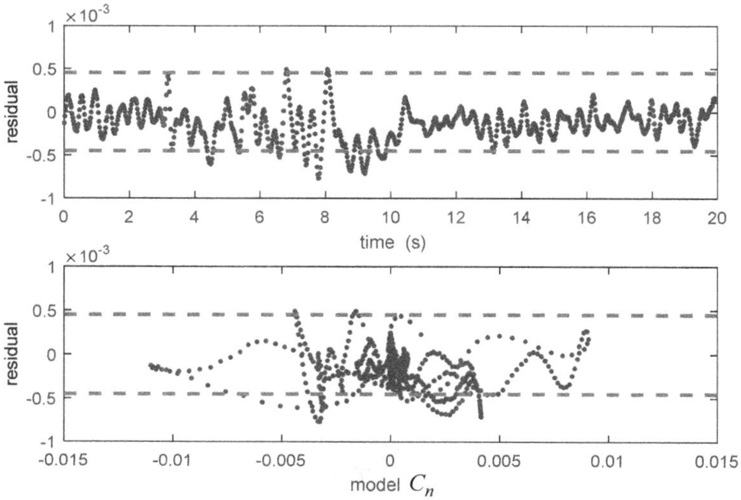

Figure 5.11 **Yawing moment coefficient prediction residuals for lateral maneuver, run 2**

Figure 5.12 shows a comparison of the residuals from Fig. 5.6, shown in the upper plot, with the random component of measured C_n obtained from the global Fourier smoother, shown in the lower plot. The magnitudes are similar.

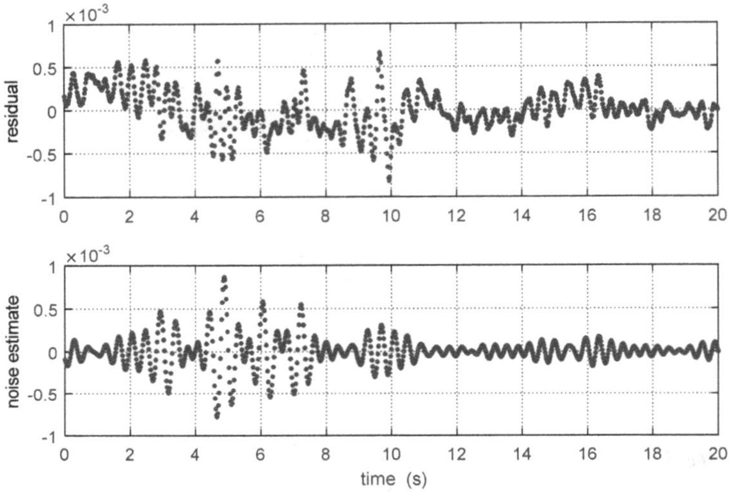

Figure 5.12 **Yawing moment coefficient residuals for lateral maneuver, run 1**

It follows that any unmodeled components in the residuals of the upper plot that are not also in the lower plot must have amplitudes within the noise level, and cannot be extracted because their signal-to-noise ratio is approximately 1 or less. This general approach can be used to assess adequacy of any assumed model structure. ■

5.2 Computational Aspects

The equation-error approach has very wide applicability. In Section 5.1, it was shown that the model terms used in equation-error modeling can be arbitrarily nonlinear, as long as the model output is linearly related to the unknown model parameters. In addition, the least-squares solution is a closed-form algebraic expression that requires no iteration [cf. Eq. (5.10)], which means that the solution can be computed very rapidly. Integration of the equations of motion is not required, because the equation-error approach identifies a model that matches derivative information in the equations of motion. Consequently, the method can be applied regardless of whether the underlying dynamic system is stable or unstable, including inherently unstable aircraft that must be stabilized by continuously-operating feedback control.

Other important practical advantages emerge because the equation-error method computes parameter estimates that solve a set of simultaneous algebraic equations in a least-squares sense. Individual equations (corresponding to individual data points) can be included or excluded, allowing the removal of data dropouts without requiring any other modifications to the procedure. This capability is important because any modeling technique based on least squares will give more attention to adjusting the model parameters to reduce the largest individual residuals, so that a data dropout that is not removed can seriously compromise the accuracy of the model. Data sampled at irregular time intervals can be used. It is also possible to combine data points from different maneuvers, or even from different flights, into a single nondimensional analysis to identify a single aerodynamic model. This idea is explored later in this chapter in association with the data partitioning approach, but the idea is also very useful when the objective is to identify a more general model, valid over a larger range of flight conditions or using a wider variety of inputs.

Often there is a desire to estimate aerodynamic model parameters using data from more than one flight test maneuver. This is useful when individual maneuvers have good information content for estimating some parameters but not others. Sometimes multiple maneuvers taken together can provide more complete information for parameter estimation. Equation-error allows concatenation of the data from various maneuvers by simply stacking the data in the measured output vector z and the regressor matrix X in Eq. (5.6). The implicit assumptions made in doing this, in addition to the usual equation-error assumptions, are: 1) the noise ν is of the same character for all concatenated maneuvers, and 2) the experimental conditions of the concatenated maneuvers are sufficiently similar that the model parameters can be expected to be the same. Violating the first assumption leads to inaccuracies in the estimated parameter uncertainties. If the second assumption is violated, the parameter estimation results are some weighted average of the results that would have been obtained from each of the concatenated maneuvers analyzed individually. Because the parameter estimates from the equation-error method are independent of the ordering of the data points, it does not matter how the data points (i.e., the rows) are ordered in z and X as far as the parameter estimates are concerned; however, the ordering is important for accurately computing parameter standard errors, as will be discussed later in this section.

In the equation-error approach, a typical linear model for the nondimensional aerodynamic pitching moment would be

$$C_m = C_{m_\alpha}\alpha + C_{m_q}\frac{\bar{c}}{2V_o}q + C_{m_\delta}\delta + C_{m_o} \tag{5.79}$$

If this expansion is interpreted as a truncated Taylor series model, then the modeling functions or regressors should be made from perturbation quantities.

If instead the measured physical quantities are used for the regressors, Eq. (5.79) becomes

$$C_m = C_{m_\alpha}\left(\alpha_o + \Delta\alpha\right) + C_{m_q}\frac{\bar{c}}{2V_o}\left(q_o + \Delta q\right) + C_{m_\delta}\left(\delta_o + \Delta\delta\right) + C_{m_o} \tag{5.80}$$

where the Δ quantities are perturbations from the reference values with subscript o. The reference values are typically from initial trim conditions. Rearranging Eq. (5.80),

$$\begin{aligned} C_m &= C_{m_\alpha}\Delta\alpha + C_{m_q}\frac{\bar{c}}{2V_o}\Delta q + C_{m_\delta}\Delta\delta \\ &\quad + \left(C_{m_o} + C_{m_\alpha}\alpha_o + C_{m_q}\frac{\bar{c}}{2V_o}q_o + C_{m_\delta}\delta_o\right) \end{aligned} \tag{5.81}$$

The bias term in Eq. (5.81) includes both C_{m_o} and terms related to the steady parts of the regressors, $C_{m_\alpha}\alpha_o + C_{m_q}\frac{\bar{c}}{2V_o}q_o + C_{m_\delta}\delta_o$. Consequently, the bias term depends on whether the regressors are assembled directly from the measured data or from perturbation data where reference values are removed first. Equations (5.79) and (5.81) show that the estimates of the other model parameters (stability and control derivatives) are unaffected, regardless of whether measured physical quantities or perturbation quantities are used. Only the bias term is affected. This issue does not occur in cases when the reference values are zero for the states and controls used to assemble the regressors, which is common for lateral cases (cf. Fig. 5.4). There are practical situations, such as a flight test maneuver without a good steady trim at the start, where determining the steady values of the explanatory variables is difficult, which makes it difficult to compute appropriate perturbation quantities. In such situations, it is permissible to use measured physical quantities for the explanatory variables, but the estimated bias term will not be the correct aerodynamic bias, but rather the aerodynamic bias plus terms related to the steady parts of the regressors. The same arguments apply if the regressors are nonlinear functions of the explanatory variables.

5.2.1 Generalized Least Squares

In the previous development for ordinary least-squares linear regression, it was assumed that the measurement errors had zero mean, and were uncorrelated with equal variance. In many practical cases, the assumptions of uncorrelated measurement errors and homogenous variances are not valid. Thus, it is necessary to make the least-squares model more general by assuming that

$$z = X\theta + v \tag{5.82a}$$

$$E(v) = 0 \qquad Cov(v) = E\left(vv^T\right) \equiv V \tag{5.82b}$$

The $N \times N$ noise covariance matrix V is nonsingular and positive definite. The least-squares estimator for this modified model is obtained by minimizing

$$J_{GLS}(\theta) = \frac{1}{2}(z - X\theta)^T V^{-1}(z - X\theta) \tag{5.83}$$

The minimizing parameter vector is computed from

$$\hat{\theta}_{GLS} = \left(X^T V^{-1} X\right)^{-1} X^T V^{-1} z \tag{5.84}$$

which is called the generalized least-squares estimator. Using calculations similar to those shown earlier for ordinary least squares, it can be shown that $\hat{\theta}_{GLS}$ is asymptotically unbiased, $E\left(\hat{\theta}_{GLS}\right) = \theta$, with covariance matrix given by

$$Cov\left(\hat{\theta}_{GLS}\right) = E\left[\left(\hat{\theta}_{GLS} - \theta\right)\left(\hat{\theta}_{GLS} - \theta\right)^T\right] = \left(X^T V^{-1} X\right)^{-1} \tag{5.85}$$

Eq. (5.85) shows that the noise covariance matrix V has a significant effect on the covariance matrix. Under the assumptions (5.82), it can be shown that $\hat{\theta}_{GLS}$ is the best linear estimator of θ (see Ref. [5.3]).

If the measurement errors for the dependent variable are uncorrelated, but with different variances, then V is a diagonal matrix with unequal elements on the diagonal. Introducing $W = V^{-1}$, the elements of W are weights for each equation in the regression problem, and the parameter estimation procedure is called weighted least squares. The expressions for the parameter estimates and covariance matrix in this case are the same as for the generalized least squares, with V^{-1} now being a diagonal matrix [cf. Eqs. (5.84) and (5.85), respectively]. This approach is equivalent to applying a weighting of $\sqrt{w_{ii}}$ to both sides of the ith regression equation, where w_{ii} is the ith diagonal element of W. The idea is that the weightings should be chosen to scale the measurement noise of each equation to approximately the same magnitude. Once that is achieved, the ordinary least-squares analysis and solution apply.

5.2.2 Accuracy of Parameter Estimates

For flight test data analysis, measured data for the linear regression problem come from several recorded time series, as in Example 5.1. Physically, the same instrumentation system is being used to collect each data sample, under similar conditions and at nearly the same time, so there is generally no justification for introducing unequal weightings to model heterogeneous variances, based on the instrumentation alone. It is true that the residuals from a particular analysis usually exhibit varying magnitudes (cf. Fig. 5.6), but this is the result of model structure deficiencies, rather than a change in the measurement error variance from point to point. Using Eq. (5.10) for ordinary least-squares parameter estimation is equivalent to assuming a constant weighting for each data point in the least-squares problem.

In practice, the residuals from flight test data analysis are often significantly correlated with their adjacent neighbors, because the data are collected sequentially in time from a maneuvering aircraft and because identified models are simplifications of reality. For static experiments, such as a wind tunnel test, randomization of the test conditions is used to ensure that the residuals are uncorrelated. This cannot be done in a flight test, so there must be some correction for correlated residuals with varying magnitude. Correlated residuals have a significant effect on the estimated parameter covariance matrix.

The parameter covariance matrix for correlated residuals is derived starting with Eq. (5.11),

$$Cov(\hat{\boldsymbol{\theta}}) = E\left[(\hat{\boldsymbol{\theta}}-\boldsymbol{\theta})(\hat{\boldsymbol{\theta}}-\boldsymbol{\theta})^T\right] = (\boldsymbol{X}^T\boldsymbol{X})^{-1}\boldsymbol{X}^T E(\boldsymbol{v}\boldsymbol{v}^T)\boldsymbol{X}(\boldsymbol{X}^T\boldsymbol{X})^{-1} \quad (5.86)$$

When $\boldsymbol{v}$ is a zero mean, weakly stationary random process,

$$E(\boldsymbol{v}\boldsymbol{v}^T) = E[v(i)\ v(j)] = \mathcal{R}_{vv}(i-j) = \mathcal{R}_{vv}(j-i) \qquad i,j=1,2,\ldots,N \quad (5.87a)$$

where $\mathcal{R}_{vv}(i-j)$ is a matrix of autocorrelation functions for the residuals at different time shifts $(i-j)$,

$$\mathcal{R}_{vv}(i-j) = \begin{bmatrix} r_{vv}(0) & r_{vv}(1) & \cdots & r_{vv}(N-1) \\ r_{vv}(1) & r_{vv}(0) & \cdots & r_{vv}(N-2) \\ \vdots & \vdots & \vdots & \vdots \\ r_{vv}(N-1) & r_{vv}(N-2) & \cdots & r_{vv}(0) \end{bmatrix} \qquad i,j=1,2,\ldots,N \quad (5.87b)$$

The estimated parameter covariance matrix for correlated residuals can be computed by substituting for $E\left(\boldsymbol{v}\boldsymbol{v}^T\right)$ from Eqs. (5.87) into Eq. (5.86), using estimates $\hat{r}_{vv}(i-j)$ from Eq. (5.68),

$$\hat{r}_{vv}(k) = \frac{1}{N}\sum_{i=1}^{N-k} v(i)v(i+k) = \hat{r}_{vv}(-k) \qquad k = 0,1,2,\ldots,m \tag{5.88}$$

The index k represents the time separation of the residuals in the summation, and m is the maximum time index difference. Since only proximate residuals are significantly correlated, the value of m can be set at $N/5$ for typical flight test maneuvers and sample rates. This reduces the required computations without significantly impacting the results.

Combining the last four equations,

$$Cov\left(\hat{\boldsymbol{\theta}}\right) = \left(\boldsymbol{X}^T\boldsymbol{X}\right)^{-1}\left[\sum_{i=1}^{N}\boldsymbol{x}(i)\sum_{j=1}^{N}\hat{r}_{vv}(i-j)\boldsymbol{x}^T(j)\right]\left(\boldsymbol{X}^T\boldsymbol{X}\right)^{-1} \tag{5.89}$$

where $\boldsymbol{x}^T(i)$ is the ith row of the $\boldsymbol{X}$ matrix, containing the measured regressors at the ith data point, and $r_{vv}(i-j)$ has been replaced by its estimate from Eq. (5.88). Note that if the residuals are uncorrelated, then

$$E\left(\boldsymbol{v}\boldsymbol{v}^T\right) = \hat{r}_{vv}(0)\boldsymbol{I} = \sigma^2\boldsymbol{I} \tag{5.90}$$

and Eq. (5.86) reduces to the ordinary least-squares expression

$$Cov\left(\hat{\boldsymbol{\theta}}\right) = \sigma^2\left(\boldsymbol{X}^T\boldsymbol{X}\right)^{-1} \tag{5.91}$$

Equations (5.88) and (5.89) represent a post-processing of the residuals from an ordinary least-squares solution to account for the effect of correlated residuals on the parameter covariance matrix. The standard errors for the estimated parameters, corrected for correlated residuals, are found as the square root of the diagonal elements of the covariance matrix calculated from Eq. (5.89).

In cases where generalized least squares is appropriate, but ordinary least squares is used instead, the parameter estimates are still unbiased, i.e., $E\left(\hat{\boldsymbol{\theta}}\right) = E\left(\hat{\boldsymbol{\theta}}_{GLS}\right) = \boldsymbol{\theta}$. However, the misspecification of $E\left(\boldsymbol{v}\boldsymbol{v}^T\right)$ results in a loss of efficiency in estimating $\boldsymbol{\theta}$, and a biased $Cov\left(\hat{\boldsymbol{\theta}}\right)$. These problems are discussed in Refs. [5.1] and [5.10]. In Ref. [5.11], investigation of bias in the parameter covariance matrix found that there is a tendency to underestimate the estimated parameter variances when the ordinary least-squares expression is used in cases where the residuals do not satisfy the assumptions made in the ordinary least-squares solution. Reference [5.12] is the source for the corrected expression (5.89), which has been found effective in characterizing estimated parameter accuracy for simulation and flight test cases with many different colored noise sequences.

The following example demonstrates the covariance matrix correction for colored residuals applied to the analysis of flight test data.

Example 5.2

Returning to the analysis in Example 5.1, the same model parameters for the yawing moment coefficient are estimated using the same measured data from run 1 [cf. Eq. (5.78c) and Fig. 5.4]. Ordinary least-squares parameter estimates and the associated standard error estimates based on the white noise residual assumption appear in Table 5.3, columns 2 and 3, respectively. These columns are repeated from Table 5.1. Column 4 of Table 5.3 contains the estimated parameter standard errors using Eq. (5.89), which accounts for colored residuals. Column 5 of Table 5.3 shows the ratio of the standard errors corrected for colored residuals to those from the ordinary least-squares calculation.

In this case, the standard errors computed based on the white residual assumption are smaller by roughly a factor of three, compared to the standard errors calculated accounting for colored residuals.

The standard errors computed using the colored residual expression (5.89) are consistent with the scatter in parameter estimates from repeated maneuvers at the same flight condition. The conventional standard error calculation results in optimistic values for the estimated parameter accuracy when the residuals are colored.

Each repeated maneuver has its own specific noise sequences, corresponding to its particular realization of the random processes. The scatter in the parameter estimates from repeated maneuvers comes from the influence of the different noise sequences, which of course carries over into the parameter estimates that are based on the measured data. As discussed earlier, the parameter estimates are therefore random variables, characterized by expected values and standard deviations when the measurement noise is assumed to be Gaussian. The goal of the analysis is to estimate mean values and confidence intervals for the parameters so that the true values of the parameters lie inside the confidence intervals centered on the estimated mean values. True values of the parameters are likely to be close to the mean value of parameter estimates from repeated maneuvers. All of this assumes that an adequate model structure has been chosen for the repeated maneuvers, and that this model structure is held constant for the analysis of each maneuver.

Table 5.3 **Least-squares estimated parameters and error bounds, aerodynamic yawing moment coefficient, run 1**

Parameter	$\hat{\theta}$	$s(\hat{\theta})$ white residual Eq. (5.91)	$s(\hat{\theta})_{corr}$ colored residual Eq. (5.89)	$\dfrac{s(\hat{\theta})_{corr}}{s(\hat{\theta})}$
C_{n_β}	8.54 x 10^{-2}	3.58 x 10^{-4}	1.16 x 10^{-3}	3.2
C_{n_p}	–5.15 x 10^{-2}	1.43 x 10^{-3}	4.01 x 10^{-3}	2.8
C_{n_r}	–1.98 x 10^{-1}	1.30 x 10^{-3}	3.66 x 10^{-3}	2.8
$C_{n_{\delta_a}}$	2.34 x 10^{-3}	5.00 x 10^{-4}	1.27 x 10^{-3}	2.5
$C_{n_{\delta_r}}$	–1.31 x 10^{-1}	5.97 x 10^{-4}	1.88 x 10^{-3}	3.1
C_{n_o}	–4.60 x 10^{-4}	7.42 x 10^{-6}	2.71 x 10^{-5}	3.7
$s = \hat{\sigma}$	2.25 x 10^{-4}			
R^2 (%)	99.6			

The best approach would be to always run repeated maneuvers at every flight condition, but this is often not possible, due to economic and other practical constraints. Instead, one or two maneuvers at the same flight condition are usually available for analysis, and the colored noise correction is applied to compute a confidence interval that accurately represents the scatter in parameter estimates that would have occurred with numerous repeated maneuvers.

Column 5 of Table 5.3 shows that the factor by which the ordinary least-squares calculation is optimistic varies slightly with the parameter, and will also vary with the particular maneuver flown. This reflects differences in the information content of the regressors. The values for these factors change with the coloring of the residual sequence. Figure 5.13 shows the power spectrum of the residual sequence shown in Fig. 5.6 for the yawing moment coefficient model. Most of the noise power is concentrated in the frequency range [0,4] Hz. This is a typical moderate coloring of the residuals. ■

It is common practice to apply a constant correction factor (typically 5 or 10) to the standard errors computed from the ordinary least-squares calculation to account for colored residuals. This practice is a rough approximation, however, and can be improved by using the colored residual formula (5.89) to compute the estimated parameter covariance matrix and standard errors. The correction for colored residuals can be run as a post-processing of the residuals, after the model parameter estimates are computed using ordinary least squares.

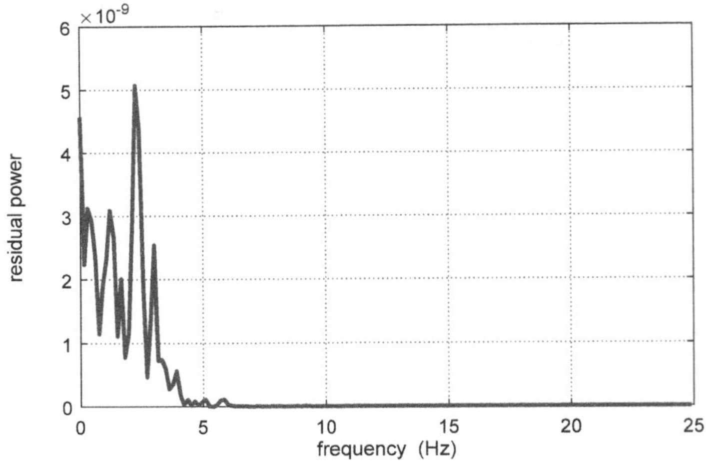

Figure 5.13 **Power spectrum of yawing moment coefficient residuals for lateral maneuver run 1**

Ideally, the generalized least-squares solution in Eqs. (5.84) and (5.85) should be used to refine the parameter estimates, using the residual sequence from ordinary least squares to estimate the weighting matrix V^{-1} from Eqs. (5.82b), (5.87), and (5.88). Then, a new residual sequence could be obtained, and the process repeated until the parameter estimates converge. Unfortunately, this procedure involves inverting an $N \times N$ weighting matrix V, which can be very large and ill-conditioned for data from typical flight test maneuvers. Consequently, computing V^{-1} takes a relatively long time, and the result can be inaccurate, which in turn causes convergence problems with the generalized least-squares parameter estimates. Because of this, practical modeling results are found using the ordinary least-squares solution with error bounds corrected for colored residuals.

5.3 Nonlinear Least Squares

In some modeling problems, the relationship between the regressors and the response variable is nonlinear in the parameters. For this case, the least-squares model was formulated in Chapter 4 as

$$z = \boldsymbol{\varphi}(\boldsymbol{\theta}) + \boldsymbol{v} \tag{5.92}$$

which is equivalent to a nonlinear regression model,

$$z(i) = f\left[\boldsymbol{x}(i), \boldsymbol{\theta}\right] + v(i) \qquad i = 1, 2, \ldots, N \tag{5.93}$$

where $\boldsymbol{x}^T(i)$ is a row vector of regressors computed from measured data at the ith data point, and f is a nonlinear function of $\boldsymbol{x}(i)$ and the parameters in the vector $\boldsymbol{\theta}$.

As before, the least-squares parameter estimate can be obtained by minimizing the sum of squared errors,

$$J(\boldsymbol{\theta})=\frac{1}{2}\sum_{i=1}^{N}\left\{z(i)-f\left[\boldsymbol{x}(i),\boldsymbol{\theta}\right]\right\}^2 \tag{5.94}$$

The minimum of this cost function is found by satisfying the normal equations,

$$\left.\frac{\partial J}{\partial \boldsymbol{\theta}}\right|_{\boldsymbol{\theta}=\hat{\boldsymbol{\theta}}}=-\sum_{i=1}^{N}\left\{z(i)-f\left[\boldsymbol{x}(i),\hat{\boldsymbol{\theta}}\right]\right\}\left.\frac{\partial f\left[\boldsymbol{x}(i),\boldsymbol{\theta}\right]}{\partial \boldsymbol{\theta}}\right|_{\boldsymbol{\theta}=\hat{\boldsymbol{\theta}}}=\mathbf{0} \tag{5.95}$$

where $\partial J/\partial \boldsymbol{\theta}$ is a row vector containing the partial derivatives of the nonlinear scalar function $J(\boldsymbol{\theta})$ with respect to the elements of $\boldsymbol{\theta}$, and $\partial f\left[\boldsymbol{x}(i),\boldsymbol{\theta}\right]/\partial \boldsymbol{\theta}$ is a row vector of model output sensitivities to changes in the model parameters.

Eq. (5.95) is a set of nonlinear algebraic equations. This means that $\hat{\boldsymbol{\theta}}$ cannot be obtained by simple matrix algebra, as in the case of a model that is linear in the parameters. Instead, an iterative nonlinear optimization technique must be used. There are many different numerical methods to solve this nonlinear minimization problem, see Ref. [5.13]. Some of these methods will be explained in Chapter 6.

5.4 Model Structure Determination

Up to this point, the data analysis techniques for linear regression problems were concerned with parameter estimation for an assumed model structure in the regression equation (5.4). The model structure refers to the number and form of the model terms in the regression equation. When the model structure is assumed known and fixed, only the constant model parameter values and their standard errors need to be estimated from the measured data. This can be done using the parameter estimation techniques discussed earlier.

Assuming a model structure also implicitly assumes that each of the terms in the model makes a significant contribution to modeling the variation in the measured response z. Choice of an adequate model structure depends heavily on how the experiment was conducted. However, in many practical cases it is not clear exactly what the model structure should be. From theory, previous experiments, or knowledge of the physical system to be modeled, candidate regressors that might be considered for the model can be postulated. Then the task is to select a subset of the candidate regressors which best model the response variable, based on the measured data. This process is called model structure determination.

In aircraft applications of system identification, the estimation of stability and control derivatives has become a standard procedure for small perturbation flight test maneuvers where the aerodynamics can be described using regressors that are linear in the explanatory variables. Interest in near- and post-stall flight regimes, and in dynamics of rapid, large amplitude maneuvers has created a need to extend aerodynamic modeling into flight regimes where nonlinear aerodynamic effects can be pronounced. Nonlinear terms can also be required to extend the validity of models to large ranges of the explanatory variables. These applications introduce the problem of determining how complex the model should be.

In Chapter 3, several nonlinear forms for the aerodynamic model were discussed, including indicial functions, splines, and multivariate polynomials in a Taylor series expansion. Splines and multivariate polynomials have been the most frequently-used modeling functions for characterizing nonlinearity, because of their relatively simple form and physical interpretation. Adding these types of terms to the linear Taylor series terms that include stability and control derivatives is a natural extension to model more complicated dependencies.

In determining the model structure, there are two conflicting objectives. On one hand, a model with many regressors might be desired so that the model can describe nearly all of the variation in the dependent variable. On the other hand, a model with as few regressors as possible is desired, because the variance of the prediction $\hat{y}$ increases as the number of regressors increases. If too many parameter estimates are sought from a given set of measured data, reduced accuracy of the estimated parameters can be expected, or, in some cases, attempts to obtain reasonable estimates of all the model parameters can fail. More informative data, which often means a greater volume of data, is needed to support accurate estimation of more model parameters.

Several algorithms for model structure determination based on measured data have been developed. Often, different methods will select different subsets of candidate regressors as the best model. There is no guarantee that any selected procedure will give the one best model, because there might be several equally good models for characterizing a particular data set, especially when nonlinearity is involved. Extensive treatments of model structure determination are given in Refs. [5.1]-[5.3], and [5.14]-[5.16].

In the following sections, the properties of a reduced linear regression model will be discussed first, followed by a description of several criteria or metrics that can be used for selecting the best model. The stepwise regression method for model structure determination will be developed next, followed by explanation of a method using multivariate orthogonal functions to determine model structure. These two model structure determination approaches have been successfully applied to flight test data for many aircraft problems. Nonlinear model structure determination will be demonstrated in an example.

5.4.1 Properties of Reduced Models

A model with one or more regressors removed relative to another model will be called a reduced model. The properties of a reduced model can be determined by comparison to the properties of a complete model with $n_p = n+1$ regressors,

$$z = X\theta + v \tag{5.96}$$

This model can be written as

$$z = X_1\theta_1 + X_2\theta_2 + v \tag{5.97}$$

where the $N \times n_p$ matrix X has been partitioned into the $N \times p$ matrix X_1 and the $N \times q$ matrix X_2, with $n_p = p + q$. The parameter vector is similarly partitioned into θ_1 and θ_2. Equation (5.97) is the same as Eq. (5.48), which was introduced to investigate the significance of $X_2\theta_2$ in the regression model.

If the model structure of Eq. (5.96) is correct, then the properties of $\hat{\theta}$ and $\hat{\sigma}^2$ as estimates of θ and σ^2 are well known. Properties of the ordinary least-squares estimates of θ_1 and σ^2 in the reduced model,

$$z = X_1\theta_1 + v \tag{5.98}$$

have been investigated by many authors. Their findings are summarized in Refs. [5.3] and [5.14]. The results motivate the use of reduced models rather than the complete model for the reasons outlined next.

Deleting regressors associated with parameters that have small numerical values and large standard errors will result in higher accuracy for the parameter estimates of the retained regressors. Deleting regressors from the model also potentially introduces bias in the estimates of θ_1 in the reduced model. However, the variance of the biased estimates $\tilde{\theta}_1$ using the reduced model will be smaller than the variance of the unbiased estimates $\hat{\theta}_1$, which is the estimate of θ_1 using the full model structure. The mean squared error of the biased estimates will also be smaller than the variance of the unbiased estimates, that is

$$
\begin{aligned}
Var\left(\tilde{\boldsymbol{\theta}}_1\right) &\le Var\left(\hat{\boldsymbol{\theta}}_1\right) \\
MSE\left(\tilde{\boldsymbol{\theta}}_1\right) &\le Var\left(\hat{\boldsymbol{\theta}}_1\right)
\end{aligned}
\tag{5.99a}
$$

where

$$
MSE\left(\tilde{\boldsymbol{\theta}}_1\right) = Var\left(\tilde{\boldsymbol{\theta}}_1\right) + \left[E\left(\tilde{\boldsymbol{\theta}}_1\right) - \boldsymbol{\theta}_1\right]^2 \tag{5.99b}
$$

Note that the mean squared error of a parameter estimate is composed of the random error variance plus the squared bias error, see Ref. [5.10]. When the parameter estimate is unbiased, as for the full model structure, then the mean squared error equals the random error variance.

The situation is illustrated in Fig. 5.14 by presenting two probability densities of a single estimated parameter θ_1 from a complete model, $p\left(\hat{\theta}_1\right)$, and from a reduced model, $p\left(\tilde{\theta}_1\right)$. The parameter estimate $\hat{\theta}_1$ is unbiased, while the parameter estimate $\tilde{\theta}_1$ is biased by $E\left(\tilde{\theta}_1\right) - \theta_1$.

If the reduced model is used for prediction, then the predicted value of the response at the ith data point is

$$
\tilde{y}_1(i) = \boldsymbol{x}_1^T(i)\tilde{\boldsymbol{\theta}}_1 \tag{5.100}
$$

with mean value

$$
E\left[\tilde{y}_1(i)\right] = \boldsymbol{x}_1^T(i)\boldsymbol{\theta}_1 + \boldsymbol{x}_1^T(i)\boldsymbol{\mathcal{A}}\boldsymbol{\theta}_2 \tag{5.101}
$$

where

$$
\boldsymbol{\mathcal{A}} = \left(\boldsymbol{X}_1^T\boldsymbol{X}_1\right)^{-1}\boldsymbol{X}_1^T\boldsymbol{X}_2 \tag{5.102}
$$

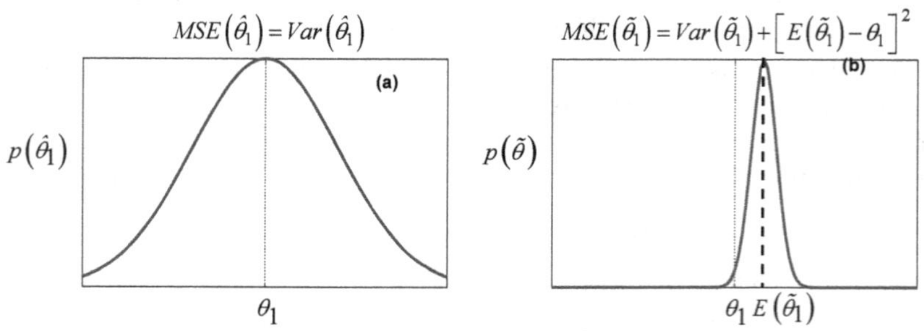

Figure 5.14 **Probability density of a parameter estimate from (a) complete model and (b) reduced model**

and the prediction interval variance is

$$Var\left[z(i)-\tilde{y}_1(i)\right]=\sigma^2\left[1+\boldsymbol{x}_1^T(i)\left(\boldsymbol{X}_1^T\boldsymbol{X}_1\right)^{-1}\boldsymbol{x}_1(i)\right] \tag{5.103}$$

The mean squared error for the prediction is given by

$$\begin{aligned} MSE\left[z(i)-\tilde{y}_1(i)\right] &= E\left\{\left[z(i)-\tilde{y}_1(i)\right]^2\right\} \\ &= \sigma^2\left[1+\boldsymbol{x}_1^T(i)\left(\boldsymbol{X}_1^T\boldsymbol{X}_1\right)^{-1}\boldsymbol{x}_1(i)\right]+\left[\boldsymbol{x}_1^T(i)\boldsymbol{\mathcal{A}}\boldsymbol{\theta}_2-\boldsymbol{x}_2^T(i)\boldsymbol{\theta}_2\right]^2 \end{aligned} \tag{5.104}$$

Detailed development of Eq. (5.104) is given in Ref. [5.14]. Combining the last two equations with Eqs. (5.97) and (5.101), the prediction mean squared error can also be expressed as

$$MSE\left(z-\tilde{y}_1\right)=Var\left(z-\tilde{y}_1\right)+\left[E\left(\tilde{y}_1\right)-y\right]^2 \tag{5.105}$$

Similarly to the properties of parameter estimates from a reduced model shown previously,

$$\begin{aligned} Var\left(z-\tilde{y}_1\right) &\leq Var\left(z-\hat{y}\right) \\ MSE\left(z-\tilde{y}_1\right) &\leq Var\left(z-\hat{y}\right) \end{aligned} \tag{5.106}$$

As the number of regressors retained in the model increases, $Var\left(z-\tilde{y}_1\right)$ will increase from its minimum value, but the bias error will decrease. This means that there will be some minimum of $MSE\left(z-\tilde{y}_1\right)$ for a certain number of regressors in the model. Figure 5.15 demonstrates this by showing $MSE\left(z-\tilde{y}_1\right)$ and its components plotted against number of regressors in the model.

The material in this section provides theoretical substantiation for the parsimony principle stated earlier and discussed further in the next section. Models with too many terms (or too few terms) can lead to reduced accuracy for estimated model parameters and degraded prediction capability. Too many terms increases the variance error, whereas too few terms increases the bias error.

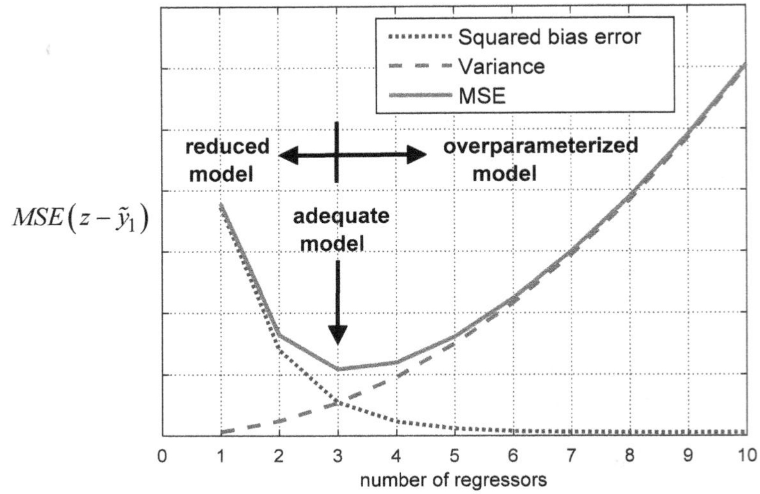

Figure 5.15 **Variation of prediction mean squared error with the number of regressors in the model**

5.4.2 Statistical Modeling Metrics and Stopping Rules

The subset of regressors selected for the model can be determined based on statistical metrics computed from the measured data. This section discusses the computation and application of various statistical modeling metrics that have been found to be useful and accurate for practical aircraft system identification. Values of these metrics provide a guide for selection of a model with good fit to the data and good prediction capability. Practical experience has shown that it is best to consider more than one statistical metric when selecting the model structure, because each statistical metric has its own characteristics. A model structure based on a consensus among several good statistical metrics improves confidence in the model selection, and this approach has been found to produce excellent results.

As mentioned earlier, there are contradictory arguments regarding the appropriate number of parameters in a model. A rule commonly used in choosing a model that would be a good predictor is known as the principle of parsimony, see Ref. [5.17]. This principle can be stated as:

Given two models fitted to the same data with nearly equal residual variances, choose the model with the fewest parameters.

Based on this principle, the objective is to include only model terms that significantly decrease the residual variance. The criterion for judging whether or not a given decrease in the residual variance is significant is some type of comparison with the variance of the measurement noise on the dependent variable. Generally speaking, if an additional model term does not characterize an effect with magnitude significantly greater than the noise level, then that term might very well be modeling noise, which is spurious and nonrepeatable. It follows that a model with too many parameters will be a poor predictor. Some statistical metrics address this problem directly by quantifying the prediction error of an identified model.

The following statistical metrics have been used successfully for aircraft aerodynamic model structure determination. Note that in this section, as in the last section, p is used to denote the number of terms in the model currently being considered, which may be different from the number of parameters in the final model, n_p.

Coefficient of Determination, R^2

The coefficient of determination was introduced earlier [cf. Eq. (5.35)],

$$R^2 = \frac{SS_R}{SS_T} = 1 - \frac{SS_E}{SS_T} = \frac{\hat{\theta}^T X^T z - N\bar{z}^2}{z^T z - N\bar{z}^2} \tag{5.107}$$

where $0 \le R^2 \le 1$. R^2 is usually given as a percentage. Adding a regressor to the model will always increase R^2. However, the more influential the added term is, the greater the change in R^2. After all influential terms are in the model, the changes in R^2 with additional terms will be small. An adequate model is achieved when R^2 is not substantially changed by adding a new term to the model. Typically, if an added model term increases R^2 by less than 0.5 percent, that term is probably not significant, and should be omitted.

Because R^2 is a measure of modeled variation about the mean to total variation about the mean, the ideal or target value of R^2 changes with signal-to-noise ratio of the dependent variable. For example, the ideal R^2 value for a dependent variable with signal-to-noise ratio (excluding the mean) equal to 3 would be 3/(3+1) = 0.75, whereas if the signal-to-noise ratio were equal to 20, the ideal R^2 would be 20/(20+1) = 0.95. Chapter 11 discusses methods for determining signal-to-noise ratio for a time series using data analysis techniques.

The R^2 metric does not account for the number of terms in the model, and therefore will always increase as model terms are added. This problem can be addressed by using the adjusted R^2 metric, defined as

$$R_{adj}^2 \equiv 1 - \frac{SS_E/(N-p)}{SS_T/(N-1)} = 1 - \frac{(N-1)}{(N-p)}\left(1-R^2\right) \tag{5.108}$$

where p is the total number of terms in the model. The R_{adj}^2 metric incorporates a model complexity penalty, and therefore will sometimes decrease as unnecessary model terms are added (p increases).

F Statistic

The *F* statistic for testing significance of a regression is [cf. Eq. (5.46)]

$$\begin{aligned} F_0 &= \frac{SS_R/n}{SS_E/(N-p)} \\ &= \frac{\hat{\boldsymbol{\theta}}^T \boldsymbol{X}^T \boldsymbol{z} - N\bar{z}^2}{ns^2} \end{aligned} \tag{5.109}$$

where n is the number of regressors in the current model, excluding the bias term, so $n = p-1$. Using Eqs. (5.107) and (5.109), the relationship between F_0 and R^2 is

$$F_0 = \frac{(N-p)}{(p-1)}\frac{SS_R}{SS_E} = \frac{(N-p)}{(p-1)}\frac{SS_R}{SS_T - SS_R} = \frac{(N-p)}{(p-1)}\frac{R^2}{1-R^2} \tag{5.110}$$

This relationship indicates that the changes in F_0 at the beginning of the model selection process can be quite pronounced. Later, however, when less influential terms are added into model, the increase in $R^2/\left(1-R^2\right)$ could be smaller than the decrease in $(N-p)/(p-1)$. As a consequence, F_0 values could reach a maximum value and then gradually decrease with more terms included in the model. Based on that behavior of F_0, it was suggested in Ref. [5.18] that $F_0 = F_{0_{max}}$ be used as a stopping rule for the model that provides the best fit to the data with the smallest number of parameters.

Predicted Sum of Squares, PRESS

The *PRESS* statistic proposed in Ref. [5.19] is defined as the sum of N residuals:

$$PRESS = \sum_{i=1}^{N} \left\{ z(i) - \hat{y}\left[i \,|\, \boldsymbol{x}(1), \boldsymbol{x}(2), \ldots, \boldsymbol{x}(i-1), \boldsymbol{x}(i+1), \ldots, \boldsymbol{x}(N)\right]\right\}^2 \tag{5.111}$$

Equation (5.111) shows that *PRESS* uses each possible subset of $N-1$ observations as the parameter estimation data set, and every observation in turn as a single prediction point. For practical computation, Ref. [5.19] shows that *PRESS* can be reformulated as

$$PRESS = \sum_{i=1}^{N} \left[\frac{v(i)}{\left(1-k_{ii}\right)} \right]^2 \tag{5.112}$$

which means that the *PRESS* metric equals the sum of squared ordinary residuals normalized by 1 minus the diagonal element of the prediction matrix,

$$k_{ii} = \boldsymbol{x}^T(i) \left(\boldsymbol{X}^T\boldsymbol{X}\right)^{-1} \boldsymbol{x}(i) \tag{5.113}$$

where $\boldsymbol{x}^T(i)$ is the *i*th row of the $\boldsymbol{X}$ matrix. The model associated with the minimum *PRESS* will be a good predictor, so minimum *PRESS* is used as a stopping rule.

The usefulness of minimum *PRESS* as a stopping rule in model structure determination can be limited in cases where $N \gg p$, which is typical of flight test data analysis. The behavior of the *PRESS* statistic for an increasing number of data points N can be examined from the limit of k_{ii} as $N \to \infty$,

$$\lim_{N\to\infty} k_{ii} = \lim_{N\to\infty} \boldsymbol{x}^T(i)\left(\boldsymbol{X}^T\boldsymbol{X}\right)^{-1} \boldsymbol{x}(i) = 0 \tag{5.114}$$

The limit of the scalar k_{ii} is zero as $N \to \infty$, because as more data are collected, the elements of the $\boldsymbol{X}^T\boldsymbol{X}$ matrix become arbitrarily large, while each single row of data $\boldsymbol{x}^T(i)$ remains fixed. From Eqs. (5.112) and (5.114), it is apparent that *PRESS* approaches the residual sum of squares SS_E as the number of data points N increases. In this situation, *PRESS* can only decrease as regression terms are added to the model. To preserve the ability of *PRESS* to reach a minimum as n_p is changed, it was recommended in Ref. [5.20] that a reduced number of data points (e.g., every tenth point) be used in computing *PRESS*, to keep N low. Another alternative is to omit and predict a number of data points, rather than a single data point, for each term in the summation used to compute *PRESS*. This just implements the idea of holding back part of the data for prediction testing. Further research is needed to find an optimal selection of data points to use with minimum *PRESS* as a stopping rule.

Predicted Square Error, PSE

Another statistical metric that can be used for model structure determination is the predicted squared error (*PSE*) defined by

$$
\begin{aligned}
PSE &\equiv \frac{1}{N}(z-\hat{y})^T(z-\hat{y}) + \sigma_{max}^2 \frac{p}{N} \\
&= MSFE + \sigma_{max}^2 \frac{p}{N}
\end{aligned}
\tag{5.115}
$$

where *MSFE* is the mean squared fit error for the modeling data,

$$
MSFE \equiv \frac{1}{N}(z-\hat{y})^T(z-\hat{y}) = \frac{1}{N}\left(v^T v\right) \tag{5.116}
$$

The *MSFE* is proportional to the ordinary least-squares cost function given in Eq. (5.8). The quantity σ_{max}^2 is a constant to be discussed next, p is the number of terms in the current model, and N is the number of data points. The *PSE* in Eq. (5.115) depends on the *MSFE* and a term proportional to the number of terms in the model, p. The *MSFE* decreases with each added model term, so that minimizing *MSFE* alone would provide no protection against overfitting the data by adding too many terms to the model. The overfit penalty term $\sigma_{max}^2\, p/N$ in the *PSE* increases with each added model term, because p increases. This term prevents overfitting the data with too many model terms, which is detrimental to model prediction accuracy, as discussed previously.

The constant σ_{max}^2 is the upper-bound estimate of the squared error between future data and the model, i.e., the upper bound mean squared error for prediction cases. The upper bound is used in the model overfit penalty term to account for the fact that *PSE* is calculated when the model structure is not correct, i.e., during the model structure determination stage. Using the upper bound is conservative in the sense that model complexity will be minimized as a result of using an upper bound for this constant in the penalty term. Because of this, the value of *PSE* computed from Eq. (5.115) for a particular model structure tends to overestimate actual prediction errors on new data. Therefore, the *PSE* metric conservatively estimates the squared error for prediction cases.

A simple estimate of σ_{max}^2 that is independent of the model structure can be obtained by considering σ_{max}^2 to be the residual variance estimate for a constant model equal to the mean of the measured response values,

$$
\sigma_{max}^2 = \frac{1}{(N-1)} \sum_{i=1}^{N} \left[z(i) - \bar{z}\right]^2 \tag{5.117}
$$

where

$$\bar{z} = \frac{1}{N}\sum_{i=1}^{N} z(i) \tag{5.118}$$

For wind tunnel testing, repeated runs at the same test conditions are often available. If $\hat{\sigma}^2$ is the measurement error variance estimated from measurements of the dependent variable for repeated runs at the same test conditions [cf. Eq. (5.22)], then σ_{max}^2 can be estimated as

$$\sigma_{max}^2 = 25\ \hat{\sigma}^2 \tag{5.119}$$

If the residuals were Gaussian, Eq. (5.119) would correspond to conservatively placing the maximum output variance at 25 times the estimated value, corresponding to a $5\hat{\sigma}$ maximum deviation. However, the estimate $\hat{\sigma}^2$ may not be very good, because of relatively few repeated runs available, or errors in the explanatory variable settings, or drift errors when duplicating test conditions for the repeat runs. The $5\hat{\sigma}$ value has been found to give accurate models in model identification algorithm tests and model structure identification based on wind tunnel data, see Refs. [5.21]-[5.22].

Correlation Coefficient

The correlation coefficient quantifies the similarity of two data vectors. The relevant characteristic is the variation of the data in the vector, which is equivalent to the waveform, if the data vector is a time series. This metric is useful for determining the similarity of the variations in selected pairs of candidate model terms, which indicates the distinct modeling capability of each term. More importantly, the correlation coefficient can be used to determine how similar the variations in a candidate model term are to the variations in the dependent variable to be modeled, which quantifies the usefulness of the candidate model term for characterizing the variations in the dependent variable.

The correlation coefficient is an inner product of two data vectors scaled to unit length with their mean values removed. The simple pairwise correlation between regressors ξ_j and ξ_k is given by

$$r_{jk} = \frac{S_{jk}}{\sqrt{S_{jj}S_{kk}}} \tag{5.120}$$

where

$$S_{jk} = \sum_{i=1}^{N} \left[\xi_j(i) - \bar{\xi}_j \right] \left[\xi_k(i) - \bar{\xi}_k \right] \tag{5.121a}$$

$$S_{jj} = \sum_{i=1}^{N} \left[\xi_j(i) - \bar{\xi}_j \right]^2 \qquad S_{kk} = \sum_{i=1}^{N} \left[\xi_k(i) - \bar{\xi}_k \right]^2 \tag{5.121b}$$

$$\bar{\xi}_j = \frac{1}{N} \sum_{i=1}^{N} \xi_j(i) \qquad \bar{\xi}_k = \frac{1}{N} \sum_{i=1}^{N} \xi_k(i) \tag{5.121c}$$

Similarly, the correlation between a regressor ξ_j and the measured dependent variable z is

$$r_{jz} = \frac{S_{jz}}{\sqrt{S_{jj} S_{zz}}} \tag{5.122}$$

where

$$\bar{z} = \frac{1}{N} \sum_{i=1}^{N} z(i) \tag{5.123a}$$

$$S_{jz} = \sum_{i=1}^{N} \left[\xi_j(i) - \bar{\xi}_j \right] \left[z(i) - \bar{z} \right] \quad S_{zz} = \sum_{i=1}^{N} \left[z(i) - \bar{z} \right]^2 \tag{5.123b}$$

Correlation coefficients lie in the range $[-1, 1]$. A value of 1 indicates data vectors with identical normalized variations, a value of -1 indicates data vectors with normalized variations that differ only by a minus sign, and a value of 0 indicates that the normalized variations are completely uncorrelated, which means the normalized data vectors are orthogonal (see Appendix A).

Other Statistical Metrics

In addition to the metrics listed earlier and their associated stopping rules, at each step of the model selection procedure, the values of the fit error and F_0 or t_0 statistic for each model parameter can be computed. The estimated fit error

$$s = \sqrt{\frac{SS_E}{N-p}} = \sqrt{\frac{\sum_{i=1}^{N} \left[z(i) - \hat{y}(i) \right]^2}{N-p}} \tag{5.124}$$

should be compared to an unbiased estimate of the standard deviation of the noise for the dependent variable. One method for obtaining such an estimate is given in Ref. [5.23]. In this approach, the random noise is separated from the deterministic part of the measured dependent variable using an optimal Fourier smoother (see Chapter 11).

The F_0 and t_0 statistics for each (*j*th) estimated parameter are obtained from Eqs. (5.59) and (5.60)

$$F_0 = \frac{\hat{\theta}_j^2}{s^2\left(\hat{\theta}_j\right)} \tag{5.125}$$

$$t_0 = \frac{\hat{\theta}_j}{s\left(\hat{\theta}_j\right)} \tag{5.126}$$

These expressions for the F_0 and t_0 statistics represent the inverse of the relative parameter variance and the inverse of the relative parameter standard error, respectively. Both of these statistics should be large for each term in an adequate model. The forms shown in Eqs. (5.125) and (5.126) make it clear that these metrics can be interpreted as tests for the value of an estimated parameter being statistically different from zero.

5.4.3 Stepwise Regression

This section explains computational techniques that can be used to evaluate subsets of a pool of candidate regressors for inclusion in a linear regression model, by adding or deleting regressors one at a time. The procedures can be classified into three basic categories: forward selection, backward elimination, and stepwise regression, which is a combination of the first two.

Forward Selection

This technique begins with the model that has no regressors, except the constant or bias term θ_o. Then one regressor at a time is added until all candidate regressors are in the model or until some selection criterion is satisfied. The first regressor selected for entry into the regression equation is the one that has the highest simple correlation with the dependent variable, adjusted for the mean value. Equation (5.122) is used to calculate the correlation coefficients for each regressor with the dependent variable, adjusted for the mean value. The regressor with the highest absolute value of correlation is also the regressor that yields the largest value of the partial F statistic for testing the significance of a single added regressor [cf. Eq. (5.59)]. This regressor enters if the partial F statistic exceeds a pre-selected value, usually called F-to-enter, or F_{in}. The second regressor selected for entry into the model is the one with the largest correlation with $z-\theta_0-\theta_1\xi_1$, which is the measured dependent variable adjusted for the effect of the mean and the first regressor in the model. This process continues until all candidate regressors are in the model, or no remaining candidate regressors pass the F_{in} criterion for entry into the model. In general, a regressor ξ_j is added to a model with p terms if [cf. Eq. (5.55)]

$$F_0 = \frac{SS_R\left(\hat{\boldsymbol{\theta}}_{p+j}\right) - SS_R\left(\hat{\boldsymbol{\theta}}_p\right)}{s^2} > F_{in} \tag{5.127}$$

where $SS_R\left(\hat{\boldsymbol{\theta}}_p\right)$ is the regression sum of squares using the p terms already in the model, and $SS_R\left(\hat{\boldsymbol{\theta}}_{p+j}\right)$ is the regression sum of squares obtained by adding the jth regressor to the original p terms. The fit error variance s^2 is computed in the usual way, assuming the jth regressor is included in the model [cf. Eq. (5.24)],

$$s^2 \equiv \hat{\sigma}^2 = \frac{\boldsymbol{v}^T\boldsymbol{v}}{(N-p-1)} = \frac{\sum_{i=1}^{N}\left[z(i)-\hat{y}(i)\right]^2}{(N-p-1)} \tag{5.128}$$

The value of F_{in} is $F(\alpha;1,N-p-1)$, where α is the selected significance level.

Backward Elimination

This technique begins with all of the candidate regressors included in the regression equation. Unnecessary regressors are then removed one at a time. At each step, the regressor with the smallest partial F-ratio, computed from the current regression, is eliminated if

$$F_0 = \min_j \frac{SS_R\left(\hat{\boldsymbol{\theta}}_p\right) - SS_R\left(\hat{\boldsymbol{\theta}}_{p-j}\right)}{s^2} < F_{out} \tag{5.129}$$

where F_{out} is $F(\alpha;1,N-p)$, and $SS_R\left(\hat{\boldsymbol{\theta}}_{p-j}\right)$ is the regression sum of squares obtained by removing the jth regressor from the original p terms. Backward elimination is the converse of forward selection, in the sense that forward selection starts with no regressors in the model and adds terms, whereas backward elimination starts with every candidate regressor in the model and removes terms.

Stepwise Regression

The stepwise regression method described here is a combination of forward selection and backward elimination, proposed in Ref. [5.24]. It modifies the forward selection by adding backward elimination to reassess the regressors entered into the model. A regressor added at an early stage may become redundant because of its relationship with regressors added subsequently to the model.

Stepwise regression starts by constructing a pool of candidate regressors, and modeling the dependent variable using only the constant or bias term, as for forward selection. Then the first regressor for the model is chosen as the one with the highest absolute value of correlation with z, adjusted for the mean value. The correlations are computed from Eq. (5.122), as in the case of forward selection. If ξ_1 is selected as the first regressor in the model, then the model

$$z = \theta_0 + \theta_1 \xi_1 + \nu \tag{5.130}$$

is used with ordinary least-squares parameter estimation to fit the measured data. The partial F statistic is computed as

$$F_0 = \frac{SS_R\left(\theta_1\right)}{s^2} \tag{5.131}$$

and compared with F_{in}. If $F_0 > F_{in}$, the model (5.131) is accepted as the model at the first step.

Next, new regressors are constructed from the remaining regressors in the pool of candidate regressors, by removing the part of each remaining candidate regressor vector that is like the terms already in the model. This can be done by modeling each regressor in the remaining pool of regressors with ξ_1 and a vector of ones (for the bias term), and treating the residuals as the new pool of regressors. This is the equivalent of removing any variation in the remaining candidate regressors that is included in terms already in the model. For example, regressor ξ_2 is modeled as

$$\xi_2 = a_0 + a_1 \xi_1 + \nu_2 \tag{5.132}$$

so that the new regressor ν_2 is given by

$$\nu_2 = \xi_2 - \hat{a}_0 - \hat{a}_1 \xi_1 \tag{5.133}$$

with the parameter estimates $\hat{a}_0$ and $\hat{a}_1$ computed from ordinary least-squares parameter estimation. The other remaining regressors in the pool of candidate regressors are similarly modified, where each remaining regressor will have its own parameter values analogous to $\hat{a}_0$ and $\hat{a}_1$ in Eq. (5.133). Note that this procedure orthogonalizes the remaining candidate regressors with respect to the regressors currently in the model [cf. Eq. (5.9c)]. In the same way, a new dependent variable $\boldsymbol{v}_z$ is computed from

$$\boldsymbol{v}_z = \boldsymbol{z} - \hat{\theta}_0 - \hat{\theta}_1 \boldsymbol{\xi}_1 \tag{5.134}$$

For the next step, a new set of correlations involving the new regressors and the new dependent variable is computed, and the process is repeated. Each stage removes the effects of model terms already chosen, so that the focus of the model structure determination is always on the model terms necessary to characterize the variation in the dependent variable that has not yet been modeled. This is helpful, because the magnitude of the early dominant terms can dwarf later terms that might be necessary for modeling smaller magnitude variations in the dependent variable. Note that only the partial correlation calculation uses quantities corrected for terms already in the model; all other calculations use the original candidate regressors and dependent variable.

At every step, the regressors incorporated into the model at previous stages and the new regressor just entering the model are examined. This is done for the regression model in the form of Eq. (5.4), but with only the selected regressors included in the model. For this part of the analysis, the selected regressors and the dependent variable are in their original form, not the conditioned form used for the forward selection step described earlier. The partial F criterion for backward elimination from Eq. (5.129) is evaluated for each regressor in the model, and compared with the value of F_{out}. Any regressor that provides a nonsignificant contribution, evidenced by a small value of F_0, is removed from the model. A regressor that may have been the best single regressor to enter at an earlier stage may, at a later stage, be redundant because of a linear or nearly linear relationship with other regressors in the model. The process of selecting and removing regressors continues until no more regressors are admitted and no more are rejected.

The pre-selected values of F_{in} and F_{out} depend on the number of data points N, the current number of model terms p, and the selected confidence level $1-\alpha$. For most practical aircraft problems, $N \gg 100$ and $p < 10$, so the effect of p on values of $F(\alpha; 1, N-p)$ is small. Therefore, for 95 percent confidence, $F(0.05; 1, N-p)$ can be taken as a constant equal to 4. Some analysts prefer to choose $F_{in} > F_{out}$, which means it will be relatively more difficult to add a regressor than to remove one.

However, using only the partial F statistic with F_{in} and F_{out} can sometimes be too restrictive. A large preselected value of F_{in} can terminate the selection procedure before all influential regressors are included in the model. On the other hand, if F_{out} is chosen small, some of the regressors with limited influence can be retained in the model.

Several stopping rules for stepwise regression applied to aircraft aerodynamic model structure determination have been proposed and tested in Ref. [5.20]. Although it is possible to automate the model selection process in stepwise regression based on statistical metrics, in practice, the best approach is often to determine the model structure based on a combination of physical insight and a consensus of several good statistical metrics, such as those presented in the previous section.

Example 5.3

This example demonstrates the use of stepwise regression for model structure determination and parameter estimation. The application is the z body axis aerodynamic force coefficient C_Z from a piloted nonlinear simulation of the F-16 aircraft. Appendix D gives a full description of the F-16 simulation, which runs in MATLAB® and is included in the SIDPAC software associated with this book. There is no automatic feedback control system included in the F-16 simulation, so that pilot stick and rudder commands move the control surfaces directly. Flaps and spoilers are fixed, which means that the elevator is the only movable longitudinal control surface. The c.g. is located at a forward position $\left(x_{cg} = 0.25\,\bar{c}\right)$ to achieve open-loop static stability, which allows the F-16 to be flown without an automatic feedback control system for stability augmentation.

Figure 5.16 shows measured time series from a piloted large-amplitude longitudinal maneuver, where the pilot applied small amplitude longitudinal stick perturbations on top of a steady pull-up and push-over. This produces data that cover large ranges of angle of attack, pitch rate, and stabilator deflection. Figure 5.17 shows cross plots of angle of attack, pitch rate, and stabilator deflection, indicating the effectiveness of this maneuver in covering a large portion of the explanatory variable space. These cross plots can also be interpreted as a graphical indication of the pairwise correlation between explanatory variables. If two explanatory variables were highly correlated, then their cross plot would approximate a straight line. The explanatory variables plotted in Fig. 5.17 show very low pairwise correlations.

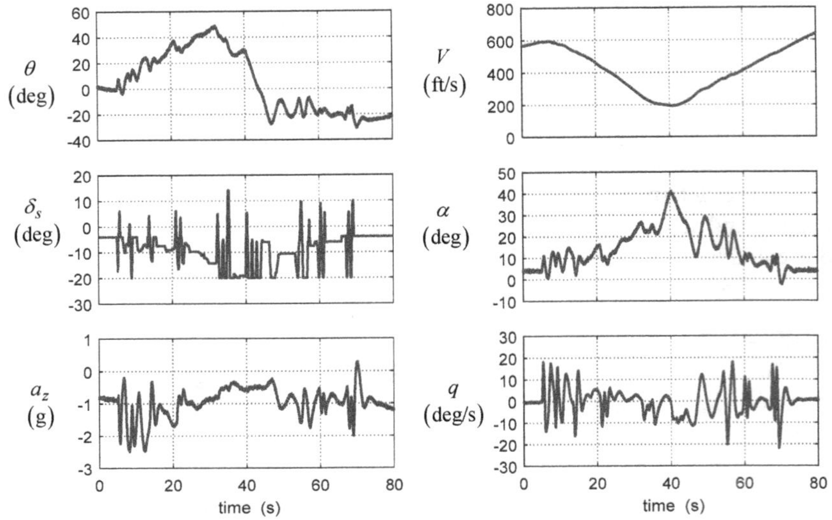

Figure 5.16 **Input and output data for a large-amplitude longitudinal maneuver**

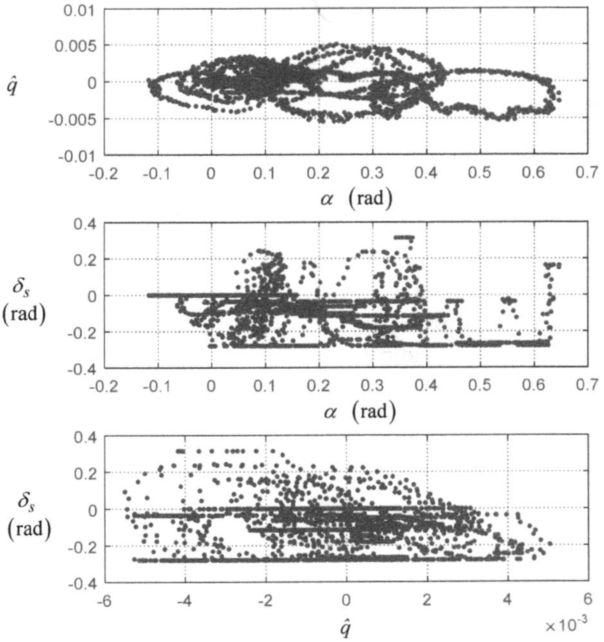

Figure 5.17 **Explanatory variable cross-plots for a large-amplitude longitudinal maneuver**

Each output variable from the nonlinear simulation was corrupted by white Gaussian noise with magnitude typical of flight test instrumentation. The resulting signal-to-noise ratio was approximately 30 for the measured response variable C_Z shown in Fig. 5.18. This signal-to-noise ratio is relatively high, and is more than adequate for good modeling results. For small perturbation maneuvers, the signal amplitudes would be much lower, while the noise levels would remain approximately the same, resulting in lower signal-to-noise ratios.

The C_Z coefficient shown in Fig. 5.18 was modeled using the stepwise regression method described earlier. Table 5.4 shows parameter estimates and statistical metrics for each step in the model structure determination. At the start, partial correlations r of each regressor with the response variable C_Z are computed. As each regressor is brought into the model, a value for the corresponding parameter estimate and F_0 statistic is computed.

At step 2, the linear q term is chosen rather than the nonlinear α^2 term, although the partial correlation for the latter term is slightly higher. In borderline cases like this, the linear term should generally be included first in the model, so that the nonlinear terms are used only to model variations that cannot be characterized with linear terms. This corresponds to adding terms of increasing complexity to the Taylor series after the simpler linear terms are already in the model. Including the linear terms first, and using the nonlinear terms to model the remaining variation, has been called modified stepwise regression.

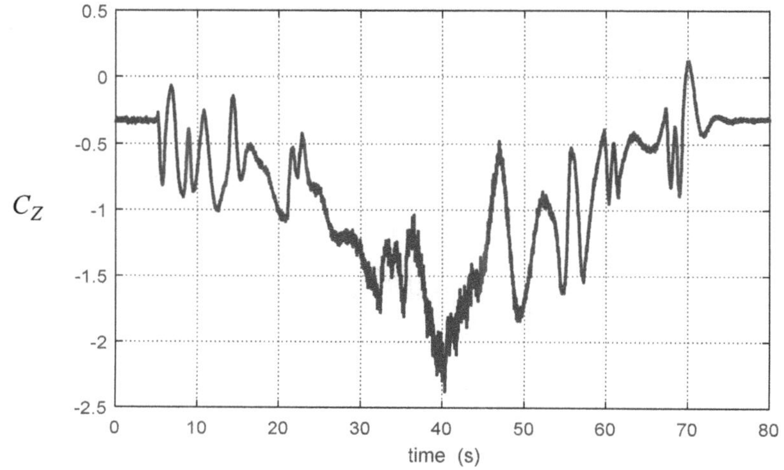

***Figure 5.18* Body-axis *Z* force coefficient for a large-amplitude longitudinal maneuver**

In the software implementation of the stepwise regression method that accompanies this book, regressors are moved in and out of the model via manual instructions from the analyst. This makes it possible to effectively handle borderline cases such as the one demonstrated in step 2 of Table 5.4, and to easily implement modified stepwise regression, if desired. The analyst is therefore involved in each decision, and can apply specialized knowledge of the physical situation or experiment to the modeling problem.

Figure 5.19 shows the model fit and residual at step 4, where the model includes the strong linear terms in α, $q\bar{c}/2V_o$, and δ_s, but not the weak linear term in Mach number, M. The residual in the lower plot of Fig. 5.19 clearly indicates some remaining deterministic content. The residual becomes random after adding the nonlinear α^2 term in step 5, see Fig. 5.20.

Statistical modeling metrics are shown at the bottom of Table 5.4 for each step in the model structure determination. The R^2 metric jumps to a large value after the initial regressor, then increases in small increments thereafter. The final value of R^2 is 99.3%, which is indicative of a good model. Both the *PRESS* and *PSE* metrics decreased for each added regressor, and the corresponding F_0 values were very high, indicating that each added regressor should be retained in the model.

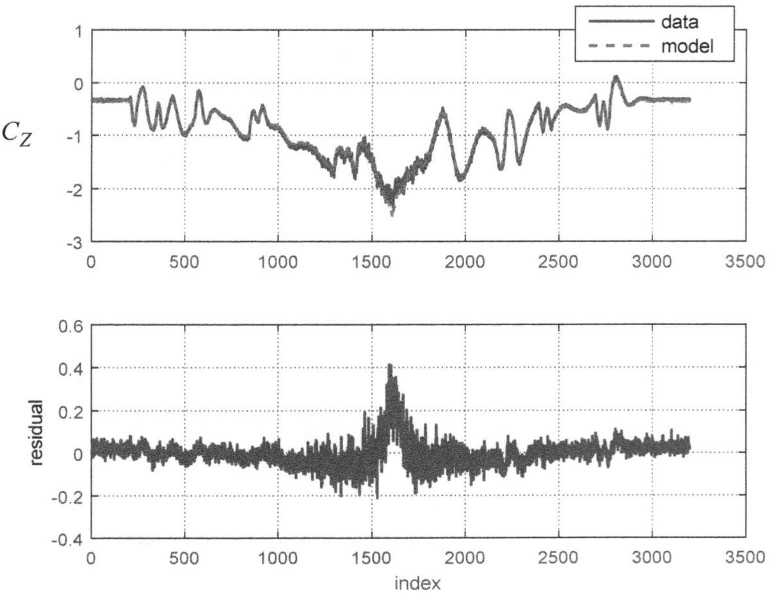

Figure 5.19 **Body-axis *Z* force coefficient modeling results after stepwise regression step 4 in Table 5.4**

Table 5.4 **Stepwise regression results, body-axis *Z* force coefficient for a large-amplitude maneuver**

Step	1			2			3			4			5		
	$\hat{\theta}$	F_0	r	$\hat{\theta}$	F_0	r	$\hat{\theta}$	F_0	r	$\hat{\theta}$	F_0	r	$\hat{\theta}$	F_0	r
C_{Z_M}	0	--	0.725	0	--	0.020	0	--	0.000	0	--	0.100	0	--	0.008
C_{Z_α}	0	--	0.973	-3.13	1.2e5	--	-3.21	1.6e5	--	-3.35	2.0e5	--	-4.17	7.0e4	--
C_{Z_q}	0	--	0.031	0	--	0.305	-27.0	1.4e3	--	-32.3	2.8e3	--	-28.1	4.0e3	--
$C_{Z_{\delta_e}}$	0	--	0.166	0	--	0.114	0	--	0.322	-0.42	1.5e3	--	-0.43	3.2e3	--
$C_{Z_{\alpha^2}}$	0	--	0.838	0	--	0.306	0	--	0.303	0	--	0.485	1.34	3.0e3	--
PRESS	852.0			23.0			16.0			10.9			5.6		
PSE	0.266			0.00735			0.00524			0.00372			0.00216		
R^2 (%)	0.0			97.3			98.1			98.7			99.3		

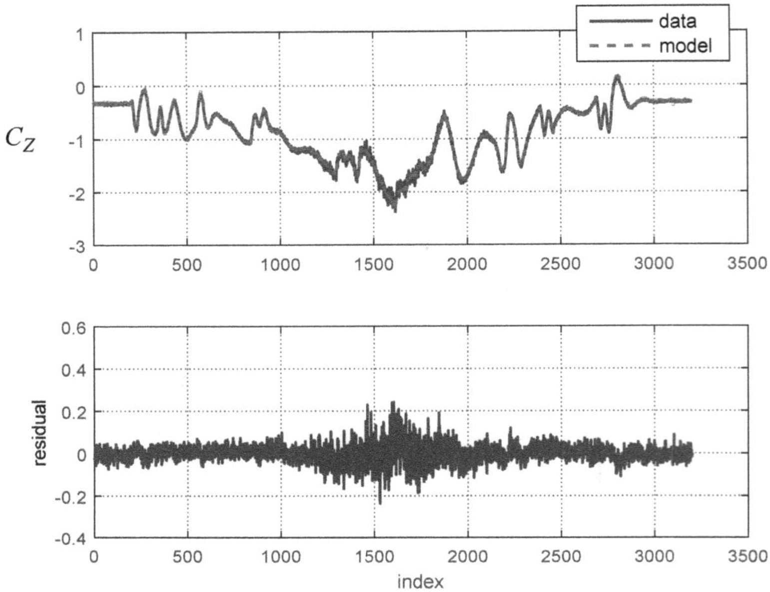

Figure 5.20 **Body-axis *Z* force coefficient modeling results after stepwise regression step 5 in Table 5.4**

The partial correlation for the Mach number regressor was never the highest among the candidate regressors not yet in the model, so Mach number was not selected for inclusion in the model. If the Mach number regressor is added at step 5 (not shown in Table 5.4), the *PSE* increases, and R^2 remains essentially unchanged, but the *PRESS* metric decreases. The value of F_0 for the Mach number regressor in the model is around 26, which is much lower than F_0 for the other regressors. These mixed messages from the statistical metrics are not uncommon, but can usually be resolved by using a majority consensus of the various metrics, as in this case.

Typically, there are more regressors in the candidate pool than shown in this example. However, given a larger pool of candidate regressors, the stepwise regression can be used to select an adequate model in a manner similar to that shown here. Sometimes there is some trial-and-error involved in assembling a good pool of candidate regressors. The stepwise regression method can only select regressors available in the candidate pool, so the modeling functions needed for an adequate model must be present in the candidate pool.

One method for identifying a good set of candidate regressors is to use the orthogonal function modeling technique described in the next section. This technique considers all multivariate polynomial combinations of the explanatory variables (up to a selected maximum order), then orthogonalizes those functions and ranks them according to their ability to reduce the mean square fit error. The ranked list of orthogonal modeling functions provides a good set of candidate regressors for stepwise regression.

Note also from Table 5.4 that the parameter estimates for terms already in the model change as each new regressor is added. This is because the regressors are correlated to some extent, and therefore are not mutually orthogonal. If the stepwise regression were done using orthogonal modeling functions, each model parameter estimate would remain unchanged as other orthogonal modeling terms were swapped in and out of the model.

Table 5.5 contains the final parameter estimation results obtained from the stepwise regression, including standard errors corrected for colored residuals. The model parameters are estimated accurately, with large t_0 statistics, and the model has a high R^2 value. This example shows that a simple nonlinear term augmentation to a linear model can be used to effectively model data from a large amplitude maneuver.

The nonlinearity highlighted in Fig. 5.19 can be characterized using other modeling functions, such as a first-order spline. Assuming a first-order spline is to be used, there is still the question of where the knot should be located. Stepwise regression can be used to help the analyst make this decision. If the pool of candidate regressors includes first-order splines with various knot locations, the best one will be indicated during the stepwise regression by the highest partial correlation after all significant linear terms are in the model.

Table 5.5 **Stepwise regression parameter estimation results, body-axis *Z* force coefficient, large-amplitude longitudinal maneuver**

Parameter	$\hat{\theta}$	$s(\hat{\theta})$	$\lvert t_0 \rvert$	$100\left[s(\hat{\theta})/\lvert\hat{\theta}\rvert\right]$
C_{Z_M}	0	--	--	--
C_{Z_α}	–4.166	0.0497	83.9	1.2
C_{Z_q}	–28.13	0.9822	28.6	3.5
$C_{Z_{\delta_e}}$	–0.4333	0.0123	35.3	2.8
$C_{Z_{\alpha^2}}$	1.344	0.0823	16.3	6.1
C_{Z_o}	–0.3229	0.0034	95.0	1.1
$s = \hat{\sigma}$	0.0418			
R^2 (%)	99.3			

Figure 5.21 shows the model fit to C_Z for this case, where a first-order spline term in angle of attack with knot location at 25 deg was selected using stepwise regression. The associated parameter estimation results appear in Table 5.6. Comparison of these results with Fig. 5.20 and Table 5.5 demonstrates that in this case the nonlinearity in C_Z can be modeled equally well with a nonlinear α^2 term or a first-order spline term. Note that parameter estimates for model terms common to the models in Tables 5.5 and 5.6 (e.g., C_{Z_α}) are different, because the regressors in the models are different and are not mutually orthogonal. Because of this, the task of characterizing variations in the response variable is shared differently among the regressors in each model. ■

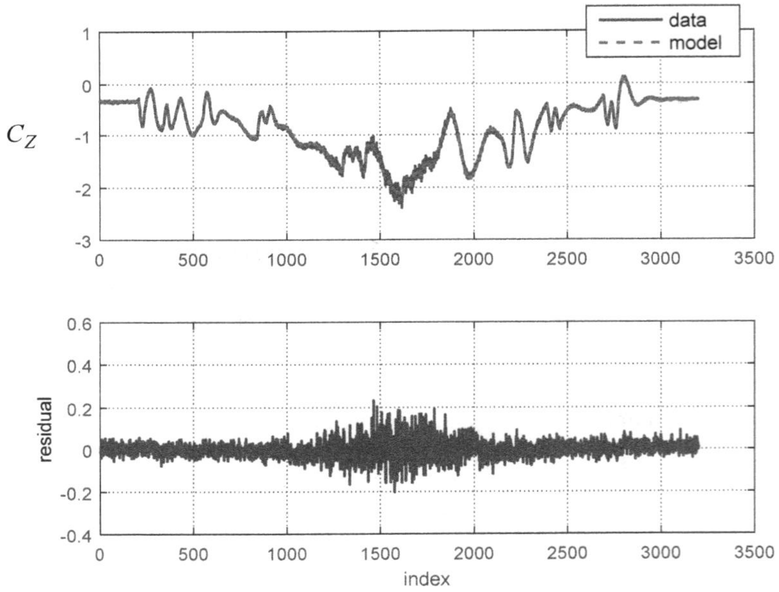

Figure 5.21 **Vertical force coefficient modeling results using a linear spline term in angle of attack to model the nonlinearity**

***Table 5.6* Stepwise regression parameter estimation results, body-axis *Z* force coefficient, large-amplitude longitudinal maneuver**

Parameter	$\hat{\theta}$	$s(\hat{\theta})$	$\lvert t_0 \rvert$	$100\left[s(\hat{\theta})/\lvert\hat{\theta}\rvert\right]$
C_{Z_M}	0	--	--	--
C_{Z_α}	–3.537	0.0160	220.7	0.5
C_{Z_q}	–29.45	0.8241	35.7	2.8
$C_{Z_{\delta_e}}$	–0.4400	0.0113	38.8	2.6
$C_{Z_{[\alpha-25(\pi/180)]_+^1}}$	1.681	0.0732	23.0	4.4
C_{Z_o}	–0.3348	0.0028	119.7	0.8
$s = \hat{\sigma}$	0.0400			
R^2 (%)	99.4			

5.4.4 Multivariate Orthogonal Function Modeling

Another method for effective model structure determination is multivariate orthogonal function modeling [5.25]-[5.26], [5.21]-[5.22]. In this approach, a pool of candidate regressors is orthogonalized first, which clarifies and simplifies the model selection process. The orthogonalization is equivalent to isolating the unique variation in each of the candidate regressors, and assigning that unique variation to individual orthogonal functions. This process results in a complete decorrelation of the modeling functions, which facilitates the selection of effective model terms. The orthogonal functions are generated in a manner that allows each of them to be decomposed without ambiguity into an expansion of ordinary multivariate functions from the candidate regressor pool. Therefore, a model identified using multivariate orthogonal functions can be subsequently converted to an expansion of ordinary multivariate functions in the explanatory variables. This latter form of the model provides physical insight into the functional dependencies.

Ordinary Least Squares using Multivariate Orthogonal Functions

The form of a multivariate orthogonal function model is

$$z = Pa + \nu \tag{5.135}$$

where

$z = [z(1) \ \ z(2) \ \ \ldots \ \ z(N)]^T = N \times 1$ response vector

$a = [a_0 \ \ a_1 \ \ \ldots \ \ a_n]^T = n_p \times 1$ vector of unknown parameters, $n_p = n+1$

$P = [p_0 \ \ p_1 \ \ \ldots \ \ p_n] = N \times n_p$ matrix of mutually orthogonal regressors

$\nu = [\nu(1) \ \ \nu(2) \ \ \ldots \ \ \nu(N)]^T = N \times 1$ modeling error vector

Equation (5.135) is the same as Eq. (5.6), except that the modeling functions are now assumed to be orthogonalized, so that

$$p_i^T p_j = 0 \qquad i \neq j \quad , \quad i, j = 0, 1, 2, \ldots, n \tag{5.136}$$

Equation (5.135) represents a mathematical model used to characterize functional dependencies in the measured data. The important questions of determining how the modeling functions p_j are computed from the explanatory variables, as well as which modeling functions should be included in Eq. (5.135), which implicitly determines n, will be addressed later. At this point, the properties of a multivariate orthogonal function model are examined.

The a_j, $j = 0,1,2,\ldots,n$ in Eq. (5.135) are constant model parameters to be determined. These parameters are different, in general, from $\theta = [\theta_0 \ \ \theta_1 \ \ \ldots \ \ \theta_n]^T$, because the modeling functions are different. The parameter vector estimate that minimizes the ordinary least-squares cost function

$$J(a) = \frac{1}{2}(z - Pa)^T (z - Pa) \tag{5.137}$$

is computed using Eq. (5.10),

$$\hat{a} = \left(P^T P\right)^{-1} P^T z \tag{5.138}$$

The estimated parameter covariance matrix is computed by Eq. (5.12),

$$Cov(\hat{a}) = E\left[(\hat{a} - a)(\hat{a} - a)^T\right] = \sigma^2 \left(P^T P\right)^{-1} \tag{5.139}$$

and the error variance σ^2 can be estimated from the residuals,

$$\boldsymbol{v} = \boldsymbol{z} - \boldsymbol{P}\hat{\boldsymbol{a}} \tag{5.140}$$

$$\hat{\sigma}^2 = \frac{1}{\left(N - n_p\right)}\left[\left(\boldsymbol{z} - \boldsymbol{P}\hat{\boldsymbol{a}}\right)^T\left(\boldsymbol{z} - \boldsymbol{P}\hat{\boldsymbol{a}}\right)\right] = \frac{\boldsymbol{v}^T\boldsymbol{v}}{\left(N - n_p\right)} \tag{5.141}$$

or σ^2 can be estimated independently, as discussed earlier. Parameter standard errors are computed as the square root of the diagonal elements of the $Cov(\hat{\boldsymbol{a}})$ matrix from Eq. (5.139), which can also be corrected for colored residuals, as described earlier. The identified model output is computed as

$$\hat{\boldsymbol{y}} = \boldsymbol{P}\hat{\boldsymbol{a}} \tag{5.142}$$

All of this looks the same as before, until the fact that the modeling functions are orthogonal is introduced. When the modeling functions are mutually orthogonal, $\boldsymbol{P}^T\boldsymbol{P}$ is a diagonal matrix with the inner product of the orthogonal functions on the main diagonal. This decouples the normal equations, so that Eq. (5.138) becomes

$$\hat{a}_j = \left(\boldsymbol{p}_j^T\,\boldsymbol{z}\right)/\left(\boldsymbol{p}_j^T\,\boldsymbol{p}_j\right) \qquad j = 0,1,2,\ldots,n \tag{5.143}$$

Using Eqs. (5.9c) and (5.136) in Eq. (5.137),

$$J(\hat{\boldsymbol{a}}) = \frac{1}{2}\left[\boldsymbol{z}^T\boldsymbol{z} - \sum_{j=0}^{n}\hat{a}_j^2\left(\boldsymbol{p}_j^T\,\boldsymbol{p}_j\right)\right] \tag{5.144}$$

or, using Eq. (5.143),

$$J(\hat{\boldsymbol{a}}) = \frac{1}{2}\left[\boldsymbol{z}^T\boldsymbol{z} - \sum_{j=0}^{n}\left(\boldsymbol{p}_j^T\,\boldsymbol{z}\right)^2/\left(\boldsymbol{p}_j^T\,\boldsymbol{p}_j\right)\right] \tag{5.145}$$

Model Structure Determination using Multivariate Orthogonal Functions

Equation (5.145) shows that when the modeling functions are orthogonal, the reduction in the least-squares cost function resulting from including the term $a_j\,\boldsymbol{p}_j$ in the model depends only on the response variable data $\boldsymbol{z}$ and the added orthogonal modeling function $\boldsymbol{p}_j$. The least-squares modeling problem is therefore decoupled, which means each orthogonal modeling function can be evaluated independently in terms of its ability to reduce the least-squares model fit to the data, regardless of which other orthogonal modeling functions are already selected for the model. When the modeling functions are instead polynomials in the explanatory variables (or any other nonorthogonal function set), the least-squares problem is coupled, and iterative analysis is required to find the subset of modeling functions for an adequate model structure.

The orthogonal modeling functions to be included in the model are chosen to minimize predicted squared error, *PSE*, defined earlier in Eq. (5.115),

$$PSE \equiv \frac{1}{N}(z-\hat{y})^T(z-\hat{y}) + \sigma_{max}^2 \frac{p}{N} \tag{5.146}$$

where p is the number of terms in the current model. The *PSE* depends on the mean squared fit error and a term proportional to the number of terms in the model, p. The latter term prevents overfitting the data with too many model terms, which is detrimental to model prediction accuracy, as discussed earlier and also in Ref. [5.27].

While the mean squared fit error must decrease with the addition of each orthogonal modeling function to the model [cf. Eq. (5.145)], the overfit penalty term $\sigma_{max}^2 \, p/N$ must increase with each added model term (p increases). Introducing the orthogonal modeling functions into the model in order of most effective to least effective in reducing the mean squared fit error (quantified by $\left(\boldsymbol{p}_j^T z\right)^2 \Big/ \left(\boldsymbol{p}_j^T \boldsymbol{p}_j\right)$ for the *j*th orthogonal modeling function $\boldsymbol{p}_j$) results in the *PSE* metric always having a single global minimum.

Figure 5.22 depicts this graphically, using actual modeling results from Ref. [5.21]. The figure shows that after the first 6 modeling functions, the added model complexity associated with an additional orthogonal modeling function is not justified by the associated reduction in mean squared fit error. This point is marked by a minimum *PSE*, which defines an adequate model structure with good predictive capability. Note that Fig. 5.22 is analogous to Fig. 5.15; however, the values shown in Fig. 5.22 are components of the *PSE* metric, computed from measured wind tunnel data. Therefore, Fig. 5.22 is a practical realization of the conceptual plot in Fig. 5.15. Reference [5.27] contains further justifying statistical arguments and analysis for the form of *PSE* given in Eqs. (5.115) and (5.146), including justification for its use in modeling problems.

Using orthogonal functions to model the response variable makes it possible to evaluate the merit of including each modeling function individually, using the predicted squared error *PSE*, because of the properties of orthogonal functions and the resultant decoupling of the associated least-squares parameter estimation problem. The goal is to select a model structure with minimum *PSE*, and the *PSE* always has a single global minimum for orthogonal modeling functions arranged according to their effectiveness in reducing mean squared error. This makes the model structure determination a well-defined and straightforward process that can be automated.

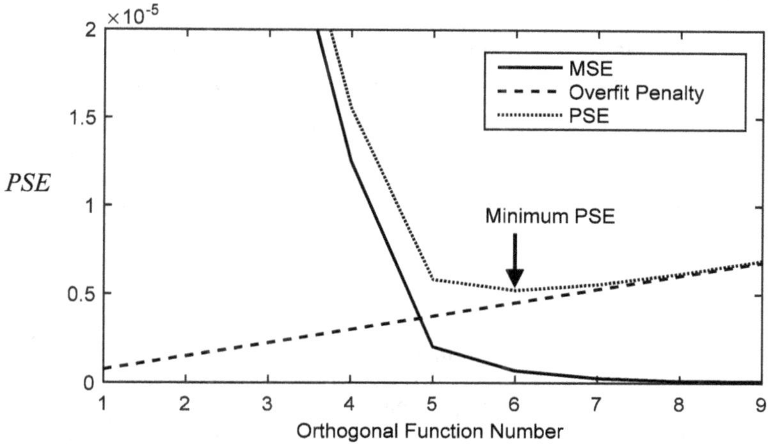

Figure 5.22 **Model structure determination using orthogonal functions and *PSE***

<u>*Generating Multivariate Orthogonal Functions*</u>

Multivariate orthogonal functions can be generated from ordinary multivariate functions in the explanatory variables using a Gram-Schmidt orthogonalization procedure. This approach is described in Refs. [5.25]-[5.26], and [5.21]-[5.22], which are the basis for the material presented here.

The process begins by choosing one of the ordinary multivariate functions as the first orthogonal function. Typically, a vector of ones (associated with the bias term in the model) is chosen as the first orthogonal function,

$$\boldsymbol{p}_0 = \boldsymbol{1} \tag{5.147}$$

In general, any function of the explanatory variables can be chosen as the first orthogonal function, without any change in the procedure. To generate the next orthogonal function, an ordinary multivariate function is made orthogonal to the preceding orthogonal function(s). Define the jth orthogonal function $\boldsymbol{p}_j$ as:

$$\boldsymbol{p}_j = \boldsymbol{\xi}_j - \sum_{k=0}^{j-1} \gamma_{kj}\, \boldsymbol{p}_k \qquad j = 1, 2, \ldots, n \tag{5.148}$$

where $\boldsymbol{\xi}_j$ is the jth ordinary multivariate function vector, and n is the total number of ordinary multivariate functions being orthogonalized, excluding the bias term, which is the selected first orthogonal function. For example, each

ξ_j could be an ordinary polynomial function of the explanatory variables, or a spline function of the explanatory variables, or some multiplicative combinations of these. The collection of all of the $\xi_j, j = 1, 2, \ldots, n$ and the bias term can be considered the pool of candidate regressors for the model. The γ_{kj} for $k = 0, 1, \ldots, j-1$ are scalars determined by multiplying both sides of Eq. (5.148) by $\boldsymbol{p}_k^T$, then invoking the mutual orthogonality of the $\boldsymbol{p}_k$, $k = 0, 1, \ldots, j$ [cf. Eq. (5.136)], and solving for γ_{kj}

$$\gamma_{kj} = \frac{\boldsymbol{p}_k^T \boldsymbol{\xi}_j}{\boldsymbol{p}_k^T \boldsymbol{p}_k} \qquad k = 0, 1, \ldots, j-1 \tag{5.149}$$

The same process can be implemented in sequence for each ordinary multivariate regressor $\xi_j, j = 1, 2, \ldots, n$. The total number of ordinary multivariate functions used as raw material for generating the multivariate orthogonal functions, including the bias term, is $n+1$. It can be seen from Eqs. (5.147)-(5.149) that each orthogonal function can be expressed exactly in terms of a linear expansion of the original multivariate functions. The orthogonal functions are generated sequentially by orthogonalizing the original multivariate functions with respect to the orthogonal functions already computed, so that each orthogonal function can be considered an orthogonalized version of an original multivariate function.

Typically, the ordinary multivariate functions to be orthogonalized will be multivariate polynomials in the explanatory variables, but that is not a requirement. These candidate regressors can in fact be arbitrary functions of the explanatory variables, such as splines or arbitrary nonlinear functions, including discontinuous functions.

The orthogonalization process described earlier can be repeated to generate orthogonal functions for multivariate polynomials of arbitrary order in the explanatory variables, subject only to limitations related to the information contained in the data. For example, it is not possible to generate an orthogonal function corresponding to α^2 if there are only two distinct values of angle of attack in the measured data. This is analogous to the requirement that at least three data points are needed to identify a quadratic model, which has three parameters. The same limit also applies to the orthogonal function corresponding to any cross term, such as $\alpha^2\delta_e$ for this example.

The multivariate orthogonal function generation normally starts by generating all possible ordinary multivariate polynomials in the explanatory variables, up to a selected maximum order. For example, when modeling the nondimensional body axis z force coefficient C_Z, the explanatory variables might be angle of attack α, nondimensional pitch rate $\hat{q} \equiv q\bar{c}/2V_o$, and stabilator deflection δ_e. If the selected maximum order is 3, then the ordinary multivariate polynomial modeling functions used as raw material for the orthogonalization process would include terms like $\alpha, \hat{q}, \delta_e, \alpha^2, \delta_e^3, \alpha\hat{q}^2, a\hat{q}\delta_e,$ etc. Note that considering any other candidate explanatory variables, such as sideslip angle β, can be done by simply including the sideslip angle among the group of explanatory variables. If it turns out that the sideslip angle is not needed to model C_Z, the model structure determination using orthogonal modeling functions will not select any orthogonal functions associated with sideslip angle. This occurs naturally and automatically in the course of the model structure determination process described earlier. Therefore, there is no harm in including explanatory variables that might not be important, except that additional computer memory and computation time will be required to identify the model structure, because additional multivariate orthogonal functions will be generated and sorted. Similarly, if the maximum order is chosen higher than necessary, and assuming the data information content can support generating the higher-order orthogonal functions, the only penalty would be the increased computer memory and computation time necessary for generating and sorting the additional orthogonal functions. The final identified model would be the same. Consequently, the choices that the analyst needs to make are easy and not critical to the quality of the final modeling results.

Conversion to Physically-Meaningful Multivariate Function Models

If the $\boldsymbol{p}_j$ vectors and the ξ_j vectors are arranged as columns of matrices $\boldsymbol{P}$ and $\boldsymbol{X}$, respectively, and the γ_{kj} are elements in the $(k+1)th$ row and $(j+1)th$ column of an upper triangular matrix $\boldsymbol{G}$ with ones on the diagonal,

$$\boldsymbol{G} = \begin{bmatrix} 1 & \gamma_{01} & \gamma_{02} & \cdots & \gamma_{0n} \\ 0 & 1 & \gamma_{12} & \cdots & \gamma_{1n} \\ 0 & 0 & 1 & \cdots & \gamma_{2n} \\ \vdots & \vdots & \vdots & \vdots & \vdots \\ 0 & 0 & 0 & \cdots & 1 \end{bmatrix} \tag{5.150}$$

Then

$$\boldsymbol{X} = \boldsymbol{P}\boldsymbol{G} \tag{5.151a}$$

which leads to

$$\boldsymbol{P} = \boldsymbol{X}\boldsymbol{G}^{-1} \tag{5.151b}$$

The columns of $\boldsymbol{G}^{-1}$ contain the coefficients for expansion of each column of $\boldsymbol{P}$ (i.e., each multivariate orthogonal function) in terms of an exact linear expansion in the original multivariate functions (candidate regressors) contained in the columns of $\boldsymbol{X}$. Equation (5.151b) can be used to express each multivariate orthogonal function in terms of the original multivariate functions. The manner in which the orthogonal functions are generated allows them to be decomposed without ambiguity into an expansion of the original multivariate functions, which have physical meaning.

For all flight test data sets and many wind tunnel data sets, the measured values of the explanatory variables are not uniformly spaced over an interval. This makes it impossible to use standard orthogonal polynomials described in many textbooks. The method described here is simple, and works regardless of the spacing of the explanatory variable measurements.

Because of the sequential nature of the method used to generate orthogonal functions from the original multivariate functions, the particular orthogonal functions generated depend on the order in which the original multivariate functions are used in the orthogonalization. This is most apparent for the case of two original multivariate functions that are highly correlated. The orthogonal function generated for the first of these original multivariate functions will have essentially all of the unique character that can be used for modeling the dependent variable. The orthogonal function for the second original multivariate function will then be close to the zero vector, because the previously-generated orthogonal function will be able to account for nearly all of the useful content of the second original multivariate function. In effect, nearly all of the useful content of the second correlated function will be removed during the orthogonalization with Eqs. (5.148) and (5.149).

The order in which the orthogonal functions are generated also affects the number of original multivariate functions that are required to represent each orthogonal function using Eq. (5.151b). In general, the later in the order that a particular orthogonal function is generated, the more original multivariate functions that will be required for its expansion. If a selected orthogonal function happens to be late in the orthogonalization order, the result is that final conversion from the selected orthogonal function back to original functions will involve small parts of original functions not needed in the final model. To avoid this, orthogonalization is done initially to determine the order of most effective to least effective orthogonal functions, then the associated original functions are re-ordered to match this order of importance, and the orthogonalization is repeated. This simple process results in the minimum number of model terms for the final model in terms of original functions.

After the model structure is determined using the multivariate orthogonal functions, the estimated model output is

$$\hat{y} = \boldsymbol{P}\,\hat{\boldsymbol{a}} \tag{5.152}$$

where the $\boldsymbol{P}$ matrix now includes only the orthogonal functions selected in the model structure determination. In general, the retained orthogonal functions will not necessarily be consecutive or the first in the orthogonalization sequence of Eq. (5.148). Each retained orthogonal function can be decomposed without error into an expansion of original functions, using the columns of $\boldsymbol{G}^{-1}$ in Eq. (5.151b) corresponding to the retained orthogonal functions. Common terms are combined using double precision arithmetic to arrive finally at a model using only original functions. Terms that contribute less than 0.1 percent of the final model root-mean-square magnitude are dropped.

The final model form is a sum of ordinary functions, with associated model parameter estimates and uncertainty. Because the conversion of orthogonal functions to ordinary functions is exact, only the uncertainty in the estimated parameter vector $\hat{\boldsymbol{a}}$, computed in the manner described earlier, is required to compute the uncertainty of the model parameters in the final model form.

When the original functions are multivariate polynomials in the explanatory variables, then the final model resembles selected terms from a multivariate Taylor series in the explanatory variables. Aircraft dynamics and control analyses are often conducted with the assumption of this form for the dependence of the nondimensional aerodynamic force and moment coefficients on explanatory variables such as angle of attack and sideslip angle. This final form of the model also allows straightforward analytic differentiation for partial derivatives of the response variable with respect to the explanatory variables, which is useful for linearization and dynamic analysis.

QR Decomposition

An alternative method for orthogonalizing the candidate modeling functions in the columns of $\boldsymbol{X}$ is the $\boldsymbol{QR}$ decomposition. An arbitrary matrix $\boldsymbol{X}$ can be decomposed into the product of a matrix of mutually orthonormal column vectors in a matrix $\boldsymbol{Q}$, and an upper triangular matrix $\boldsymbol{R}$,

$$\boldsymbol{X} = \boldsymbol{QR} \tag{5.153a}$$

where

$$\boldsymbol{Q}^T \boldsymbol{Q} = \boldsymbol{I} \tag{5.153b}$$

Several numerical analysis software packages, including MATLAB®, provide an algorithm for computing the $\boldsymbol{QR}$ decomposition. The $\boldsymbol{QR}$ decomposition is achieved with a sequential orthogonalization, analogous to the Gram-Schmidt orthogonalization. Comparing Eqs. (5.151a) and (5.153a) shows that the Gram-Schmidt orthogonalization is related to the $\boldsymbol{QR}$ decomposition, with a difference being that the columns of $\boldsymbol{P}$ are orthogonal, whereas the columns of $\boldsymbol{Q}$ are orthonormal. The $\boldsymbol{G}$ and $\boldsymbol{R}$ matrices are both upper triangular. For the $\boldsymbol{QR}$ decomposition, the scaling of the orthogonal functions is in the $\boldsymbol{R}$ matrix (because the vector lengths of the columns of $\boldsymbol{Q}$

are normalized to 1, cf. Eq. (5.153b)), whereas for the Gram-Schmidt orthogonalization, the scaling of the orthogonal functions is in the columns of $\boldsymbol{P}$ (because the diagonal elements of the $\boldsymbol{G}$ matrix are equal to 1 in Eq. (5.150)). The magnitude of each value on the diagonal of the $\boldsymbol{R}$ matrix is the vector norm of the corresponding column of the $\boldsymbol{P}$ matrix, so that each row of $\boldsymbol{G}$ can be obtained by dividing the corresponding row of $\boldsymbol{R}$ by the diagonal element in that row. This can be seen from the identity

$$\boldsymbol{X} = \boldsymbol{Q}\boldsymbol{R} = \boldsymbol{P}\boldsymbol{G} \tag{5.154a}$$

Columns of $\boldsymbol{Q}$ are the columns of $\boldsymbol{P}$ with their vector lengths normalized to 1,

$$\boldsymbol{q}_j = \boldsymbol{p}_j \Big/ \sqrt{\boldsymbol{p}_j^T \boldsymbol{p}_j} \qquad j = 0, 1, 2, \ldots, n \tag{5.154b}$$

Note that there is a sign ambiguity in the $\boldsymbol{QR}$ decomposition, because the elements in the $\boldsymbol{R}$ matrix can change sign, in which case there is a corresponding sign change in the columns of $\boldsymbol{Q}$, and consequently in Eq. (5.154b), i.e.,

$$\boldsymbol{X} = \boldsymbol{Q}\boldsymbol{R} = (-\boldsymbol{Q})(-\boldsymbol{R}) \tag{5.155}$$

However, this issue does not affect the orthogonality properties.

5.5 Data Collinearity

As noted earlier, polynomials in the explanatory variables are often used to model the functional dependence of the response variables on the explanatory variables. One difficulty that can arise with this approach is that polynomial terms can be highly correlated, because of similar values in restricted ranges. For example, linear and cubic functions near zero have a similar form. In addition, it often happens in aircraft flight testing that some explanatory variables are highly correlated because of how the aircraft normally flies, or because of high gain feedback control. For example, the pitch rate and time derivative of the angle of attack are often quite similar when the dominant aircraft response is longitudinal short period motion. For aircraft with active pitch rate feedback to stabilize the aircraft in pitch, or to improve the short period damping, movements of the longitudinal pitch control surface, typically the elevator, are often highly correlated with pitch rate.

Correlation among regressors in a proposed model structure causes problems in parameter estimation, simply because the parameter estimation algorithm cannot assign accurate parameter values for the regressors in the model when some of the regressors are similar. Stated another way, the parameter estimator cannot determine which model term should be used to model a particular variation in the response variable if there are several model terms that could fill the same role nearly as well. This problem manifests itself mathematically in poor conditioning of the $X^T X$ matrix, which affects the parameter estimates and error bounds through inaccuracies in the $X^T X$ matrix inversion [cf. Eqs. (5.10) and (5.12)].

This problem is addressed automatically in the multivariate orthogonal function approach to model structure determination, because any correlated regressors after the first one will be transformed to a zero (or near zero) vector during the orthogonalization, then dropped in the model structure determination process, because of its consequent low impact on reducing mean squared fit error. Regressor correlation problems are also apparent when using stepwise regression, because when candidate regressors are highly correlated, including any one of them makes the others appear insignificant, and the model can be made almost equally good by including any of the correlated regressors.

If a regressor in a linear regression model such as Eq. (5.6) is equal to a linear combination of one or more of the other regressors, then all the involved regressors are said to be linearly dependent. This makes the parameter estimation problem ill-conditioned. In practice, regressors might be almost linearly dependent, but not exactly so. But the fundamental difficulty remains the same, with its severity becoming worse as regressors get closer to being perfectly correlated, i.e., as $\left|r_{jk}\right|$, $j,k \in (1,2,\ldots,n)$, from Eq. (5.120) approaches 1. Any situation where regressors are correlated at a high enough level to cause problems in the parameter estimation is called data collinearity. Operationally, when data collinearity is present, the parameter estimation routines will produce inaccurate parameter estimates with large variances, or in severe cases the parameter estimation routine may fail.

Collinearity is a data problem, and is not specific to any particular data analysis or modeling method. There are at least three different sources of collinearity: 1) design of the experiment, 2) constraints, and 3) model specification.

If the experiment is conducted in a manner such that the data for two or more of the regressors are mostly changed proportionally, then collinearity can occur. The problem can also arise when the changes in the regressors are not sufficient to excite the response variable above the noise level. The key point is that the experiment must involve significant, independent changes in any regressor for which an accurate, independent estimate of influence (i.e., parameter estimate) is sought (see Chapter 9).

Constraints in the data could be caused by an inherent property or mode of operation for the system being tested. For example, an aircraft control system can deflect various control surfaces in proportion to measured body-axis angular rates, for stability augmentation, or can deflect several control surfaces proportionally for improved control authority. This causes near-linear dependency among the regressors, and data collinearity occurs. The regressors are never perfectly correlated for the aircraft problem, because the control system has some delay between measuring the angular rates and implementing the control surface deflections, and there is also some noise on all the measured signals, which is different for each signal. However, the absolute value of pairwise regressor correlations can exceed 0.9 for highly augmented aircraft such as jet fighters.

Collinearity can also be caused by specifying a model with terms that are not warranted. For example, if ξ_1 is small, then the regressors ξ_1^2 and $\xi_1\xi_2$ might have very little influence on the response variable, because these regressors will be close to zero. This shows up as a data collinearity, where ξ_1^2 and $\xi_1\xi_2$ are roughly the same regressor, namely a zero vector.

The remainder of this section describes methods for assessing data collinearity and some approaches for getting good parameter estimation results when data collinearity exists. For this analysis, it is convenient to use standardized regressors, described next.

5.5.1 Standardized Regressors

For assessment of data collinearity , it is advantageous to work with scaled versions of the regressors and response variable. There are several scaling techniques; however, only the one known as unit length scaling will be introduced here. The equations for the transformed regressors and response variable are the same as those used to compute correlation coefficients earlier,

$$\xi_j^*(i) = \frac{\xi_j(i) - \bar{\xi}_j}{\sqrt{S_{jj}}} \qquad j = 1,2,\ldots,n \quad , \quad i = 1,2,\ldots,N \tag{5.156a}$$

$$z^*(i) = \frac{z(i) - \bar{z}}{\sqrt{S_{zz}}} \qquad i = 1,2,\ldots,N \tag{5.156b}$$

where

$$S_{jj} = \sum_{i=1}^{N} \left[\xi_j(i) - \bar{\xi}_j \right]^2 \tag{5.156c}$$

is the centered sum of squares for the regressor ξ_j, and

$$S_{zz} = \sum_{i=1}^{N} \left[z(i) - \bar{z} \right]^2 \tag{5.157}$$

is the centered sum of squares for the response z. With this scaling, each new regressor ξ_j^* has a mean value equal to zero,

$$\bar{\xi}_j^* = 0 \qquad j = 1,2,\ldots,n \tag{5.158a}$$

and unit length

$$\| \xi_j^* \| = \sqrt{\sum_{i=1}^{N} \left[\xi_j^*(i) \right]^2} = 1 \qquad j = 1,2,\ldots,n \tag{5.158b}$$

The regression model with scaled regressors takes the form

$$z^*(i) = \theta_1^* \xi_1^*(i) + \theta_2^* \xi_2^*(i) + \ldots + \theta_n^* \xi_n^*(i) + v(i) \qquad i = 1,2,\ldots,N \tag{5.159}$$

where the constant bias term is omitted because the regressors and the response variable are now centered, with their mean values removed. Using vector and matrix notation as before, the least-squares estimator is

$$\hat{\boldsymbol{\theta}}^* = \left(\boldsymbol{X}^{*T} \boldsymbol{X}^* \right)^{-1} \boldsymbol{X}^{*T} \boldsymbol{z}^* \tag{5.160a}$$

with covariance matrix

$$Cov\left(\hat{\boldsymbol{\theta}}^*\right) = \left(\sigma^*\right)^2 \left(\boldsymbol{X}^{*T} \boldsymbol{X}^* \right)^{-1} \tag{5.160b}$$

where $\boldsymbol{X}^* = \left[\xi_1^* \;\; \xi_2^* \;\; \ldots \;\; \xi_n^* \right]$ and $\left(\sigma^*\right)^2$ is the error variance for z^*. The $\boldsymbol{X}^{*T} \boldsymbol{X}^*$ matrix then takes the form of a correlation matrix,

$$\boldsymbol{X}^{*T} \boldsymbol{X}^* = \begin{bmatrix} 1 & r_{12} & r_{13} & \cdots & r_{1n} \\ r_{21} & 1 & r_{23} & \cdots & r_{2n} \\ \vdots & & & & \vdots \\ r_{n1} & r_{n2} & r_{n3} & \cdots & 1 \end{bmatrix} \tag{5.161}$$

where

$$r_{jk} = \frac{\sum_{i=1}^{N} \left[\xi_j(i) - \bar{\xi}_j \right] \left[\xi_k(i) - \bar{\xi}_k \right]}{\sqrt{S_{jj} \, S_{kk}}} \qquad j,k = 1,2,\ldots,n \tag{5.162}$$

is the simple pairwise correlation between ξ_j and ξ_k [cf. Eq. (5.120)]. Similarly,

$$\boldsymbol{X}^{*T} \boldsymbol{z}^* = \left[r_{1z} \;\; r_{2z} \;\; \cdots \;\; r_{nz} \right]^T \tag{5.163}$$

where

$$r_{jz} = \frac{\sum_{i=1}^{N}\left[\xi_j(i)-\bar{\xi}_j\right]\left[z(i)-\bar{z}\right]}{\sqrt{S_{jj}\,S_{zz}}} \qquad j=1,2,\ldots,n \qquad (5.164)$$

is the simple correlation between ξ_j^* and z^*. The parameters in the vector $\boldsymbol{\theta}^*$ are usually called standardized parameters. They are dimensionless and are related to the original parameters by

$$\theta_j = \theta_j^* \sqrt{\frac{S_{zz}}{S_{jj}}} \qquad j=1,2,\ldots,n \qquad (5.165)$$

$$\hat{\theta}_0 = \bar{z} - \sum_{j=1}^{n} \hat{\theta}_j^* \bar{\xi}_j \qquad (5.166)$$

The matrix $\boldsymbol{X}^{*T}\boldsymbol{X}^*$ is the $n \times n$ matrix of pairwise regressor correlation coefficients. If $\boldsymbol{\xi}_j^{*T}\boldsymbol{\xi}_k^* = 0,\ j \neq k$, the regressors are orthogonal, and the $\boldsymbol{X}^{*T}\boldsymbol{X}^*$ matrix is a diagonal matrix equal to the identity matrix. The vectors $\xi_1^*,\ \xi_2^*,\ldots\xi_n^*$ are linearly dependent if there is a set of constants c_j, not all zero, such that

$$\sum_{j=1}^{n} c_j \xi_j^* = \mathbf{0} \qquad (5.167)$$

In that case, the rank of $\boldsymbol{X}^{*T}\boldsymbol{X}^*$ is less than n, $det\left(\boldsymbol{X}^{*T}\boldsymbol{X}^*\right)=0$, and $\left(\boldsymbol{X}^{*T}\boldsymbol{X}^*\right)^{-1}$ does not exist.

In many applications of linear regression, Eq. (5.167) is only approximately true. This indicates near-linear dependency among the columns of $\boldsymbol{X}^*$, and the problem of data collinearity exists. In such a case, the $\boldsymbol{X}^{*T}\boldsymbol{X}^*$ matrix is called ill-conditioned. This causes computational problems, and reduces the accuracy of the parameter estimates, because of the use of the matrix $\left(\boldsymbol{X}^{*T}\boldsymbol{X}^*\right)^{-1}$ in both the parameter estimation and estimated parameter covariance calculations [cf. Eq. (5.160)].

5.5.2 Detection and Assessment of Data Collinearity

Many procedures have been developed to detect collinearity. Some of these are discussed in Refs. [5.3] and [5.28]. In this section, three methods will be considered: 1) examination of the regressor correlation matrix, 2) eigensystem analysis and singular value decomposition, and 3) parameter variance decomposition. The use of each method will be demonstrated in an example.

Examination of the Regressor Correlation Matrix

The simplest and most straightforward procedure for assessing collinearity is to examine the regressor correlation matrix. A high correlation coefficient between two regressors can point to a possible collinearity problem. A good practical rule of thumb is that collinearity might cause problems in the parameter estimation for any pairwise regressor correlation with absolute value greater than 0.9. However, this rule of thumb is only approximate. It has been found in practice that the level of correlation associated with collinearity decreases with increasing noise levels on the regressors, and vice versa. In effect, regressors can be distinguished better when noise levels are lower and the noise obscures smaller differences.

The absence of high pairwise correlation cannot be viewed as a guarantee of no collinearity problems. The correlation matrix is unable to reveal the presence of correlation among more than two regressors. This was demonstrated by Goldberg [5.11].

The aforementioned shortcoming in using the regressor correlation matrix $\boldsymbol{X}^{*T}\boldsymbol{X}^{*}$ as a diagnostic measure of collinearity limits the usefulness of its inverse in a similar way. The diagonal elements of $\left(\boldsymbol{X}^{*T}\boldsymbol{X}^{*}\right)^{-1}$ are often called the variance inflation factors, VIF, which can be expressed as

$$VIF_j = \frac{1}{1-R_j^2} \qquad j = 1, 2, \ldots, n \tag{5.168}$$

for the jth regressor, where R_j^2 is the coefficient of determination for the regressor ξ_j modeled as a linear function of the other regressors in the model (see Ref. [5.29]). The quantity VIF_j is the jth diagonal element of $\left(\boldsymbol{X}^{*T}\boldsymbol{X}^{*}\right)^{-1}$. The name "variance inflation factor" comes from a relationship with the jth parameter variance. As shown in Ref. [5.29], this relationship is

$$Var\left(\theta_j\right) = \frac{\sigma^2}{\xi_j^{*T}\xi_j^{*}} VIF_j \qquad j = 1, 2, \ldots, n \tag{5.169}$$

The diagnostic value of VIF_j is evident from Eq. (5.168). For data with no collinearity at all (i.e., orthogonal regressors), the variance inflation factors VIF_j are equal to 1, since $R_j^2 = 0$ for every *j*. Large values of VIF_j indicate an R_j^2 near unity, which points to collinearity. The weakness of this diagnostic measure is its inability to distinguish among several coexisting near-dependencies, and the lack of a specific value for VIF_j that can be considered an acceptable upper bound for low data collinearity. According to Ref. [5.2], a VIF_j larger than 10 indicates a serious problem with collinearity.

Eigensystem Analysis and Singular Value Decomposition

The matrix $\boldsymbol{X}^T\boldsymbol{X}$ can be decomposed as follows:

$$\boldsymbol{X}^T\boldsymbol{X} = \boldsymbol{T}\,\boldsymbol{\Lambda}\,\boldsymbol{T}^T \tag{5.170}$$

where $\boldsymbol{\Lambda}$ is an $n \times n$ diagonal matrix whose diagonal elements are the eigenvalues λ_j, $j = 1,2,\ldots,n$, of $\boldsymbol{X}^T\boldsymbol{X}$, and $\boldsymbol{T}$ is an $n \times n$ orthonormal matrix whose columns are the eigenvectors of $\boldsymbol{X}^T\boldsymbol{X}$. The $N \times n$ matrix $\boldsymbol{X}$ considered here can be scaled and/or centered. For diagnostic purposes, Ref. [5.28] recommends scaling the columns of $\boldsymbol{X}$ to unit length. The matrix $\boldsymbol{X}$ should not be centered if the role of the intercept term in near-linear dependencies is being investigated.

One or more small eigenvalues imply that there are near-linear dependencies among the columns of $\boldsymbol{X}$. The severity of these dependencies is indicated by how small the eigenvalues are, compared with the maximum eigenvalue. Near-linear dependencies among the columns of the regressor matrix $\boldsymbol{X}$ can be measured by the condition index

$$\frac{\lambda_{max}}{\lambda_j} \geq 1 \qquad j = 1,2,\ldots,n \tag{5.171}$$

The largest value $\lambda_{max}/\lambda_{min}$ is known as the condition number of the $\boldsymbol{X}^T\boldsymbol{X}$ matrix. According to Ref. [5.28], any condition index from Eq. (5.171) in the range of 100 to 1000 indicates moderate to strong collinearity. However, experience from the analysis of flight data using linear regression has revealed that in some cases the parameter estimates could be affected by data collinearity even if the condition number of the $\boldsymbol{X}^T\boldsymbol{X}$ matrix is less than 100.

Reference [5.28] recommends an approach using singular-value decomposition for diagnosing collinearity. This approach is based on a stable numerical decomposition of the matrix $\boldsymbol{X}$ as

$$\boldsymbol{X} = \boldsymbol{U}\,\boldsymbol{D}\,\boldsymbol{T}^T \tag{5.172}$$

where $\boldsymbol{U}$ is an $N \times n$ matrix with mutually orthonormal columns, and $\boldsymbol{T}$ is an $n \times n$ matrix with mutually orthonormal columns, so that $\boldsymbol{U}^T\boldsymbol{U} = \boldsymbol{T}^T\boldsymbol{T} = \boldsymbol{I}$. The matrix $\boldsymbol{D}$ is an $n \times n$ diagonal matrix with nonnegative elements on the diagonal, μ_j, $j = 1,2,\ldots,n$, which are called the singular values of $\boldsymbol{X}$. The singular-value decomposition, *SVD*, is closely related to the concept of eigenvalues and eigenvectors, since from Eqs. (5.170) and (5.172),

$$\boldsymbol{X}^T\boldsymbol{X} = \boldsymbol{T}\,\boldsymbol{D}^2\,\boldsymbol{T}^T = \boldsymbol{T}\,\boldsymbol{\Lambda}\,\boldsymbol{T}^T \tag{5.173}$$

The diagonal elements of $\boldsymbol{D}^2$ are therefore the eigenvalues of $\boldsymbol{X}^T\boldsymbol{X}$ and the columns of $\boldsymbol{T}$ in Eq. (5.172) are the eigenvectors of $\boldsymbol{X}^T\boldsymbol{X}$ associated with the n nonzero eigenvalues. The severity of ill-conditioning in the $\boldsymbol{X}^T\boldsymbol{X}$ matrix is indicated by how small the singular values are relative to the maximum singular value. Therefore, a condition index for the matrix $\boldsymbol{X}$ is proposed as

$$\frac{\mu_{max}}{\mu_j} = \sqrt{\frac{\lambda_{max}}{\lambda_j}} \qquad j = 1,2,\ldots,n \tag{5.174}$$

The largest value μ_{max}/μ_{min} is the condition number of the $\boldsymbol{X}$ matrix.

The *SVD* of the matrix $\boldsymbol{X}$ provides similar information to that given by the eigensystem of $\boldsymbol{X}^T\boldsymbol{X}$. However, the *SVD* is generally preferred, because of greater numerical stability in its computing algorithm, compared to that for the eigensystem analysis of $\boldsymbol{X}^T\boldsymbol{X}$.

Parameter Variance Decomposition

In general, pairwise correlation coefficients between regressors cannot be used to diagnose collinearity among more than two regressors. Collinearity among more than two regressors is fairly common in practice, and can have a significant impact on modeling results.

An approach for detecting multiple collinearity using parameter variance decomposition was proposed in Ref. [5.28]. It is based on the covariance matrix of parameter estimates $\hat{\boldsymbol{\theta}}$, which can be formulated as

$$Cov\left(\hat{\boldsymbol{\theta}}\right) = \sigma^2\left(\boldsymbol{X}^T\boldsymbol{X}\right)^{-1} = \sigma^2\,\boldsymbol{T}\boldsymbol{\Lambda}^{-1}\boldsymbol{T}^T \tag{5.175}$$

where the last equality uses Eq. (5.173) and the fact that $\boldsymbol{T}$ is an orthonormal matrix. The variance of each parameter is then equal to

$$Var\left(\hat{\theta}_k\right) = \sigma^2 \sum_{j=1}^{n} \frac{t_{kj}^2}{\lambda_j} = \sigma^2 \sum_{j=1}^{n} \frac{t_{kj}^2}{\mu_j^2} \tag{5.176}$$

where t_{kj} is the *k*th element of the *j*th eigenvector associated with λ_j, and the *j*th eigenvector is the *j*th column of $\boldsymbol{T}$. The expression (5.176) shows that the variance for each estimated parameter can be decomposed into a sum of components, each corresponding to one of the n eigenvalues λ_j, $j = 1, 2, \ldots, n$. For each term in the summation of Eq. (5.176), the eigenvalue appears in the denominator, so one or more small eigenvalues can substantially increase the variance of $\hat{\theta}_k$. A high proportion of the variance for two or more estimated parameters from the same small eigenvalue can provide evidence that the corresponding near-dependency is causing problems. Introducing

$$\phi_{kj} \equiv \frac{t_{kj}^2}{\lambda_j} \quad \text{and} \quad \phi_k \equiv \sum_{j=1}^{n} \frac{t_{kj}^2}{\lambda_j} = \sum_{j=1}^{n} \phi_{kj} \tag{5.177}$$

The k, j variance proportion π_{kj} is defined as the proportion of the variance of the *k*th estimated parameter associated with the *j*th component of the decomposition in Eq. (5.176),

$$\pi_{kj} \equiv \frac{\phi_{kj}}{\phi_k} \qquad j = 1, 2, \ldots, n \tag{5.178}$$

Because two or more regressors are required to create near-dependency, two or more variances will be adversely affected by high variance proportions associated with each small singular value. Variance proportions greater than 0.5 are recommended guidance for possible collinearity problems. High values of the variance proportions π_{kj}, $j = 1, 2, \ldots, n$ indicate which of the regressors are involved in a multiple collinearity.

The variance proportions complement the other diagnostics in assessing the effect of data collinearity on the estimated parameter variances. Diagnosis for data collinearity should involve the condition indices to assess the severity of a particular dependency, the variance proportions to indicate which regressors are involved in the dependency and to what extent, and the *VIF* to aid in determining the damage to individual parameter estimates and variances.

5.5.3 Adverse Effects of Data Collinearity

The presence of collinearity in the data for a linear regression problem results in various unwanted properties of the least-squares parameter estimates and variances. The increased variance in the estimates is apparent from Eqs. (5.168), (5.169), and (5.176). Collinearity also tends to produce least-squares estimates that are large in absolute value. This can be demonstrated using Eq. (4.11) for the mean squared error in the vector $\hat{\boldsymbol{\theta}}$ with no bias error from model structure misspecification,

$$MSE\left(\hat{\boldsymbol{\theta}}\right) \equiv E\left[\left(\hat{\boldsymbol{\theta}}-\boldsymbol{\theta}\right)^T\left(\hat{\boldsymbol{\theta}}-\boldsymbol{\theta}\right)\right] \tag{5.179}$$

which can also be expressed as

$$\begin{aligned} MSE\left(\hat{\boldsymbol{\theta}}\right) &= \sum_{j=1}^{n} E\left[\left(\hat{\theta}_j-\theta_j\right)^2\right] \\ &= \sigma^2\, Tr\left[\left(\boldsymbol{X}^T\boldsymbol{X}\right)^{-1}\right] = \sigma^2\, Tr\left[\left(\boldsymbol{T}\boldsymbol{\Lambda}\boldsymbol{T}^T\right)^{-1}\right] \\ &= \sigma^2 \sum_{j=1}^{n} \lambda_j^{-1} \end{aligned} \tag{5.180}$$

Eq. (5.180) shows that the distance from $\hat{\boldsymbol{\theta}}$ to $\boldsymbol{\theta}$ may be large if at least one of the λ_j is small.

Reference [5.28] addresses the problem of decreased accuracy of parameters estimated from collinear data, in connection with insufficient information in the data. For this case, the linear regression model can be formulated as

$$\begin{aligned} \boldsymbol{z} &= \boldsymbol{X}\boldsymbol{\theta} + \boldsymbol{v} \\ &= \boldsymbol{X}_1\boldsymbol{\theta}_1 + \boldsymbol{X}_2\boldsymbol{\theta}_2 + \boldsymbol{v} \end{aligned} \tag{5.181}$$

where $\boldsymbol{X}$ has been partitioned into two matrices $\boldsymbol{X}_1$ and $\boldsymbol{X}_2$, with dimensions $N \times n_1$ and $N \times n_2$, respectively. The matrix $\boldsymbol{X}_2$ contains regressors involved in n_2 near-dependencies with regressors in $\boldsymbol{X}_1$. Modeling the regressors in $\boldsymbol{X}_2$ as a linear combination of the regressors in $\boldsymbol{X}_1$ gives

$$\boldsymbol{X}_2 = \boldsymbol{X}_1\boldsymbol{C} + \boldsymbol{V}_2 \tag{5.182}$$

where $\boldsymbol{C}$ is a matrix of parameters with each column containing the parameters for the model of the corresponding column of $\boldsymbol{X}_2$ in terms of $\boldsymbol{X}_1$, and $\boldsymbol{V}_2$ is the associated matrix of residuals. Substituting Eq. (5.182) into (5.181) results in the regression model

$$z = \boldsymbol{X}_1\left(\boldsymbol{\theta}_1 + \boldsymbol{C}\boldsymbol{\theta}_2\right) + \boldsymbol{V}_2\boldsymbol{\theta}_2 + \boldsymbol{\nu} \tag{5.183}$$

Since $\boldsymbol{X}_1^T\boldsymbol{V}_2 = \boldsymbol{V}_2^T\boldsymbol{X}_1 = \boldsymbol{0}$ from the normal equations associated with Eq. (5.182), it is possible to estimate $\boldsymbol{\theta}_1 + \boldsymbol{C}\boldsymbol{\theta}_2$ and $\boldsymbol{\theta}_2$ using separate regressions of z as a linear function of $\boldsymbol{X}_1$ and $\boldsymbol{V}_2$, respectively. Further, because of the near-linear dependency between $\boldsymbol{X}_1$ and $\boldsymbol{X}_2$, $det\left(\boldsymbol{V}_2^T\boldsymbol{V}_2\right) \ll det\left(\boldsymbol{X}_1^T\boldsymbol{X}_1\right)$, which means that most of the information in the data will be applied to the estimation of $\boldsymbol{\theta}_1 + \boldsymbol{C}\boldsymbol{\theta}_2$, and only a small amount will be available for the estimation of $\boldsymbol{\theta}_2$. Thus, collinearity can create a situation where there is insufficient information in the data for highly accurate estimation of all model parameters.

5.5.4 Least Squares with Prior Information

As discussed previously, the ordinary least-squares technique provides an unbiased linear estimator which has minimum variance in the class of unbiased linear estimators. However, there is no guarantee that the variance will be small. It was seen in the last section that the application of ordinary least squares to data with collinearity problems can result in large estimated parameters with large variances. Figure 5.14 showed that biased parameter estimates can have smaller variance than unbiased parameter estimates, so that for small bias errors, the mean squared error for biased parameter estimates can be smaller than the variance of parameter estimates from an unbiased estimator. This possibility has inspired the development of various biased parameter estimation techniques. Some of these techniques are reviewed in Ref. [5.14]. In the following, only one biased parameter estimation technique, known as the mixed estimator, will be described and applied to experimental data. This technique incorporates prior information about the model parameters to improve the accuracy of the parameter estimates.

The mixed estimator is described in Ref. [5.10] as a Bayes-like technique that augments the measured data with prior information about the model parameters. For the linear regression model

$$\begin{gathered} z = \boldsymbol{X}\boldsymbol{\theta} + \boldsymbol{\nu} \\ E(\boldsymbol{\nu}) = \boldsymbol{0} \quad \text{and} \quad E\left(\boldsymbol{\nu}\boldsymbol{\nu}^T\right) = \sigma^2 \boldsymbol{I} \end{gathered} \tag{5.184}$$

it is assumed that $m \le n_p$ prior constraints on the elements of $\boldsymbol{\theta}$ are available.

These constraints are formulated as

$$d = B\theta + \varsigma \tag{5.185}$$

In Eq. (5.185), B is an $m \times n_p$ matrix with known constant elements, d is an $m \times 1$ vector of values that can be specified, and ς is a random vector with

$$E(\varsigma) = 0 \qquad E\left(\nu\varsigma^T\right) = 0 \qquad E\left(\varsigma\varsigma^T\right) = V \tag{5.186}$$

where V is a known matrix.

Combining Eqs. (5.184)-(5.186), the mixed model is given as

$$\begin{bmatrix} z \\ d \end{bmatrix} = \begin{bmatrix} X \\ B \end{bmatrix}\theta + \begin{bmatrix} \nu \\ \varsigma \end{bmatrix}$$
$$E\left\{\begin{bmatrix} \nu \\ \varsigma \end{bmatrix}\right\} = 0 \quad \text{and} \quad E\left\{\begin{bmatrix} \nu \\ \varsigma \end{bmatrix}\begin{bmatrix} \nu^T & \varsigma^T \end{bmatrix}\right\} = \begin{bmatrix} \sigma^2 & 0 \\ 0 & V \end{bmatrix} \tag{5.187}$$

The least-squares cost function to be minimized is

$$J_{ME} = \frac{1}{2}\begin{bmatrix} (z - X\theta) \\ (d - B\theta) \end{bmatrix}^T \begin{bmatrix} \sigma^2 & 0 \\ 0 & V \end{bmatrix}^{-1} \begin{bmatrix} (z - X\theta) \\ (d - B\theta) \end{bmatrix}$$

$$\begin{aligned} J_{ME} &= \frac{1}{2\sigma^2}(z - X\theta)^T (z - X\theta) + \frac{1}{2}(d - B\theta)^T V^{-1} (d - B\theta) \\ &= \frac{1}{2\sigma^2}(z - X\theta)^T (z - X\theta) \\ &\quad + \frac{1}{2}\left[\theta - B^T\left(BB^T\right)^{-1} d\right]^T B^T V^{-1} B \left[\theta - B^T\left(BB^T\right)^{-1} d\right] \end{aligned} \tag{5.188}$$

which is the form of the cost function for the Bayesian estimator, cf. Eq. (4.28) with $\theta_p = B^T\left(BB^T\right)^{-1} d$ and $\Sigma_p^{-1} = B^T V^{-1} B$. Applying ordinary least-squares parameter estimation to the cost function in (5.188) results in the mixed estimator

$$\hat{\theta}_{ME} = \left(X^T X/\sigma^2 + B^T V^{-1} B\right)^{-1} \left(X^T z/\sigma^2 + B^T V^{-1} d\right) \tag{5.189}$$

with covariance matrix

$$Cov\left(\hat{\boldsymbol{\theta}}_{ME}\right)=\left(\boldsymbol{X}^T\boldsymbol{X}/\sigma^2+\boldsymbol{B}^T V^{-1}\boldsymbol{B}\right)^{-1} \tag{5.190}$$

where σ^2 can be estimated using data smoothing methods from Chapter 11 or using the estimation method described previously, applied to the regression problem without prior information.

Introducing the augmented variables,

$$\boldsymbol{z}_a \equiv \begin{bmatrix} \boldsymbol{z} \\ \boldsymbol{d} \end{bmatrix} \qquad \boldsymbol{X}_a \equiv \begin{bmatrix} \boldsymbol{X} \\ \boldsymbol{B} \end{bmatrix} \qquad \boldsymbol{v}_a \equiv \begin{bmatrix} \boldsymbol{v} \\ \boldsymbol{\varsigma} \end{bmatrix} \tag{5.191}$$

the mixed model can also be written as

$$\begin{gathered} \boldsymbol{z}_a = \boldsymbol{X}_a\boldsymbol{\theta}+\boldsymbol{v}_a \\ E\left(\boldsymbol{v}_a\right)=\boldsymbol{0} \quad \text{and} \quad E\left(\boldsymbol{v}_a\boldsymbol{v}_a^T\right)=\begin{bmatrix} \sigma^2 & \boldsymbol{0} \\ \boldsymbol{0} & V \end{bmatrix}\equiv \boldsymbol{V}_a \end{gathered} \tag{5.192}$$

The model given in (5.192) is a generalized least-squares problem, with the solution

$$\hat{\boldsymbol{\theta}}_{ME}=\left(\boldsymbol{X}_a^T\boldsymbol{V}_a^{-1}\boldsymbol{X}_a\right)^{-1}\boldsymbol{X}_a^T\boldsymbol{V}_a^{-1}\boldsymbol{z}_a \tag{5.193}$$

and covariance matrix

$$Cov\left(\hat{\boldsymbol{\theta}}_{ME}\right)=\left(\boldsymbol{X}_a^T\boldsymbol{V}_a^{-1}\boldsymbol{X}_a\right)^{-1} \tag{5.194}$$

The quantity $\hat{\boldsymbol{\theta}}_{ME}$ computed from Eq. (5.189) or (5.193) is an unbiased linear estimator of $\boldsymbol{\theta}$, because the constraint equations (5.185) were assumed to hold exactly.

In practical applications of mixed estimation, the *a priori* constraints on the parameters are usually not known exactly. In this case, Eq. (5.185) is changed to

$$\boldsymbol{d}=\boldsymbol{B}\boldsymbol{\theta}+\boldsymbol{b}+\boldsymbol{\varsigma} \tag{5.195}$$

where $\boldsymbol{b}$ is an unknown vector. The mixed estimate associated with the condition (5.195) will be denoted $\tilde{\boldsymbol{\theta}}_{ME}$. The expected value of $\tilde{\boldsymbol{\theta}}_{ME}$ is obtained by substituting from Eqs. (5.184) and (5.195) into (5.189),

$$E\left(\tilde{\boldsymbol{\theta}}_{ME}\right)=\left(\boldsymbol{X}^T\boldsymbol{X}/\sigma^2+\boldsymbol{B}^T\boldsymbol{V}^{-1}\boldsymbol{B}\right)^{-1}E\Big[\left(\boldsymbol{X}^T\boldsymbol{X}\boldsymbol{\theta}+\boldsymbol{X}^T\boldsymbol{\nu}\right)/\sigma^2$$

$$+\boldsymbol{B}^T\boldsymbol{V}^{-1}\boldsymbol{B}\boldsymbol{\theta}+\boldsymbol{B}^T\boldsymbol{V}^{-1}\boldsymbol{b}+\boldsymbol{B}^T\boldsymbol{V}^{-1}\boldsymbol{\varsigma}\Big]$$

$$=\left(\boldsymbol{X}^T\boldsymbol{X}/\sigma^2+\boldsymbol{B}^T\boldsymbol{V}^{-1}\boldsymbol{B}\right)^{-1}\Big[\left(\boldsymbol{X}^T\boldsymbol{X}/\sigma^2+\boldsymbol{B}^T\boldsymbol{V}^{-1}\boldsymbol{B}\right)E(\boldsymbol{\theta})$$

$$+\boldsymbol{X}^T E(\boldsymbol{\nu})/\sigma^2+\boldsymbol{B}^T\boldsymbol{V}^{-1}\boldsymbol{b}+\boldsymbol{B}^T\boldsymbol{V}^{-1}E(\boldsymbol{\varsigma})\Big]$$

$$=\boldsymbol{\theta}+\boldsymbol{M}^{-1}\boldsymbol{B}^T\boldsymbol{V}^{-1}\boldsymbol{b} \tag{5.196}$$

where

$$\boldsymbol{M}\equiv\boldsymbol{X}^T\boldsymbol{X}/\sigma^2+\boldsymbol{B}^T\boldsymbol{V}^{-1}\boldsymbol{B} \tag{5.197}$$

From Eq. (5.196), the estimate $\tilde{\boldsymbol{\theta}}_{ME}$ is biased by the quantity $\boldsymbol{M}^{-1}\boldsymbol{B}^T\boldsymbol{V}^{-1}\boldsymbol{b}$. The covariance matrix for the mixed estimator $\tilde{\boldsymbol{\theta}}_{ME}$ is

$$Cov\left(\tilde{\boldsymbol{\theta}}_{ME}\right)=\left(\boldsymbol{X}^T\boldsymbol{X}/\sigma^2+\boldsymbol{B}^T\boldsymbol{V}^{-1}\boldsymbol{B}\right)^{-1}=\boldsymbol{M}^{-1} \tag{5.198}$$

The difference between the covariance of the ordinary least-squares estimator and the mixed estimator is

$$Cov\left(\hat{\boldsymbol{\theta}}\right)-Cov\left(\tilde{\boldsymbol{\theta}}_{ME}\right)=\left(\boldsymbol{X}^T\boldsymbol{X}/\sigma^2\right)^{-1}-\left(\boldsymbol{X}^T\boldsymbol{X}/\sigma^2+\boldsymbol{B}^T\boldsymbol{V}^{-1}\boldsymbol{B}\right)^{-1} \tag{5.199}$$

As shown in Ref. [5.10], the right side of Eq. (5.199) is a nonnegative definite matrix. Adding prior information to ordinary least-squares regression therefore results in a reduction in the estimated parameter variances, compared to the ordinary least squares. Considering the mean squared error of the estimated parameter vector as a measure of the accuracy of an estimator, then for the mixed estimator,

$$MSE\left(\tilde{\boldsymbol{\theta}}_{ME}\right)=Tr\left(\boldsymbol{M}^{-1}\right)+\left(\boldsymbol{\theta}-\boldsymbol{M}^{-1}\boldsymbol{B}^T\boldsymbol{V}^{-1}\boldsymbol{b}\right)^T\left(\boldsymbol{\theta}-\boldsymbol{M}^{-1}\boldsymbol{B}^T\boldsymbol{V}^{-1}\boldsymbol{b}\right) \tag{5.200}$$

Thus, more accurate *a priori* values of $\boldsymbol{d}$ (i.e., smaller $\boldsymbol{b}$) will result in a more accurate estimate $\tilde{\boldsymbol{\theta}}_{ME}$.

The constraints on the parameters given by Eq. (5.185) can take several forms. The most common are:

1) A separate prior estimate of $\boldsymbol{\theta}$ exists. If this estimate is called $\boldsymbol{\theta}_p$, with covariance matrix $\boldsymbol{\Sigma}_p$, then Eq. (5.185) becomes

$$\boldsymbol{\theta}_p = \boldsymbol{\theta} + \boldsymbol{\varsigma} \tag{5.201a}$$

and the expression for the mixed estimator is

$$\hat{\boldsymbol{\theta}}_{ME} = \left(\boldsymbol{X}^T \boldsymbol{X}/\sigma^2 + \boldsymbol{\Sigma}_p^{-1}\right)^{-1} \left(\boldsymbol{X}^T \boldsymbol{z}/\sigma^2 + \boldsymbol{\Sigma}_p^{-1} \boldsymbol{\theta}_p\right) \tag{5.201b}$$

with covariance matrix

$$Cov\left(\hat{\boldsymbol{\theta}}_{ME}\right) = \left(\boldsymbol{X}^T \boldsymbol{X}/\sigma^2 + \boldsymbol{\Sigma}_p^{-1}\right)^{-1} \tag{5.201c}$$

Equations (5.201) can be used to estimate parameters and error bounds for combined data from the experiment that generated $\boldsymbol{X}$ and $\boldsymbol{z}$, and a previous experiment, where information from the previous experiment is included through the *a priori* parameter estimate $\boldsymbol{\theta}_p$ and covariance matrix $\boldsymbol{\Sigma}_p$. This idea can be used repeatedly to estimate parameters based on combined data from more than one flight maneuver and/or wind tunnel test. Furthermore, the technique can be used if prior information exists for only some of the elements in the $\boldsymbol{\theta}$ vector. For the elements of $\boldsymbol{\theta}$ without prior information, the corresponding elements of $\boldsymbol{\theta}_p$ can be set to zero, and the corresponding diagonal element of $\boldsymbol{\Sigma}_p^{-1}$ also set to zero. This implements zero prior information for those elements of the $\boldsymbol{\theta}$ vector.

2) The prior information is given as a statement that particular parameters lie in a certain range $\left(d_{min}, d_{max}\right)$. For a parameter θ_j, the prior value can be placed at the center of the range, with the appropriate uncertainty,

$$\theta_{p_j} = \frac{1}{2}\left(d_{min_j} + d_{max_j}\right) = \theta_j + \varsigma_j \qquad j = 1,2,\ldots,n_p \tag{5.202}$$

The situation is then identical to having *a priori* parameter estimates, as in the preceding case.

3) Setting $\boldsymbol{d} = \boldsymbol{0}$, $\boldsymbol{B} = \boldsymbol{I}$, and $E\left(\boldsymbol{\varsigma}\boldsymbol{\varsigma}^T\right) = \sigma^2 \boldsymbol{V} = k_R^{-1} \boldsymbol{I}$, the mixed estimator becomes a ridge estimator,

$$\tilde{\boldsymbol{\theta}}_R = \left(\boldsymbol{X}^T\boldsymbol{X} + k_R\boldsymbol{I}\right)^{-1}\boldsymbol{X}^T\boldsymbol{z} \tag{5.203}$$

where k_R is a scalar biasing parameter. Methods for choosing k_R and an extension that allows separate biasing parameters for each regressor are described in Ref. [5.3]. The ridge regression is frequently applied to collinear data when *a priori* values for the parameters are unavailable. Note that this approach can be interpreted as a variation of case 1), with prior estimates of the parameters in $\boldsymbol{\theta}_p$ equal to zero, and the prior covariance matrix $\boldsymbol{\Sigma}_p$ equal to $\sigma^2\left(k_R\boldsymbol{I}\right)^{-1}$. It follows that adjusting k_R to larger positive values drives the parameter estimates toward zero, which will drop out model terms for which there is little data information.

Example 5.4

This example is based on the material in Ref. [5.30]. The test vehicle is the X-29 aircraft, which is a single-engine, single-seat fighter-type research aircraft with forward-swept wings. Figure 5.23 is a photograph of the X-29 in flight. The aircraft has relaxed static longitudinal stability in subsonic and transonic regimes, and near-neutral stability in supersonic regimes. Because of this, the aircraft has a full-time automatic flight control system for stability augmentation. For longitudinal control, deflections of canard, wing flaps (flaperons), and fuselage strakes were used. Lateral control was provided by the rudder and asymmetric deflection of the flaperons.

Figure 5.23 **X-29 aircraft**

In addition to pilot stick and rudder inputs, the aircraft responses could also be excited using the concept of a remotely augmented vehicle (RAV). The RAV arrangement employed a ground computer to augment commands from the onboard control system. This capability could be used to introduce a command to the pilot controls (pitch stick, roll stick, or rudder pedals), or to control surfaces (symmetric flaperons, strakes, canard, rudder, or asymmetric flaperons). The RAV commands, usually a pulse or doublet, were summed with the already existing commands to independently move pilot controls or control surfaces. A detailed description of the aircraft and its control system can be found in Ref. [5.31].

For this example, the following two sets of data were used to estimate aircraft parameters:

1) Longitudinal maneuvers implemented by a pilot, at Mach numbers ranging from 0.5 to 1.4.

2) Longitudinal maneuvers implemented by computer-generated inputs using the RAV system, at Mach numbers from 0.6 to 1.3.

During the data analysis, several problems associated with the inherent instability, high augmentation, and sometimes insufficient excitation of the aircraft responses had to be addressed. The main issues were: 1) parameter estimation for an aircraft that is open-loop unstable, 2) adverse effects of data collinearity on parameter identifiability and accuracy, and 3) insufficient information content in the data.

To address these problems, the RAV system was used in the experiment, and both linear regression and mixed estimation were used in the analysis. Figures 5.24 and 5.25 show measurements of the pitch stick η, and longitudinal control surface deflections of the canard δ_c, symmetric flaperons δ_f, and strake δ_s, along with the associated longitudinal response variables, for a piloted pitch stick input maneuver and a maneuver using a sequence of computer-generated RAV inputs.

For the pilot pitch stick input maneuver in Fig. 5.24, all control surface deflections have similar forms, suggesting that data collinearity exists. In general, regressor time series must be dissimilar in form for good parameter estimation, because when any regressor can be scaled to approximately match another, there is an indeterminacy in how variations in the dependent variable can be modeled, and data collinearity exists.

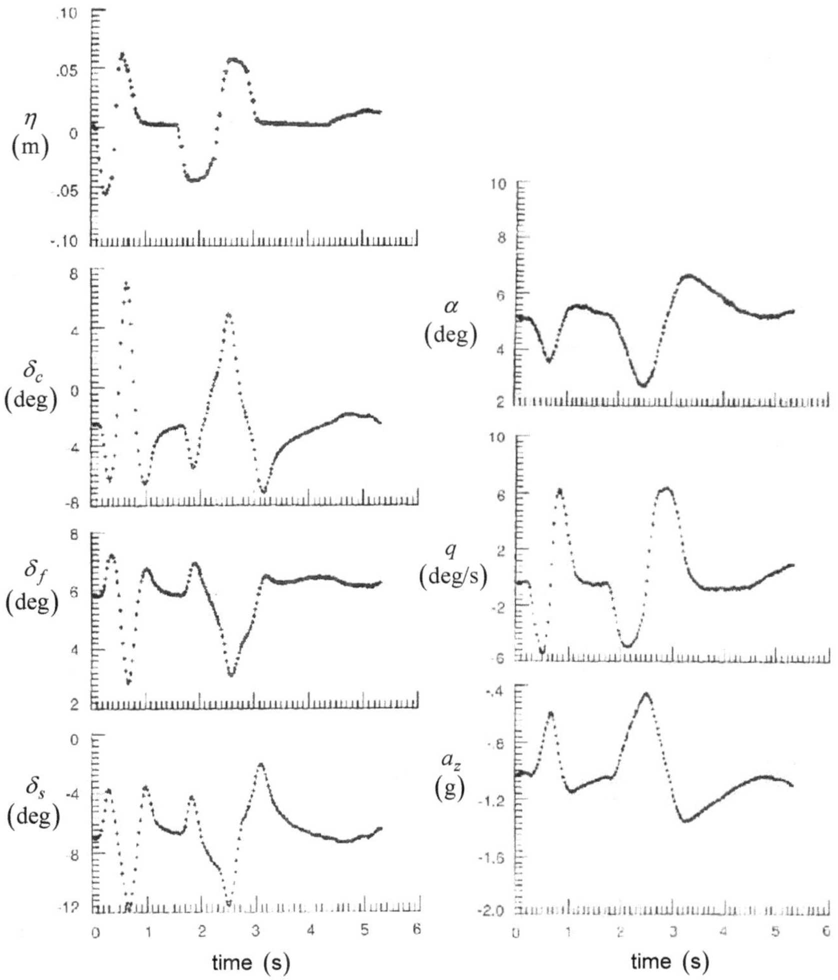

Figure 5.24 **Pilot pitch stick input maneuver, X-29 aircraft**

Figure 5.25 demonstrates the change in data collinearity that results from replacing the pilot pitch stick input with a computer-generated sequence of doublets on the symmetric flaperons, strake, canard, and pitch stick. For both maneuvers, data collinearity was assessed by examining pairwise correlations between regressors, and the parameter variance proportions.

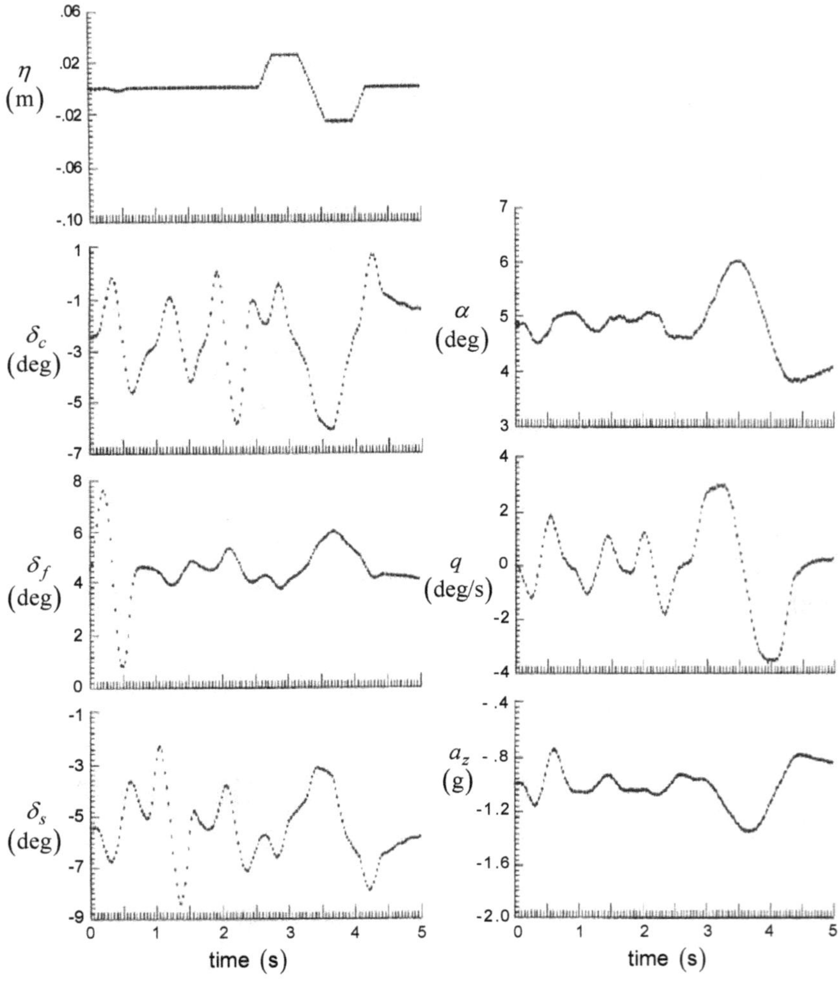

Figure 5.25 **RAV input sequence maneuver, X-29 aircraft**

The aerodynamic model equations for the z body axis force coefficient C_Z and the pitching moment coefficient C_m were postulated as:

$$C_Z = C_{Z_o} + C_{Z_\alpha} \Delta\alpha + C_{Z_q} \frac{q\bar{c}}{2V_o} + C_{Z_{\delta_s}} \Delta\delta_s + C_{Z_{\delta_f}} \Delta\delta_f + C_{Z_{\delta_c}} \Delta\delta_c$$

$$C_m = C_{m_o} + C_{m_\alpha} \Delta\alpha + C_{m_q} \frac{q\bar{c}}{2V_o} + C_{m_{\delta_s}} \Delta\delta_s + C_{m_{\delta_f}} \Delta\delta_f + C_{m_{\delta_c}} \Delta\delta_c$$

where Δ indicates perturbation relative to the trim value.

Correlation matrices for the two sets of data are given in Tables 5.7 and 5.8, respectively. The results for the pilot input show high correlation between $q\bar{c}/2V_o$ and δ_c, and between δ_c and δ_s. High correlation is defined as a pairwise correlation with absolute value greater than or equal to 0.9. The RAV data do not exhibit any high correlations between regressors, and generally have lower pairwise correlations overall.

Table 5.7 **Correlation matrix for the regressors, pilot pitch stick input.**

	α	$q\bar{c}/2V_o$	δ_s	δ_f	δ_c
α	1.000	0.151	0.753	0.711	–0.795
$q\bar{c}/2V_o$		1.000	0.318	–0.341	–0.980
δ_s			1.000	0.643	–0.928
δ_f				1.000	–0.844
δ_c					1.000

Table 5.8 **Correlation matrix for the regressors, RAV input sequence.**

	α	$q\bar{c}/2V_o$	δ_s	δ_f	δ_c
α	1.000	0.275	0.673	0.328	–0.772
$q\bar{c}/2V_o$		1.000	0.345	–0.375	–0.089
δ_s			1.000	0.098	–0.578
δ_f				1.000	–0.303
δ_c					1.000

Condition indices and parameter variance proportions for the two data sets are given in Tables 5.9 and 5.10. The maximum condition index (i.e., the condition number) for the pilot input data is 174, whereas the maximum condition index for the RAV input sequence is 14, indicating a reduced spread of eigenvalues when the RAV system was used. The variance proportions in Table 5.9 for the largest condition index show strong collinearity among the bias term, strake effectiveness, and canard effectiveness. The same quantities in Table 5.10 indicate only a moderate collinearity between the bias term and canard effectiveness.

Table 5.9 **Collinearity diagnostics, pilot pitch stick input.**

Eigenvalue	Condition Index	Variance Proportions (scaled regressors)					
		1	α	$q\bar{c}/2V_o$	δ_s	δ_f	δ_c
3.310	1	0.0000	0.0092	0.4806	0.0017	0.0006	0.0001
1.278	3	0.0099	0.0001	0.0397	0.0720	0.0058	0.0001
1.033	3	0.0001	0.2572	0.1604	0.0000	0.0342	0.0002
0.261	13	0.0504	0.1140	0.0991	0.0762	0.1887	0.0245
0.098	34	0.1220	0.6194	0.2143	0.0041	0.7123	0.0548
0.019	174	0.8176	0.0001	0.0061	0.8460	0.0585	0.9204

Table 5.10 **Collinearity diagnostics, RAV input sequence.**

Eigenvalue	Condition Index	Variance Proportions (scaled regressors)					
		1	α	$q\bar{c}/2V_o$	δ_s	δ_f	δ_c
2.532	1	0.0003	0.0017	0.8594	0.0003	0.0000	0.0018
1.434	2	0.0905	0.0002	0.0049	0.0008	0.0132	0.1901
1.003	3	0.0156	0.2543	0.0098	0.1481	0.1685	0.2565
0.474	5	0.2173	0.0355	0.0171	0.0178	0.4886	0.0172
0.380	7	0.0460	0.6463	0.0220	0.4605	0.0094	0.0506
0.177	14	0.6302	0.0621	0.0867	0.3724	0.4720	0.7147

Table 5.11 contains identifiability diagnostics for estimating pitching moment coefficient parameters from each data set. The table lists the increment in coefficient of determination ΔR^2, t statistic, and variance inflation factor for each regressor. The values of ΔR^2 represent the amount of variation in the pitching moment coefficient that was modeled by the individual terms. The t statistic can be considered a measure of significance of the individual parameters. Analysis of the pilot input data revealed that $C_{m_{\delta_c}}\delta_c$ is the most influential term in the pitching moment equation, with a limited possibility for accurate estimates of parameters $C_{m_{\delta_s}}$ and $C_{m_{\delta_f}}$, and that the significance of the $C_{m_q}\,q\bar{c}/2V_o$ term is almost zero. The RAV experiment improved the identifiability of parameters C_{m_α}, $C_{m_{\delta_s}}$, and $C_{m_{\delta_f}}$,

and showed $C_{m_{\delta_c}}\,\delta_c$ as still the dominant term. The chance for accurate estimation of C_{m_q} remains small.

Table 5.11 **Identifiability diagnostics, pilot pitch stick input, Fig. 5.24, and RAV input sequence, Fig. 5.25.**

Parameter	Pilot input			RAV input sequence		
	ΔR^2	$\lvert t_0 \rvert$	VIF	ΔR^2	$\lvert t_0 \rvert$	VIF
C_{m_α}	5.1	28.0	3.3	9.1	64.8	3.6
C_{m_q}	0.0	0.0	3.2	1.1	13.7	1.6
$C_{m_{\delta_s}}$	1.0	9.2	31.8	4.5	42.7	2.7
$C_{m_{\delta_f}}$	0.9	9.2	14.0	8.1	46.4	1.5
$C_{m_{\delta_c}}$	91.5	12.3	14.9	75.7	99.1	2.0

Estimates of two parameters C_{m_α} and $C_{m_{\delta_c}}$, which contribute the most to the pitching moment model, are plotted against Mach number in Fig. 5.26, along with wind tunnel results. The estimates were obtained from data generated by pilot pitch stick inputs using ordinary least squares and mixed estimation, and from RAV inputs using only ordinary least squares. The wind tunnel values for the strake effectiveness were used as *a priori* values in the mixed estimation. The accuracies of the *a priori* values were determined from repeated tests in two different wind tunnels. As can be seen from Fig. 5.26, the scatter in the parameter estimates was reduced by either applying the mixed estimator to pilot input data, or using data from the RAV experiment. ■

Further examples of least squares parameter estimation using prior information can be found in Ref. [5.32]. Another important application is updating the aerodynamic database in a nonlinear aircraft simulation based on flight data. This involves localized multivariate orthogonal function modeling of data extracted from the aerodynamic data tables, and the use of these local models as *a priori* information for flight data analysis. The method has been developed and applied successfully using flight data in both the time domain [5.33] and the frequency domain [5.34].

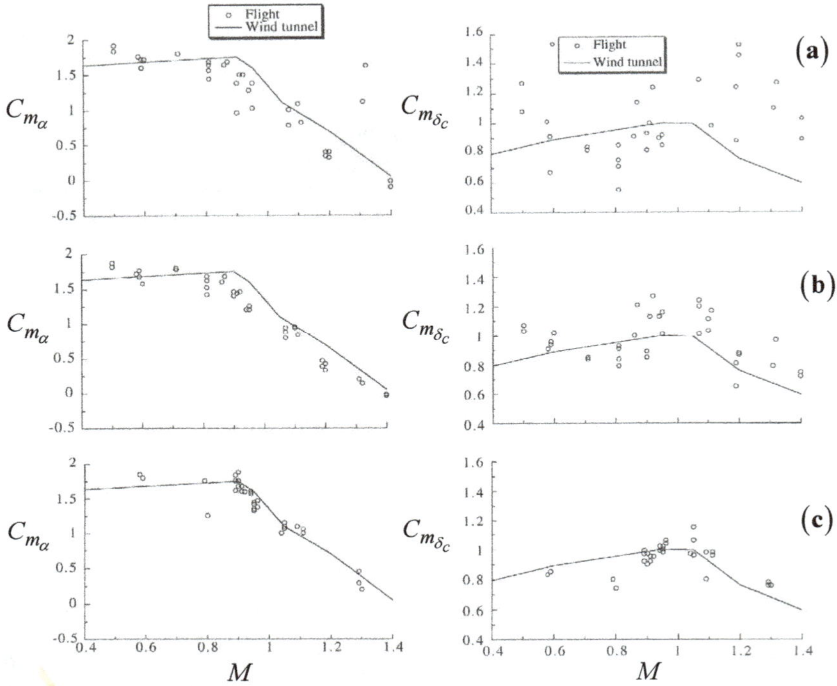

Figure 5.26 **Pitching moment parameters estimated from flight data: (a) pilot input using ordinary least squares, (b) pilot input using mixed estimation, and (c) RAV input sequence using ordinary least squares**

5.6 Data Partitioning

In most practical applications of system identification to aircraft problems, parameter estimation methods are applied to measured data from small amplitude maneuvers executed about a trimmed flight condition. For these maneuvers, a linear aerodynamic model can be assumed. However, results from this analysis are only valid locally, that is, near the flight condition where the maneuver was performed. Information about the aerodynamic characteristics over a wider range of conditions can be obtained by analyzing maneuvers involving large variations in angle of attack, sideslip angle, and control deflections.

Analysis of large amplitude maneuvers requires postulation of a model which might involve a relatively large number of parameters, e.g., including higher-order terms in a multivariate Taylor series expansion. Very often the increased model complexity cannot be supported by the information content in the data. This can result in parameter estimates with low accuracy, or the parameter estimation might fail.

To overcome these problems, a procedure known as data partitioning can be used (cf. Ref. [5.35]). The idea is to divide the data points from a maneuver or set of maneuvers covering a large range of some important explanatory variables into partitions, where each partition contains the data points with values of important explanatory variables that lie within small ranges. This converts a modeling problem that might require a complicated nonlinear model structure and many model parameters into a series of simpler problems which require only linear model structures and just a few model parameters for each of the simpler problems.

Data partitioning with respect to a single variable, e.g., angle of attack α, involves dividing the measured data into m subsets, where each subset contains the data points which satisfy the condition:

$$\alpha_k < \alpha \le \alpha_{k+1} \qquad k = 1,2,\ldots,m \tag{5.204}$$

and the boundary values for the angle of attack partitioning, α_k, $k = 1,2,\ldots,m+1$, are selected by the analyst. The median value of α in the kth subset is

$$\bar{\alpha}_k = \frac{\alpha_k + \alpha_{k+1}}{2} \qquad k = 1,2,\ldots,m \tag{5.205}$$

The interval $\Delta\alpha_k = |\alpha_{k+1} - \alpha_k|$ should be small enough so that the assumption

$$\alpha \approx \bar{\alpha}_k = \text{constant} \tag{5.206}$$

can be made. In this way, dependence on α is simplified or removed for each data subset, so that the modeling problem for each individual subset is simplified. The model for each subset is

$$z_k = \boldsymbol{X}_k \boldsymbol{\theta}_k + \boldsymbol{v}_k \qquad k = 1,2,\ldots,m \tag{5.207}$$

where the subscript k indicates that the data points are selected according to the value of α using Eq. (5.204), and the parameter estimates apply only for that restricted range of α. All of the $\boldsymbol{\theta}_k$, $k = 1,2,\ldots,m$, considered as a function of median α, $\bar{\alpha}_k$, together give a global view of the aerodynamic characteristics for data covering a large range in angle of attack.

In the following example, an aircraft performed many lateral maneuvers over a wide range of angle of attack α. For these maneuvers taken together, the lateral coefficients might depend on α in a nonlinear way,

$$C_n = C_n(\alpha, \beta, p, r, \delta_a, \delta_r) \tag{5.208}$$

and similarly for C_l and C_Y. To simplify the modeling and parameter estimation, the measured data are partitioned into subsets with respect to α. For each subset, the model is changed to

$$C_n(\bar{\alpha}_k) = C_n(\beta, p, r, \delta_a, \delta_r)_{\alpha_k < \alpha \le \alpha_{k+1}} \qquad k = 1,2,\ldots,m \quad (5.209)$$

and similarly for C_l and C_Y. In general, each subset can include data from more than one section of time during the maneuver, or even from different maneuvers or different flights.

In the data partitioning process, each partition must contain a sufficient number of data points, and the data for the regressors must exhibit substantial variation, to allow accurate parameter estimation for each partition. Model structure determination might still be necessary for each partition, since large variations may have occurred in variables other than the one on which the partitioning was based. Some trial and error is usually necessary to select appropriate boundaries for the partitions. In addition to Ref. [5.35], examples of data partitioning and subsequent model formulation and parameter estimation can be found in Refs. [5.20] and [5.36]-[5.38].

Example 5.5

Measured data from 56 low-speed lateral maneuvers of the X-29 aircraft, described earlier in Example 5.4, were assembled into one set with 50,201 data points. The data was then partitioned into 41 one-degree α subsets, and 1 three-degree α subset. The distribution of data points in these subsets is shown in Fig. 5.27.

Half of the lateral maneuvers were analyzed as individual maneuvers using stepwise regression. The possibility of data collinearity in the measured data was investigated by procedures explained in Example 5.4. Application of stepwise regression to the partitioned data resulted in model structures for the lateral aerodynamic coefficients and least-squares estimates of parameters for each data partition. For data subsets with $\alpha \le 40$ deg, models with linear stability and control derivatives were adequate. For data at $\alpha > 40$ deg, models for the lateral force and yawing moment coefficients included some of the nonlinear terms β^2, β^3, $\alpha\beta$, $\beta\delta_c$, or δ_r^3. However, the parameter estimates for these additional terms did not provide any consistent comprehensive information about aerodynamic nonlinearities or effects of longitudinal variables on the lateral aerodynamic coefficients.

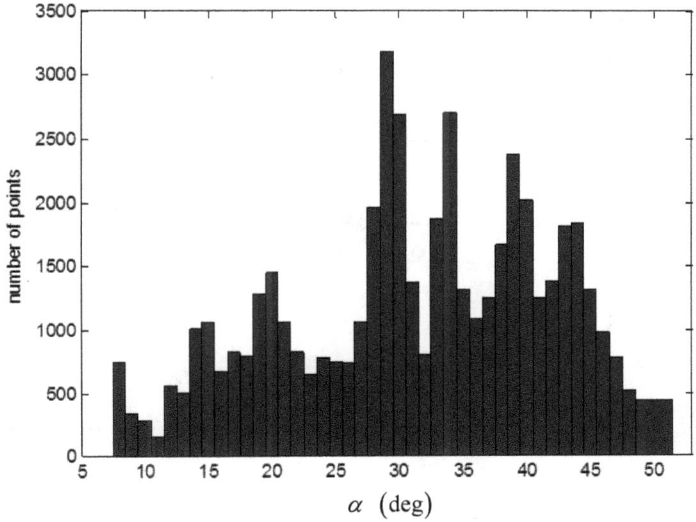

Figure 5.27 **Data partitioning for 56 lateral maneuvers, X-29 aircraft**

The estimated parameters (stability and control derivatives) from the 43 data partitions were fitted by quadratic polynomial splines in angle of attack. A comparison of estimates for the directional static stability derivative C_{n_β} from partitioned data and single maneuvers is shown in Fig. 5.28. In addition, the 95% confidence limits (±2-sigma limits) for prediction using the fitted quadratic spline model, computed from Eq. (5.43), are shown as dashed lines. Most of the estimates from single maneuvers lie within the prediction intervals shown. Similar behavior was observed for the other lateral aerodynamic parameters. It follows that the single maneuver results were consistent with those obtained using partitioned data. This suggests that the modeling and data analysis were done properly for the many maneuvers involved in the analysis, and that combining the data from different maneuvers and using data partitioning produced good modeling results for global nonlinear aerodynamics. ■

The general idea of combining local linear models to create a global nonlinear model is well-established in aircraft system identification practice. Other manifestations of this general approach differ mainly in how the local regions are defined, how the linear models are identified, and how the local linear models are combined or blended to form a global nonlinear model. Examples include Refs. [5.39]-[5.41].

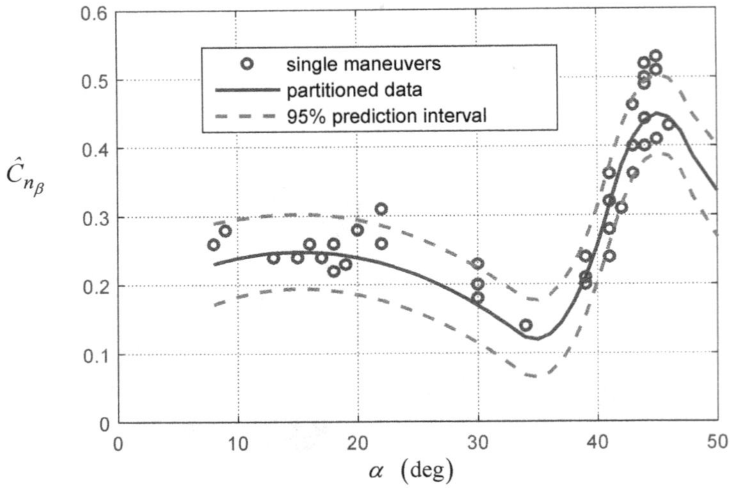

Figure 5.28 **Directional static stability parameter for the X-29 aircraft estimated from partitioned data and single maneuvers**

5.7 Summary and Concluding Remarks

The techniques discussed in this chapter belong to a group of methods known as regression analysis, which is probably the most frequently used of all data analysis approaches. This chapter was concerned in particular with linear regression, in which the model equation relates a dependent variable to a sum of model terms called regressors, and each regressor is multiplied by an unknown constant parameter to be determined from measured data. The main topics covered were linear regression, model structure determination, and data collinearity.

Details were given for the calculations required to estimate unknown parameters in a linear regression model using the least-squares principle. The resulting parameter estimates are unbiased, efficient, and consistent. The calculations also provide standard errors of the estimated parameters as a measure of their accuracy, along with measures of the quality of the model fit to the data, such as fit error and the coefficient of determination. By assuming a normal distribution for measurement errors, it was possible to construct confidence intervals for the estimated parameters, estimated output, and predicted output.

Least-squares parameter estimation using linear regression assumes that the form of the model is known. In practice, however, it is often unclear what terms should be included in the model. This uncertainty may result in a reduced model or in a model with too many terms, neither of which will be a good predictor for other similar data. Therefore, there is a need to find an adequate model which fits the data well and is a good predictor. Statistical metrics and model structure determination techniques were introduced for identifying which model terms are significant, based on the measured data, and should therefore be retained in the model. This makes it possible to identify an adequate model structure based on measured data. Two methods that have been found useful for this purpose, stepwise regression and orthogonal function modeling, were discussed in detail.

Historically, linear regression was developed for the situation where the variables that will be used to formulate the model, called the explanatory variables, are set to selected values by the experimenter, and therefore are assumed to be known without error. The response variable is measured directly, and assumed to be corrupted by random noise, which is the combination of measurement errors and unknown influences. This situation corresponds to a typical wind tunnel experiment, and linear regression methods are used for this problem on a routine basis.

For flight test data, the situation is different in some important respects. Dependent variables, typically nondimensional aerodynamic force and moment coefficients, are not measured directly, but rather are derived from other measurements. Explanatory variables can no longer be set independently, because the values of these quantities change dynamically as the aircraft flies. The explanatory variables are typically aircraft states and control surface deflections. Although measurements of control surface deflections typically have very low noise levels and can be assumed to be measured without error, aircraft state measurements are usually noisy. This violates the assumption that explanatory variables are measured without error, with the result that the least-squares parameter estimation results are biased and inefficient. However, these effects can be mitigated or eliminated with good instrumentation, careful data handling, frequency-domain techniques discussed in Chapter 7, and/or data smoothing techniques described in Chapter 11.

There are significant advantages to using linear regression methods for aerodynamic modeling based on flight data. The modeling can be done using individual equations, one at a time, in an equation error formulation, so that the complete problem of estimating aerodynamic model parameters for an aircraft can be done by solving a series of smaller problems. Linear regression model parameters are usually stability and control derivatives that multiply aircraft states or controls; however, model parameters can also multiply nonlinear modeling terms, such as polynomials or polynomial splines. This allows the modeling to be extended to nonlinear dependencies. Least-squares parameter estimation is relatively simple, and the solution does not require iteration. This efficiency is needed when identifying the model structure, which typically involves evaluating many different candidate model terms by estimating the parameters for each candidate model term and using statistical modeling metrics to choose among them. In the equation-error approach, the model generally matches state time-derivative information in the dynamic system equations. Consequently, there is no need to integrate equations of motion, because the matching is done directly in the equations of motion (hence the name "equation-error"). An important practical consequence is that the equation-error method can be applied equally well to data from inherently unstable aircraft flying with active automatic feedback control. Chapter 7 shows that the same least-squares parameter estimation equations developed in this chapter can be applied to frequency-domain data as well. Because of the model structure determination tools available for linear regression methods, and the capability to incorporate nonlinear functions as modeling terms, the linear regression approach is very general and useful for modeling complicated nonlinear aerodynamic dependencies, which are common for maneuvers that include large ranges or amplitudes of the aerodynamic angles, high angular rates, or unsteady aerodynamic effects.

Residuals from linear regression models based on flight data are typically correlated in time, or colored. The theory assumes the residuals are uncorrelated in time, or white. For experiments such as wind tunnel tests, the test points are randomized to enforce the white residuals assumption. For flight data, this is not possible because the data points are collected sequentially as the aircraft flies. The difference between the assumption and reality when using flight data results in low values of parameter standard errors when white residuals are assumed in the calculations done with colored residuals. For that reason, the theory must be modified to account for colored residuals, by re-formulating the expression for the parameter covariance matrix estimate to include the autocorrelation of the residuals. Parameter estimates are practically unaffected by the residual coloring, so the modified parameter covariance matrix and standard errors are computed after the parameters are estimated by ordinary least squares.

Another practical problem, which is not limited to linear regression modeling, is data collinearity, where model terms have similar forms, quantified by high correlation coefficients. This can occur because of high-gain automatic feedback control, control surfaces that move simultaneously for increased control authority, and also the natural relationship among various quantities for an aircraft in flight. Procedures for detection and assessment of collinearity in linear regression were discussed in this chapter. These include evaluation of the regressor correlation matrix and its inverse, eigensystem analysis or singular value decomposition, and parameter variance decomposition.

One way of dealing with data collinearity is to use different estimation techniques from ordinary least squares. One of these techniques is the mixed estimator, which is a Bayes-like method applied to measured data augmented by prior information. The parameter estimates are biased but can have lower mean squared error than parameter estimates from ordinary least squares, when high data collinearity exists. An example demonstrated that the proposed procedure for dealing with data collinearity can be useful for estimating parameters of a highly-augmented aircraft from flight data.

Finally, data partitioning was introduced as a data-handling strategy that can be used with linear regression methods to identify a model from data covering a wide range of important explanatory variables, such as angle of attack. Examples presented throughout the chapter demonstrated this and other modeling techniques.

Linear regression methods are often applied initially to measured flight data, because of their simplicity and generality. The next chapter describes additional practical methods for aircraft system identification, based on the principle of maximum likelihood.

References

[5.1] Draper, N.R. and Smith, H., *Applied Regression Analysis*, 2nd Ed., John Wiley & Sons, Inc., New York, NY, 1981.

[5.2] Myers, R.H., *Classical and Modern Regression with Applications*, Duxbury Press, Boston, MA, 1986.

[5.3] Montgomery, D.C., Peck, E.A., and Vining, G.G., *Introduction to Linear Regression Analysis*, 3rd Ed., John Wiley & Sons, Inc., New York, NY, 2001.

[5.4] Chatterjee, S. and Hadi, A.S., *Sensitivity Analysis in Linear Regression*, John Wiley & Sons, Inc., New York, NY, 1988.

[5.5] Bendat, J.S. and Piersol, A.G., *Random Data Analysis and Measurement Procedures*, 2nd Ed., John Wiley & Sons, New York, NY, 1986.

[5.6] Box, G.E.P. and Jenkins, G.M., *Time Series Analysis: forecasting and control*, Holden-Day, Inc., San Francisco, CA, 1976.

[5.7] Grauer, J.A. and Morelli, E.A. "Dynamic Modeling Accuracy Dependence on Errors in Sensor Measurements, Mass Properties, and Aircraft Geometry," AIAA-2013-0949, *51st AIAA Aerospace Sciences Meeting*, Grapevine, TX, January 2013.

[5.8] Grauer, J.A. and Morelli, E.A. "Dependence of Dynamic Modeling Accuracy on Sensor Measurements, Mass Properties, and Aircraft Geometry," NASA/TM-2013-218056, November 2013.

[5.9] Morelli, E.A., "Practical Aspects of the Equation-Error Method for Aircraft Parameter Estimation," AIAA-2006-6144, *AIAA Atmospheric Flight Mechanics Conference*, Keystone, CO, August 2006.

[5.10] Toutenburg, H., *Prior Information in Linear Models*, John Wiley & Sons, Ltd., London, UK, 1982.

[5.11] Goldberg, A.S., *Estimation Theory*, John Wiley & Sons, Ltd., London, UK, 1964.

[5.12] Morelli, E.A. and Klein, V., "Accuracy of Aerodynamic Model Parameters Estimated from Flight Test Data," *Journal of Guidance, Control, and Dynamics*, Vol. 20, No. 1, 1997, pp. 74-80.

[5.13] Speedy, C.B., Brown, R.F., and Goodwin, G.C., *Control Theory: Identification and Optimal Control*, Oliver and Boyd, Edinburgh, UK, 1970.

[5.14] Hocking, R.R., "The Analysis and Selection of Variables in Linear Regression," *Biometrics*, Vol. 32, 1976, pp. 1-49.

[5.15] Klein, V., "On the Adequate Model for Aircraft Parameter Estimation," Cranfield Report Aero No. 28, Cranfield Institute of Technology, Cranfield, UK, 1975.

[5.16] Gupta, N.K., Hall, W.E., Jr., and Trankle, T.L., "Advanced Methods of Model Structure Determination from Test Data," *Journal of Guidance and Control*, Vol. 1, No. 3, 1978, pp. 197-204

[5.17] Kashyap, R.L., "A Bayesian Comparison of Different Classes of Dynamic Models Using Empirical Data," *IEEE Transactions on Automatic Control*, Vol. AC-22, No. 5, 1977, pp. 715-727.

[5.18] Hall, W.E., Gupta, N.K., and Smith, R.G., "Identification of Aircraft Stability and Control Coefficients for the High Angle-of-Attack Regime," Engineering Technical Report No. 2, Systems Control, Inc., Palo Alto, CA, 1974.

[5.19] Allen, D.M., "The Prediction Sum of Squares as a Criterion for Selecting Predictor Variables," Technical Report No. 23, University of Kentucky, 1971.

[5.20] Klein, V., Batterson, J.G., and Murphy, P.C., "Determination of Airplane Model Structure From Flight Data by Using Modified Stepwise Regression," NASA TP-1916, 1981.

[5.21] Morelli, E.A. and DeLoach, R., "Wind Tunnel Database Development using Modern Experiment Design and Multivariate Orthogonal Functions," AIAA Paper 2003-0653, *41st AIAA Aerospace Sciences Meeting and Exhibit*, Reno, NV, 2003.

[5.22] Morelli, E.A. and DeLoach, R., "Response Surface Modeling using Multivariate Orthogonal Functions," AIAA Paper 2001-0168, *39th AIAA Aerospace Sciences Meeting and Exhibit*, Reno, NV, 2001.

[5.23] Morelli, E.A., "Estimating Noise Characteristics from Flight Test Data using Optimal Fourier Smoothing," *Journal of Aircraft*, Vol. 32, No. 4, 1995, pp. 689-695.

[5.24] Efroymson, M.A., "Multiple Regression Analysis," *Mathematical Methods for Digital Computer*, edited by A. Ralston and H.S. Wilf, John Wiley & Sons, Inc., New York, NY, 1960.

[5.25] Morelli, E.A., "Global Nonlinear Aerodynamic Modeling using Multivariate Orthogonal Functions," *Journal of Aircraft*, Vol. 32, No. 2, 1995, pp. 270-77.

[5.26] Morelli, E.A., "Efficient Global Aerodynamic Modeling from Flight Data," AIAA-2012-1050, *50th AIAA Aerospace Sciences Meeting*, Nashville, TN, January 2012.

[5.27] Barron, A.R., "Predicted Squared Error : A Criterion for Automatic Model Selection," *Self-Organizing Methods in Modeling*, edited by S.J. Farlow, Marcel Dekker, Inc., New York, NY, 1984, pp. 87-104.

[5.28] Belsley, D.A., Kuh, E., and Welsh, R.E., *Regression Diagnostics: Identifying Influential Data and Sources of Collinearity*, John Wiley & Sons, Inc., New York, NY, 1980.

[5.29] Theil, H., *Principles of Econometrics*, John Wiley & Sons, Inc., New York, NY, 1971.

[5.30] Klein, V. and Murphy, P.C., "Aerodynamic Parameters of High Performance Aircraft Estimated from Wind Tunnel and Flight Test Data," *System Identification for Integrated Aircraft Development and Flight Testing*, RTO-MP-11, Paper 18, 1998.

[5.31] Gera, J., "Dynamic and Controls Flight Testing of the X-29A Airplane," NASA TM-86803, 1986.

[5.32] Klein, V., "Two Biased Estimation Techniques in Linear Regression – Application to Aircraft," NASA TM 100649, 1988.

[5.33] Morelli, E.A. and Ward, D.G., "Automated Simulation Updates based on Flight Data," AIAA-2007-6714, *AIAA Atmospheric Flight Mechanics Conference*, Hilton Head, SC, August 2007.

[5.34] Morelli, E.A. and Cooper, J., "Frequency-Domain Method for Automated Simulation Updates based on Flight Data," *Journal of Aircraft*, Vol. 52, No. 6, November-December 2015, pp. 1995-2008.

[5.35] Batterson, J.G., "Estimation of Airplane Stability and Control Derivatives From Large Amplitude Longitudinal Maneuvers," NASA TM 83185, 1981.

[5.36] Klein, V., Ratvasky, T.R., and Cobleigh, B.R., "Aerodynamic Parameters of High-Angle-of-Attack Research Vehicle (HARV) Estimated From Flight Data," NASA TM-102692, 1990.

[5.37] Klein, V., Batterson, J.G., and Murphy, P.C., "Airplane Model Structure Determination From Flight Data," *Journal of Aircraft*, Vol. 20, No. 5, May 1983, pp. 469-474.

[5.38] Batterson, J.G. and Klein, V., "Partitioning of Flight. Data for Aerodynamic Modeling of Aircraft at High Angles of Attack," *Journal of Aircraft*, Vol. 26, No. 4, 1989, pp. 334-339

[5.39] Seher-Weiss, S. "Identification of Nonlinear Aerodynamic Derivatives Using Classical and Extended Local Model Networks," *Aerospace Science and Technology*, Vol. 15, Issue 1, January-February 2011, pp. 33-44.

[5.40] Brandon, J. and Morelli, E.A. "Nonlinear Aerodynamic Modeling From Flight Data Using Advanced Piloted Maneuvers and Fuzzy Logic," NASA/TM-2012-217778, October 2012.

[5.41] Tobias, E.L. and Tischler, M.B., "A Model Stitching Architecture for Continuous Full Flight-Envelope Simulation of Fixed-Wing Aircraft and Rotorcraft from Discrete-Point Linear Models," Special Report RDMR-AF-16-01, AMRDEC, Redstone Arsenal, AL, April 2016.

Problems

5.1 Answer the following, using words, graphs, equations, and/or diagrams:

a) Discuss the similarities and differences between orthogonal function modeling and stepwise regression modeling.

b) Explain data collinearity and why it is important.

c) Using time-domain least-squares linear regression, show why model parameter estimates do not depend on whether or not the residuals are colored, but the error bound estimates do.

d) Discuss the advantages of analyzing multiple maneuvers at a particular flight condition.

e) What are the advantages and disadvantages of estimating dimensional derivatives, compared to estimating nondimensional derivatives?

f) Discuss the significance of orthogonal modeling functions in equation-error modeling.

g) Explain how stepwise regression works.

h) Discuss methods for model validation.

5.2 Fly the F-16 at a steady level flight condition of your choice and execute a maneuver consisting of a sequence of doublets on the stabilator, rudder, and aileron, in that order.

a) Use equation-error parameter estimation to identify models for nondimensional stability and control derivatives for vertical force and pitching moment.

b) Use equation-error parameter estimation to identify models for nondimensional stability and control derivatives for the side force, rolling moment, and yawing moment.

5.3 Repeat problem **5.2** using the SIDPAC Graphical User Interface (GUI), accessed by typing `sid` at the MATLAB® prompt.. Reconcile any differences in the results.

5.4 A longitudinal control law has the structure shown in the block diagram below. Flight data are available for a piloted maneuver with the control system operating. There is interest in the control system gains $K_\alpha, K_q,$ and K_η, but the organization that designed the control system will not reveal the values of these constants, nor how they were determined. Your boss asks: "Is there any way to extract those control system constants from the flight data?" Write your answer.

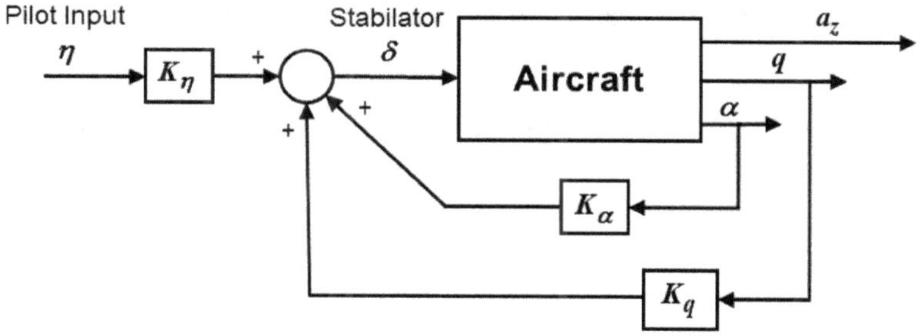

5.5 For the Twin Otter flight data in the file totter_demo_data.mat, compute force and moment coefficients using compfc.m and compmc.m.

a) Compute equation-error estimates of stability and control derivatives for lift coefficient using the least-squares solution directly or lesq.m.

b) Plot the measured and model values for the lift coefficient. Then plot the difference between the measured and modeled values. Is your model adequate? How do you know? What would you do if it were not?

c) Repeat parts a) and b) with the regressors centered about the initial condition. Which parameters are different and why?

d) Repeat parts a), b), and c) for the pitching moment coefficient. Comment on the results.

e) Compute equation-error estimates of stability and control derivatives for pitching moment coefficient using both dimensional and nondimensional parameters. Are the estimates consistent, accounting for the estimated uncertainties? Discuss your results, including any ideas you have on why the estimates might differ.

5.6 Repeat the previous problem using the SIDPAC GUI, accessed by typing `sid` at the MATLAB® prompt.

5.7 Investigate the change in aileron control effectiveness and roll damping with increasing trim angle of attack for the F-16. Use the nominal $x_{cg} = 0.25\bar{c}$, and 10,000 ft altitude. Apply simple multi-step inputs to the aileron, such as a doublet, doublet sequence, or a 2-1-1-2 (see mksqw.m). Use the roll mode approximation, equation-error for nondimensional parameters, and use 5% Gaussian noise on the measured responses from the F-16 simulation (this noise level is implemented in fly_f16.m, for example). Make plots of parameter estimates and uncertainties, and plot the true aileron control effectiveness for the F-16 simulation from the aerodynamic data tables inside the F-16 simulation. Compare your estimation results with actual values and discuss the comparison.

5.8 For the F-16 flight data in the file f16_lon_001_data.mat, use the stepwise regression software swr.m to determine a model for the lift coefficient.

a) Purposely include a nonsignificant term in the model and compute the parameter estimates and standard errors. What is different about the results for the nonsignificant model term? How do you interpret this?

b) Determine the model structure again, this time using spline functions generated with splgen.m and/or other functions that you think might be good candidates for the model. Use plots and cmpsigs.m to suggest possible model terms. Note that nonlinearity can usually be modeled almost equally well in more than one way. Sometimes the manner in which a nonlinearity is modeled depends on which final model form would be the most useful.

c) Repeat parts a) and b) using the SIDPAC GUI, accessed by typing `sid` at the MATLAB® prompt.

5.9 Trim the F-16 at a nominal flight condition of 20 deg angle of attack and 10,000 ft altitude with $x_{cg} = 0.25\bar{c}$. Inside gen_f16_model.m, set initial airspeed to 200 ft/s, angle of attack and pitch angle to 20 deg, initial stabilator deflection to −15 deg, and initial throttle to 0.5, to achieve trim at 20 deg angle of attack. Fly a lateral maneuver on the F-16 at this flight condition, using rudder and aileron inputs with moderate amplitudes.

a) Apply stepwise regression and orthogonal function modeling to identify models for nondimensional aerodynamic rolling moment and yawing moment. Are the models from stepwise regression different than those from orthogonal function modeling? Why or why not?

b) Apply data partitioning to the same data to obtain another model form for the nondimensional aerodynamic rolling moment and yawing moment.

c) Which models would you put forward as the best, and why? How could you check your answer?

5.10 For the example data in ch5_prob10_data.mat, identify a model for the output **z** based on the inputs **u1**, **u2**, and **u3**.

5.11 For the simulated F-16 flight data in the file f16_05A_lon_doublet_data.mat, apply stepwise regression and multivariate orthogonal function modeling to determine adequate model structures and parameter estimates for the nondimensional vertical force and pitching moment coefficients. Do both methods arrive at the same model structures? Why or why not?

5.12 For the flight data in ch5_prob12_data.mat (all angular data are in deg or deg/s):

a) Identify a model for the nondimensional aerodynamic yawing moment coefficient. Model terms must be monomials of order 2 or lower, and splines are allowed.

b) Discuss any techniques, analysis, arguments, or considerations that could be used to convince others that your identified model is good.

Chapter 6

Maximum Likelihood Methods

In Chapter 4, the maximum likelihood estimator was developed for the Fisher model, formulated as

$$z = X\theta + v \qquad \text{or} \qquad z = \varphi(\theta) + v \tag{6.1}$$

where θ is a vector of unknown constant parameters and v is $\mathbb{N}(0, R)$.

It was shown that for a linear measurement equation $z = X\theta + v$, the maximum likelihood estimator reduces to a least squares estimator with weighting equal to the inverse of the noise covariance matrix. This chapter starts with the development of the maximum likelihood estimator for a stochastic dynamic system described by differential equations with process noise. In this case, the measurements are a nonlinear function of the parameters, as shown in Chapter 4. The general form of the relevant measurement equation is therefore $z = \varphi(\theta) + v$.

In general, model parameter estimates are found by maximizing a likelihood function, which involves minimizing the weighted least squares difference between measured outputs and model outputs. The solution combines a state estimator represented by a Kalman filter and a nonlinear parameter estimator. The state estimator is necessary because the presence of process noise in the dynamic equations means that the states are random variables. A nonlinear parameter estimator is required because of the nonlinear connection between model parameters and model outputs, mentioned earlier. Deterministic inputs are still assumed to be measured without error, as in Chapter 5. Because the maximum likelihood estimator includes a Kalman filter to estimate the states, and the outputs are computed from the resulting state estimates, this algorithm for parameter estimation is often called the filter-error method.

The combined estimation of states and parameters constitutes a difficult nonlinear estimation problem. For minimization of the cost function, the Newton-Raphson method is adopted, although other nonlinear optimization methods could also be used. For practical implementation, three simplifications are made:

1) The Newton-Raphson optimization scheme is replaced by a simplified version known as the Gauss-Newton method or the modified Newton-Raphson method.

2) For linear dynamic models, the general Kalman filter is replaced by its steady-state form.

3) The unknowns are divided into three subsets: states, model parameters, and noise covariance matrix elements.

Presentation of the filter-error estimation algorithm and its various forms is followed by a summary of the properties of the maximum likelihood estimates.

For practical application, the additional assumption of no process noise is made. When the process noise is neglected, the states can be computed deterministically by direct numerical integration. The maximum likelihood cost function again involves weighted squared differences between measured and computed outputs. The resulting maximum likelihood estimator is known as the output-error method. This method is a generalization of the nonlinear least squares estimator presented in Chapter 5.

For data collection, the standard procedure is to flight test on days with calm air, using maneuvers designed so that an assumed aerodynamic model structure (usually linear) will be adequate to characterize the data. This eliminates the need to include process noise in the model, so that the simpler output-error method can be used in the data analysis and modeling.

Because the output-error method is one of the most frequently used techniques for aircraft parameter estimation, attention is given to the computational aspects of the solution. These include computing the sensitivities, inverting the Fisher information matrix, and computing the accuracy of the parameter estimates. Practical application of the output-error method is demonstrated with examples.

Finally, it is shown that assuming the state variables are measured without error leads to an equation-error formulation of the maximum likelihood estimator that is identical to linear regression.

6.1 Dynamic System with Process Noise

A wide range of practical system identification problems can be characterized by discrete-time measurements made on a continuous-time dynamic system. The aircraft dynamic modeling problem is in this category.

A stochastic forcing term is sometimes added to the linear dynamic equations to model gusts or to account for inadequacy of the linear model structure. The model equations are then

$$\dot{\boldsymbol{x}}(t) = \boldsymbol{A}\boldsymbol{x}(t) + \boldsymbol{B}\boldsymbol{u}(t) + \boldsymbol{B}_W \boldsymbol{w}(t) \tag{6.2a}$$

$$\boldsymbol{y}(t) = \boldsymbol{C}\boldsymbol{x}(t) + \boldsymbol{D}\boldsymbol{u}(t) \tag{6.2b}$$

$$\boldsymbol{z}(i) = \boldsymbol{y}(i) + \boldsymbol{v}(i) \qquad i = 1, 2, \ldots, N \tag{6.2c}$$

where

$$E\left[x(0)\right]=\bar{x}_0 \qquad E\left\{\left[x(0)-\bar{x}_0\right]\left[x(0)-\bar{x}_0\right]^T\right\}=P_0 \tag{6.2d}$$

The random vectors $w(t)$ and $v(i)$ are assumed to be white with

$$\begin{aligned} E\left[w(t)\right]&=0 \qquad & E\left[w(t_i)w^T(t_j)\right]&=Q(t_i)\delta(t_i-t_j) \\ E\left[v(i)\right]&=0 \qquad & E\left[v(i)v^T(j)\right]&=R(i)\delta_{ij} \end{aligned} \tag{6.2e}$$

Elements of the vector of unknown parameters $\boldsymbol{\theta}$, in general, appear in the elements of the $A, B, B_w, C, D, P_0, Q,$ and R matrices, and the initial state vector $\bar{x}_0$. The $P_0, Q,$ and R matrices are typically diagonal. The state x is a vector of random variables because of the stochastic forcing term $B_w w(t)$ in the dynamic equations.

Following the development in Chapter 4, the likelihood function for a sequence of measurements $Z_N=\left[z(1) \quad z(2) \quad \ldots \quad z(N)\right]^T$ will be denoted by $\mathbb{L}[Z_N;\boldsymbol{\theta}]$. By successive applications of Bayes's rule (see Appendix B), the expression for the likelihood function is

$$\begin{aligned} \mathbb{L}[Z_N;\boldsymbol{\theta}]&=\mathbb{L}\left[z(1),z(2),\ldots z(N);\boldsymbol{\theta}\right] \\ &=\mathbb{L}\left[z(N)\,|\,Z_{N-1};\boldsymbol{\theta}\right]\mathbb{L}[Z_{N-1};\boldsymbol{\theta}] \\ &=\mathbb{L}\left[z(N)\,|\,Z_{N-1};\boldsymbol{\theta}\right]\mathbb{L}\left[z(N-1)\,|\,Z_{N-2};\boldsymbol{\theta}\right]\mathbb{L}[Z_{N-2};\boldsymbol{\theta}] \\ &= \vdots \\ &=\prod_{i=1}^{N}\mathbb{L}\left[z(i)\,|\,Z_{i-1};\boldsymbol{\theta}\right] \end{aligned} \tag{6.3}$$

For computational purposes, it is advantageous to minimize the negative logarithm of the likelihood function, rather than maximize the likelihood function. This is permissible because the likelihood function is nonnegative, and the logarithm is a monotonic function of nonnegative arguments. Using the negative logarithm simplifies the expressions shown next, and converts the problem from maximization to minimization, which is consistent with the operation of most numerical optimization routines. The maximum likelihood estimator can be expressed in the form

$$\hat{\boldsymbol{\theta}} = \max_{\boldsymbol{\theta}} \mathbb{L}[\boldsymbol{Z}_N;\boldsymbol{\theta}]$$

$$= \max_{\boldsymbol{\theta}} \prod_{i=1}^{N} \mathbb{L}\left[z(i) \mid \boldsymbol{Z}_{i-1};\boldsymbol{\theta}\right] \tag{6.4}$$

$$= \min_{\boldsymbol{\theta}} \sum_{i=1}^{N} -ln\left\{\mathbb{L}\left[z(i) \mid \boldsymbol{Z}_{i-1};\boldsymbol{\theta}\right]\right\}$$

If $\boldsymbol{w}(t)$ and $\boldsymbol{v}(i)$ are independent and normally distributed, then $z(i)$ will also have these properties. Then,

$$\mathbb{L}\left[z(i) \mid \boldsymbol{Z}_{i-1};\boldsymbol{\theta}\right] \equiv \mathbb{L}\left[z(i);\boldsymbol{\theta}\right] \tag{6.5}$$

will be uniquely determined by the mean and covariance. By definition,

$$E\left[z(i);\boldsymbol{\theta}\right] \equiv \hat{y}(i \mid i-1)$$

$$Cov\left[z(i);\boldsymbol{\theta}\right] \equiv E\left\{\left[z(i)-\hat{y}(i \mid i-1)\right]\left[z(i)-\hat{y}(i \mid i-1)\right]^T\right\} \tag{6.6}$$

$$= E\left[\boldsymbol{\upsilon}(i)\boldsymbol{\upsilon}^T(i)\right] \equiv \boldsymbol{\mathcal{B}}(i)$$

where

$$\boldsymbol{\upsilon}(i) = z(i) - \hat{y}(i \mid i-1) \tag{6.7}$$

is the $(n_o \times 1)$ vector of innovations and $\boldsymbol{\mathcal{B}}(i)$ is the innovation covariance matrix.

As the number of data samples increases, the probability density of the innovations approach a Gaussian distribution (see Ref. [6.1]). Thus, for a sufficient number of data points, the likelihood function can be written as

$$\mathbb{L}\left[z(i);\boldsymbol{\theta}\right] = (2\pi)^{-n_o/2} \left|\boldsymbol{\mathcal{B}}(i)\right|^{-1/2} exp\left[-\frac{1}{2}\boldsymbol{\upsilon}^T(i)\,\boldsymbol{\mathcal{B}}^{-1}(i)\,\boldsymbol{\upsilon}(i)\right] \tag{6.8}$$

with the negative log-likelihood function for all of the measured data equal to

$$-ln\left[\mathbb{L}(\boldsymbol{Z}_N;\boldsymbol{\theta})\right] = \frac{1}{2}\sum_{i=1}^{N}\left[\boldsymbol{\upsilon}^T(i)\boldsymbol{\mathcal{B}}^{-1}(i)\boldsymbol{\upsilon}(i) + ln\left|\boldsymbol{\mathcal{B}}(i)\right|\right] + \frac{Nn_o}{2}ln(2\pi) \tag{6.9}$$

where n_o is the number of outputs. The term on the far right is a constant that has no effect on the optimization problem, so it can be dropped. This leaves

$$-ln\left[\mathbb{L}(\boldsymbol{Z}_N;\boldsymbol{\theta})\right] = \frac{1}{2}\sum_{i=1}^{N}\left[\boldsymbol{\upsilon}^T(i)\,\boldsymbol{\mathcal{B}}^{-1}(i)\,\boldsymbol{\upsilon}(i) + ln\left|\boldsymbol{\mathcal{B}}(i)\right|\right] \tag{6.10}$$

The problem of determining the negative log-likelihood function is thus reduced to finding the mean and covariance of the innovations $\upsilon(i)$. These two statistics can be obtained from a Kalman filter. As shown in Chapter 4, the Kalman filter recursively processes measurements one at a time. At each point, the filter produces minimum variance estimates of the state and output vector, based on all data measured up to that point. The filtering is a two-step procedure, consisting of prediction and measurement update. The corresponding sets of equations are given below, and are based on Eqs. (4.61) for a linear, time-varying system. The equations are modified for constant parameters, which results in a time-invariant system.

Initial Conditions

$$\begin{aligned} \boldsymbol{x}(0) &= E[\boldsymbol{x}(0)] = \bar{\boldsymbol{x}}_0 \\ \boldsymbol{P}(0) &= E\left\{[\boldsymbol{x}(0)-\bar{\boldsymbol{x}}_0][\boldsymbol{x}(0)-\bar{\boldsymbol{x}}_0]^T\right\} = \boldsymbol{P}_0 \end{aligned} \tag{6.11a}$$

Prediction

$$\frac{d}{dt}\left[\hat{\boldsymbol{x}}(t|i-1)\right] = \boldsymbol{A}\,\hat{\boldsymbol{x}}(t|i-1) + \boldsymbol{B}\boldsymbol{u}(t) \tag{6.11b}$$

$$\frac{d}{dt}\left[\boldsymbol{P}(t|i-1)\right] = \boldsymbol{A}\,\boldsymbol{P}(t|i-1) + \boldsymbol{P}(t|i-1)\boldsymbol{A}^T + \boldsymbol{B}_w\,\boldsymbol{Q}\,\boldsymbol{B}_w^T \tag{6.11c}$$

$$\text{for } (i-1)\Delta t \le t \le i\Delta t$$

Measurement Update

$$\hat{\boldsymbol{x}}(i|i) = \hat{\boldsymbol{x}}(i|i-1) + \boldsymbol{K}(i)\,\boldsymbol{\upsilon}(i) \tag{6.11d}$$

$$\boldsymbol{K}(i) = \boldsymbol{P}(i|i-1)\,\boldsymbol{C}^T\mathcal{B}^{-1}(i|i-1) \tag{6.11e}$$

$$\boldsymbol{P}(i|i) = \left[\boldsymbol{I} - \boldsymbol{K}(i)\boldsymbol{C}\right]\boldsymbol{P}(i|i-1) \tag{6.11f}$$

where

$$\mathcal{B}(i|i-1) = \boldsymbol{C}\,\boldsymbol{P}(i|i-1)\,\boldsymbol{C}^T + \boldsymbol{R} \tag{6.11g}$$

$$\boldsymbol{\upsilon}(i) = \boldsymbol{z}(i) - \boldsymbol{C}\hat{\boldsymbol{x}}(i|i-1) - \boldsymbol{D}\boldsymbol{u}(i) \tag{6.11h}$$

6.1.1 Optimization Algorithm

Maximum likelihood parameter estimates are obtained by minimizing the negative log-likelihood function from Eq. (6.10),

$$J(\boldsymbol{\theta}) = \frac{1}{2}\sum_{i=1}^{N} \boldsymbol{v}^T(i)\,\boldsymbol{\mathcal{B}}^{-1}(i)\,\boldsymbol{v}(i) + \frac{1}{2}\sum_{i=1}^{N} ln\left|\boldsymbol{\mathcal{B}}(i)\right| \tag{6.12}$$

subject to the constraints imposed by Eqs. (6.11).

There are several optimization techniques that could be applied to this nonlinear optimization problem. A comparison of different methods in Ref. [6.2] showed that the Newton-Raphson scheme has a very good convergence rate. This approach requires first- and second-order gradients of the cost function, which appear in the Taylor series expansion of $J(\boldsymbol{\theta})$. Assuming the vector $\boldsymbol{\theta}$ can be expressed as a small perturbation $\Delta\boldsymbol{\theta}$ from a nominal parameter estimate $\boldsymbol{\theta}_o$,

$$J(\boldsymbol{\theta}_o + \Delta\boldsymbol{\theta}) = J(\boldsymbol{\theta}_o) + \Delta\boldsymbol{\theta}^T \left.\frac{\partial J}{\partial \boldsymbol{\theta}}\right|_{\boldsymbol{\theta}=\boldsymbol{\theta}_o} + \frac{1}{2}\Delta\boldsymbol{\theta}^T \left.\frac{\partial^2 J}{\partial \boldsymbol{\theta}\partial \boldsymbol{\theta}^T}\right|_{\boldsymbol{\theta}=\boldsymbol{\theta}_o} \Delta\boldsymbol{\theta} + \dots \tag{6.13}$$

where

$\Delta\boldsymbol{\theta}$ = vector of changes in $\boldsymbol{\theta}$

$\partial J/\partial \boldsymbol{\theta}$ = vector of gradients $\partial J/\partial \theta_j$, $j = 1,2,\dots,n_p$

$\dfrac{\partial^2 J}{\partial \boldsymbol{\theta}\partial \boldsymbol{\theta}^T}$ = second-order gradient matrix, called the Hessian matrix, with elements $\partial^2 J/\partial \theta_j \partial \theta_k$, $j,k = 1,2,\dots,n_p$

Using the second-order expansion in Eq. (6.13) as an approximation for $J(\boldsymbol{\theta}_o + \Delta\boldsymbol{\theta})$,

$$J(\boldsymbol{\theta}_o + \Delta\boldsymbol{\theta}) \approx J(\boldsymbol{\theta}_o) + \Delta\boldsymbol{\theta}^T \left.\frac{\partial J}{\partial \boldsymbol{\theta}}\right|_{\boldsymbol{\theta}=\boldsymbol{\theta}_o} + \frac{1}{2}\Delta\boldsymbol{\theta}^T \left.\frac{\partial^2 J}{\partial \boldsymbol{\theta}\partial \boldsymbol{\theta}^T}\right|_{\boldsymbol{\theta}=\boldsymbol{\theta}_o} \Delta\boldsymbol{\theta} \tag{6.14}$$

The necessary condition for $J(\boldsymbol{\theta}_o + \Delta\boldsymbol{\theta})$ to be a minimum is

$$\frac{\partial}{\partial \boldsymbol{\theta}}\left[J(\boldsymbol{\theta}_o + \Delta\boldsymbol{\theta})\right] = \mathbf{0} \tag{6.15}$$

Combining the last two equations,

$$\frac{\partial}{\partial \boldsymbol{\theta}}\left[J(\boldsymbol{\theta}_o + \Delta\boldsymbol{\theta})\right] = \left.\frac{\partial J}{\partial \boldsymbol{\theta}}\right|_{\boldsymbol{\theta}=\boldsymbol{\theta}_o} + \left.\frac{\partial^2 J}{\partial \boldsymbol{\theta}\partial \boldsymbol{\theta}^T}\right|_{\boldsymbol{\theta}=\boldsymbol{\theta}_o} \Delta\boldsymbol{\theta} = \mathbf{0} \tag{6.16}$$

The solution of the last equation provides an estimate for the vector of parameter changes,

$$\Delta\hat{\boldsymbol{\theta}} = -\left[\left.\frac{\partial^2 J}{\partial \boldsymbol{\theta}\partial \boldsymbol{\theta}^T}\right|_{\boldsymbol{\theta}=\boldsymbol{\theta}_o}\right]^{-1} \left.\frac{\partial J}{\partial \boldsymbol{\theta}}\right|_{\boldsymbol{\theta}=\boldsymbol{\theta}_o} \tag{6.17}$$

assuming that the Hessian matrix is nonsingular. Since the first and second gradients of the cost function are computed at a nominal value of the parameters $\boldsymbol{\theta}_o$, the updated parameter estimate $\hat{\boldsymbol{\theta}}$ is computed from

$$\hat{\boldsymbol{\theta}} = \boldsymbol{\theta}_o + \Delta\hat{\boldsymbol{\theta}} \tag{6.18}$$

Because of the approximation to $J(\boldsymbol{\theta})$ in Eq. (6.14), it is necessary to repeat the estimation procedure by taking the estimated parameter vector $\hat{\boldsymbol{\theta}}$ as the new nominal value $\boldsymbol{\theta}_o$, i.e., set $\boldsymbol{\theta}_o = \hat{\boldsymbol{\theta}}$ for the next iteration. The reason for this iteration is that the output depends nonlinearly on the parameters, so the cost dependence on the parameters is more complicated than quadratic. In the linear regression problem of Chapter 5, the output depended linearly on the parameters, so that the cost was a quadratic function of the parameters, and Eq. (6.14) in that case was exact, not an approximation. The parameter estimates could then be obtained in one iteration, corresponding to the solution of the normal equations [cf. Eq. (5.10)], which is equivalent to Eq. (6.17). In the present case, repeated quadratic approximations to the nonlinear dependence of the cost on the parameters are used to iteratively arrive at the solution. The iterative process is completed when selected convergence criteria are satisfied. Convergence criteria that have been found useful in practice are given in Section 6.3.

A complete set of expressions for the first- and second-order gradients in Eq. (6.17) can be found in Refs. [6.3] and [6.4]. A block diagram of the computing algorithm for parameter estimation based on Eqs. (6.11), (6.17), and (6.18) is given in Fig. 6.1. As mentioned earlier, this algorithm for parameter estimation is called the filter-error method, because the maximum likelihood estimator shown in Fig. 6.1 includes a Kalman filter for output estimation.

Minimization of the negative log-likelihood function for the filter-error problem is generally a very difficult optimization problem. The second-order gradients in the Hessian matrix required for the Newton-Raphson algorithm are computationally expensive to obtain, and are susceptible to numerical error. Because of this, practical applications require the use of a simplified approach known as the modified Newton-Raphson method. Details of this simplified approach to nonlinear optimization appear in Section 6.2.

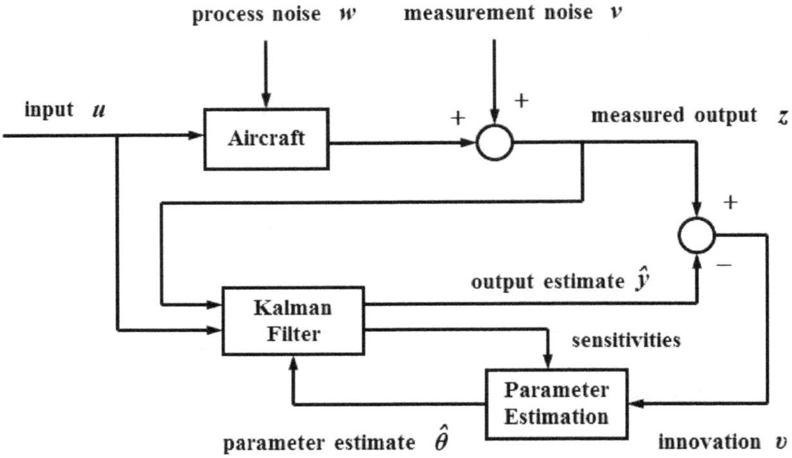

Figure 6.1 **Block diagram for filter-error parameter estimation**

The estimation algorithm described earlier can be generalized to a nonlinear dynamic system with process noise and nonzero initial conditions (see Ref. [6.5]). The system can be described by

$$\dot{x}(t) = f\left[x(t), u(t), \theta\right] + B_w\, w(t) \tag{6.19a}$$

$$y(t) = g\left[x(t), u(t), \theta\right] \tag{6.19b}$$

$$z(i) = y(i) + v(i) \qquad i = 1, 2, \ldots, N \tag{6.19c}$$

where

$$E\left[x(0)\right] = \bar{x}_0 \quad \text{and} \quad E\left\{\left[x(0) - \bar{x}_0\right]\left[x(0) - \bar{x}_0\right]^T\right\} = P_0 \tag{6.19d}$$

The unknown parameters appear in the nonlinear vector functions f and h, as well as in the matrices B_w, P_0, Q, R, and in the initial state vector $\bar{x}_0$. The Kalman filter is replaced by the extended Kalman filter outlined in Section 4.3.4.

6.1.2 Simplified Algorithms

The maximum likelihood estimation algorithm for a time-invariant stochastic linear dynamic system can be simplified by replacing the Kalman filter equations (6.11) with their steady-state forms. The equations follow from Eqs. (4.61) as

Initial Conditions

$$E\left[\boldsymbol{x}(0)\right] = \bar{\boldsymbol{x}}_0 \qquad E\left\{\left[\boldsymbol{x}(0) - \bar{\boldsymbol{x}}_0\right]\left[\boldsymbol{x}(0) - \bar{\boldsymbol{x}}_0\right]^T\right\} = \boldsymbol{P}_0 \tag{6.20a}$$

Prediction

$$\frac{d}{dt}\left[\hat{\boldsymbol{x}}(t \mid i-1)\right] = \boldsymbol{A}\,\hat{\boldsymbol{x}}(t \mid i-1) + \boldsymbol{B}\,\boldsymbol{u}(t) \tag{6.20b}$$

$$\text{for } (i-1)\Delta t \le t \le i\Delta t$$

Measurement Update

$$\hat{\boldsymbol{x}}(i \mid i) = \hat{\boldsymbol{x}}(i \mid i-1) + \boldsymbol{K}\boldsymbol{\upsilon}(i) \tag{6.20c}$$

$$\boldsymbol{K} = \boldsymbol{P}\,\boldsymbol{C}^T \boldsymbol{\mathcal{B}}^{-1} \tag{6.20d}$$

where

$$\boldsymbol{\mathcal{B}} = \boldsymbol{C}\,\boldsymbol{P}\,\boldsymbol{C}^T + \boldsymbol{R} \tag{6.20e}$$

$$\boldsymbol{\upsilon}(i) = \boldsymbol{z}(i) - \boldsymbol{C}\hat{\boldsymbol{x}}(i \mid i-1) - \boldsymbol{D}\boldsymbol{u}(i) \tag{6.20f}$$

As suggested in Ref. [6.5], the steady-state discrete-time form of the Riccati equation can be solved for $\boldsymbol{P}$. Eq. (4.55) is one of the forms of the Riccati equation,

$$\boldsymbol{P} = \boldsymbol{\Phi}\left[\boldsymbol{P} - \boldsymbol{P}\boldsymbol{C}^T\left(\boldsymbol{C}\boldsymbol{P}\boldsymbol{C}^T + \boldsymbol{R}\right)^{-1}\boldsymbol{C}\boldsymbol{P}\right]\boldsymbol{\Phi}^T + \boldsymbol{\Gamma}_w\,\boldsymbol{Q}\,\boldsymbol{\Gamma}_w^T \tag{6.20g}$$

For the continuous-discrete form of the filter equations, $\boldsymbol{\Phi}$ and $\boldsymbol{\Gamma}_w$ are obtained from Eq. (2.21) as

$$\boldsymbol{\Phi} \equiv e^{\boldsymbol{A}\Delta t} \qquad \boldsymbol{\Gamma}_w \equiv \left[\boldsymbol{A}^{-1}\left(e^{\boldsymbol{A}\Delta t} - \boldsymbol{I}\right)\right]\boldsymbol{B}_w \tag{6.20h}$$

In the simplified formulation, the unknown parameters are elements of matrices $\boldsymbol{A}, \boldsymbol{B}, \boldsymbol{B}_w, \boldsymbol{C}, \boldsymbol{D}, \boldsymbol{Q},$ and $\boldsymbol{R}$. The steady-state filter brings simplification to the equation for the first-order gradients of the cost function, because the partial derivatives of the covariance matrix $\boldsymbol{P}$ with respect to the parameters are obtained from a linear matrix equation (see Ref. [6.5]). Despite this simplification, the estimation algorithm can still have difficulties, because the elements of both noise covariance matrices $\boldsymbol{Q}$ and $\boldsymbol{R}$ are treated as unknowns. These matrices appear indirectly in the likelihood function through $\boldsymbol{\mathcal{B}}$, which is a complicated function of $\boldsymbol{A}, \boldsymbol{B}_w, \boldsymbol{C}, \boldsymbol{Q},$ and $\boldsymbol{R}$. In addition, the convergence of the algorithm can be a problem, as illustrated in Ref. [6.5] on a scalar case.

In the general form given here, this estimation algorithm is very difficult to use in practice. Consequently, there have been only a few applications of this algorithm to flight data. In Ref. [6.6], the problem of estimating the noise covariance matrices was not addressed, and in Ref. [6.7] only the process noise covariance matrix was estimated, and not the measurement noise covariance matrix.

Further simplification to the algorithm was suggested in Ref. [6.8]. The vector of unknown parameters was recast to include elements of $\boldsymbol{K}$ and $\boldsymbol{\mathcal{B}}$, rather than elements of the $\boldsymbol{Q}$ and $\boldsymbol{R}$ noise covariance matrices. The gradient of the cost function with respect to $\boldsymbol{\mathcal{B}}$ can be computed by writing the cost function as

$$\begin{aligned} J(\boldsymbol{\theta}) &= \frac{1}{2}\sum_{i=1}^{N} \boldsymbol{\upsilon}^T(i)\boldsymbol{\mathcal{B}}^{-1}\boldsymbol{\upsilon}(i) + \frac{N}{2} ln|\boldsymbol{\mathcal{B}}| \\ &= \frac{1}{2} Tr\left[\boldsymbol{\mathcal{B}}^{-1}\sum_{i=1}^{N}\boldsymbol{\upsilon}(i)\boldsymbol{\upsilon}^T(i)\right] + \frac{N}{2} ln|\boldsymbol{\mathcal{B}}| \end{aligned} \tag{6.21}$$

The scalar cost can then be differentiated with respect to the matrix $\boldsymbol{\mathcal{B}}$ (see Appendix A) to obtain

$$\frac{\partial J}{\partial \boldsymbol{\mathcal{B}}} = -\frac{1}{2}\boldsymbol{\mathcal{B}}^{-1}\sum_{i=1}^{N}\boldsymbol{\upsilon}(i)\boldsymbol{\upsilon}^T(i)\boldsymbol{\mathcal{B}}^{-1} + \frac{N}{2}\boldsymbol{\mathcal{B}}^{-1} \tag{6.22}$$

Setting the gradient equal to zero and solving for $\boldsymbol{\mathcal{B}}$ results in the estimator for the covariance matrix of the innovations,

$$\hat{\boldsymbol{\mathcal{B}}} = \frac{1}{N}\sum_{i=1}^{N}\boldsymbol{\upsilon}(i)\boldsymbol{\upsilon}^T(i) \tag{6.23}$$

Then, for a given estimate $\hat{\boldsymbol{\mathcal{B}}}$, the cost function becomes

$$J(\boldsymbol{\theta}) = \frac{1}{2}\sum_{i=1}^{N} \boldsymbol{\upsilon}^T(i)\hat{\boldsymbol{\mathcal{B}}}^{-1}\boldsymbol{\upsilon}(i) \tag{6.24}$$

This cost is minimized with respect to the unknown parameters in matrices $\boldsymbol{A}, \boldsymbol{B}, \boldsymbol{C}, \boldsymbol{D},$ and $\boldsymbol{K}$ using a nonlinear optimization technique, such as the

Newton-Raphson method described above. Once the estimates of the unknown parameters are obtained, a new sequence of residuals is computed, from which the estimate $\hat{\mathcal{B}}$ can be updated using Eq. (6.23). The procedure continues until selected convergence criteria for the entire set of unknown parameters are satisfied. The unknowns in the problem are estimated in subsets, where the parameters contained in $\boldsymbol{A}, \boldsymbol{B}, \boldsymbol{C}, \boldsymbol{D},$ and $\boldsymbol{K}$ are held constant while the parameters in $\mathcal{B}$ are estimated, and then the parameters contained in $\mathcal{B}$ are held constant while the parameters in $\boldsymbol{A}, \boldsymbol{B}, \boldsymbol{C}, \boldsymbol{D},$ and $\boldsymbol{K}$ are estimated using a nonlinear optimizer. This approach to estimating the full set of unknown parameters is often referred to as a relaxation technique.

The estimate of the measurement noise covariance matrix can be easily calculated from Eqs. (6.20d) and (6.20e) as

$$\hat{\boldsymbol{R}} = \left(\boldsymbol{I} - \hat{\boldsymbol{C}}\hat{\boldsymbol{K}}\right)\hat{\mathcal{B}} \tag{6.25}$$

The computation of $\boldsymbol{B}_w$ and $\boldsymbol{Q}$ from Eqs. (6.20d), (6.20e), (6.20g), and (6.20h) is much more difficult. In general, these equations do not have a unique solution. In order to find one, it is necessary to impose further constraints on $\boldsymbol{B}_w$ and $\boldsymbol{Q}$ (see Refs. [6.8] and [6.9]).

Estimating $\mathcal{B}$ and $\boldsymbol{K}$ directly simplifies the estimation problem considerably. Unfortunately, there are only a few references where this estimation algorithm was applied to flight data. In these examples, identifiability problems were detected, and the algorithm resulted in poor estimates of $\boldsymbol{K}$ (see Refs. [6.10] and [6.11]). Maine and Iliff [6.5] point out a theoretical problem associated with the estimation of $\boldsymbol{K}$. Because of the requirement that $\boldsymbol{P} \geq \boldsymbol{0}$, elements of $\boldsymbol{K}$ cannot take on arbitrary values, but instead must be constrained to lie within certain boundaries. This makes the estimation problem into a constrained nonlinear optimization.

Because of the difficulties with the two simplified versions of the general estimator, a new formulation was proposed in Ref. [6.5] and reiterated in Ref. [6.12]. It takes advantage of the previous approaches by specifying the unknowns as elements in the matrices $\boldsymbol{A}, \boldsymbol{B}, \boldsymbol{B}_w, \boldsymbol{C}, \boldsymbol{D}, \boldsymbol{Q},$ and $\mathcal{B}$. The algorithm proceeds as follows:

1) Using nominal values of the unknown parameters, assemble the system and covariance matrices $\boldsymbol{A}, \boldsymbol{B}, \boldsymbol{B}_w, \boldsymbol{C}, \boldsymbol{D}, \boldsymbol{Q},$ and $\boldsymbol{R}$, and compute the innovations $\upsilon(i)$ for $i = 1, 2, \ldots, N$.

2) Compute $\hat{\mathcal{B}}$ from Eq. (6.23).

3) Find estimates for the unknown parameters in $\boldsymbol{A}, \boldsymbol{B}, \boldsymbol{B}_w, \boldsymbol{C}, \boldsymbol{D}$, and $\boldsymbol{Q}$ using a nonlinear optimization technique to minimize the cost in Eq. (6.24).

4) For $\mathcal{B} = \hat{\mathcal{B}}$, compute $\hat{\boldsymbol{K}}$ and $\hat{\boldsymbol{R}}$ from Eqs. (6.20d) and (6.25).

5) Update the nominal values of the unknown parameters.

6) Repeat steps 1 through 5 until convergence criteria are satisfied.

Another practical approach developed by Grauer and Morelli [6.13] involves separately computing $\hat{\boldsymbol{R}}$ by numerically smoothing the measurements first, then estimating the unknown parameters in $\boldsymbol{A}, \boldsymbol{B}, \boldsymbol{B}_w, \boldsymbol{C}$, and $\boldsymbol{D}$ using a nonlinear optimizer to minimize the cost in Eq. (6.24), computing $\hat{\mathcal{B}}$ from Eq. (6.23), then using a simple estimator for $\boldsymbol{Q}$, in a relaxation technique. This approach accurately assigns the physical measurement noise to $\hat{\boldsymbol{R}}$, which reduces the number of unknown parameters and better represents the real physical situation. Accurate modeling results can be obtained without convergence or identifiability problems, and without requiring analyst judgment or tuning parameter adjustments. Successful application was demonstrated using flight data in turbulence.

The filter-error method just detailed is the most general estimation method used in practical aircraft system identification, in that it allows for the existence of both process noise and measurement noise in the model. The main practical problems with this method are the large number of parameters to be estimated, and the complexity of applying both a Kalman filter and a nonlinear optimizer. This often leads to difficulties with identifiability and insufficient information in the data for accurate parameter estimation.

6.1.3 Properties of Maximum Likelihood Parameter Estimates

The accuracy of the estimated parameters is related to the properties of the estimates. These have been discussed in numerous references, e.g., Refs. [6.14] and [6.15]. Using the definitions in Section 4.1, the properties of the maximum likelihood estimates can be summarized as follows:

1) Maximum likelihood estimates of dynamic system parameters are asymptotically unbiased,

$$E\left(\hat{\boldsymbol{\theta}}\right) \rightarrow \boldsymbol{\theta} \quad \text{as} \quad N \rightarrow \infty$$

The mean of the distribution for the random vector $\hat{\boldsymbol{\theta}}$ approaches the true parameter vector $\boldsymbol{\theta}$ as the number of data points increases.

2) Maximum likelihood estimates are consistent,

$$\hat{\boldsymbol{\theta}} \rightarrow \boldsymbol{\theta} \quad \text{as} \quad N \rightarrow \infty$$

The estimate $\hat{\boldsymbol{\theta}}$ approaches the true value $\boldsymbol{\theta}$ as the number of data points increases.

3) Maximum likelihood estimates are asymptotically efficient,

$$Cov\left(\hat{\boldsymbol{\theta}}\right) \rightarrow \boldsymbol{M}^{-1} \quad \text{for} \quad N \rightarrow \infty$$

where $\boldsymbol{M}$ is the Fisher information matrix defined as [cf. Eq. (4.13) and Appendix B]

$$\boldsymbol{M} \equiv E\left\{\left[\frac{\partial \ln \mathbb{L}(\boldsymbol{Z}_N;\boldsymbol{\theta})}{\partial \boldsymbol{\theta}}\right]\left[\frac{\partial \ln \mathbb{L}(\boldsymbol{Z}_N;\boldsymbol{\theta})}{\partial \boldsymbol{\theta}}\right]^T\right\} = -E\left[\frac{\partial^2 \ln \mathbb{L}(\boldsymbol{Z}_N;\boldsymbol{\theta})}{\partial \boldsymbol{\theta}\partial \boldsymbol{\theta}^T}\right]$$

The main diagonal elements of the inverse information matrix provide the lower bounds on the parameter variances, called the Cramér-Rao bounds. Thus, the diagonal elements of $\boldsymbol{M}^{-1}$ represent the achievable accuracy for the estimated parameters. In practice, this achievable accuracy can be closely approached for values of N associated with typical flight test maneuver lengths and sampling rates.

4) Maximum likelihood estimates are asymptotically normal, i.e., the distribution of the estimates asymptotically approaches a normal distribution with mean $\boldsymbol{\theta}$ and variance $\boldsymbol{M}^{-1}$, so that $\hat{\boldsymbol{\theta}}$ is $\mathbb{N}\left(\boldsymbol{\theta},\boldsymbol{M}^{-1}\right)$.

6.2 Output-Error Method

The maximum likelihood parameter estimation method can be substantially simplified when applied to a deterministic dynamic system, which has no process noise. In practice, a deterministic linear dynamic system is commonly used, which can be described by

$$\dot{\boldsymbol{x}}(t) = \boldsymbol{A}\boldsymbol{x}(t) + \boldsymbol{B}\boldsymbol{u}(t) \qquad \boldsymbol{x}(0) = \boldsymbol{x}_o \tag{6.26a}$$

$$\boldsymbol{y}(t) = \boldsymbol{C}\boldsymbol{x}(t) + \boldsymbol{D}\boldsymbol{u}(t) \tag{6.26b}$$

$$\boldsymbol{z}(i) = \boldsymbol{y}(i) + \boldsymbol{v}(i) \qquad i = 1,2,\ldots,N \tag{6.26c}$$

$$\boldsymbol{v} \text{ is } \mathbb{N}(\boldsymbol{0},\boldsymbol{R}) \tag{6.26d}$$

$$Cov\left[\boldsymbol{v}(i)\right] = E\left[\boldsymbol{v}(i)\boldsymbol{v}^T(j)\right] = \boldsymbol{R}\,\delta_{ij} \tag{6.26e}$$

where model parameters $\boldsymbol{\theta}$ are included in the system matrices $\boldsymbol{A}, \boldsymbol{B}, \boldsymbol{C}, \boldsymbol{D}$.

Because there is no process noise, $\boldsymbol{Q}$ is zero, and the state equations (6.26a) are deterministic. The Kalman gain $\boldsymbol{K}$ is also zero, and the Kalman filter is replaced by a simple integration of the state equations. For a given parameter vector estimate $\boldsymbol{\theta}$, the innovations become the output errors,

$$\boldsymbol{v}(i) = \boldsymbol{z}(i) - \boldsymbol{y}(i) = \boldsymbol{z}(i) - \boldsymbol{C}\boldsymbol{x}(i) - \boldsymbol{D}\boldsymbol{u}(i) \qquad i = 1,2,\ldots,N \qquad (6.27)$$

The negative log-likelihood function takes the form

$$-ln\mathbb{L}(\boldsymbol{Z}_N;\boldsymbol{\theta}) = \frac{1}{2}\sum_{i=1}^{N}\boldsymbol{v}^T(i)\,\boldsymbol{R}^{-1}\,\boldsymbol{v}(i) + \frac{N}{2}ln|\boldsymbol{R}| + \frac{N\,n_o}{2}ln(2\pi) \quad (6.28)$$

For the linear dynamic system specified by Eqs. (6.26), the unknown parameters are the elements of $\boldsymbol{\theta}$ and $\boldsymbol{R}$, and the initial condition vector $\boldsymbol{x}_o$.

Elements of the initial condition vector can be considered as unknown parameters and estimated along with the other unknown quantities, but typically the initial condition vector is estimated separately first, by applying local numerical smoothing methods to measured data, cf. Chapter 11. This reduces the number of unknown parameters and improves the conditioning of the modeling problem.

Optimizing the right side of Eq. (6.28) with respect to $\boldsymbol{R}$ is done by differentiating with respect to $\boldsymbol{R}$, setting the result equal to zero, and solving for $\boldsymbol{R}$, which gives [cf. Eq. (6.23)],

$$\hat{\boldsymbol{R}} = \frac{1}{N}\sum_{i=1}^{N}\boldsymbol{v}(i)\,\boldsymbol{v}^T(i) \qquad (6.29)$$

Usually only the diagonal elements of the $\boldsymbol{R}$ matrix are estimated from Eq. (6.29), enforcing an assumption that the measurement noise sequences for the n_o measured outputs are uncorrelated with one another. This is a good assumption in practice, and a diagonal $\hat{\boldsymbol{R}}$ simplifies the calculations. Retaining the full $\boldsymbol{R}$ matrix could be done without conceptual difficulty, but the small difference in the results usually does not warrant the extra computation involved.

For a given $\hat{\boldsymbol{R}}$, the negative log-likelihood cost function $J(\boldsymbol{\theta})$ becomes

$$\begin{aligned} J(\boldsymbol{\theta}) &= \frac{1}{2}\sum_{i=1}^{N}\boldsymbol{v}^T(i)\,\hat{\boldsymbol{R}}^{-1}\,\boldsymbol{v}(i) \\ &= \frac{1}{2}\sum_{i=1}^{N}\left[\boldsymbol{z}(i) - \boldsymbol{y}(i)\right]^T\hat{\boldsymbol{R}}^{-1}\left[\boldsymbol{z}(i) - \boldsymbol{y}(i)\right] \end{aligned} \qquad (6.30)$$

where the last two terms in Eq. (6.28) are dropped because they do not depend on the unknown model parameter vector $\boldsymbol{\theta}$, which includes all unknown parameters except the elements of the $\boldsymbol{R}$ matrix. Note that the $\boldsymbol{R}^{-1}$ matrix serves the practical purposes of accounting for different scaling or units among the multiple outputs, and also accounts for different noise levels on the outputs. Since the innovations $\boldsymbol{v}(i)$ in the cost function are output errors, this approach is called the output-error method.

The negative log-likelihood cost function is minimized using a relaxation technique, by computing $\hat{\boldsymbol{R}}$ from Eq. (6.29) for a given fixed $\boldsymbol{\theta}$, then fixing $\boldsymbol{R} = \hat{\boldsymbol{R}}$, and minimizing the cost in Eq. (6.30) with respect to $\boldsymbol{\theta}$. The idea is that optimization with respect to the complete set of unknown parameters in $\boldsymbol{\theta}$ and $\boldsymbol{R}$ is more well-conditioned if $\boldsymbol{\theta}$ and $\boldsymbol{R}$ are adjusted alternately, with one being allowed to vary while the other is held constant. The two steps are repeated until convergence criteria are satisfied. There is no general proof that this sequence will converge, but extensive practical experience has shown that the sequence does in fact converge.

Optimizing the cost in Eq. (6.30) can be done using the Newton-Raphson method, as discussed earlier. The gradient of the cost function is obtained as

$$\begin{aligned}\frac{\partial J(\boldsymbol{\theta})}{\partial \boldsymbol{\theta}} &= \sum_{i=1}^{N} \frac{\partial \boldsymbol{v}^T(i)}{\partial \boldsymbol{\theta}} \hat{\boldsymbol{R}}^{-1} \boldsymbol{v}(i) \\ &= -\sum_{i=1}^{N} \frac{\partial \boldsymbol{y}^T(i)}{\partial \boldsymbol{\theta}} \hat{\boldsymbol{R}}^{-1} \boldsymbol{v}(i)\end{aligned} \tag{6.31}$$

which is a vector with elements

$$\frac{\partial J(\boldsymbol{\theta})}{\partial \theta_j} = -\sum_{i=1}^{N} \frac{\partial \boldsymbol{y}^T(i)}{\partial \theta_j} \hat{\boldsymbol{R}}^{-1} \boldsymbol{v}(i) \qquad j = 1,2,\ldots,n_p \tag{6.32}$$

The second equality in Eq. (6.31) follows from Eq. (6.27) and the fact that the measurements $\boldsymbol{z}(i)$ do not depend on the model parameters $\boldsymbol{\theta}$. The elements of the second-order gradient matrix are

$$\frac{\partial^2 J(\boldsymbol{\theta})}{\partial \theta_j \partial \theta_k} = \sum_{i=1}^{N} \frac{\partial \boldsymbol{y}^T(i)}{\partial \theta_j} \hat{\boldsymbol{R}}^{-1} \frac{\partial \boldsymbol{y}(i)}{\partial \theta_k} - \sum_{i=1}^{N} \frac{\partial^2 \boldsymbol{y}(i)}{\partial \theta_j \partial \theta_k} \hat{\boldsymbol{R}}^{-1} \boldsymbol{v}(i) \tag{6.33}$$

$$j,k = 1,2,\ldots,n_p$$

If the second-order partial derivative term in Eq. (6.33) is neglected, the resulting optimization algorithm is called Gauss-Newton or modified Newton-Raphson. This simplification is made for practical reasons, because the second-order gradient is computationally expensive to obtain and susceptible to numerical error due to the higher-order differentiation. Since the second-order gradient term is multiplied by the residual $\boldsymbol{v}(i)$, the approximation gets better as the estimated parameter vector approaches the solution, and is very good near the solution.

Using the approximate second-order gradient matrix, the estimate for the parameter vector change is [cf. Eq. (6.17)]

$$\Delta\hat{\boldsymbol{\theta}} = \left[\sum_{i=1}^{N} \frac{\partial \boldsymbol{y}^T(i)}{\partial \boldsymbol{\theta}} \hat{\boldsymbol{R}}^{-1} \frac{\partial \boldsymbol{y}(i)}{\partial \boldsymbol{\theta}}\right]_{\boldsymbol{\theta}=\boldsymbol{\theta}_o}^{-1} \left[\sum_{i=1}^{N} \frac{\partial \boldsymbol{y}^T(i)}{\partial \boldsymbol{\theta}} \hat{\boldsymbol{R}}^{-1} \boldsymbol{v}(i)\right]_{\boldsymbol{\theta}=\boldsymbol{\theta}_o} \tag{6.34}$$

Elements of the $n_o \times n_p$ matrix $\partial \boldsymbol{y}/\partial \boldsymbol{\theta}$ are called output sensitivities. The output sensitivities quantify the change in the outputs due to changes in the parameters. Since $\hat{\boldsymbol{R}}^{-1}$ is typically diagonal, Eq. (6.34) shows that the output sensitivities must be linearly independent and nonzero for good matrix inversion and a reasonable $\Delta\hat{\boldsymbol{\theta}}$. When the output sensitivities are linearly independent and nonzero, each model parameter has a unique and significant impact on the model outputs, so that minimizing the output error will be a well-conditioned problem leading to accurate values for the unknown parameters.

Using the approximation for the second-order gradient of the cost function, and assuming a given constant $\boldsymbol{R} = \hat{\boldsymbol{R}}$, the Fisher information matrix is simplified to

$$\boldsymbol{M} \equiv -E\left[\frac{\partial^2 \ln \mathbb{L}(\boldsymbol{Z}_N;\boldsymbol{\theta})}{\partial \boldsymbol{\theta}\,\partial \boldsymbol{\theta}^T}\right] = \sum_{i=1}^{N} \frac{\partial \boldsymbol{y}^T(i)}{\partial \boldsymbol{\theta}} \hat{\boldsymbol{R}}^{-1} \frac{\partial \boldsymbol{y}(i)}{\partial \boldsymbol{\theta}} \tag{6.35}$$

The maximum likelihood parameter estimator can be expressed as

$$\hat{\boldsymbol{\theta}} = \boldsymbol{\theta}_o - \boldsymbol{M}_{\boldsymbol{\theta}=\boldsymbol{\theta}_o}^{-1} \left[\frac{\partial J(\boldsymbol{\theta})}{\partial \boldsymbol{\theta}}\right]_{\boldsymbol{\theta}=\boldsymbol{\theta}_o} \tag{6.36}$$

and the parameter covariance matrix satisfies

$$Cov\left(\hat{\boldsymbol{\theta}}\right) \geq \boldsymbol{M}_{\boldsymbol{\theta}=\hat{\boldsymbol{\theta}}}^{-1} \tag{6.37}$$

Equation (6.37) is the Cramér-Rao inequality, indicating the lower bound for the parameter covariance matrix (see Appendix B). A block diagram of the output-error method is shown in Fig. 6.2.

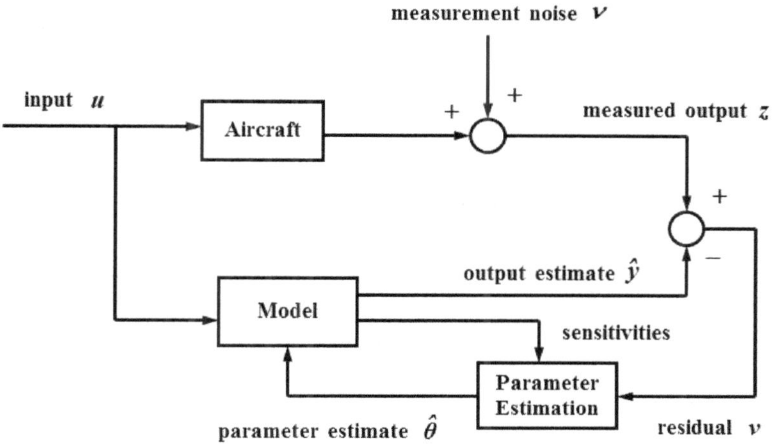

Figure 6.2 **Block diagram for output-error parameter estimation**

Equation (6.34) for the parameter vector estimate update can also be derived by replacing the model output in the cost function of Eq. (6.30) with a Taylor series in $\boldsymbol{\theta}$, expanded about a nominal parameter vector value $\boldsymbol{\theta}_0$, and truncated after the linear term, i.e.,

$$y(i) \approx y(i)\Big|_{\theta=\theta_0} + \frac{\partial y(i)}{\partial \boldsymbol{\theta}}\Bigg|_{\theta=\theta_0} (\boldsymbol{\theta} - \boldsymbol{\theta}_0) \qquad i = 1,2,\ldots,N \quad (6.38)$$

Equation (6.38) is a linear approximation for $y(i)$ in the neighborhood of a nominal starting value for the parameter vector, $\boldsymbol{\theta}_0$. Substituting this linear approximation into Eq. (6.30), setting the gradient $\partial J / \partial \boldsymbol{\theta}$ equal to zero, and solving for $\Delta\boldsymbol{\theta} \equiv \boldsymbol{\theta} - \boldsymbol{\theta}_0$ results in the modified Newton-Raphson parameter vector update expression in Eq. (6.34).

The model output is a nonlinear function of the parameter vector, due to the time integration involved in solving the dynamic system equations, regardless of whether or not the dynamic system is linear. It follows that the dynamic system model can in fact be an arbitrary nonlinear function of the parameter vector without any change in the procedure described in this section. Consequently, the output-error method can be used for arbitrarily nonlinear models. In particular, the full nonlinear aircraft equations of motion can be used as the dynamic system model equations, without any change to the output-error cost formulation or the nonlinear optimization. Similarly, the nonlinear least squares problem formulated in Chapter 5 can be solved using the same nonlinear optimization technique described here, among others.

6.3 Computational Aspects

Aspects of computing maximum likelihood estimates are presented here in connection with the output-error algorithm. Most of the material is also applicable to the more general case where the dynamic system model includes process noise.

As pointed out in Ref. [6.16], practical application of the maximum likelihood estimator to flight data sometimes leads to difficulties. For example, the shape of the log-likelihood function can be far from quadratic, as assumed by many nonlinear optimization methods, and can have multiple minima or discontinuities. The surface of the log-likelihood function can be flat or wavy in places. These properties can prevent finding the global minimum of the negative log-likelihood function (equivalently, the global maximum of the likelihood function), which is necessary in order to achieve unbiased and efficient parameter estimates. Furthermore, the information matrix can be singular or nearly singular, creating problems with its inversion.

The parameter estimation method for the output-error method using modified Newton-Raphson optimization explained earlier can be summarized as

$$\hat{\boldsymbol{\theta}} = \boldsymbol{\theta}_o + \Delta\hat{\boldsymbol{\theta}} \tag{6.39a}$$

$$\Delta\hat{\boldsymbol{\theta}} = -\boldsymbol{M}^{-1}_{\theta=\theta_o}\,\boldsymbol{g}_{\theta=\theta_o} \tag{6.39b}$$

$$Cov\left(\hat{\boldsymbol{\theta}}\right) \ge \boldsymbol{M}^{-1}_{\theta=\hat{\theta}} \tag{6.39c}$$

where

$$\boldsymbol{M}_{\theta=\theta_o} = \sum_{i=1}^{N}\left[\boldsymbol{S}^T(i)\,\hat{\boldsymbol{R}}^{-1}\,\boldsymbol{S}(i)\right]_{\theta=\theta_o} \tag{6.40a}$$

$$\boldsymbol{g}_{\theta=\theta_o} = \sum_{i=1}^{N}\left[\boldsymbol{S}^T(i)\,\hat{\boldsymbol{R}}^{-1}\,\boldsymbol{v}(i)\right]_{\theta=\theta_o} \tag{6.40b}$$

$$\boldsymbol{S}(i) = \left[s_{jk}(i)\right] = \left[\frac{\partial y_j(i,\boldsymbol{\theta})}{\partial \theta_k}\right] \qquad \begin{aligned} j &= 1,2,\ldots,n_o \\ k &= 1,2,\ldots,n_p \end{aligned} \tag{6.40c}$$

$$\boldsymbol{v}(i) = \boldsymbol{z}(i) - \hat{\boldsymbol{y}}(i,\boldsymbol{\theta}) \tag{6.40d}$$

The information matrix $\boldsymbol{M}$ has dimensions $n_p \times n_p$, and the sensitivity matrix $\boldsymbol{S}(i)$ has dimensions $n_o \times n_p$. To arrive at the global maximum of the likelihood function using the modified Newton-Raphson method, the nominal parameter vector estimate $\boldsymbol{\theta}_o$ must be near the value of $\boldsymbol{\theta}$ corresponding to the minimum of the cost function $J(\boldsymbol{\theta})$ for a given $\hat{\boldsymbol{R}}$. The reason for this is simply that the modified Newton-Raphson method is based on the assumption that $\boldsymbol{\theta}_o$ is close to the solution. If the starting value $\boldsymbol{\theta}_o$ is not in the same valley of the cost function as the global solution, then the result will be either convergence to a local minimum, or divergence, the latter being the more common result for flight test data analysis.

Fortunately, excellent starting values for $\boldsymbol{\theta}_o$ can be obtained using the linear regression methods described in Chapter 5. Alternatively, Section 6.3.1 includes a description of a method that can be used to eliminate the need to compute starting values $\boldsymbol{\theta}_o$. When there is some doubt as to the model structure, model structure determination methods discussed in Chapter 5 can be used to identify both an adequate model structure and good starting values for the parameters.

An initial estimate of $\boldsymbol{R}$ can be obtained using the starting values $\boldsymbol{\theta}_o$ and Eq. (6.29), or from numerically smoothing the measurements (see Chapter 11), or $\hat{\boldsymbol{R}}$ can be initially set to the identity matrix, $\hat{\boldsymbol{R}} = \boldsymbol{I}$. The updates of $\hat{\boldsymbol{R}}$ using Eq. (6.29) should occur after each optimization of $J(\boldsymbol{\theta})$ in Eq. (6.30) satisfies the convergence criteria. The combined optimization is finished when convergence criteria for both $\hat{\boldsymbol{\theta}}$ and $\hat{\boldsymbol{R}}$ are satisfied. Typical convergence criteria involve one or more of the following:

1) Absolute value of the elements of $\Delta\boldsymbol{\theta}$ are sufficiently small.
2) Changes in the cost $J(\hat{\boldsymbol{\theta}})$ are sufficiently small for consecutive iterations.
3) Absolute values of the elements of the cost gradient $\boldsymbol{g}$ are sufficiently close to zero.
4) Changes in the elements of $\hat{\boldsymbol{R}}$ are sufficiently small.

The following convergence criteria have been found to work well for aircraft parameter estimation problems:

$$\left|\left(\hat{\theta}_j\right)_k - \left(\hat{\theta}_j\right)_{k-1}\right| < 1.0 \times 10^{-5} \quad \forall j, \quad j = 1,2,\ldots,n_p \tag{6.41a}$$

or alternatively,

$$\frac{\left\|\hat{\boldsymbol{\theta}}_k - \hat{\boldsymbol{\theta}}_{k-1}\right\|}{\left\|\hat{\boldsymbol{\theta}}_{k-1}\right\|} < 0.001 \tag{6.41b}$$

where k denotes the kth iteration and $\|\ \|$ indicates the Euclidean norm, or the square root of the sum of squares of the vector elements. At the same time,

$$\left|\frac{J(\hat{\boldsymbol{\theta}}_k) - J(\hat{\boldsymbol{\theta}}_{k-1})}{J(\hat{\boldsymbol{\theta}}_{k-1})}\right| < 0.001 \tag{6.41c}$$

$$\left|\left(\frac{\partial J(\boldsymbol{\theta})}{\partial \theta_j}\right)_{\boldsymbol{\theta}=\hat{\boldsymbol{\theta}}_k}\right| < 0.05 \quad \forall j, \quad j = 1,2,\ldots,n_p \tag{6.41d}$$

$$\left|\frac{(\hat{r}_{jj})_k - (\hat{r}_{jj})_{k-1}}{(\hat{r}_{jj})_{k-1}}\right| < 0.05 \quad \forall j \quad j = 1,2,\ldots,n_o \tag{6.41e}$$

where $\hat{r}_{jj}$ denotes the estimate of the jth diagonal element of $\hat{\boldsymbol{R}}$.

In some cases, varying the magnitude of $\Delta\hat{\boldsymbol{\theta}}$ in the direction computed from Eq. (6.39b) might provide an additional decrease in the cost function for each iteration. This can be done by multiplying the expression for $\Delta\hat{\boldsymbol{\theta}}$ by a constant c,

$$\Delta\hat{\boldsymbol{\theta}} = -\,c\,\boldsymbol{M}_{\boldsymbol{\theta}=\boldsymbol{\theta}_o}^{-1}\,\boldsymbol{g}_{\boldsymbol{\theta}=\boldsymbol{\theta}_o} \tag{6.42}$$

This approach is discussed in Ref. [6.17]. Reference [6.3] gives the formula for computing c using quadratic interpolation. However, this approach is computationally expensive because it requires three computed values of the cost for each iteration. For practical aircraft parameter estimation problems, this refinement of the modified Newton-Raphson algorithm is probably not worth the trouble, because modern computers are so fast that the difference in convergence time is small.

6.3.1 Computing Sensitivities

Two methods can be used to compute the sensitivities $\partial y/\partial \theta_j \;\; j = 1,2,\ldots,n_p$ appearing in the equations of the preceding section. The methods are sometimes called the analytical approach and the numerical approach.

In the analytical approach, equations for the output sensitivities $\partial y/\partial \theta_j \;\; j=1,2,\ldots,n_p$ are generated by taking partial derivatives of the output equation with respect to the unknown parameters $\theta_j \;\; j=1,2,\ldots,n_p$, and solving the resulting equations. For the linear dynamic system in Eqs. (6.26), the result is

$$\frac{\partial y}{\partial \theta_j} = \boldsymbol{C}\frac{\partial \boldsymbol{x}}{\partial \theta_j} + \frac{\partial \boldsymbol{C}}{\partial \theta_j}\boldsymbol{x} + \frac{\partial \boldsymbol{D}}{\partial \theta_j}\boldsymbol{u} \qquad j=1,2,\ldots,n_p \tag{6.43}$$

Equations (6.43) are called the output sensitivity equations. Since $\boldsymbol{C}$ and $\boldsymbol{D}$ are constant matrices containing unknown parameters, $\partial \boldsymbol{C}/\partial \theta_j$ and $\partial \boldsymbol{D}/\partial \theta_j$ are also constant matrices. The state $\boldsymbol{x}$ is computed from the state equations for the dynamic system

$$\dot{\boldsymbol{x}} = \boldsymbol{A}\boldsymbol{x} + \boldsymbol{B}\boldsymbol{u} \qquad \boldsymbol{x}(0) = \boldsymbol{x}_o \tag{6.44}$$

The state sensitivities $\partial \boldsymbol{x}/\partial \theta_j \;\; j=1,2,\ldots,n_p$ are computed by solving the state sensitivity equations, which are obtained by differentiating the state equation (6.44) with respect to the unknown parameters,

$$\frac{d}{dt}\left(\frac{\partial \boldsymbol{x}}{\partial \theta_j}\right) = \boldsymbol{A}\frac{\partial \boldsymbol{x}}{\partial \theta_j} + \frac{\partial \boldsymbol{A}}{\partial \theta_j}\boldsymbol{x} + \frac{\partial \boldsymbol{B}}{\partial \theta_j}\boldsymbol{u} \qquad \frac{\partial \boldsymbol{x}(0)}{\partial \theta_j} = \boldsymbol{0} \qquad j=1,2,\ldots,n_p \tag{6.45}$$

On the left side of Eq. (6.45), the order of differentiation with respect to time and θ_j was switched, which can be done under the assumption that $\boldsymbol{x}$ is analytic. The initial conditions for the state sensitivity equations are zero when the initial condition $\boldsymbol{x}_o$ is known, as assumed here. The initial state vector $\boldsymbol{x}_0$ can be obtained from locally smoothed initial measurements (see Chapter 11). Equations (6.44) and (6.45) are integrated numerically, using a Runge-Kutta method, for example.

For n_s states, there are $n_s n_p$ differential equations to be solved for the state sensitivities, plus n_s for the dynamic system equations, for a total of $n_s\left(n_p+1\right)$ differential equations to solve. A scheme for reducing the sensitivity calculations is presented in Ref. [6.18].

Alternatively, discrete formulas can be used with the state transition matrix. The discrete formulas follow from Eqs. (2.20) and (2.21) as

$$\boldsymbol{x}(i) = \boldsymbol{\Phi}\boldsymbol{x}(i-1) + \boldsymbol{\Gamma}\frac{\left[\boldsymbol{u}(i)+\boldsymbol{u}(i-1)\right]}{2} \qquad \boldsymbol{x}(0) = \boldsymbol{x}_0 \tag{6.46a}$$

where

$$\boldsymbol{\Phi} \equiv e^{A\Delta t} \qquad \boldsymbol{\Gamma} \equiv \int_0^{\Delta t} e^{A\tau} d\tau\, \boldsymbol{B} = \left[\boldsymbol{A}^{-1}\left(e^{A\Delta t} - \boldsymbol{I}\right)\right]\boldsymbol{B} \tag{6.46b}$$

so that

$$\boldsymbol{x}(i) = e^{A\Delta t}\boldsymbol{x}(i-1) + \boldsymbol{A}^{-1}\left(e^{A\Delta t} - \boldsymbol{I}\right)\boldsymbol{B}\frac{\left[\boldsymbol{u}(i)+\boldsymbol{u}(i-1)\right]}{2} \qquad \boldsymbol{x}(0) = \boldsymbol{x}_0 \tag{6.46c}$$

The quantity $\left[\boldsymbol{u}(i)+\boldsymbol{u}(i-1)\right]/2$ is an averaging of the input values at the endpoints of the time interval, which improves the accuracy of the solution for the discrete-time dynamic equation. Because of the similarity of Eqs. (6.44) and (6.45), the discrete-time state sensitivity equations can be written by analogy to Eq. (6.46c),

$$\begin{aligned} \frac{\partial \boldsymbol{x}(i)}{\partial \theta_j} &= e^{A\Delta t}\frac{\partial \boldsymbol{x}(i-1)}{\partial \theta_j} + \boldsymbol{A}^{-1}\left(e^{A\Delta t} - \boldsymbol{I}\right) \\ &\quad \cdot \left[\frac{\partial \boldsymbol{A}}{\partial \theta_j}\frac{\left[\boldsymbol{x}(i)+\boldsymbol{x}(i-1)\right]}{2} + \frac{\partial \boldsymbol{B}}{\partial \theta_j}\frac{\left[\boldsymbol{u}(i)+\boldsymbol{u}(i-1)\right]}{2}\right] \\ \frac{\partial \boldsymbol{x}(0)}{\partial \theta_j} &= \boldsymbol{0} \end{aligned} \tag{6.46d}$$

The forcing functions for the state sensitivities in Eq. (6.45) are the states and controls. Taylor and Iliff [6.19] recommend computing the state sensitivities initially using measured states, rather than states obtained from the dynamic system equations. This has the effect of generating sensitivities that are close enough to the final sensitivities that the parameter change computed from Eq. (6.34) using these initial sensitivities will put the estimated parameter vector in close proximity to the solution, regardless of the starting point. Therefore, the starting value $\boldsymbol{\theta}_o$ can be a zero vector, as long as the initial sensitivities are computed using measured states in the state sensitivity equations. This approach works well in practice, and avoids the need to assemble a good initial parameter vector estimate from equation-error methods or wind tunnel results. The price for this convenience is some additional programming to make the first iteration of the state sensitivity calculations different from all subsequent calculations, along with measuring the states.

The numerical approach to computing output sensitivities uses an approximation to the definition of numerical partial derivatives,

$$\frac{\partial y}{\partial \theta_j} = \frac{y\left(\boldsymbol{\theta}_o + \delta\boldsymbol{\theta}_j\right) - y\left(\boldsymbol{\theta}_o\right)}{\left|\delta\boldsymbol{\theta}_j\right|} \qquad j = 1, 2, \ldots, n_p \tag{6.47a}$$

or

$$\frac{\partial y}{\partial \theta_j} = \frac{y(\boldsymbol{\theta}_o + \delta\boldsymbol{\theta}_j) - y(\boldsymbol{\theta}_o - \delta\boldsymbol{\theta}_j)}{2|\delta\boldsymbol{\theta}_j|} \qquad j = 1,2,\ldots,n_p \qquad (6.47b)$$

where $\delta\boldsymbol{\theta}_j$ denotes a vector with all zero elements except for the jth element, which contains the perturbation for parameter θ_j, and $|\delta\boldsymbol{\theta}_j|$ is the scalar magnitude of $\delta\boldsymbol{\theta}_j$. The size of the perturbation for each parameter can be chosen according to how that parameter influences the outputs, but in practice using perturbations that are fixed at one percent of the nominal parameter value works well for many applications. If the nominal parameter value is near zero, then the perturbation can be set to 0.01. The output $\boldsymbol{y}$ is computed by numerically solving the dynamic equations, using the nominal and perturbed values of the parameter vector.

Equation (6.47a) implements forward finite differences, and Eq. (6.47b) implements central finite differences. Forward finite differences requires $(n_p + 1)$ solutions of the dynamic equations, whereas central finite differences requires $2n_p$ solutions. The central finite difference approach is much more accurate, since its error in approximating the partial derivatives is $O\left(|\delta\boldsymbol{\theta}_j|^2\right)$, whereas the errors for forward finite differences are $O\left(|\delta\boldsymbol{\theta}_j|\right)$. Note that the sensitivities must be calculated for every iteration of the modified Newton-Raphson method, so the difference in the amount of computation can be large. However, the central finite difference calculation is preferred, because the number of iterations of Eqs. (6.39) will generally be lower with higher accuracy sensitivities.

When the dynamic system model is nonlinear,

$$\dot{\boldsymbol{x}}(t) = \boldsymbol{f}(\boldsymbol{x},\boldsymbol{u},\boldsymbol{\theta}) \qquad \boldsymbol{x}(0) = \boldsymbol{x}_o \qquad (6.48a)$$

$$\boldsymbol{y} = \boldsymbol{g}(\boldsymbol{x},\boldsymbol{u},\boldsymbol{\theta}) \qquad (6.48b)$$

the resulting sensitivity equations are

$$\frac{d}{dt}\left(\frac{\partial \boldsymbol{x}}{\partial \theta_j}\right) = \frac{\partial \boldsymbol{f}}{\partial \boldsymbol{x}}\frac{\partial \boldsymbol{x}}{\partial \theta_j} + \frac{\partial \boldsymbol{f}}{\partial \theta_j} \qquad \frac{\partial \boldsymbol{x}(0)}{\partial \theta_j} = \boldsymbol{0} \qquad j = 1,2,\ldots,n_p \; (6.49a)$$

$$\frac{\partial \boldsymbol{y}}{\partial \theta_j} = \frac{\partial \boldsymbol{g}}{\partial \boldsymbol{x}}\frac{\partial \boldsymbol{x}}{\partial \theta_j} + \frac{\partial \boldsymbol{g}}{\partial \theta_j} \qquad j = 1,2,\ldots,n_p \qquad (6.49b)$$

In this case, the numerical approach is used more often, because the analytic derivatives of the nonlinear functions $\boldsymbol{f}$ and $\boldsymbol{g}$ can be complicated and are different for every problem.

A generalization of the finite difference method is the local surface approximation method of Ref. [6.20]. In this technique, the output sensitivities are computed using the slope information from a local surface approximation of $\boldsymbol{y}(\boldsymbol{\theta})$. The approximations are made near $\boldsymbol{y}(\boldsymbol{\theta}_o)$. According to Ref. [6.20], this approach requires less computational effort than either a finite difference method or integration of state and output sensitivity equations.

Example 6.1

Flight test data from Example 5.1, run 1, are used for this example. The data are from a lateral maneuver of the NASA Twin Otter aircraft. Figure 5.4 shows the measured inputs and outputs. In Chapter 5, this data was analyzed using linear regression. In this example, the output-error method will be used to obtain parameter estimates from the same data.

In the output-error method, the equations of motion are used together, so the lateral aerodynamic model parameters are estimated all at once. This contrasts with the equation-error method, where a separate analysis is done for each dynamic equation to estimate the parameters associated with each lateral aerodynamic coefficient C_Y, C_l, and C_n.

The maneuver shown in Fig. 5.4 is a small perturbation maneuver at low angle of attack, so a linearized dynamic model can be used, with the aerodynamic model equations given by Eq. (5.78). This results in the following dynamic model equations [cf. Eqs. (3.111) and (3.116)],

$$\dot{\beta} = \frac{\bar{q}_o S}{m V_o}\left(C_{Y_\beta}\beta + C_{Y_r}\frac{rb}{2V_o} + C_{Y_{\delta_r}}\delta_r \right) + p\, sin\alpha_o - r\, cos\alpha_o + \frac{g\cos\theta_o}{V_o}\phi + b_{\dot{\beta}} \tag{6.50a}$$

$$I_x\dot{p} - I_{xz}\dot{r} = \bar{q}_o S b\left(C_{l_\beta}\beta + C_{l_p}\frac{pb}{2V_o} + C_{l_r}\frac{rb}{2V_o} + C_{l_{\delta_a}}\delta_a + C_{l_{\delta_r}}\delta_r \right) + b_{\dot{p}} \tag{6.50b}$$

$$I_z\dot{r} - I_{xz}\dot{p} = \bar{q}_o S b\left(C_{n_\beta}\beta + C_{n_p}\frac{pb}{2V_o} + C_{n_r}\frac{rb}{2V_o} + C_{n_{\delta_a}}\delta_a + C_{n_{\delta_r}}\delta_r \right) + b_{\dot{r}} \tag{6.50c}$$

$$\dot{\phi} = p + tan\theta_o r + b_{\dot{\phi}} \tag{6.50d}$$

$$a_y = \frac{\bar{q}_o S}{mg}\left(C_{Y_\beta}\beta + C_{Y_r}\frac{rb}{2V_o} + C_{Y_{\delta_r}}\delta_r \right) + b_{a_Y} \tag{6.51}$$

Equations (6.50) are linear dynamic equations for the lateral states $\beta, p, r,$ and ϕ, which are also measured outputs, and Eq. (6.51) is an output equation for lateral acceleration. Perturbation quantities are used in these linearized equations of motion, and the aerodynamic model structure is the same as for Example 5.1. Values with subscript o are constants equal to measured trim values at the start of the maneuver, or mean values computed from the measured data. In this example, mean values are used. Mass and inertia quantities are also mean values, although, for simplicity, this is not shown in the notation.

Note that all of the dynamic equations (6.50) include a bias parameter. This is to account for any measurement biases not removed by data compatibility analysis (see Chapter 9) or biases that result from using perturbation quantities. For example, if the initial measured value of rudder deflection is biased or noisy, then using that initial value to generate the perturbation quantity δ_r will artificially induce a bias in δ_r. Estimating a bias parameter in each dynamic equation will remove such effects. Without the bias parameters for each dynamic equation, integration of the dynamic equations can result in a drift in the computed state time histories. The bias term in the output equation for lateral acceleration is introduced to account for biases from inaccurate trim values (as described earlier) and bias in the measured perturbation output a_y.

Equations (6.50) and (6.51) can be written in state-space form as

$$\begin{bmatrix} 1 & 0 & 0 & 0 \\ 0 & I_{xx} & -I_{xz} & 0 \\ 0 & -I_{xz} & I_{zz} & 0 \\ 0 & 0 & 0 & 1 \end{bmatrix}\begin{bmatrix} \dot{\beta} \\ \dot{p} \\ \dot{r} \\ \dot{\phi} \end{bmatrix} = \begin{bmatrix} k_1 C_{Y_\beta} & \sin\alpha_o & \left(k_1 C_{Y_r} k_2 - \cos\alpha_o\right) & \dfrac{g\cos\theta_o}{V_o} \\ k_3 C_{l_\beta} & k_3 C_{l_p} k_2 & k_3 C_{l_r} k_2 & 0 \\ k_3 C_{n_\beta} & k_3 C_{n_p} k_2 & k_3 C_{n_r} k_2 & 0 \\ 0 & 1 & \tan\theta_o & 0 \end{bmatrix}\begin{bmatrix} \beta \\ p \\ r \\ \phi \end{bmatrix}$$

$$+\begin{bmatrix} 0 & k_1 C_{Y_{\delta_r}} & b_{\dot{\beta}} \\ k_3 C_{l_{\delta_a}} & k_3 C_{l_{\delta_r}} & b_{\dot{p}} \\ k_3 C_{n_{\delta_a}} & k_3 C_{n_{\delta_r}} & b_{\dot{r}} \\ 0 & 0 & b_{\dot{\phi}} \end{bmatrix}\begin{bmatrix} \delta_a \\ \delta_r \\ 1 \end{bmatrix}$$

(6.52a)

$$\begin{bmatrix} \beta \\ p \\ r \\ \phi \\ a_y \end{bmatrix} = \begin{bmatrix} 1 & 0 & 0 & 0 \\ 0 & 1 & 0 & 0 \\ 0 & 0 & 1 & 0 \\ 0 & 0 & 0 & 1 \\ k_4 C_{Y_\beta} & 0 & k_4 C_{Y_r} k_2 & 0 \end{bmatrix} \begin{bmatrix} \beta \\ p \\ r \\ \phi \end{bmatrix} + \begin{bmatrix} 0 & 0 & 0 \\ 0 & 0 & 0 \\ 0 & 0 & 0 \\ 0 & 0 & 0 \\ 0 & k_4 C_{Y_{\delta_r}} & b_{a_y} \end{bmatrix} \begin{bmatrix} \delta_a \\ \delta_r \\ 1 \end{bmatrix} \tag{6.52b}$$

where

$$k_1 = \frac{\bar{q}_o S}{m V_o} \qquad k_2 = \frac{b}{2V_o} \qquad k_3 = \bar{q}_o S b \qquad k_4 = \frac{\bar{q}_o S}{mg}$$

Equations (6.50b) and (6.50c) (equivalently, the second and third rows of Eq. (6.52a)) must be combined to produce equations with only a single state derivative on the left hand side. This is the form required for numerical integration techniques such as Runge-Kutta methods. The required calculations are simple, and were done in Chapter 3 to arrive at Eqs. (3.41). Equations (6.50b) and (6.50c) are transformed as follows prior to numerical integration:

$$\dot{p} = c_3 L + c_4 N + b_{\dot{p}}' \tag{6.53a}$$

$$\dot{r} = c_4 L + c_9 N + b_{\dot{r}}' \tag{6.53b}$$

where

$$L = \bar{q}_o S b \left(C_{l_\beta} \beta + C_{l_p} \frac{pb}{2V_o} + C_{l_r} \frac{rb}{2V_o} + C_{l_{\delta_a}} \delta_a + C_{l_{\delta_r}} \delta_r \right)$$

$$N = \bar{q}_o S b \left(C_{n_\beta} \beta + C_{n_p} \frac{pb}{2V_o} + C_{n_r} \frac{rb}{2V_o} + C_{n_{\delta_a}} \delta_a + C_{n_{\delta_r}} \delta_r \right)$$

$$\Gamma = I_x I_z - I_{xz}^2 \qquad c_3 = I_z / \Gamma \qquad c_4 = I_{xz} / \Gamma \qquad c_9 = I_x / \Gamma$$

$$b_{\dot{p}}' = c_3 b_{\dot{p}} + c_4 b_{\dot{r}}$$

$$b_{\dot{r}}' = c_4 b_{\dot{p}} + c_9 b_{\dot{r}}$$

The bias parameters exist only to correct for biases in the measured perturbation quantities. Because of this, it is acceptable to re-define the bias terms as shown above. Such parameters, sometimes called nuisance parameters, must be estimated along with the others, although their estimates are not of direct interest for the problem. When the identified model is used for analysis or control system design, these nuisance parameters are omitted from the model.

Modified Newton-Raphson optimization [cf. Eqs. (6.39)-(6.40)] was used to minimize the cost function in Eq. (6.30), with the measurement noise covariance matrix estimate computed from Eq. (6.29). Output sensitivities were computed numerically using central finite differences [cf. Eq. (6.47b)]. The vector of parameters to be estimated was:

$$\boldsymbol{\theta} = \Big[\; C_{Y_\beta} \;\; C_{Y_r} \;\; C_{Y_{\delta_r}} \;\; b_{\dot{\beta}} \;\; C_{l_\beta} \;\; C_{l_p} \;\; C_{l_r} \;\; C_{l_{\delta_a}} \;\; C_{l_{\delta_r}} \;\; b_{\dot{p}}' \;\; C_{n_\beta} \;\; C_{n_p} \;\; C_{n_r} \;\; C_{n_{\delta_a}} \;\; C_{n_{\delta_r}} \;\; b_{\dot{r}}' \;\; b_{\dot{\phi}} \;\; b_{a_y} \Big]^T$$

Figure 6.3 shows the model fit to the measured outputs $\beta, p, r, \phi,$ and a_y. The plots show excellent model fit to the measured data, using the assumed model structure. The solution took 41 modified Newton-Raphson iterations, and 5 updates of $\hat{\boldsymbol{R}}$. Criteria in Eq. (6.41) were used to define convergence.

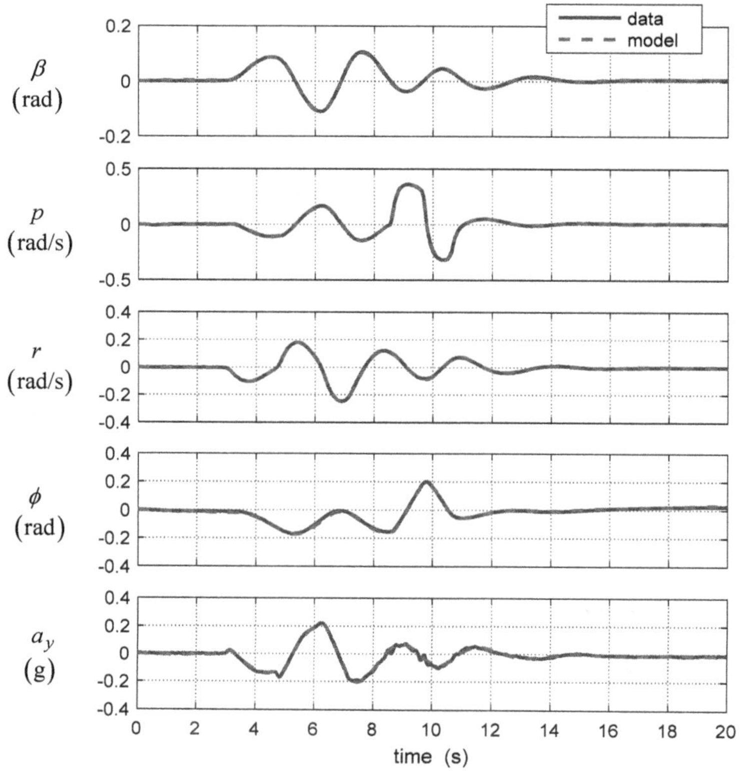

Figure 6.3 **Output-error model fit to measured perturbation outputs for lateral maneuver, run 1**

Parameter estimation results from the output-error method are shown in Table 6.1, together with results obtained using the equation-error method demonstrated in Example 5.1. The estimated parameter covariance matrix for the output-error method was computed from Eq. (6.39c) as the Cramér-Rao lower bound, $Cov(\hat{\boldsymbol{\theta}}) = \boldsymbol{M}^{-1}_{\boldsymbol{\theta}=\hat{\boldsymbol{\theta}}}$. In this calculation, the residuals are assumed to be white. The next example in this chapter demonstrates the changes to the estimated parameter errors when the residual coloring is taken into account.

Table 6.1 **Parameter estimation results for lateral maneuver, run 1**

	Output-error		Equation-error	
Parameter	$\hat{\theta}$	$s(\hat{\theta})$	$\hat{\theta}$	$s(\hat{\theta})$
C_{Y_β}	-8.65×10^{-1}	2.83×10^{-3}	-8.45×10^{-1}	7.41×10^{-3}
C_{Y_r}	9.33×10^{-1}	1.67×10^{-2}	9.63×10^{-1}	5.53×10^{-2}
$C_{Y_{\delta_r}}$	3.80×10^{-1}	8.07×10^{-3}	3.43×10^{-1}	2.89×10^{-2}
C_{l_β}	-1.20×10^{-1}	6.07×10^{-4}	-1.05×10^{-1}	2.54×10^{-3}
C_{l_p}	-5.86×10^{-1}	2.61×10^{-3}	-5.28×10^{-1}	9.93×10^{-3}
C_{l_r}	1.89×10^{-1}	1.83×10^{-3}	1.93×10^{-1}	8.33×10^{-3}
$C_{l_{\delta_a}}$	-2.28×10^{-1}	9.21×10^{-4}	-2.12×10^{-1}	3.39×10^{-3}
$C_{l_{\delta_r}}$	4.22×10^{-2}	7.48×10^{-4}	2.99×10^{-2}	3.86×10^{-3}
C_{n_β}	8.64×10^{-2}	2.15×10^{-4}	8.54×10^{-2}	3.58×10^{-4}
C_{n_p}	-6.44×10^{-2}	9.83×10^{-4}	-5.15×10^{-2}	1.43×10^{-3}
C_{n_r}	-1.91×10^{-1}	6.25×10^{-4}	-1.98×10^{-1}	1.30×10^{-3}
$C_{n_{\delta_a}}$	-2.98×10^{-3}	3.62×10^{-4}	2.34×10^{-3}	5.00×10^{-4}
$C_{n_{\delta_r}}$	-1.36×10^{-1}	2.70×10^{-4}	-1.31×10^{-1}	5.97×10^{-4}

It is also possible to use the measured inputs and outputs directly in the output-error method, rather than using perturbation quantities. The only difference is that the bias parameters then also take on the task of estimating the constants that would have been subtracted to produce perturbation quantities. For example, in Eq. (6.50a), the bias term $b_{\dot{\beta}}$ would now also include the term $-(\bar{q}_o S/mV_o)C_{Y_{\delta_r}}\delta_{r_o}$ (among others), where δ_{r_o} is trim value of the rudder. These terms were already removed when perturbation quantities were used in the analysis. Since the biases are nuisance parameters anyway, their role can be changed in this way without any effect on the useful results.

Using measured inputs and outputs in the analysis puts the burden of estimating the trim value to be removed into the bias term estimate. The disadvantage is that all of the trim value estimates for each equation are lumped into one parameter, so it is not clear what term was effectively subtracted for each variable. The advantage is that the analysis is done using the measured physical magnitudes of the measured inputs and outputs, and it is not necessary to determine the trim value that should be subtracted for the perturbations, since those values are estimated by the bias term. There is not much difference either way for maneuvers at low angles of attack where most of the trim quantities are small. However, the issue increases in prominence as the trim angle of attack for the maneuver increases and the trim values become larger, for both longitudinal and lateral maneuvers.

Figure 6.4 shows measured and predicted outputs for a different lateral maneuver at a similar flight condition, run 2. This is the same prediction maneuver used in Example 5.1. For the prediction case, the bias parameters $b_{\dot{\beta}}$, $b'_{\dot{p}}$, $b'_{\dot{r}}$, $b_{\dot{\phi}}$, and b_{a_y} were re-estimated using output-error parameter estimation and the prediction data, to account for small bias errors in the measurements for the prediction maneuver. All other model parameter were held fixed at the estimated values given in Table 6.1. The plots show that the identified model is a good predictor for a maneuver with different inputs at a similar flight condition. ■

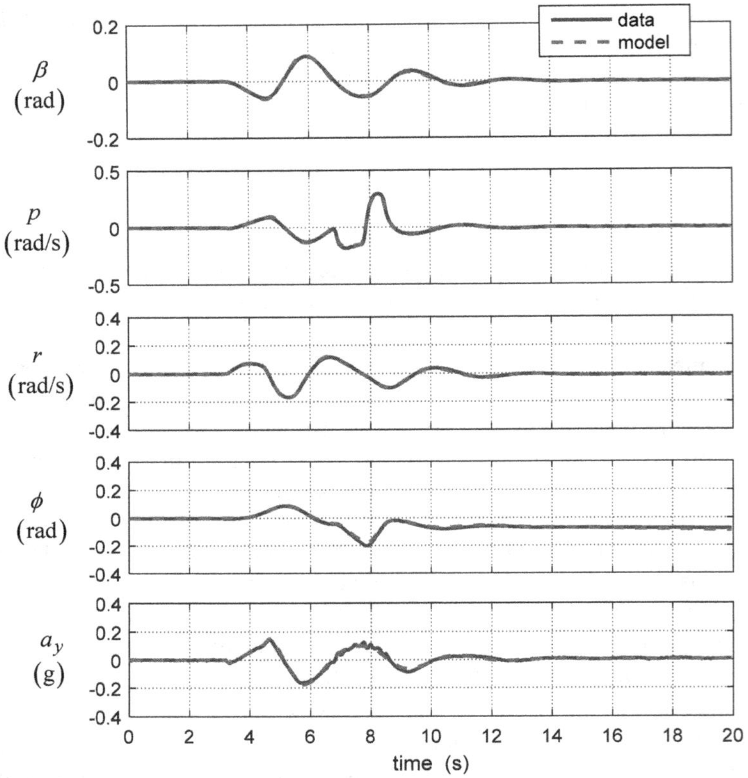

Figure 6.4 **Model prediction for lateral maneuver, run 2**

6.3.2 Nearly Singular Information Matrix

The problem of a nearly singular information matrix, also called an ill-conditioned information matrix, has already been discussed in connection with properties of the $\boldsymbol{X}^T\boldsymbol{X}$ matrix for linear regression. In that case, the sensitivities of the model output to the parameters were just the columns of the $\boldsymbol{X}$ matrix. For the case where the output is a vector nonlinear function of the parameters, the analog of the $\boldsymbol{X}$ matrix is the matrix of output sensitivities $\partial \boldsymbol{y}/\partial \boldsymbol{\theta}$, which are weighted in the information matrix using the noise covariance matrix [cf. Eq. (6.40)]. This arrangement includes contributions to the information matrix from each element of the output vector, using signal-to-noise ratio for the modeling problem, roughly speaking.

A common reason for an ill-conditioned information matrix is having too many unknown model parameters, also known as overparameterization. When this happens, the information content in the data can be too low for the number of estimated parameters required, so that the estimator cannot produce results with sufficient accuracy, or the estimation process can fail completely. Another cause is misspecification of the model, where changes in more than one model parameter produce nearly equivalent changes in the outputs, or perhaps one or more model parameters have little or no effect on the outputs. A third cause is insufficient information content in the data, where there is so little movement in the outputs that it appears that one or more parameters have little or no effect on the outputs; i.e., the corresponding output sensitivities are near zero. Another form of this problem occurs when a model parameter is associated with a quantity that is constant or nearly constant, which causes the associated parameter to be confounded with the bias parameter. The latter problem could also be classified as model misspecification.

There are two main consequences of a nearly singular information matrix:

1) The information matrix $\boldsymbol{M}$ can be negative definite, resulting in a cost increase $J\left(\hat{\boldsymbol{\theta}}_{k+1}\right) > J\left(\hat{\boldsymbol{\theta}}_k\right)$ for the modified Newton-Raphson step.

2) The modified Newton-Raphson step size $\Delta\hat{\boldsymbol{\theta}}$ can be large in one or more directions.

The second problem can be seen from the decomposition of the inverse $\boldsymbol{M}$ matrix as

$$\boldsymbol{M}^{-1} = \sum_{j=1}^{n_p} \frac{1}{\lambda_j} \boldsymbol{t}_j \boldsymbol{t}_j^T \tag{6.54}$$

as already discussed in Chapter 5. In Eq. (6.54), λ_j are the eigenvalues of $\boldsymbol{M}$ corresponding to the eigenvectors $\boldsymbol{t}_j$. Then the modified Newton-Raphson step in the direction of $\boldsymbol{t}_j$ is found by combining the previous expression with Eq. (6.39b),

$$\Delta\hat{\boldsymbol{\theta}}_{\lambda_j} = -\frac{1}{\lambda_j}\left(\boldsymbol{t}_j^T \boldsymbol{g}\right)\boldsymbol{t}_j \tag{6.55}$$

This step will be large for small λ_j, in a direction $\boldsymbol{t}_j$ for which little information is available. The parameter step $\Delta\hat{\boldsymbol{\theta}}$ is therefore dominated by bad information, resulting in a parameter vector update does not approach the minimum of the cost function.

There are several techniques for dealing with an ill-conditioned information matrix. Three of them associated with the modified Newton-Raphson method will be described here.

Rank Deficient Method

The rank deficient method is based on the reduced-order inverse of $\boldsymbol{M}$ using the singular value decomposition (SVD),

$$\boldsymbol{M} = \boldsymbol{U}\,\boldsymbol{D}\,\boldsymbol{T}^T \tag{6.56}$$

discussed in Chapter 5. The formula for the reduced-order inverse is

$$\boldsymbol{M}^{-1} = \boldsymbol{T}\begin{bmatrix} 1/\mu_1 & & & & & \\ & 1/\mu_2 & & 0 & & \\ & & \ddots & & & \\ & & & 1/\mu_{n_p-m} & & \\ & 0 & & & 0 & \\ & & & & & \ddots \\ & & & & & & 0 \end{bmatrix}\boldsymbol{U}^T = \sum_{j=1}^{n_p-m} \frac{1}{\mu_j}\boldsymbol{t}_j\boldsymbol{u}_j^T \tag{6.57}$$

where $\boldsymbol{t}_j$ and $\boldsymbol{u}_j$ are the jth columns of the $\boldsymbol{T}$ and $\boldsymbol{U}$ matrices, respectively, and the singular values are arranged from largest to smallest in magnitude,

$$\mu_1 > \mu_2 > \ldots > \mu_{n_p-m} > \mu_{n_p-m+1} > \ldots > \mu_{n_p} \tag{6.58}$$

The terms associated with the smallest m singular values are dropped in computing the matrix inverse. The criterion is to drop any singular value for which

$$\frac{\mu_j}{\mu_{max}} < N\varepsilon \tag{6.59}$$

where N is the number of data points, and ε is the computing machine precision. This implements the reduced-order matrix inverse and gives the inverse of $\boldsymbol{M}$ with rank reduced from n_p to $n_p - m$. The rank deficient method is, in principle, similar to principal component regression for reducing the adverse effects of data collinearity (see Refs. [6.21] and [6.22]).

Levenberg-Marquardt Method

The Levenberg-Marquardt method [6.23] augments the information matrix to improve its conditioning and thereby produce a more reasonable inverse. The formula for the matrix inverse is

$$\boldsymbol{M}^{-1} = \left(\boldsymbol{M}_o + k\boldsymbol{A}\right)^{-1} \tag{6.60}$$

where M_o is the original information matrix, which is presumably ill-conditioned, k is a positive nonzero scalar parameter, and A is a positive definite matrix. Usually A is taken as the identity matrix, so

$$M^{-1} = \left(M_o + kI\right)^{-1} \tag{6.61}$$

The scalar k can be obtained by an iterative procedure proposed by Theil [6.22]. The initial value for k is recommended as $k = 0.01$. As k increases, the Levenberg-Marquardt inverse of the information matrix causes the modified Newton-Raphson step to follow the cost gradient vector more closely [cf. Eq. (6.39b)]. The Levenberg-Marquardt method resembles the ridge regression method mentioned in Chapter 5.

Bayes-Like Method

The Bayes-like estimation method improves the conditioning of the M matrix by combining the measured data with prior estimates of some or all of the unknown parameters in the model. The cost function for the parameter estimation is similar to that for the Bayesian estimator [cf. Eq. (4.28)],

$$J(\theta) = \frac{1}{2}\sum_{i=1}^{N} v^T(i)\, R^{-1} v(i) + \frac{1}{2}\left(\theta - \theta_p\right)^T \Sigma_p^{-1}\left(\theta - \theta_p\right) \tag{6.62}$$

where θ_p is the vector of prior parameter values, and Σ_p is a positive semi-definite covariance matrix that reflects the confidence in the parameter values θ_p.

Using the modified Newton-Raphson minimization technique, $\Delta\hat{\theta}$ is computed from

$$\Delta\hat{\theta} = -\left(M_{\theta=\theta_o} + \Sigma_p^{-1}\right)^{-1}\left[g + \Sigma_p^{-1}\left(\theta - \theta_p\right)\right]_{\theta=\theta_o} \tag{6.63}$$

This estimator is similar to the mixed estimator introduced in Chapter 5, although its development followed a different path. The Σ_p matrix is usually diagonal, with each diagonal element representing the variance of the corresponding element of θ_p.

From Eq. (6.63), it can be seen that prior values can help particularly with parameters for which there is little or no information in the data. In a sense, the prior information can fill information deficiencies in the measured data, and thereby regularize the matrix inversion. This is a good example of a general principle in modeling, which is that information about the model parameters is either supplied to the estimator by some mechanism like prior estimates and their variances, or extracted from the measured data.

The estimates obtained from all three techniques just mentioned will be biased; however, in practical cases, the bias is small in comparison to the inaccuracy that would be introduced if $\boldsymbol{M}$ was nearly singular and nothing was done to regularize the estimation.

In some cases, the information matrix $\boldsymbol{M}$ is nearly singular only at certain points in the progression toward a solution using the modified Newton-Raphson method. In those cases, the methods described earlier can be used temporarily to get past the difficult points, and then omitted as the sequence of parameter estimates $\hat{\boldsymbol{\theta}}_k$ approaches the solution.

The rank deficiency method described earlier is easy to use in this way. To implement the approach, the inverse of the information matrix is always calculated using the *SVD* (which is a superior method for calculating the inverse anyway, see Ref. [6.24]), and the singular value ratios are checked against the criterion in Eq. (6.59). If any singular values are too small, their corresponding terms are dropped from the inverse using Eq. (6.57), and the modified Newton-Raphson method proceeds in the usual way. This approach is a simple method for addressing the near-singularity of the information matrix and works well in practice.

6.3.3 Accuracy of Parameter Estimates

In Chapter 5, the focus was on linear unbiased estimators and their accuracy. The accuracy of a parameter estimate was determined by the computed variance or standard error. With the assumption of Gaussian measurement noise, it was possible to construct a confidence interval for each parameter, or a confidence region for more than one parameter.

For a nonlinear estimator, the evaluation of parameter accuracy is much more difficult. As in the linear estimation problem, for a finite number of data samples, the parameter estimates are biased. Because of the nonlinear dependence of the model outputs on the parameters, the parameter covariance matrix represents only an approximation to the lower bounds for the parameter covariances. Furthermore, normality of the measurement noise does not guarantee normal distribution of the parameter estimates, because the parameter estimates are a nonlinear function of the measurements. Nevertheless, the usual statement regarding accuracy of the parameters for a nonlinear maximum likelihood estimator quotes the Cramér-Rao lower bound on the parameter variances, which are obtained from the main diagonal of the inverse information matrix [cf. Eq. (6.39c)].

Practical experience with aircraft parameter estimation reveals large discrepancies between the Cramér-Rao lower bound and the ensemble variance of parameter estimates obtained from analysis of repeated maneuvers at the same flight condition [6.25]-[6.27]. On the other hand, for simulated data, the Cramér-Rao lower bound has been found to be a very good estimator for the variance of an ensemble of parameter estimates.

Several approaches have been developed to explain the discrepancy: 1) development of a technique for determining bias and mean square error for an arbitrary parameter estimator, 2) changing the assumptions about the measurement noise, and 3) computation of confidence intervals by considering the likelihood function for the model, rather than its quadratic approximation.

A technique for determining accuracy for a general parameter estimator was proposed in Ref. [6.28]. The technique is based on higher-order sensitivity analysis of an estimator for a general dynamic system. First- and second-order approximations of the expectation $E(\boldsymbol{Z}_N;\boldsymbol{\theta})$ resulted in expressions for biases and mean squared error, *MSE*. When applied to a maximum likelihood estimator, it was found that, to first order, the bias and *MSE* are obtained as

$$\boldsymbol{b} \equiv E(\hat{\boldsymbol{\theta}}) - \boldsymbol{\theta} = \boldsymbol{0}$$

$$MSE \equiv E\left[(\hat{\boldsymbol{\theta}}-\boldsymbol{\theta})(\hat{\boldsymbol{\theta}}-\boldsymbol{\theta})^T\right] = \left[\sum_{i=1}^{N} \frac{\partial \boldsymbol{v}^T(i)}{\partial \boldsymbol{\theta}} \boldsymbol{R}^{-1} \frac{\partial \boldsymbol{v}(i)}{\partial \boldsymbol{\theta}}\right]^{-1} \tag{6.64}$$

These results are identical to those obtained using the modified Newton-Raphson algorithm. For better assessment of the parameter accuracy, second-order expressions for the biases and *MSE* have to be computed. The expressions for these quantities become very complex, requiring computation of the second-order gradient $\partial^2 \boldsymbol{v}(i)/\partial\boldsymbol{\theta}\partial\boldsymbol{\theta}^T$. The sensitivity analysis provides a very general tool which can, in theory, be used in the error analysis for any estimator, including the effects of any misspecification of the postulated model. Unfortunately, the complexity of the analysis makes this approach impractical for the multidimensional dynamic systems used to model aircraft dynamics.

In the development of the maximum likelihood estimator, it was assumed that the measurement noise $\boldsymbol{v}(i)$ came from a random white noise process that was zero-mean and Gaussian. This means that the power of the noise should be evenly spread over the frequency range $[0, f_N]$ Hz, where $f_N = 1/(2\Delta t)$ is the Nyquist frequency. The relation between the one-sided power spectral density of the noise vector and the noise covariance matrix is therefore

$$\boldsymbol{R} = \boldsymbol{PSD}\, f_N = \frac{\boldsymbol{PSD}}{2\Delta t} \tag{6.65}$$

where ***PSD*** is a diagonal matrix with the power spectral densities of the elements of the noise vector on the main diagonal. From before, the

Cramér-Rao inequality for the parameter covariance matrix of the output-error estimator is

$$Cov\left(\hat{\boldsymbol{\theta}}\right) \geq \left[\sum_{i=1}^{N} \boldsymbol{S}^T(i)\boldsymbol{R}^{-1}\boldsymbol{S}(i)\right]^{-1} \tag{6.66}$$

or, in terms of the power spectral density,

$$Cov\left(\hat{\boldsymbol{\theta}}\right) \geq \frac{1}{2\Delta t}\left[\sum_{i=1}^{N} \boldsymbol{S}^T(i)(\boldsymbol{PSD})^{-1}\boldsymbol{S}(i)\right]^{-1} \tag{6.67}$$

In many practical cases, analysis of the residuals from output-error estimation indicates that the residuals are colored. This means that the power spectral densities of the residuals show more power concentrated in a particular frequency band, compared to the relatively constant power out to the Nyquist frequency that is characteristic of white noise. For aircraft parameter estimation, this concentration of noise power occurs in the low frequency band corresponding to the rigid-body aircraft dynamics. The phenomena is usually caused by inadequacies in the proposed model structure. For example, it could be that the true model parameters are not really constant, or that the data include some nonlinear effects not accounted for by the proposed linear model structure.

The theoretical development of the maximum likelihood estimator assumes that the residuals are white, because the power spectral densities of the residuals are not known *a priori*, and using the assumption of white residuals makes the theoretical development more straightforward. The assumption of white residuals is equivalent to assuming that the proposed model structure is completely adequate to characterize the measured data. In simulation, this assumption is perfectly true, because the same model structure is used both to generate the data and to analyze it, and the added noise sequence is typically white and Gaussian. For this case, the application matches the theoretical development, and the usual calculation of the parameter variances provides an excellent measure of the parameter accuracy.

To handle the colored residuals encountered in practice, Ref. [6.29] suggested approximating the colored noise with band-limited white noise of the same power. The approximation is depicted graphically in Fig. 6.5 for one element of the residual vector, where the one-sided power spectral density (PSD) of wide-band white noise is included for comparison.

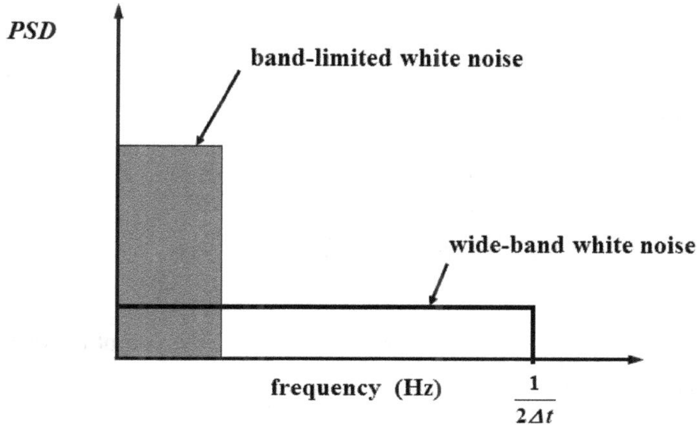

Figure 6.5 **Approximation of colored noise with band-limited white noise**

For the power spectral density of band-limited white noise with one-sided bandwidth B, the analog of Eq. (6.65) is

$$\boldsymbol{PSD} = \frac{\boldsymbol{R}}{B} \tag{6.68}$$

Thus, Eq. (6.67) changes to

$$Cov(\hat{\theta}) \geq \frac{1}{2\Delta t\, B}\left[\sum_{i=1}^{N} \boldsymbol{S}^T(i)\boldsymbol{R}^{-1}\boldsymbol{S}(i)\right]^{-1} \tag{6.69}$$

which reduces to Eq. (6.66) when $B = f_N = 1/(2\Delta t)$. Expression (6.69) means that for colored noise, the Cramér-Rao bounds can be underestimated by a factor of $1/(2\Delta t\, B) = f_N / B$.

For example, if the bandwidth of the rigid-body dynamics and the colored residuals is 1 Hz, and the sampling rate is 50 Hz, then $f_N = 25$ Hz, and the Cramér-Rao bounds for the parameter covariance matrix elements would be too low by a factor of 25. This translates into parameter standard error bounds that are too low by a factor of 5. These results provide an explanation for the discrepancies between Cramér-Rao bound estimates computed from simulation versus flight data, and between Cramér-Rao bounds computed from a single flight maneuver versus an ensemble variance estimate computed from repeated maneuvers. More details can be found in Refs. [6.26] and [6.29].

Further improvement in the Cramér-Rao bound estimate can be obtained by replacing the white residual assumption by assuming the residuals are colored, and including an estimate of the residual coloring in the expression for the Cramér-Rao bound estimate. This approach has already been mentioned in Chapter 5 for the least squares estimator. In Refs. [6.27] and [6.30], the approach was generalized for the output-error method, resulting in an expression for the parameter covariance matrix for colored residuals,

$$Cov(\hat{\boldsymbol{\theta}}) \geq \boldsymbol{M}^{-1}\left[\sum_{i=1}^{N} \boldsymbol{S}^T(i)\boldsymbol{R}^{-1}\sum_{j=1}^{N} \mathcal{R}_{vv}(i-j)\boldsymbol{R}^{-1}\boldsymbol{S}(j)\right]\boldsymbol{M}^{-1} \qquad (6.70)$$

where $\mathcal{R}_{vv}(i-j)$ is the autocorrelation matrix for the output residual vector. This quantity can be estimated using

$$\hat{\mathcal{R}}_{vv}(k) = \frac{1}{N}\sum_{i=1}^{N-k} \boldsymbol{v}(i)\boldsymbol{v}^T(i+k) = \hat{\mathcal{R}}_{vv}(-k) \qquad k = 0,1,2,\ldots,m \quad (6.71)$$

The estimate in Eq. (6.71) usually only includes the diagonal terms, since the off-diagonal terms are close to zero for the same reasons given in the discussion of estimating the noise covariance matrix $\boldsymbol{R}$ from Eq. (6.29). The value of m in Eq. (6.71) determines the maximum time difference used in the computation of the residual autocorrelation estimate. Small values of m correspond to computing residual coloring at low frequency. Since the residual coloring that occurs in practice is at low frequencies associated with the rigid-body dynamic modes, the value of m need not run all the way up to $N-1$, but instead can be stopped at about $N/5$ for typical flight test maneuvers and sample rates. This reduces the calculations required and has no discernable effect on the quality of the results.

The values for $\boldsymbol{S}$, $\boldsymbol{M}$, and $\boldsymbol{R}$ in Eq. (6.70) are from the conventional maximum likelihood estimation calculations, so they are the same as the quantities appearing in Eqs. (6.29), (6.40), and (6.66). Consequently, Eqs. (6.70) and (6.71) can be used as a residual post-processor after conventional maximum likelihood parameter estimation. The residual post-processing corrects the parameter covariance matrix to account for colored residuals. Parameter variances computed from Eqs. (6.70) and (6.71) have been shown to be consistent with ensemble estimates of the variance of parameter estimates from repeated maneuvers, using simulated data with many different realistic colored noise sequences, as well as flight data (see Refs. [6.27] and [6.30]).

For a linear unbiased estimate, the parameter confidence interval can be easily obtained, and is symmetric about the value of the estimated parameter. The confidence level associated with this interval expresses the probability of finding the parameter true value within the interval. The confidence interval for a single parameter can be extended to a parameter confidence region for multiple parameters, defined by a confidence ellipsoid (see Ref. [6.26]).

For nonlinear parameter estimation, the confidence ellipsoid is only an approximation to the confidence region. This region could have an irregular shape quite different from an ellipsoid. Furthermore, because the parameter estimates in a nonlinear case do not have Gaussian distribution, the confidence level based on the F-distribution is again an approximation.

A method for finding confidence intervals for parameter estimates obtained by maximum likelihood estimation was presented in Ref. [6.20], for both linear and nonlinear estimation cases. The construction of the confidence interval for a one-dimensional case is shown in Fig. 6.6. The confidence interval for a linear estimate is given by the interval for which the change in the maximum likelihood cost function is a constant selected value ΔJ_L,

$$\Delta J_L = n_p\, F\left(\alpha; n_p, N - n_p\right) \tag{6.72}$$

where $F\left(\alpha; n_p, N - n_p\right)$ is the value of the F-distribution with confidence level α. For the nonlinear estimate,

$$\Delta J_{NL} = n_p\, F\left(\alpha; n_p, N - n_p\right)\left[1 - \frac{N\left(n_p + 2\right)}{\left(N - n_p\right) n_p} N_\phi\right] \tag{6.73}$$

where N_ϕ is the nondimensional intrinsic nonlinearity measure introduced in Ref. [6.31].

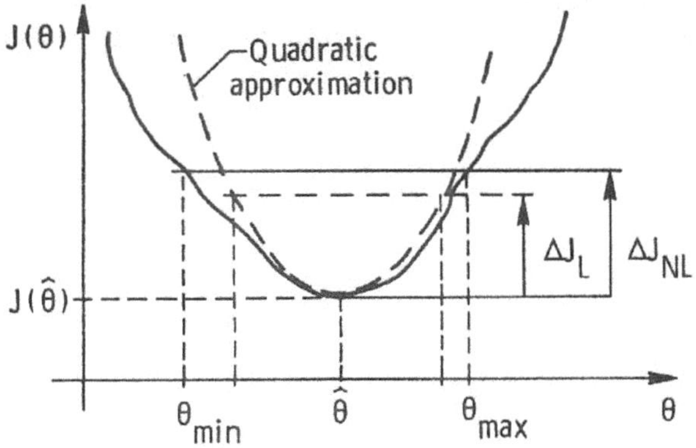

Figure 6.6 **Confidence interval determination by cost increments (from Ref. [6.20])**

For computing the limits of the confidence interval $\hat{\theta}_{min}$ and $\hat{\theta}_{max}$, the search algorithm developed in Ref. [6.32] can be used. This algorithm finds the contour boundaries of the cost function by testing a series of randomly-selected points in and around the confidence region. Through many iterations, the limits of the confidence region can be determined, and confidence intervals for each of the parameters are obtained. This method has not been widely used, probably because the search algorithm is computationally very demanding, even for a system with a small number of parameters.

Example 6.2

In this example, data from Example 6.1 are used to demonstrate the estimated parameter covariance corrections for colored residuals. The method is similar to that shown in Example 5.2 for linear regression, with the main difference being that the number of model outputs for the regression case was one, whereas the number of model outputs for the output-error case is generally more than one.

Figure 6.7 shows the model fit to the measured output data, with model parameters estimated using output-error method. The same model structure as before was used [cf. Eqs. (6.52) and (6.53)]. In this case, the output-error parameter estimation was done using the measured input-output data directly, i.e., the measured time series were not converted to perturbations about initial trim values. This can be seen by comparing the initial values shown in Figs. 6.3 and 6.7, particularly for a_y. The output-error parameter estimation results are identical to those found in Example 6.1, except for the bias parameters.

Figure 6.8 shows the output-error residuals, which are the difference between the data and model time series shown in Fig. 6.7. The output-error residuals are obviously colored, because they do not resemble random sequences.

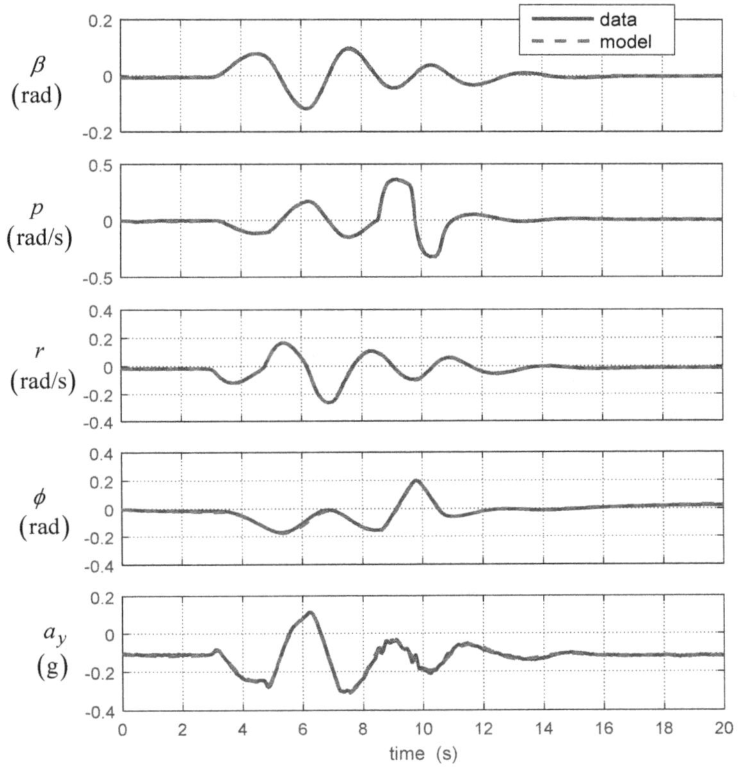

Figure 6.7 **Output-error model fit to measured outputs for lateral maneuver, run 1**

Table 6.2 contains the output-error parameter estimation results, including estimated parameter standard errors using both the white residual assumption [Eq. (6.66)], and accounting for colored residuals [Eq. (6.70)]. In Fig. 6.9, equation-error parameter estimates from Examples 5.1 and 5.2 are plotted along with output-error parameter estimates based on the same data, using the same model structure. The error bounds shown on all the parameter estimates are 95% confidence intervals, accounting for colored residuals. The figure shows that the equation-error and output-error parameter estimates are in good agreement.

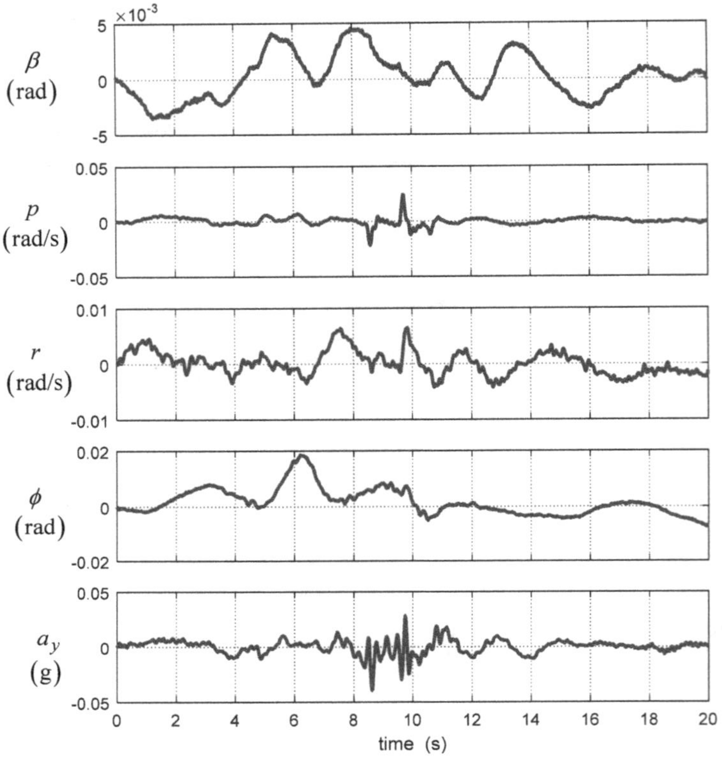

Figure 6.8 **Output-error residuals for lateral maneuver, run 1**

The equation-error estimates are biased as a result of the regressors being noisy, (and not deterministic, as assumed in the theory), as well as from any deterministic model structure error, which should be very small for this perturbation maneuver at low angle of attack. The bias in the output-error parameter estimates due to deterministic model structure error should likewise be small. Estimates from either method could be biased because of the finite number of measurements, since in theory the estimation methods are only asymptotically unbiased. More accurate parameter estimates can be obtained by averaging the estimates from repeated maneuvers at the same flight condition. ■

Table 6.2 **Output-error parameter estimation results for lateral maneuver, run 1**

Parameter	$\hat{\theta}$	$s(\hat{\theta})$ white residual Eq. (6.66)	$s(\hat{\theta})_{corr}$ colored residual Eq. (6.70)	$\frac{s(\hat{\theta})_{corr}}{s(\hat{\theta})}$
C_{Y_β}	-8.66×10^{-1}	2.80×10^{-3}	8.82×10^{-3}	3.15
C_{Y_r}	9.31×10^{-1}	1.65×10^{-2}	6.04×10^{-2}	3.65
$C_{Y_{\delta_r}}$	3.75×10^{-1}	7.98×10^{-3}	3.15×10^{-2}	3.95
C_{l_β}	-1.19×10^{-1}	5.60×10^{-4}	1.89×10^{-3}	3.37
C_{l_p}	-5.84×10^{-1}	2.39×10^{-3}	8.23×10^{-3}	3.44
C_{l_r}	1.88×10^{-1}	1.71×10^{-3}	8.89×10^{-3}	5.19
$C_{l_{\delta_a}}$	-2.28×10^{-1}	8.43×10^{-4}	3.03×10^{-3}	3.60
$C_{l_{\delta_r}}$	3.84×10^{-2}	7.12×10^{-4}	6.03×10^{-3}	8.47
C_{n_β}	8.65×10^{-2}	2.07×10^{-4}	1.53×10^{-3}	7.36
C_{n_p}	-6.39×10^{-2}	9.49×10^{-4}	6.41×10^{-3}	6.76
C_{n_r}	-1.92×10^{-1}	6.07×10^{-4}	4.06×10^{-3}	6.68
$C_{n_{\delta_a}}$	-2.73×10^{-3}	3.50×10^{-4}	2.23×10^{-3}	6.35
$C_{n_{\delta_r}}$	-1.36×10^{-1}	2.66×10^{-4}	1.86×10^{-3}	7.00

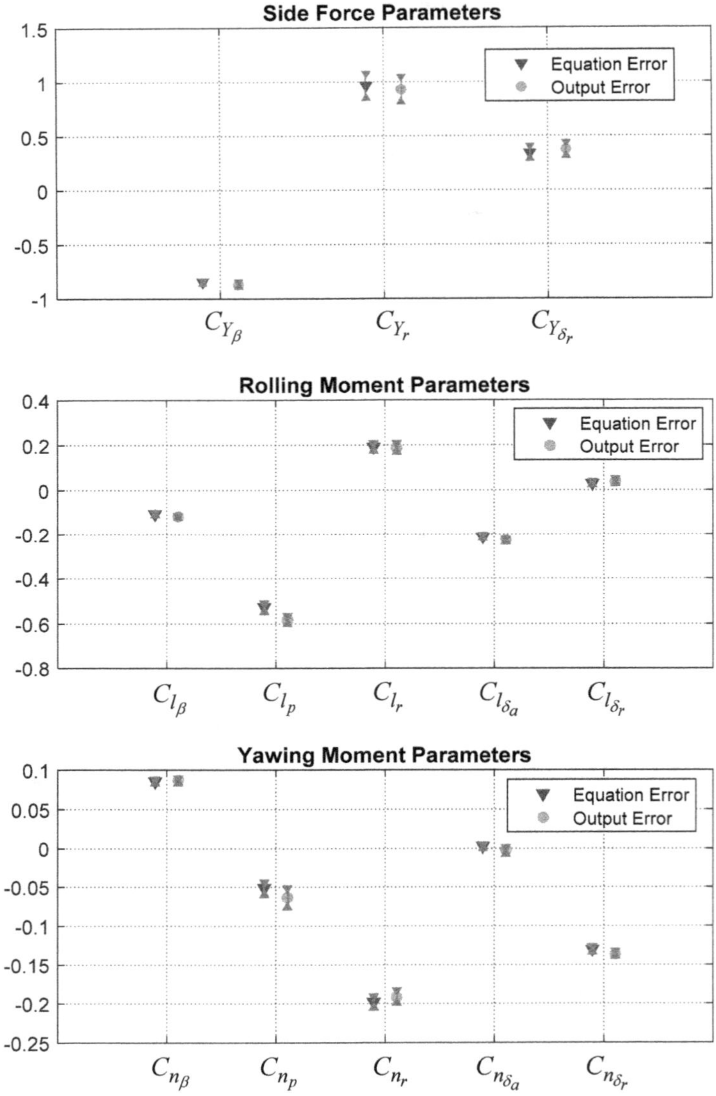

Figure 6.9 **Comparison of parameter estimates for lateral maneuver, run 1**

6.4 Equation-Error Method

Another variation of the maximum likelihood parameter estimation method is obtained by assuming:

1) The dynamic model is in linear state space form with process noise.
2) The model parameters to be estimated are dimensional stability and control derivatives.
3) All the states are measured without error, in addition to the inputs.

Under these assumptions, the model equations are formed as

$$\dot{\boldsymbol{x}}(t) = \boldsymbol{A}\boldsymbol{x}(t) + \boldsymbol{B}\boldsymbol{u}(t) \qquad \boldsymbol{x}(0) = \boldsymbol{x}_o \tag{6.74a}$$

$$\boldsymbol{z}(i) = \boldsymbol{A}\boldsymbol{x}(i) + \boldsymbol{B}\boldsymbol{u}(i) + \boldsymbol{v}(i) \quad i = 1,2,\ldots,N \tag{6.74b}$$

Note that the equation error $\boldsymbol{v}(i) = \boldsymbol{z}(i) - \boldsymbol{A}\boldsymbol{x}(i) - \boldsymbol{B}\boldsymbol{u}(i)$ can include measurement errors for the state derivative, as well as process noise that may be present in the state equations. Typically, the state derivatives are not measured directly, but instead are found from smoothed numerical differentiation of the measured states (see Chapter 11). If the equation errors $\boldsymbol{v}(i)$ are assumed to be Gaussian with

$$E\left[\boldsymbol{v}(i)\right] = \boldsymbol{0} \qquad E\left[\boldsymbol{v}(i)\boldsymbol{v}^T(j)\right] = \boldsymbol{R}\delta_{ij} \tag{6.75}$$

then the negative log-likelihood function has the form

$$-\ln \mathbb{L}(\boldsymbol{Z}_N;\boldsymbol{\theta}) = \frac{1}{2}\sum_{i=1}^{N}\left[\boldsymbol{z}(i) - \boldsymbol{A}\boldsymbol{x}(i) - \boldsymbol{B}\boldsymbol{u}(i)\right]^T \boldsymbol{R}^{-1}\left[\boldsymbol{z}(i) - \boldsymbol{A}\boldsymbol{x}(i) - \boldsymbol{B}\boldsymbol{u}(i)\right] + \frac{N}{2}\ln|\boldsymbol{R}| + \frac{N n_o}{2}\ln(2\pi) \tag{6.76}$$

For known $\hat{\boldsymbol{R}}$, the cost function to be minimized is

$$J(\boldsymbol{\theta}) = \frac{1}{2}\sum_{i=1}^{N}\left[\boldsymbol{z}(i) - \boldsymbol{A}\boldsymbol{x}(i) - \boldsymbol{B}\boldsymbol{u}(i)\right]^T \hat{\boldsymbol{R}}^{-1}\left[\boldsymbol{z}(i) - \boldsymbol{A}\boldsymbol{x}(i) - \boldsymbol{B}\boldsymbol{u}(i)\right] \tag{6.77}$$

The measurement equation (6.74b) and the cost function in Eq. (6.77) indicate that the parameter estimation has been simplified to that of linear regression, where the innovations are the equation errors,

$$\boldsymbol{v}(i) = \boldsymbol{z}(i) - \boldsymbol{A}\boldsymbol{x}(i) - \boldsymbol{B}\boldsymbol{u}(i) \quad i = 1,2,\ldots,N \tag{6.78}$$

If the noise covariance matrix estimate $\hat{\boldsymbol{R}}$ is assumed diagonal, minimizing the cost function in Eq. (6.77) can be done by minimizing the cost associated with each individual row in the state space model. Therefore, the unknown parameters in matrices $\boldsymbol{A}$ and $\boldsymbol{B}$ can be estimated in subsets corresponding to their individual equation in the state space model. For example, the lift parameters can be estimated from the $\dot{\alpha}$ equation, the pitching moment parameters from the $\dot{q}$ equation, and so on.

It is just as simple to use output equations instead of dynamic equations for this approach. Equation (6.74b) can be changed to use measured output equations in the state space formulation as follows:

$$\boldsymbol{z}(i) = \boldsymbol{C}\boldsymbol{x}(i) + \boldsymbol{D}\boldsymbol{u}(i) + \boldsymbol{v}(i) \qquad i = 1,2,\ldots,N \tag{6.79}$$

Elements in the relevant rows of the $\boldsymbol{C}$ and $\boldsymbol{D}$ matrices must contain model parameters to be estimated. Consequently, the measurement equations for translational accelerometers can be used in this way. A block diagram of the equation-error method is shown in Fig. 6.10.

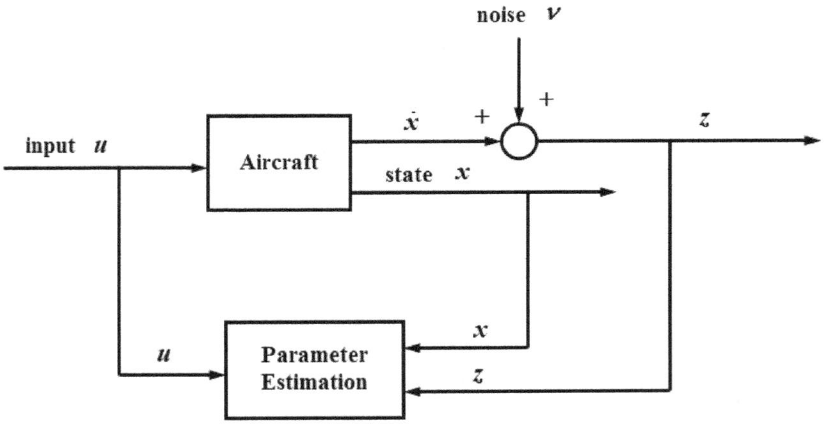

Figure 6.10 **Block diagram for equation-error parameter estimation**

In any practical situation, the measured states and inputs are corrupted by measurement errors. Regression of the elements of $\dot{\boldsymbol{x}}$ or $\boldsymbol{z}$ on measured $\boldsymbol{x}$ and $\boldsymbol{u}$ will normally result in biased estimates of the unknown model parameters in $\boldsymbol{A}, \boldsymbol{B}, \boldsymbol{C},$ and $\boldsymbol{D}$, (see Ref. [6.33] and Section 5.1.5). However, the bias errors in the parameter estimates can be eliminated by smoothing the regressors in the time domain or the frequency domain (see Ref. [6.34]).

Example 6.3

Measured flight test data for this example comes from a longitudinal perturbation maneuver of the NASA Twin Otter aircraft, pictured in Figure 5.3. Figure 6.11 shows measured input-output data for the maneuver. The equation-error method described in the previous section was used to estimate dimensional stability and control derivatives.

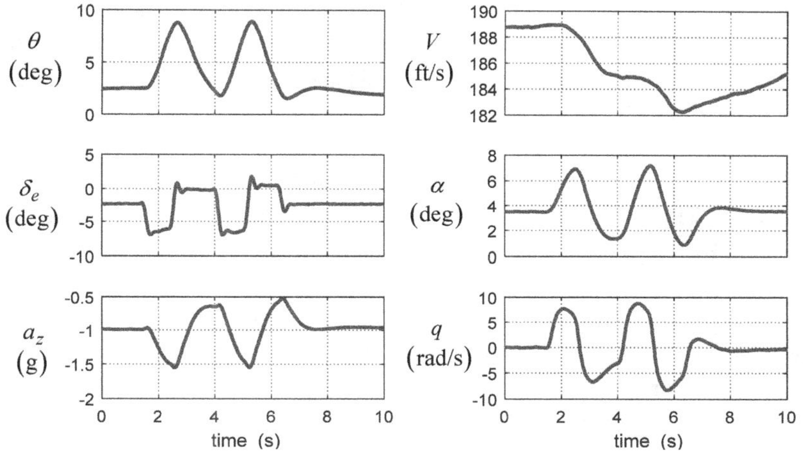

Figure 6.11 **Measured input and output data for a longitudinal maneuver**

The linearized longitudinal equations in state space form are [cf. Eqs. (3.128)],

$$\dot{\alpha} = Z_\alpha \alpha + q + Z_{\delta_e} \delta_e \tag{6.80a}$$

$$\dot{q} = M_\alpha \alpha + M_q q + M_{\delta_e} \delta_e \tag{6.80b}$$

$$a_z = \frac{V_o}{g}\left(Z_\alpha \alpha + Z_{\delta_e} \delta_e\right) \tag{6.80c}$$

where all states and controls are perturbation quantities, which is consistent with the linear model structure. The maneuver was executed at a low trim angle of attack, so the lift force is nearly along the z body axis, which allows use of the form shown for the $\dot{\alpha}$ equation. At low angles of attack, $Z_q \approx 0$, so the associated term does not appear in Eqs. (6.80a) and (6.80c). The stepwise regression technique described in Chapter 5 was used to verify that the Z_q term should not be included in the model structure.

The dynamic equation for angle of attack α is typically dominated by two terms: the inertial term q and the $Z_\alpha \alpha$ term. The output equation for z body-axis acceleration a_z does not include the inertial term, so all of the information in the a_z measurement can be used for aerodynamic parameter estimation. In addition, the a_z measurement is directly applicable to estimating the dimensional parameters associated with the Z force, whereas using the dynamic equation for α involves an additional approximation, namely $-L = Z \cos\alpha \approx Z$. As a result, the output equation for a_z usually gives better estimates for the aerodynamic parameters associated with the body-axis force Z. The output equation for a_z, Eq. (6.80c), was used to estimate the Z aerodynamic force parameters, and the dynamic equation for pitch rate q, Eq. (6.80b), was used to estimate the pitching moment parameters. This also means that only $\dot{q}$ needed to be computed numerically, and not $\dot{\alpha}$. The state derivative $\dot{q}$ was computed using smoothed local numerical differentiation described in Chapter 11.

Table 6.3 gives the equation-error estimates of the dimensional stability and control derivatives found by minimizing the cost function of Eq. (6.77) twice – once for the Z force parameters using Eq. (6.80c) and once for the pitching moment parameters using Eq. (6.80b). Parameter estimates and associated error bounds were found in the usual manner for linear regression problems, as described in Chapter 5. Estimated parameter error bounds were corrected for colored residuals in linear regression problems, as in Chapter 5. The measured outputs to be fit by the models were a_z and $\dot{q}$, respectively. Figure 6.12 shows the model fit to the data, using estimated parameters from Table 6.3. The parameter estimates given in Table 6.3 are good, useful estimates, and also provide excellent starting values for the iterative output-error method. The dimensional derivative estimates in Table 6.3 can be nondimensionalized using Eqs. (3.129), if necessary. The main advantage of this approach is that useful stability and control information can be obtained rapidly with very few measurements.

Table 6.3 **Equation-error dimensional parameter estimation results for a longitudinal maneuver**

Parameter	$\hat{\theta}$	$s(\hat{\theta})$	$\lvert t_0 \rvert$	$100\left[s(\hat{\theta})/\lvert\hat{\theta}\rvert\right]$
Z_α	–1.589	0.026	61.5	1.6
Z_{δ_e}	–0.038	0.016	2.4	41.2
Z_o	–0.406	0.014	29.9	3.4
M_α	–5.245	0.050	104.8	1.0
M_q	–2.598	0.047	55.4	1.8
M_{δ_e}	–7.852	0.082	95.6	1.0
M_o	–0.003	0.006	0.6	165.1

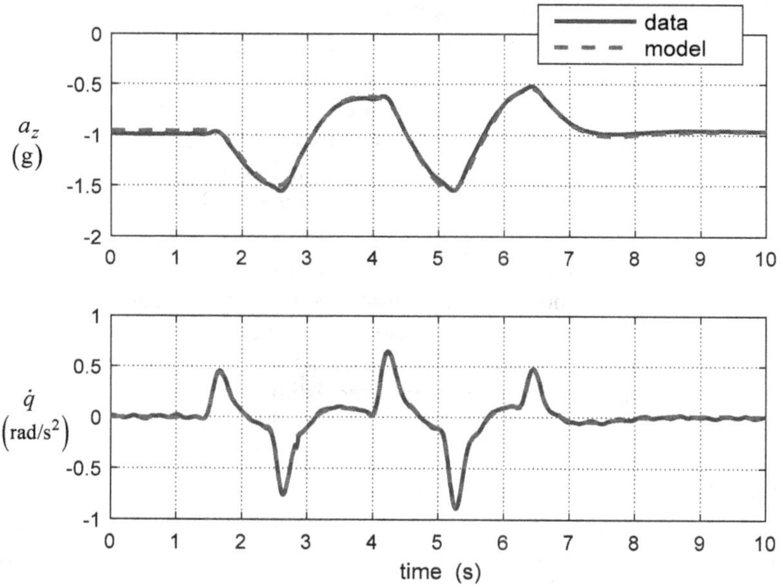

Figure 6.12 **Equation-error model fit to a_z and $\dot{q}$ for a longitudinal maneuver**

Finally, the expressions in Eq. (3.129) show that the estimated dimensional parameters include the dynamic pressure $\bar{q} = \rho V^2/2$ and airspeed V. Because of this, changes in airspeed and altitude must be relatively small during the maneuver used for estimating dimensional stability and control derivatives. This applies to any method where dimensional stability and control derivatives are being estimated. The reason is that the model parameters are assumed to be constants in the analysis, but dimensional parameters will vary with airspeed and altitude changes, because dimensional derivatives include dynamic pressure and airspeed. If the dynamic pressure and/or airspeed vary significantly during the maneuver, then varying quantities (the dimensional stability and control derivatives) will be modeled using constant model parameters, which introduces modeling error. Similar statements apply for the mass / inertia properties, but these quantities rarely vary significantly over the time period typical of a stability and control maneuver. A good rule of thumb is that dimensional parameters can be used when the airspeed variation is less than 10 percent of the mean value. Maximum allowable variation in altitude can be defined as the altitude change that induces a similar change in the magnitude of the dynamic pressure. ■

6.5 Summary and Concluding Remarks

The maximum likelihood method for parameter estimation is based on a relatively simple idea: assuming the outcome of an experiment can be characterized by a model that depends on unknown parameters, find the values of those parameters which make the measured values the most likely to occur. This concept is realized by maximization of a likelihood function, which equals the probability density function for measured values, given the parameters.

In aircraft applications, the measured data are time series of input and output variables for a nonlinear dynamic system with process noise. It is usually assumed that the inputs are measured without errors, and the measured outputs are corrupted by measurement noise which is Gaussian, white, and zero mean. The cost function for parameter estimation includes innovations weighted by their covariance matrix. The innovations are the differences between measured and computed outputs, where the computed outputs are based on states estimated using an extended Kalman filter.

The parameter estimation as stated represents a difficult optimization problem. Several optimization schemes are available, but the modified Newton-Raphson method is usually used in practice. A simplification of the estimation problem is achieved when the dynamic system model is linear. Then the extended Kalman filter is replaced by a Kalman filter, or its simplified version, the steady-state Kalman filter. The estimation includes both the state of the dynamic system and the model parameters. This approach is called the filter-error method.

The resulting parameter estimates have favorable statistical properties: they are asymptotically unbiased, efficient, consistent, and normally distributed. Despite the generality of the problem formulation, this full version of the maximum likelihood estimator is rarely used in practice because of the complexity of the algorithm and possible identifiability problems caused by low information content relative to the number of unknowns to be estimated. Good treatments of the maximum likelihood method with examples can be found in reports by Stepner and Mehra [6.4], Hall et al. [6.35], Jategaonkar and Plaetschke [6.12] and [6.36], and Maine and Iliff [6.5] and [6.37].

Substantial simplification of the algorithm for the maximum likelihood method results from assuming there is no process noise. Then, the state estimation reduces to integration of the state equations, which are then deterministic. The maximum likelihood method becomes the output-error method, which is, in principle, identical to a nonlinear regression. Excellent explanation of the output-error method with many examples is given by Maine and Iliff [6.38] and Maine [6.39].

If the states are also assumed to be measured without errors, and the dynamic system model is linear with dimensional stability and control derivatives, then the maximum likelihood method reduces to a linear regression.

Because the output-error method is widely used in aircraft parameter estimation, some practical aspects of computing the estimates were presented. These include computing sensitivities, inverting a nearly singular information matrix, and estimating parameter accuracy. Several examples were included to demonstrate applications to aircraft modeling problems.

References

[6.1] Kailath, T., "A General Likelihood-Ratio Formula for Random Signals in Gaussian Noise," *IEEE Transactions on Information Theory*, Vol. IT-15, No. 3, 1969, pp. 350-361.

[6.2] Taylor, L.W., Iliff, K.W., and Powers, B.G., "A Comparison of Newton-Raphson and Other Methods for Determining Stability Derivatives from Flight Data," AIAA Paper 69-315, March 1969.

[6.3] Speedy, C.B., Brown, R.F., and Goodwin, G.C., *Control Theory, Identification, and Optimal Control*, Oliver and Boyd, Edinburgh, UK, 1970.

[6.4] Stepner, D.E. and Mehra, R.K., "Maximum Likelihood Identification and Optimal Input Design for Identifying Aircraft Stability and Control Derivatives," NASA CR-2200, 1973.

[6.5] Maine, R.E. and Iliff, K.W., "Formulation and Implementation of a Practical Algorithm for Parameter Estimation with Process and Measurement Noise," *SIAM Journal of Applied Mathematics*, Vol. 41, No. 3, 1981, pp. 558-579.

[6.6] Schultz, G., "Maximum Likelihood Identification Using Kalman Filter – Least Squares Estimation: A Comparison for the Estimation of Stability Derivatives Considering Gust Disturbances," European Space Agency Technical Translation, ESA TT-258, 1976.

[6.7] Iliff, K.W., "Identification and Stochastic Control with Application to Flight Control in Turbulence," UCLA-ENG-7340, School of Engineering and Applied Sciences, University of California, Los Angeles, CA, 1973.

[6.8] Mehra, R.K., "Identification of Stochastic Linear Dynamic System Using Kalman Filter Representation," *AIAA Journal*, Vol. 9, No. 1, 1971, pp. 28-31.

[6.9] Mehra, R.K., "On the Identification of Variances and Adaptive Kalman Filtering," *IEEE Transactions on Automatic Control*, Vol. AC-15, No. 2, 1970, pp. 175-184.

[6.10] Bach, R.E., Jr., "A User's Manual for AMES (A Parameter Estimation Program), Rep. NAS2-7397, NASA Ames Research Center, Moffett Field, CA, 1974.

[6.11] Klein, V., "Maximum Likelihood Method for Estimating Airplane Stability and Control Parameters From Flight Data in Frequency Domain," NASA TP-1637, 1980.

[6.12] Jategaonkar, R. and Plaetschke, E., "Maximum Likelihood Estimation of Parameters in Linear Systems with Process and Measurement Noise," DFVLR Forschungsbericht 87-20, 1987.

[6.13] Grauer, J.A. and Morelli, E.A. "A New Formulation of the Filter-Error Method for Aerodynamic Parameter Estimation in Turbulence," AIAA 2015-2704, *AIAA Aviation 2015 Conference*, Dallas, TX, June 2015.

[6.14] Åström, K.J., and Eykhoff, P., "System Identification – A Survey," *Automatica*, Vol. 7, March 1971, pp. 123-162.

[6.15] Eykhoff, P., *System Identification, Parameter and State Estimation*, John Wiley & Sons, New York, NY, 1974.

[6.16] Iliff, K.W. and Taylor, L.W., "Determination of Stability Derivatives from Flight Data Using a Newton-Raphson Minimization Technique," NASA TN D-6579, 1971.

[6.17] Myers, R.H., *Classical and Modern Regression with Applications*, Duxbury Press, Boston, MA, 1986.

[6.18] Gupta, N.K. and Mehra, R.K., "Computational Aspects of Maximum Likelihood Estimation and Reduction in Sensitivity Function Calculations," *IEEE Transactions on Automatic Control*, Vol. AC-19, No. 6, 1974, pp. 774-783.

[6.19] Taylor, L.W. and Iliff, K.W., "Systems Identification Using a Modified Newton-Raphson Method: A FORTRAN Program," NASA TN D-6734, 1972.

[6.20] Murphy, P.C., "A Methodology for Airplane Parameter Estimation and Confidence Interval Determination in Nonlinear Estimation Problems," NASA RP-1153, 1986.

[6.21] Montgomery, D.C., Peck, E.A., and Vining, G.G., *Introduction to Linear Regression Analysis*, 3rd Ed., John Wiley & Sons, Inc., New York, NY, 2001.

[6.22] Theil, H., *Principles of Econometrics*, John Wiley & Sons, Inc., New York, NY, 2001.

[6.23] Marquardt, D.W., "An Algorithm for Least Squares Estimation of Nonlinear Parameters," *SIAM Journal of Numerical Analysis*, Vol. 11, No. 2, 1963, pp. 431-441.

[6.24] Press, W.H., Teukolsky, S.A., Vettering, W.T., and Flannery, B.R., *Numerical Recipes in FORTRAN: The Art of Scientific Computing*, 2nd Ed., Cambridge University Press, New York, NY, 1992.

[6.25] Klein, V., "Determination of Stability and Control Parameters of a Light Airplane From Flight Data Using Two Estimation Methods," NASA TP-1306, 1979.

[6.26] Maine, R.E. and Iliff, K.W., "The Theory and Practice of Estimating the Accuracy of Dynamic Flight-Determined Coefficients," NASA RP 1077, 1981.

[6.27] Morelli, E.A. and Klein, V., "Determining the Accuracy of Maximum Likelihood Parameter Estimates with Colored Residuals," NASA CR 194893, 1994.

[6.28] Gupta, N.K., "Bias and Mean Square Error Properties of General Estimators," *Proceedings of the 12th Conference on Decision and Control*, December 1976, pp. 624-628.

[6.29] Balakrishnan, A.V. and Maine, R.E., "Improvements in Aircraft Extraction Programs," NASA CR-145090, 1976.

[6.30] Morelli, E.A. and Klein, V., "Accuracy of Aerodynamic Model Parameters Estimated from Flight Test Data," *Journal of Guidance, Control, and Dynamics*, Vol. 20, No. 1, 1997, pp. 74-80.

[6.31] Beale, E.M.L., "Confidence Regions in Non-linear Estimation," *Journal of the Royal Statistical Society*, Series B, Vol. 22, No. 1, 1960, pp. 41-88.

[6.32] Mereau, P. and Raymond, J., "Computation of the Intervals of Uncertainties About the Parameters Found for Identification," NASA TM-76978, 1982.

[6.33] Klein, V., "Identification Evaluation Methods," *Parameter Identification*, AGARD-LS-104, Paper 2, 1979.

[6.34] Morelli, E.A., "Practical Aspects of the Equation-Error Method for Aircraft Parameter Estimation," AIAA-2006-6144, *AIAA Atmospheric Flight Mechanics Conference*, Keystone, CO, August 2006.

[6.35] Hall, W.E., Gupta, N.K., and Smith, R.G., "Identification of Aircraft Stability and Control Coefficients for the High Angle-of-Attack Regime," Engineering Technical Report No. 2, Systems Control, Inc., Palo Alto, CA, 1974.

[6.36] Jategaonkar, R. and Plaetschke, E., "Estimation of Aircraft Parameters Using Filter Error Methods and Extended Kalman Filter," DFVLR Forschungsbericht 88-15, 1988.

[6.37] Maine, R.E. and Iliff, K.W. "Identification of Dynamic Systems, Theory and Formulation," NASA RP 1138, 1985.

[6.38] Maine, R.E. and Iliff, K.W., "Application of Parameter Estimation to Aircraft Stability and Control, The Output Error Approach," NASA RP 1168, 1986.

[6.39] Maine, R.E., "Programmer's Manual for MMLE3, A General FORTRAN Program for Maximum Likelihood Parameter Estimation," NASA TP-1690, 1981.

Problems

6.1 Answer the following, using words, graphs, equations, and/or diagrams:

a) Compare and contrast the solution methods for linear and nonlinear parameter estimation problems.

b) Explain why it is necessary to correct estimated parameter errors for colored residuals in the time domain.

c) Compare and contrast equation-error parameter estimation with output-error parameter estimation.

d) Explain why output-error methods are not used for model structure determination.

e) Explain the difference between Newton-Raphson optimization and modified Newton-Raphson optimization. Why is this difference important?

f) Explain why equation-error can be applied to one equation at a time, while output-error must use several equations at once.

g) Explain why equation-error can be applied to unstable systems, while output-error generally cannot.

6.2 Fly a simple roll doublet maneuver using the F-16 simulation, or use the data in f16_roll_data.mat.

a) Use oe.m with the model file tlatss_roll.m (which implements the roll mode approximation) to find output-error parameter estimates.

b) Compute the roll mode time constant for the F-16 based on the results from part a).

c) Make a Bode plot using cmpbode.m, with no time lag, `tau=0`.

d) Repeat parts a), b), and c) at a significantly different angle of attack.

6.3 Repeat the previous problem using the SIDPAC GUI, accessed by typing `sid` at the MATLAB® prompt.

6.4 The short-period equations of motion for an aircraft can be written in terms of dimensional derivatives as follows:

$$\dot{\alpha} = -L_{\alpha}\,\alpha + q - L_{\delta_e}\,\delta_e$$

$$\dot{q} = M_{\alpha}\,\alpha + M_q\,q + M_{\delta_e}\,\delta_e$$

a) Assuming that the output is only pitch rate q, write the output sensitivity equation for each unknown parameter. Assume that all quantities are analytic, so that the order of differentiation with respect to different quantities can be switched. Control is implemented by the pilot and/or automatic control, and therefore does not depend on parameters, whereas all states depend on all parameters, because of the coupled state equations.

b) Referring to the dynamic equations, your answer from part a), and the analytic expression for estimated parameter errors using output-error parameter estimation, discuss how the estimated parameter errors can be affected by the experiment. Assume that the noise covariance matrix is known.

6.5 For the Twin Otter flight data in the file totter_demo_data.mat, compute time-domain equation-error estimates of stability and control derivatives for pitching moment coefficient using both dimensional and nondimensional parameters. Then repeat using output error. Correct the error bounds for colored residuals.

a) Are the estimates consistent, accounting for the estimated uncertainties?

b) Discuss your results, including any ideas you have on why the estimates might differ.

6.6 Derive the output sensitivity equation for equation-error linear regression using Eq. (5.5). Explain how this affects the solution method, compared to the solution method required for output-error parameter estimation.

Chapter 7

Frequency Domain Methods

Many methods for data analysis and modeling can be formulated in the frequency domain. Frequency domain analysis has certain advantages, including physical insight in terms of frequency content, direct applicability to control system design, and a smaller number of data points for parameter estimation, among others. The basis for frequency domain methods is the finite Fourier transform, which is the mechanism for transforming time-domain data to the frequency domain. Any errors in the transformation from time to frequency domain affect the accuracy of the data in the frequency domain, which in turn impacts data analysis and modeling results. This chapter begins by presenting a method for accurately evaluating the finite Fourier transform for sampled time-domain data, and continues with a discussion of spectral densities and frequency response computed from measured input-output data.

For parametric modeling in the frequency domain, two different models for uncertainty are considered, the Fisher model and the least-squares model. The Bayesian model is not discussed because of difficulties in formulating probability densities for complex random variables, and the consequent very limited use of this model in aircraft system identification. The two models for frequency domain analysis have the following forms:

Fisher Model

1) $\tilde{z} = \tilde{\boldsymbol{\varphi}}(\boldsymbol{\theta}) + \tilde{\boldsymbol{v}}$

2) $\boldsymbol{\theta}$ is a vector of unknown constant real parameters.

3) $\tilde{\boldsymbol{v}}$ is a complex random vector with probability density $p(\tilde{\boldsymbol{v}})$.

Least Squares Model:

1) $\tilde{z} = \tilde{\boldsymbol{X}}\boldsymbol{\theta} + \tilde{\boldsymbol{v}}$

2) $\boldsymbol{\theta}$ is a vector of unknown constant real parameters.

3) $\tilde{\boldsymbol{v}}$ is a complex random vector of measurement noise.

where the $\tilde{}$ notation indicates the Fourier transform. In the measurement equations, $\tilde{\boldsymbol{\varphi}}(\boldsymbol{\theta})$ is a known complex vector function and $\tilde{\boldsymbol{X}}$ is a known complex matrix. Both $\tilde{\boldsymbol{\varphi}}(\boldsymbol{\theta})$ and $\tilde{\boldsymbol{X}}$ are typically functions of the transformed states and controls, $\tilde{\boldsymbol{x}}$ and $\tilde{\boldsymbol{u}}$, respectively. The measurement vector $\tilde{\boldsymbol{z}}$ is the Fourier transform of the time-domain vector $\boldsymbol{z}$. Properties of the complex random sequence $\tilde{\boldsymbol{v}}$ are presented in Ref. [7.1]. These properties were developed from the assumed properties of the real random sequence $\boldsymbol{v}$.

The two models just introduced lead to maximum likelihood estimators for a dynamic system and to an ordinary least squares estimator. In the development of maximum likelihood estimators, a linear dynamic system with process noise is considered first. Then, in a manner similar to the time-domain approach, estimation algorithms for simplified models with no process noise or no measurement noise are presented. For the least squares model, a general form of complex linear regression is described. Applications of frequency domain techniques for parameter estimation from flight and wind tunnel data are demonstrated in examples.

7.1 Transforming Measured Data to the Frequency Domain

The transformation of measured data from the time domain to the frequency domain is based on the Fourier integral. For a given continuous function in the time domain $x(t)$, the corresponding continuous function in the frequency domain $\tilde{x}(\omega)$ is

$$\mathcal{F}\left[x(t)\right] \equiv \tilde{x}(\omega) = \int_{-\infty}^{\infty} x(t)\, e^{-j\omega t}\, dt \tag{7.1}$$

where $j = \sqrt{-1}$ and ω is the angular frequency in radians per second. The quantity $\tilde{x}(\omega)$ defined in the above equation is called the Fourier transform of $x(t)$. The $\tilde{x}(\omega)$ function can be transformed back to the time domain using the inverse Fourier transform,

$$x(t) = \frac{1}{2\pi} \int_{-\infty}^{\infty} \tilde{x}(\omega)\, e^{j\omega t}\, d\omega \tag{7.2}$$

The functions $x(t)$ and $\tilde{x}(\omega)$ are related by Eqs. (7.1) and (7.2), which are known as the Fourier transform pair. Transformation properties are discussed in detail in many references, e.g. [7.2].

The Fourier transform of a continuous scalar time function $x(t)$ on a finite time interval $[0,T]$ is called the finite Fourier transform, defined by

$$\tilde{x}(\omega) \equiv \int_{0}^{T} x(t)\, e^{-j\omega t}\, dt \tag{7.3}$$

or

$$\tilde{x}(f) \equiv \int_0^T x(t)\, e^{-j2\pi ft}\, dt \tag{7.4}$$

where

$$\omega = 2\pi f \tag{7.5}$$

The above relationships show that the finite Fourier transform $\tilde{x}(\omega)$ can be interpreted as a coefficient in an expansion of $x(t)$ in terms of the basis function $e^{-j\omega t} = cos\,\omega t - j\,sin\,\omega t$, for each frequency ω.

When the time function $x(t)$ is sampled at discrete regular time intervals Δt, the finite Fourier transform can be approximated by

$$\tilde{x}(f) \approx \Delta t \sum_{i=0}^{N-1} x(i)\, e^{-j2\pi f i \Delta t} \tag{7.6}$$

where the time length of the measured data is $T = (N-1)\Delta t$ and

$$\begin{aligned} &\Delta t = T/(N-1) \\ t_i = i\Delta t \qquad &x(i) \equiv x(t_i) = x(i\Delta t) \qquad i = 0,1,2,\ldots,N-1 \end{aligned} \tag{7.7}$$

Note that the total number of data points in the time domain is N, and the time index i starts at 0.

For a conventional finite Fourier transform, the frequencies are chosen as

$$f_k = \frac{k}{N\Delta t} \qquad k = 0,1,2,\ldots,N-1 \tag{7.8}$$

or

$$\omega_k = 2\pi f_k = 2\pi \frac{k}{N\Delta t} \qquad k = 0,1,2,\ldots,N-1 \tag{7.9}$$

Using the discrete frequencies defined in Eqs. (7.8) and (7.9), the approximation to the finite Fourier transform in Eq. (7.6) becomes

$$\tilde{x}(k) = \Delta t \sum_{i=0}^{N-1} x(i)\; e^{-j(2\pi k/N)\, i} \qquad k = 0,1,2,\ldots,N-1 \tag{7.10}$$

and the inverse is

$$x(i) = \frac{1}{N\Delta t} \sum_{k=0}^{N-1} \tilde{x}(k)\; e^{j(2\pi k/N)\, i} \qquad i = 0,1,2,\ldots,N-1 \tag{7.11}$$

Often Eqs. (7.10) and (7.11) are written for the normalized sampling interval $\Delta t = 1$. The result is called the discrete Fourier transform, defined by

$$X(k) \equiv \sum_{i=0}^{N-1} x(i)\, e^{-j(2\pi k/N)\, i} \qquad k = 0,1,2,\ldots,N-1 \quad (7.12)$$

which has the inverse

$$x(i) = \frac{1}{N} \sum_{k=0}^{N-1} X(k)\, e^{j(2\pi k/N)\, i} \qquad i = 0,1,2,\ldots,N-1 \quad (7.13)$$

Eqs. (7.12) and (7.13) are called the discrete Fourier transform (DFT) pair.

In general, the *DFT* coefficients are complex numbers, with real and imaginary parts. If $x(i),\ i = 0,1,2,\ldots,N-1$, is a sequence of real numbers, as is the case for flight test data, it can be seen from Eq. (7.12) that

$$X(N-k) = X^*(k) \qquad (7.14)$$

where $X^*(k)$ is the complex conjugate of $X(k)$. It follows that the transformed data has conjugate symmetry about the value of k corresponding to the Nyquist frequency $1/(2\Delta t)$. Because of this, it suffices to compute the finite Fourier transform for only the first M frequencies in Eq. (7.9), where

$$M = \begin{cases} N/2+1 & \text{for } N \text{ even} \\ (N+1)/2 & \text{for } N \text{ odd} \end{cases} \qquad (7.15)$$

The values given for M in Eq. (7.15) represent the fundamental limitation that information in a sampled time series must fall within the frequency band $[0, f_N]$, where $f_N = 1/(2\Delta t)$ is the Nyquist frequency, defined as half the sampling frequency.

Direct computation of the *DFT* using Eq. (7.12) is not economical computationally. A numerically efficient method for computing the *DFT* is the Fast Fourier Transform (FFT) algorithm, proposed by Cooley and Tukey [7.3]. Bendat and Piersol [7.2] give a more general form of this algorithm, which is used as the basis for many Fourier transform computations. A method based on the chirp *z*-transform that can be used to compute the finite Fourier transform defined in Eq. (7.3) with very high accuracy at arbitrarily selected frequencies within the frequency band $[0, f_N]$ was developed in Ref. [7.4]. This method is described in detail in Chapter 11.

Often, data from more than one maneuver or data record must be combined for a single analysis. In the frequency domain, this can be done by simply adding the Fourier transforms from individual maneuvers at corresponding frequencies. Each maneuver to be combined must be transformed into the frequency domain using the same frequencies. This is easily done when the frequencies for the transformation can be selected arbitrarily, as in the procedure described in Chapter 11.

If frequency-domain data from two maneuvers are to be combined to represent the situation where the associated time series are concatenated in time, the task can be done in the frequency domain by multiplying the frequency domain data from the second maneuver by $e^{-j\omega T_1}$, where T_1 is the time length of the first maneuver, then adding the frequency-domain data at corresponding frequencies ω. This is equivalent to a pure time-shift of the data in the second maneuver by T_1 seconds.

7.2 Frequency Response

In Section 2.1, the nonparametric relation between the input and output vectors of a linear, time-invariant system was formed as

$$y(t)=\int_0^t \boldsymbol{H}(t-\tau)\,\boldsymbol{u}(\tau)\,d\tau \tag{7.16}$$

assuming zero initial conditions. The Laplace transform of the above equation is

$$\tilde{y}(s)=\boldsymbol{H}(s)\,\tilde{\boldsymbol{u}}(s) \tag{7.17}$$

where $\boldsymbol{H}(s)$ is the transfer function matrix (see Appendix A). Replacing the Laplace transform variable s with $j\omega$ results in the frequency response matrix $\boldsymbol{H}(\omega)$ with elements

$$H_{jk}(\omega)=\frac{\tilde{y}_j(\omega)}{\tilde{u}_k(\omega)} \qquad \begin{array}{l} j=1,2,\ldots,n_o \\ k=1,2,\ldots,n_i \end{array} \tag{7.18}$$

Individual frequency response functions can be determined experimentally from measured input-output data as

$$\hat{H}_{jk}(\omega)=\frac{\tilde{z}_j(\omega)}{\tilde{u}_k(\omega)} \qquad \begin{array}{l} j=1,2,\ldots,n_o \\ k=1,2,\ldots,n_i \end{array} \tag{7.19}$$

where $\tilde{z}_j(\omega)$ and $\tilde{u}_k(\omega)$ are finite Fourier transforms of the relevant measured input and output. The frequency response matrix is therefore composed of elements that are frequency response functions for a scalar input and a scalar output. Simplifying the notation by dropping the subscripts indicating the particular input and output, an individual frequency response function can be estimated from

$$\hat{H}(\omega) = \frac{\tilde{z}(\omega)}{\tilde{u}(\omega)} \tag{7.20}$$

For each selected frequency, the frequency response function is, in general, a complex number,

$$\begin{aligned} H(\omega) &= Re\left[H(\omega)\right] + j\,Im\left[H(\omega)\right] \\ &= R(\omega)\, e^{j\phi(\omega)} \end{aligned} \tag{7.21}$$

where

$$\begin{aligned} R(\omega) &= \sqrt{\left\{Re\left[H(\omega)\right]\right\}^2 + \left\{Im\left[H(\omega)\right]\right\}^2} \\ \phi(\omega) &= tan^{-1}\left\{\frac{Im\left[H(\omega)\right]}{Re\left[H(\omega)\right]}\right\} \end{aligned} \tag{7.22}$$

A collection of frequency response functions for many different frequencies is called the frequency response, which can be presented graphically, usually by displaying the amplitude ratio $R(\omega)$ and the phase angle $\phi(\omega)$ between the input and output, as a function of frequency. Such plots are called frequency response curves. Plotting $20\, log_{10}\, R(\omega)$ and phase $\phi(\omega)$ versus $log_{10}\, \omega$ results in a Bode plot, shown in the lower portion Fig. 7.1 for a simple transfer function.

In the special case where the input is a simple harmonic function, the steady-state output of a linear, time-invariant system is also a simple harmonic function, with the same frequency but modified amplitude and a phase shift. An example can be seen in Fig. 7.1, where the input to the transfer function is a pure sine wave with unit amplitude and frequency $\omega = 1.5$ rad/sec . The steady-state amplitude ratio and phase angle are estimated as

$$\begin{aligned} \hat{R}(\omega) &= |z|/|u| \\ \hat{\phi}(\omega) &= \tau\omega \end{aligned} \tag{7.23}$$

where $|u|$ and $|z|$ are the steady-state input and output amplitudes, respectively, and τ is the time delay from input to output, which can be evaluated at a zero crossing in the time domain, for example.

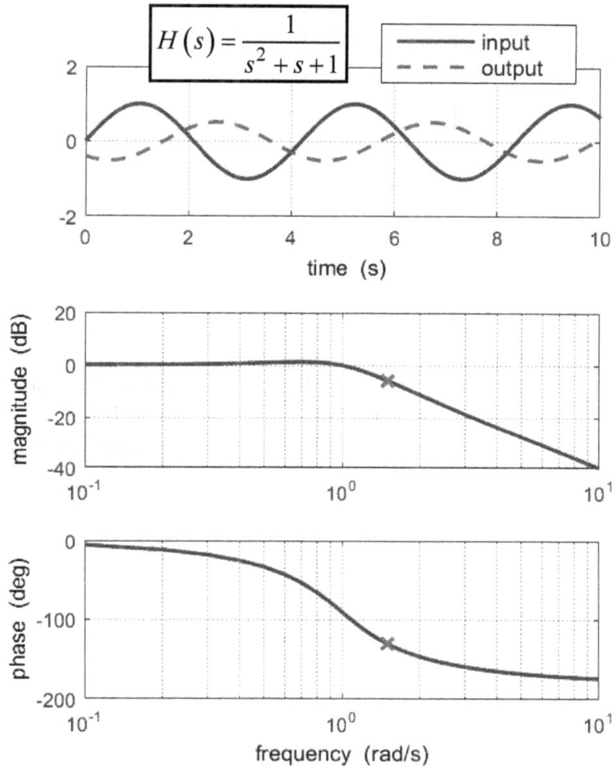

Figure 7.1 **Frequency response for a linear system**

Frequency response is based on the linear relationship between input and output. Frequency response functions are often estimated from spectral densities, which are defined as the Fourier transform of the autocorrelation function. The autocorrelation functions for the zero-mean time series input u and output y are (see Appendix B)

$$\mathcal{R}_{uu}(\tau) = \lim_{T\to\infty} \frac{1}{T}\int_0^T u(t)\,u(t+\tau)\,dt \tag{7.24a}$$

$$\mathcal{R}_{yy}(\tau) = \lim_{T\to\infty} \frac{1}{T}\int_0^T y(t)\,y(t+\tau)\,dt \tag{7.24b}$$

and the cross correlation is

$$\mathcal{R}_{uy}(\tau) = \lim_{T\to\infty} \frac{1}{T}\int_0^T u(t)\,y(t+\tau)\,dt \tag{7.24c}$$

Applying the Fourier transform to the correlation functions results in power spectral densities

$$S_{uu}(\omega) \equiv \int_{-\infty}^{\infty} \mathcal{R}_{uu}(\tau)\, e^{-j\omega\tau} \; d\tau \tag{7.25a}$$

$$S_{uu}(\omega) \equiv \int_{-\infty}^{\infty} \mathcal{R}_{uu}(\tau)\, e^{-j\omega\tau} \; d\tau \tag{7.25b}$$

$$S_{uy}(\omega) \equiv \int_{-\infty}^{\infty} \mathcal{R}_{uy}(\tau)\, e^{-j\omega\tau} \; d\tau \tag{7.25c}$$

The quantities in Eqs. (7.25a) and (7.25b) are also called auto spectral densities, and the quantity in Eq. (7.25c) is also called the cross spectral density. From Eqs. (7.24), the autocorrelation functions have the properties

$$\mathcal{R}_{uu}(\tau) = \mathcal{R}_{uu}(-\tau) \qquad \mathcal{R}_{yy}(\tau) = \mathcal{R}_{yy}(-\tau) \tag{7.26a}$$

$$\mathcal{R}_{uy}(\tau) = \mathcal{R}_{yu}(-\tau) \tag{7.26b}$$

Combining Eqs. (7.26) and (7.25), the spectral densities have the properties

$$S_{uu}(\omega) = S_{uu}(-\omega) \qquad S_{yy}(\omega) = S_{yy}(-\omega) \tag{7.27a}$$

$$S_{uy}(\omega) = S_{uy}^{*}(-\omega) = S_{yu}(-\omega) \tag{7.27b}$$

If the input and output variables for a linear, time-invariant dynamic system are expressed in terms of correlation functions, as is done for stochastic systems, then the cross correlation can be expressed as (see Appendix B)

$$\mathcal{R}_{uy}(\tau) = \int_{0}^{\infty} H(\mu)\, \mathcal{R}_{uu}(\tau - \mu)\, d\mu \tag{7.28}$$

Since convolution in the time domain is equivalent to multiplication in the frequency domain (see Appendix A), the Fourier transform of Eq. (7.28) is

$$S_{uy}(\omega) = H(\omega) S_{uu}(\omega) \tag{7.29}$$

The spectral densities $S_{uu}(\omega)$, $S_{yy}(\omega)$, and $S_{uy}(\omega)$ are called two-sided spectral densities, because the frequency ω can be positive or negative. For practical work, the one-sided spectral densities use only positive frequencies (see Appendix B), and satisfy a relationship similar to Eq. (7.29),

$$G_{uy}(\omega) = H(\omega) G_{uu}(\omega) \tag{7.30}$$

Estimates of the one-sided spectral densities can be computed from (see Refs. [7.2], [7.5]-[7.9]),

$$\hat{G}_{uu}(\omega) = \frac{2}{T}\tilde{u}^*(\omega,T)\,\tilde{u}(\omega,T) \qquad \omega>0 \qquad (7.31a)$$

$$\hat{G}_{yy}(\omega) = \frac{2}{T}\tilde{z}^*(\omega,T)\,\tilde{z}(\omega,T) \qquad \omega>0 \qquad (7.31b)$$

$$\hat{G}_{uy}(\omega) = \frac{2}{T}\tilde{u}^*(\omega,T)\,\tilde{z}(\omega,T) \qquad \omega>0 \qquad (7.31c)$$

where the fact that the finite Fourier transforms depend on the time length T is shown explicitly in the notation, and the measured output z is used.

Chapter 11 includes a discussion of some practical methods for computing the spectral densities required in Eq. (7.30) to calculate an empirical estimate of the frequency response using

$$\hat{H}(\omega) = \frac{\hat{G}_{uy}(\omega)}{\hat{G}_{uu}(\omega)} \qquad (7.32)$$

Computing the frequency response from Eq. (7.32), as opposed to Eq. (7.20), is done when there is uncertainty in the distribution of the input and output signal energy with respect to frequency. In that case, spectral density estimates are used. This issue is highlighted in Chapter 11 in connection with a technique for estimating frequency responses using multisine inputs.

The coherence between the input and output is a real-valued quantity defined in terms of the spectral densities,

$$\gamma_{uy}^2(\omega) = \frac{\left|G_{uy}(\omega)\right|^2}{G_{uu}(\omega)\,G_{yy}(\omega)} \qquad (7.33)$$

$$0 \le \gamma_{uy}^2(\omega) \le 1$$

An estimate of the coherence $\gamma_{uy}^2(\omega)$ is obtained using estimates of the spectral densities. The coherence is a measure of linearity between the input and output, so $\gamma_{uy}^2(\omega)=1$ indicates that the input and output are related perfectly by a linear dynamic system. The coherence is often used as a metric for evaluating the suitability of frequency-domain data for a linear model. The coherence can be less than 1 for any of the following reasons:

1) Measurement noise is present in the data.
2) The dynamic system is nonlinear.

3) Other inputs or disturbances influence the output.

More information on the coherence and its extension for a general multivariable system can be found in Ref. [7.2].

The use of frequency response data for aircraft system identification involves some important practical considerations:

1) The frequency response characterizes a linear dynamic relationship between a single input u and a single output y. Aircraft are nonlinear dynamic systems with multiple inputs and multiple outputs. Because of this mismatch, using frequency responses for aircraft system identification generally requires significant flight test time. The main reasons are that only one control can be moved at a time (alternatively, multiple moving inputs must be uncorrelated), and each frequency response can only characterize a range of flight conditions and responses where linearity assumptions are valid.

2) The expressions in Eq. (7.31) for estimating one-sided frequency responses from a single data record of length T can produce spectral density estimates with random error close to 100% of the estimated values (see Ref. [7.8]), regardless of the time length T. The random error can be reduced using some form of averaging, which is commonly done by windowing the data and averaging results from each window at corresponding frequencies. However, windowing splits the data into subsets, and this raises the lowest frequency available and makes the frequency resolution more coarse [cf. Eq. (7.8)]. Overlapping the data windows by up to 50 percent of the window width can mitigate the problem. But the fundamental trade-off between reduced random error (for an increased number of smaller windows) and reduced frequency range and resolution (for fewer larger windows) remains. This problem has been addressed by analyzing the data using various window sizes, then combining the results using an optimization based on coherence (see Ref. [7.6]). Another way to address the problem is to simply increase the maneuver time length, but that approach requires more flight test time and can be impractical for flight conditions that are difficult to maintain for a long period of time.

3) Although the frequency response can be used to predict transient response, it cannot be identified well from transient data. The frequency response is the steady-state, forced response for a linear, time-invariant system to a single input, as can be seen from the preceding development. Frequency response is clearly exhibited in the data after the transient response has died out, but is not as clearly identifiable from transient data. As a consequence, the flight test maneuvers typically used for frequency response estimation are sine sweeps with slowly-varying

frequency (see Chapter 8), which require more flight test time than other approaches.

4) Spectral densities quantify the signal power as a function of frequency, based on data collected over a finite time length T. This situation requires special data handling to avoid a phenomenon called leakage, which can adversely impact spectral density function estimates. The issue is discussed in Chapter 11, and also in Refs. [7.2], [7.5]-[7.9].

5) The frequency response is a nonparametric model, and therefore does not require a parametric model structure. The frequency response is equivalent to fitting the system dynamics with a linear, time-invariant, dynamic system model, without specifying the order of the model. Such models have been referred to as describing functions [7.10]. The frequency response therefore has no inherent capability to model nonlinearity, apart from combining multiple local linear models.

References [7.2], [7.5]-[7.9] contain more information on frequency responses, including how they can be computed accurately from measured data, and how they can be used for aircraft system identification. The frequency response data form requires significant effort both in conducting the flight test experiments and in processing the data.

The remainder of the chapter shows how finite Fourier transform data can be used directly in formulations analogous to those already discussed in the time domain, including practical application examples. Using the Fourier transform data directly, instead of estimating spectral density functions, avoids the problems in the list just given, while retaining many important advantages of working in the frequency domain.

7.3 Maximum Likelihood Estimator

To develop the maximum likelihood estimator in the frequency domain, it is necessary to postulate a model for the dynamic system, and then transform this model into the frequency domain. The next step is to formulate the likelihood function of the innovations, and then to maximize the likelihood function with respect to the unknown parameters. When the estimation problem is nonlinear, the modified Newton-Raphson algorithm is usually selected for the optimization, as in the time-domain case.

Model equations for a linear continuous-time dynamic system with process noise and discrete-time measurements are considered,

$$\dot{\boldsymbol{x}}(t) = \boldsymbol{A}\,\boldsymbol{x}(t) + \boldsymbol{B}\,\boldsymbol{u}(t) + \boldsymbol{B}_w\,\boldsymbol{w}(t) \tag{7.31a}$$

$$\boldsymbol{z}(i) = \boldsymbol{C}\,\boldsymbol{x}(i) + \boldsymbol{D}\,\boldsymbol{u}(i) + \boldsymbol{v}(i) \qquad i = 1,2,\ldots,N \tag{7.31b}$$

$$E\left[x(0)\right]=\mathbf{0} \qquad E\left[x(0)x^T(0)\right]=P_0 \tag{7.31c}$$

where $w(t)$ and $v(i)$ are assumed to be independent white Gaussian noise sequences with

$$\begin{aligned} E\left[w(t)\right]&=\mathbf{0} \qquad E\left[w(t_i)w^T(t_j)\right]=Q\,\delta(t_i-t_j) \\ E\left[v(i)\right]&=\mathbf{0} \qquad E\left[v(i)v^T(j)\right]=R\,\delta_{ij} \end{aligned} \tag{7.31d}$$

In the general case, the unknown parameters will occur in the matrices $A, B, B_w, C, D, P_0, Q,$ and R. As discussed in Section 6.1, estimation of all of these parameters can be extremely difficult because of the algorithm complexity and possible identifiability problems. For these reasons, the problem will be simplified by using a steady-state Kalman filter and estimating the unknown parameters in this formulation. Returning to Eqs. (6.20) and (6.6), the filter equations take the form

$$\frac{d}{dt}\left[\hat{x}(t\,|\,i-1)\right]=A\,\hat{x}(t\,|\,i-1)+B\,u(t) \tag{7.32a}$$

for $(i-1)\Delta t \le t \le i\Delta t$,

$$\hat{x}(i\,|\,i)=\hat{x}(i\,|\,i-1)+K\upsilon(i) \tag{7.32b}$$

where the innovations

$$\upsilon(i)=z(i)-C\,\hat{x}(i\,|\,i-1)-D\,u(i) \tag{7.32c}$$

form a sequence of independent Gaussian vectors with

$$E\left[\upsilon(i)\right]=\mathbf{0} \qquad E\left[\upsilon(i)\upsilon^T(j)\right]=\mathcal{B}\,\delta_{ij} \tag{7.32d}$$

Based on the suggestion of Mehra [7.11], the unknown parameters are selected as the elements of the matrices $A, B, C, D, K,$ and $\mathcal{B}$.

In the next step, the time functions in Eqs. (7.32) are written in the form of their Fourier series expansions. The Fourier series components of $x(i)$ are defined by Eq. (7.10) as

$$\tilde{x}(k)=\Delta t\sum_{i=0}^{N-1} x(i)\; e^{-j(2\pi k/N)\,i} \qquad k=0,1,2,\ldots,N-1 \tag{7.33}$$

Similar expressions apply for $z(i)$, $u(i)$, and $\upsilon(i)$. As stated in Ref. [7.1], the Fourier series expansion of random variables holds in the mean-squared sense, so that

$$E\left\{\left|\upsilon(i)-\frac{1}{N\Delta t}\sum_{k=0}^{N-1}\tilde{\upsilon}(k)e^{j(2\pi k/N)i}\right|^2\right\}=\mathbf{0} \tag{7.34}$$

and similarly for the other transformed quantities. The transform in Eq. (7.33) represents an Euler approximation to the finite Fourier integral, which can be replaced by a high-accuracy calculation mentioned earlier and described in Chapter 11. Now Eqs. (7.32) transformed into the frequency domain become

$$j\omega_k\,\tilde{x}(k)=A\tilde{x}(k)+B\tilde{u}(k)+K\tilde{\upsilon}(k) \tag{7.35a}$$

or

$$\tilde{x}(k)=\left(j\omega_k I-A\right)^{-1}B\tilde{u}(k)+\left(j\omega_k I-A\right)^{-1}K\tilde{\upsilon}(k) \tag{7.35b}$$

$$\tilde{z}(k)=C\tilde{x}(k)+D\tilde{u}(k)+\tilde{\upsilon}(k) \qquad k=0,1,2,\ldots,N-1 \tag{7.35c}$$

where ω_k is the kth frequency, and $\omega_k=2\pi k/N$ when the transform in Eq. (7.33) is used. The transformed innovations $\tilde{\upsilon}(k)$ are uncorrelated Gaussian random variables (cf. Ref. [7.1]), with

$$E\left[\tilde{\upsilon}(k)\right]=\mathbf{0} \qquad E\left[\tilde{\upsilon}(k)\tilde{\upsilon}^{\dagger}(k)\right]=\frac{S_{\upsilon\upsilon}}{N} \tag{7.35d}$$

where $S_{\upsilon\upsilon}$ is a real diagonal matrix with diagonal elements equal to the power spectral densities of the elements of $\upsilon(i)$, and $\tilde{\upsilon}^{\dagger}(k)$ is the complex conjugate transpose of $\tilde{\upsilon}(k)$.

The left side of Eq. (7.35a) requires that the bias and trend be removed from $x(t)$, so that $x(T)=x(0)=0$ prior to the Fourier transformation, since

$$\begin{aligned}\mathcal{F}\left[\dot{x}(t)\right]&=\int_0^T \dot{x}(t)\,e^{-j\omega t}dt\\&=j\omega\int_0^T x(t)\,e^{-j\omega t}dt+x(T)\,e^{-j\omega T}-x(0)\\&=j\omega\tilde{x}(\omega)+x(T)\,e^{-j\omega T}-x(0)\end{aligned} \tag{7.36}$$

The bias and trend should be removed from any time series prior to Fourier transformation, for practical reasons discussed in Section 7.7.2. It is also possible to eliminate the effects of nonzero endpoints using a modulating function approach based on harmonically-related sinusoids. This approach is discussed in Ref. [7.12] and also in Section 7.7.3.

From Eqs. (7.35),

$$\begin{aligned}\tilde{z}(k) &= \left[C(j\omega_k I - A)^{-1} B + D\right]\tilde{u}(k) + \left[C(j\omega_k I - A)^{-1} K + I\right]\tilde{\upsilon}(k) \\ &= H_1(k,\theta)\tilde{u}(k) + H_2(k,\theta)\tilde{\upsilon}(k)\end{aligned} \tag{7.37}$$

where H_1 and H_2 are the system transfer function matrices defined as

$$H_1(k,\theta) = C(j\omega_k I - A)^{-1} B + D \tag{7.38a}$$

$$H_2(k,\theta) = C(j\omega_k I - A)^{-1} K + I \tag{7.38b}$$

and θ is the vector of unknown parameters in matrices A, B, C, D, K, and $S_{\upsilon\upsilon}$. Since the Kalman filter representation in Eqs. (7.32) is invertible [7.13], H_2 is nonsingular, and the innovations can be expressed as

$$\tilde{\upsilon}(k) = H_2^{-1}\tilde{z}(k) - H_2^{-1} H_1 \tilde{u}(k) \tag{7.39}$$

To develop the likelihood function, the quantity V_N is introduced, which consists of all innovations up to and including the frequency for which $k = N-1$,

$$V_N \equiv \left[\tilde{\upsilon}(0) \;\; \tilde{\upsilon}(1) \;\; \ldots \;\; \tilde{\upsilon}(N-1)\right]^T \tag{7.40}$$

Assuming that $\tilde{\upsilon}(k)$ has a probability density $p(\tilde{\upsilon})$, then it follows from the definition of probability densities that

$$p(V_N) = p\left[\tilde{\upsilon}(N-1) \mid V_{N-1}\right] p(V_{N-1}) \tag{7.41}$$

Repeated use of Eq. (7.41) gives the expression for the likelihood function,

$$\begin{aligned}\mathbb{L}(V_N;\theta) &= p(V_N \mid \theta) \\ &= p\left[\tilde{\upsilon}(N-1) \mid V_{N-1}\right] p\left[\tilde{\upsilon}(N-2) \mid V_{N-2}\right] \cdots p\left[\tilde{\upsilon}(1) \mid V_1\right] p\left[\tilde{\upsilon}(0)\right]\end{aligned} \tag{7.42}$$

Since the distribution of $\tilde{\upsilon}(k)$ is Gaussian, the distribution of $\tilde{\upsilon}(k)$ given V_k is also Gaussian. The probability density for this complex Gaussian vector is

$$p\left[\tilde{\upsilon}(k) \mid V_k\right] = \frac{N^{n_o}}{\pi^{n_o}\left|\boldsymbol{S}_{\upsilon\upsilon}\right|} exp\left[-N\,\tilde{\upsilon}^{\dagger}(k)\,\boldsymbol{S}_{\upsilon\upsilon}^{-1}\,\tilde{\upsilon}(k)\right] \tag{7.43}$$

See Refs. [7.1] and [7.14] for more details. The power spectral density $\boldsymbol{S}_{\upsilon\upsilon}$ appears in Eq. (7.43) because of the definitions in Eq. (7.35d). Combining Eqs. (7.42) and (7.43), the negative logarithm of $\mathbb{L}(V_N;\boldsymbol{\theta})$ is

$$-ln\left[\mathbb{L}(V_N;\boldsymbol{\theta})\right] \equiv J(\boldsymbol{\theta}) = N\sum_{k=0}^{N-1}\tilde{\upsilon}^{\dagger}(k)\,\boldsymbol{S}_{\upsilon\upsilon}^{-1}\,\tilde{\upsilon}(k) + N\,ln\left|\boldsymbol{S}_{\upsilon\upsilon}\right| + N n_o\,ln\left(\frac{\pi}{N}\right) \tag{7.44}$$

Excluding the constant term which is unaffected by $\boldsymbol{\theta}$, the cost function to be minimized is

$$J(\boldsymbol{\theta}) = N\sum_{k=0}^{N-1}\tilde{\upsilon}^{\dagger}(k)\,\boldsymbol{S}_{\upsilon\upsilon}^{-1}\,\tilde{\upsilon}(k) + N\,ln\left|\boldsymbol{S}_{\upsilon\upsilon}\right| \tag{7.45}$$

The maximum likelihood parameter vector estimate is the value of $\boldsymbol{\theta}$ that minimizes the cost function in Eq. (7.45). Optimizing $J(\boldsymbol{\theta})$ for the parameters in $\boldsymbol{S}_{\upsilon\upsilon}$ gives

$$\hat{\boldsymbol{S}}_{\upsilon\upsilon} = \sum_{k=0}^{N-1}\tilde{\upsilon}(k,\hat{\boldsymbol{\theta}})\,\tilde{\upsilon}^{\dagger}(k,\hat{\boldsymbol{\theta}}) = \sum_{k=0}^{N-1}\left[\tilde{z}(k)-\tilde{y}(k,\hat{\boldsymbol{\theta}})\right]\left[\tilde{z}(k)-\tilde{y}(k,\hat{\boldsymbol{\theta}})\right]^{\dagger} \tag{7.46}$$

where $\hat{\boldsymbol{\theta}}$ is an estimate of the vector of parameters $\boldsymbol{\theta}$, which appear in the dynamic system matrices. The initial estimate of $\boldsymbol{S}_{\upsilon\upsilon}$ is made with $\hat{\boldsymbol{\theta}} = \boldsymbol{\theta}_o$, where $\boldsymbol{\theta}_o$ is a nominal starting value of $\boldsymbol{\theta}$.

The estimates of the remaining parameters are found using the modified Newton-Raphson technique for nonlinear optimization. For a given $\hat{\boldsymbol{S}}_{\upsilon\upsilon}$, the cost function

$$J(\boldsymbol{\theta}) = N\sum_{k=0}^{N-1}\tilde{\upsilon}^{\dagger}(k)\,\hat{\boldsymbol{S}}_{\upsilon\upsilon}^{-1}\,\tilde{\upsilon}(k) \tag{7.47}$$

is minimized by iteratively computing

$$\hat{\boldsymbol{\theta}} = \boldsymbol{\theta}_o + \Delta\hat{\boldsymbol{\theta}} \tag{7.48}$$

where the parameter vector update $\Delta\hat{\boldsymbol{\theta}}$ is given by

$$\Delta\hat{\boldsymbol{\theta}} = -\,\boldsymbol{M}_{\boldsymbol{\theta}=\boldsymbol{\theta}_o}^{-1}\left[\frac{\partial J(\boldsymbol{\theta})}{\partial\boldsymbol{\theta}}\right]_{\boldsymbol{\theta}=\boldsymbol{\theta}_o} \tag{7.49}$$

and $\boldsymbol{M}$ is the Fisher information matrix,

$$\boldsymbol{M} = -E\left[\frac{\partial^2 \ln\mathbb{L}(V_N;\boldsymbol{\theta})}{\partial\boldsymbol{\theta}\,\partial\boldsymbol{\theta}^T}\right] = E\left[\frac{\partial^2 J(\boldsymbol{\theta})}{\partial\boldsymbol{\theta}\,\partial\boldsymbol{\theta}^T}\right] \tag{7.50}$$

Because $\Delta\hat{\boldsymbol{\theta}}$ is a vector with only real elements, and the log-likelihood function is real, the expressions for the first- and second-order gradients of $J(\boldsymbol{\theta})$ are also real,

$$\begin{aligned}\frac{\partial J(\boldsymbol{\theta})}{\partial\boldsymbol{\theta}} &= 2N\,Re\left[\sum_{k=0}^{N-1}\tilde{\boldsymbol{v}}^{\dagger}(k)\,\hat{\boldsymbol{S}}_{\boldsymbol{vv}}^{-1}\frac{\partial\tilde{\boldsymbol{v}}(k)}{\partial\boldsymbol{\theta}}\right] \\ &= -2N\,Re\left[\sum_{k=0}^{N-1}\tilde{\boldsymbol{v}}^{\dagger}(k)\,\hat{\boldsymbol{S}}_{\boldsymbol{vv}}^{-1}\frac{\partial\tilde{\boldsymbol{y}}(k)}{\partial\boldsymbol{\theta}}\right]\end{aligned} \tag{7.51}$$

$$\begin{aligned}\frac{\partial^2 J(\boldsymbol{\theta})}{\partial\boldsymbol{\theta}\partial\boldsymbol{\theta}^T} &= 2N\,Re\left[\sum_{k=0}^{N-1}\frac{\partial\tilde{\boldsymbol{v}}^{\dagger}(k)}{\partial\boldsymbol{\theta}}\,\hat{\boldsymbol{S}}_{\boldsymbol{vv}}^{-1}\frac{\partial\tilde{\boldsymbol{v}}(k)}{\partial\boldsymbol{\theta}}\right] \\ &= 2N\,Re\left[\sum_{k=0}^{N-1}\frac{\partial\tilde{\boldsymbol{y}}^{\dagger}(k)}{\partial\boldsymbol{\theta}}\,\hat{\boldsymbol{S}}_{\boldsymbol{vv}}^{-1}\frac{\partial\tilde{\boldsymbol{y}}(k)}{\partial\boldsymbol{\theta}}\right]\end{aligned} \tag{7.52}$$

where the expression in Eq. (7.52) is the simplified second gradient used in the modified Newton-Raphson method.

After estimating $\boldsymbol{S}_{\boldsymbol{vv}}$ from Eq. (7.46), the vector of remaining unknown parameters $\boldsymbol{\theta}$ is composed of elements of the $\boldsymbol{A}, \boldsymbol{B}, \boldsymbol{C}, \boldsymbol{D},$ and $\boldsymbol{K}$ matrices. The final estimated parameters for the model of Eqs. (7.35) have the properties given in Section 6.1.3.

7.4 Output-Error Method

As in the time-domain analysis, the maximum likelihood estimator in the frequency domain is substantially simplified when there is no process noise. In that case, the dynamic system is deterministic,

$$j\omega_k \, \tilde{x}(k) = A\tilde{x}(k) + B\tilde{u}(k) \tag{7.53a}$$

$$\tilde{y}(k) = C\tilde{x}(k) + D\tilde{u}(k) \tag{7.53b}$$

$$\tilde{z}(k) = H(k,\theta)\,\tilde{u}(k) + \tilde{v}(k) \tag{7.53c}$$

$$E\left[\tilde{v}(k)\right] = 0 \qquad E\left[\tilde{v}(k)\tilde{v}^{\dagger}(k)\right] = \frac{S_{vv}}{N} \tag{7.53d}$$

where

$$H(k,\theta) \equiv H_1(k,\theta) = C\left(j\omega_k I - A\right)^{-1} B + D \tag{7.54a}$$

$$\omega_k = 2\pi k/T \qquad k = 0,1,2,\ldots,N-1 \tag{7.54b}$$

and S_{vv} is the spectral density of $v(i)$. For these model equations, the innovations are reduced to output errors

$$\tilde{v}(k,\theta) = \tilde{z}(k) - \tilde{y}(k,\theta) = \tilde{z}(k) - H(k,\theta)\tilde{u}(k) \tag{7.55}$$

The output-error cost function is the negative log-likelihood function, excluding the constant term,

$$J(\theta) = N\sum_{k=0}^{N-1} \tilde{v}^{\dagger}(k,\theta)\, S_{vv}^{-1}\, \tilde{v}(k,\theta) \; + N\,ln\left|S_{vv}\right| \tag{7.56}$$

and the parameter estimation process is similar to that presented earlier for the more general model equations. The estimates for the parameters in S_{vv} are obtained from

$$\hat{S}_{vv} = \sum_{k=0}^{N-1} \tilde{v}(k,\hat{\theta})\,\tilde{v}^{\dagger}(k,\hat{\theta}) = \sum_{k=0}^{N-1}\left[\tilde{z}(k) - \tilde{y}(k,\hat{\theta})\right]\left[\tilde{z}(k) - \tilde{y}(k,\hat{\theta})\right]^{\dagger} \tag{7.57}$$

To estimate the unknown parameters in matrices $A, B, C,$ and D, the modified Newton-Raphson algorithm is used again. The expressions for the gradient of the negative log-likelihood function and the information matrix can be derived from the output-error cost function,

$$J(\theta) = N\sum_{k=0}^{N-1} \tilde{v}^{\dagger}(k,\theta)\, \hat{S}_{vv}^{-1}\, \tilde{v}(k,\theta) \tag{7.58}$$

as

$$\frac{\partial J(\boldsymbol{\theta})}{\partial \boldsymbol{\theta}} = -2N\, Re\left[\sum_{k=0}^{N-1} \boldsymbol{S}^{\dagger}(k)\, \hat{\boldsymbol{S}}_{vv}^{-1}\, \tilde{\boldsymbol{v}}(k,\boldsymbol{\theta})\right] \tag{7.59}$$

$$\boldsymbol{M}(\boldsymbol{\theta}) = \frac{\partial^2 J(\boldsymbol{\theta})}{\partial \boldsymbol{\theta} \partial \boldsymbol{\theta}^T} = 2N\, Re\left[\sum_{k=0}^{N-1} \boldsymbol{S}^{\dagger}(k)\, \hat{\boldsymbol{S}}_{vv}^{-1}\, \boldsymbol{S}(k)\right] \tag{7.60}$$

where $\boldsymbol{S}(k)$ is the $n_o \times n_p$ output sensitivity matrix, given by

$$\boldsymbol{S}(k) = \frac{\partial \boldsymbol{H}(k,\boldsymbol{\theta})\tilde{\boldsymbol{u}}(k)}{\partial \boldsymbol{\theta}} \tag{7.61}$$

The modified Newton-Raphson step is given by Eqs. (7.48) and (7.49), as before.

For a state-space model, $\boldsymbol{H}(k,\boldsymbol{\theta})$ is computed from Eq. (7.54a); however, $\boldsymbol{H}(k,\boldsymbol{\theta})$ can also be a matrix of transfer functions with transfer function coefficients as the unknown parameters in $\boldsymbol{\theta}$.

The output-error parameter estimation algorithm can be easily modified for measured data in the form of frequency response curves. In this case, the cost function takes the form

$$J(\boldsymbol{\theta}) = N\sum_{k=0}^{N-1} \tilde{\boldsymbol{u}}(k)^{\dagger}\left[\boldsymbol{H}_E(k) - \boldsymbol{H}(k,\boldsymbol{\theta})\right]^{\dagger}\, \boldsymbol{S}_{vv}^{-1}\left[\boldsymbol{H}_E(k) - \boldsymbol{H}(k,\boldsymbol{\theta})\right]\tilde{\boldsymbol{u}}(k) \tag{7.62}$$

where $\boldsymbol{H}(k,\boldsymbol{\theta})$ is a matrix with transfer functions as elements, and $\boldsymbol{H}_E(k)$ is a matrix of experimentally-determined frequency responses, which can be found using spectral estimates, as described earlier and in Chapter 11.

Cost functions in Eq. (7.58) or (7.62) can be minimized with respect to unknown parameters in $\boldsymbol{A}, \boldsymbol{B}, \boldsymbol{C},$ and $\boldsymbol{D}$ or with respect to transfer function coefficients in the elements of $\boldsymbol{H}(k,\boldsymbol{\theta})$. The parameter estimates are obtained from Eqs. (7.48)-(7.52). Spectral densities of the residuals are estimated from Eq. (7.57).

For a dynamic system with a single input, the output-error cost function with measured frequency response curves is defined as

$$J(\boldsymbol{\theta}) = N\sum_{k=0}^{N-1} \tilde{u}^*(k)\left[\boldsymbol{H}_E(k) - \boldsymbol{H}(k,\boldsymbol{\theta})\right]^{\dagger}\, \boldsymbol{S}_{vv}^{-1}\left[\boldsymbol{H}_E(k) - \boldsymbol{H}(k,\boldsymbol{\theta})\right]\tilde{u}(k) \tag{7.63}$$

In this formulation, the scalar variable $\tilde{u}(k)$ can be interpreted as a weighting function quantifying the importance of the measured data in the frequency domain according to the harmonic content of the input.

Another approach is to formulate the cost function without the input weighting,

$$J(\boldsymbol{\theta}) = N\sum_{k=0}^{N-1} \left[\boldsymbol{H}_E(k) - \boldsymbol{H}(k,\boldsymbol{\theta})\right]^{\dagger} \boldsymbol{S}_{vv}^{-1}\left[\boldsymbol{H}_E(k) - \boldsymbol{H}(k,\boldsymbol{\theta})\right] \quad (7.64)$$

When the cost functions in Eqs. (7.63) and (7.64) are used for a single-input, single-output case, the factor NS_{vv}^{-1} can be omitted from the cost without affecting the parameter estimation results.

A variation of Eq. (7.64) is to introduce relative weighting on each term (i.e., for each frequency), according to the coherence. The idea is to weight the frequency-domain data with higher coherence more heavily in the cost function. The coherence is a measure of the linear relationship between input and output, so that data with high coherence are good in the sense of being compatible with a linear model structure and not overly corrupted by noise.

Sometimes the cost function is partitioned into separate parts, quantifying the model fit to the magnitude and phase angle for a Bode plot. To implement this for a single-input, single-output case, the cost function is

$$\begin{aligned} J(\boldsymbol{\theta}) = &\sum_{k=0}^{N-1} \left[20\,log_{10}\left|H_E(k)\right| - 20\,log_{10}\left|H(k,\boldsymbol{\theta})\right|\right]^2 \\ &+ w\sum_{k=0}^{N-1} \left\{arg\left[H_E(k)\right] - arg\left[H(k,\boldsymbol{\theta})\right]\right\}^2 \end{aligned} \quad (7.65)$$

where typically the weighting w is set to 0.0175, to balance the contributions of the magnitude and phase error components of the cost function. The cost formulation in Eq. (7.65), with coherence weighting, is the basis of the commercially-available FORTRAN software called CIFER® (see Refs. [7.6], [7.15]-[7.17]), which has been used extensively in practice, particularly for rotorcraft and V/STOL aircraft.

Although the summations over the frequency index k in the above expressions are written to include all frequencies that would be calculated by the discrete Fourier transform, it is not necessary to include all these frequencies. Using selected frequencies would only change the summation indices to correspond to the selected frequencies. Arbitrary frequencies can be used when the transform is done using the high-accuracy Fourier transform algorithm described in Chapter 11. For selected frequencies, there is a difference is in interpretation of the results, because the model and spectral density estimate $\boldsymbol{S}_{vv}$ will apply only for signal components at the selected frequencies, and will not include all of the power in the time-domain data. Transforming the time-domain data using a small frequency band automatically discards some of the power in the time-domain data. This is equivalent to a zero-lag, low-pass filtering operation, which improves the convergence and accuracy of the parameter estimation.

Finally, the weighting matrix $\boldsymbol{S}_{vv}$ in the cost function is the spectral density of the residuals, which ideally should be estimated as a function of frequency index k [cf. Eq. (7.60)]. However, in most practical cases, the frequencies used for the analysis correspond to the bandwidth of the dynamic system modes, which is also the bandwidth where most of the residual power resides when the residuals are colored. Therefore, using a constant estimate of $\boldsymbol{S}_{vv}$ over the frequencies used in the analysis [cf. Eq. (7.60), Figs. 5.13 and 6.5] is a good representation of the spectral density of the residual power in the frequency domain. Consequently, the noise assumptions in the theory match well with reality, and there is no need to correct the estimated error bounds for colored residuals when the parameter estimation is done in the frequency domain.

7.5 Equation-Error Method

Further simplification of the parameter estimation can be made by assuming that all state variables are measured without errors. Including this assumption results in the equation-error method. The dynamic model equation in the frequency domain is

$$j\omega_k\,\tilde{\boldsymbol{x}}(\omega_k) = \boldsymbol{A}\tilde{\boldsymbol{x}}(\omega_k) + \boldsymbol{B}\tilde{\boldsymbol{u}}(\omega_k) + \boldsymbol{B}_w\,\tilde{\boldsymbol{w}}(\omega_k) \tag{7.66}$$

where process noise is included, but there is no measurement noise because both the inputs and states are assumed to be measured without errors. Considering the state derivatives as the measured outputs at discrete frequencies $\omega_k = 2\pi k/T$,

$$\tilde{\boldsymbol{z}}(k) = j\omega_k\,\tilde{\boldsymbol{x}}(k) \equiv j\omega_k\,\tilde{\boldsymbol{x}}(\omega_k) \tag{7.67}$$

$$\omega_k = 2\pi k/T \qquad k = 0,1,2,\ldots,N-1 \tag{7.68}$$

The model equation then becomes

$$\tilde{\boldsymbol{z}}(k) = \boldsymbol{A}\tilde{\boldsymbol{x}}(k) + \boldsymbol{B}\tilde{\boldsymbol{u}}(k) + \tilde{\boldsymbol{v}}(k) \qquad k = 0,1,2,\ldots,N-1 \tag{7.69}$$

where $\tilde{\boldsymbol{v}}(k) \equiv \boldsymbol{B}_w\,\tilde{\boldsymbol{w}}(k)$ are the equation errors, which are Gaussian with

$$E\left[\tilde{\boldsymbol{v}}(k)\right]=\boldsymbol{0} \qquad E\left[\tilde{\boldsymbol{v}}(k)\tilde{\boldsymbol{v}}^{\dagger}(k)\right]=\frac{\boldsymbol{S}_{vv}}{N} \tag{7.70}$$

The cost function for equation-error parameter estimation is

$$J(\boldsymbol{\theta})=N\sum_{k=0}^{N-1}\left[\tilde{\boldsymbol{z}}(k)-\boldsymbol{A}\tilde{\boldsymbol{x}}(k)-\boldsymbol{B}\tilde{\boldsymbol{u}}(k)\right]^{\dagger}\boldsymbol{S}_{vv}^{-1}\left[\tilde{\boldsymbol{z}}(k)-\boldsymbol{A}\tilde{\boldsymbol{x}}(k)-\boldsymbol{B}\tilde{\boldsymbol{u}}(k)\right] \tag{7.71}$$

where the innovations

$$\tilde{\boldsymbol{v}}(k)=\tilde{\boldsymbol{z}}(k)-\boldsymbol{A}\tilde{\boldsymbol{x}}(k)-\boldsymbol{B}\tilde{\boldsymbol{u}}(k) \tag{7.72}$$

are equation errors.

Parameter estimates are obtained by minimizing the cost in Eq. (7.71) with respect to the unknown parameters in matrices $\boldsymbol{A}$ and $\boldsymbol{B}$. The minimization is applied separately to each state equation, corresponding to each row in the vector equation (7.66). This assumes that the errors in the equations are mutually uncorrelated, so that $\boldsymbol{S}_{vv}$ is a diagonal matrix. This is generally a very good assumption in practice. Since the equation error is minimized for one equation at a time, the weighting $N\,S_{vv}^{-1}$ can be omitted from the least squares optimization; however, this weighting must be included to compute the error bounds properly. As in the time-domain case, the equation-error method can be used to estimate dimensional or nondimensional aerodynamic parameters, by using the appropriate model equations.

The equation for the measured outputs of the dynamic system,

$$\tilde{\boldsymbol{z}}(k)=\boldsymbol{C}\tilde{\boldsymbol{x}}(k)+\boldsymbol{D}\tilde{\boldsymbol{u}}(k)+\tilde{\boldsymbol{v}}(k) \qquad k=0,1,2,\ldots,N-1 \tag{7.73}$$

can also be used to estimate aerodynamic parameters appearing in the $\boldsymbol{C}$ and $\boldsymbol{D}$ matrices. In this case, the actual measured outputs of the dynamic system are used for $\tilde{\boldsymbol{z}}(k)$, and

$$\tilde{\boldsymbol{v}}(k)=\tilde{\boldsymbol{z}}(k)-\boldsymbol{C}\tilde{\boldsymbol{x}}(k)-\boldsymbol{D}\tilde{\boldsymbol{u}}(k) \tag{7.74}$$

For example, output equations for measured accelerometer outputs can be used to estimate aerodynamic force parameters.

When the model is formulated as a transfer function,

$$\begin{aligned}\tilde{z}(k)&=H(k,\boldsymbol{\theta})\tilde{u}(k)+\tilde{v}(k)\\&=\frac{num(k,\boldsymbol{\theta})}{den(k,\boldsymbol{\theta})}\tilde{u}(k)+\tilde{v}(k)\end{aligned}$$

$$\tilde{z}(k) = \frac{num_1(k,\boldsymbol{\theta})}{(j\omega_k)^n + den_1(k,\boldsymbol{\theta})}\tilde{u}(k) + \tilde{v}(k) \tag{7.75}$$

where $den_1(k)$ and $num_1(k)$ are obtained by dividing $den(k)$ and $num(k)$ by the coefficient of the highest order term, $(j\omega_k)^n$, in $den(k)$. The least-squares cost function is

$$J(\boldsymbol{\theta}) = N\sum_{k=0}^{N-1} \tilde{v}^*(k)\, S_{vv}^{-1}\, \tilde{v}(k) \tag{7.76}$$

where

$$\tilde{v}(k) = (j\omega)^n\, \tilde{z}(k) + den_1(k,\boldsymbol{\theta})\tilde{z}(k) - num_1(k,\boldsymbol{\theta})\tilde{u}(k) \tag{7.77}$$

In Eq. (7.77), the model parameters are transfer function coefficients, which are estimated as the values that minimize the cost function (7.76).

The main disadvantage of using the cost function in Eq. (7.76) with Eq. (7.77) is the appearance of factors of $j\omega_k$ for each power of s in the numerator and denominator of the transfer function, corresponding to each time derivative. This causes frequency-domain data at higher frequencies to be given increased weighting in the least squares problem, because factors like $(j\omega_k)^2$ are larger for high frequencies than for low frequencies.

To fix this problem, the residuals in the cost function can be normalized by factors of $(j\omega_k)$. One approach that works well in practice is to divide the residuals by a factor of $(j\omega_k)^{n-1}$, where n is the order of the denominator of the transfer function, which is also the number of states in the dynamic system model. This effectively removes the high-frequency weighting, which is an artifact of the cost formulation, and therefore has no physical substantiation. The revised cost function is still

$$J(\boldsymbol{\theta}) = N\sum_{k=0}^{N-1} \tilde{v}^*(k)\, S_{vv}^{-1}\, \tilde{v}(k) \tag{7.78}$$

but the residuals are changed to

$$\begin{aligned}\tilde{v}(k) = (j\omega_k)\tilde{z}(k) &+ \left[den_1(k,\boldsymbol{\theta})/(j\omega_k)^{n-1}\right]\tilde{z}(k) \\ &- \left[num_1(k,\boldsymbol{\theta})/(j\omega_k)^{n-1}\right]\tilde{u}(k)\end{aligned} \tag{7.79}$$

The above modification to the cost function is equivalent to scaling the kth equation to be solved in the least squares sense by the factor $1/(j\omega_k)^{n-1}$. This is a weighted least squares formulation, as discussed in Chapter 5.

As in the case of the output-error method, selected frequencies can be used to generate the finite Fourier transforms for equation-error parameter estimation in the frequency domain. This results in different limits for the summations in the above expressions, changes the number of frequencies to $M \le N$, and makes the results applicable for only the frequencies included. Specifying frequencies near the modal frequencies of the physical system results in good signal-to-noise ratios and automatic smoothing, which produces accurate parameter estimates with very small bias errors [7.18].

For both the state-space and transfer function model forms, the resulting parameter estimation is a least squares linear regression problem with complex numbers. The next section gives the solution for this problem.

The frequency-domain output-error and equation-error methods presented here are based on the material in Refs. [7.1], [7.18]-[7.20]. Practical applications of the methods can be found in Refs. [7.1], [7.18]-[7.29].

7.6 Complex Linear Regression

Linear regression in the frequency domain follows the same approach used in Section 5.1 for time-domain data. The general form of the regression equation is

$$\tilde{z} = \tilde{X}\theta + \tilde{v} \tag{7.80}$$

where

$\tilde{z}$ = $N \times 1$ vector of transformed dependent variable measurements

θ = $n_p \times 1$ vector of unknown parameters

$\tilde{X}$ = $N \times n_p$ matrix of vectors of ones and transformed regressors, $n_p = n+1$

$\tilde{v}$ = $N \times 1$ vector of complex measurement errors

and the properties of $\tilde{v}$ are

$$E(\tilde{v}) = 0 \qquad E\left(\tilde{v}\tilde{v}^{\dagger}\right) = \sigma^2 I \tag{7.81}$$

where the quantity σ^2 is a real number representing the squared magnitude of the complex residual.

The least squares estimate of $\boldsymbol{\theta}$ given $\tilde{\boldsymbol{z}}$ follows from the minimization of

$$J(\boldsymbol{\theta})=\left(\tilde{\boldsymbol{z}}-\tilde{\boldsymbol{X}}\boldsymbol{\theta}\right)^{\dagger}\left(\tilde{\boldsymbol{z}}-\tilde{\boldsymbol{X}}\boldsymbol{\theta}\right)=\tilde{\boldsymbol{v}}^{\dagger}\tilde{\boldsymbol{v}}=|\tilde{\boldsymbol{v}}|^{2} \tag{7.82}$$

which results in the least squares estimate

$$\hat{\boldsymbol{\theta}}=\left[\tilde{\boldsymbol{X}}^{\dagger}\tilde{\boldsymbol{X}}\right]^{-1}\tilde{\boldsymbol{X}}^{\dagger}\tilde{\boldsymbol{z}} \tag{7.83}$$

The covariance matrix of the parameter estimates is

$$Cov\left(\hat{\boldsymbol{\theta}}\right)=E\left[\left(\hat{\boldsymbol{\theta}}-\boldsymbol{\theta}\right)\left(\hat{\boldsymbol{\theta}}-\boldsymbol{\theta}\right)^{\dagger}\right]=\sigma^{2}\left[\tilde{\boldsymbol{X}}^{\dagger}\tilde{\boldsymbol{X}}\right]^{-1} \tag{7.84}$$

In the last two expressions, $\tilde{\boldsymbol{X}}^{\dagger}\tilde{\boldsymbol{X}}$ is a Hermitian matrix that is nonsingular and positive definite.

The vector of estimated dependent variables is

$$\hat{\tilde{\boldsymbol{y}}}=\tilde{\boldsymbol{X}}\hat{\boldsymbol{\theta}} \tag{7.85}$$

and the unbiased estimate of the fit error variance σ^2 is

$$s^{2}\equiv\hat{\sigma}^{2}=\frac{\tilde{\boldsymbol{v}}^{\dagger}\tilde{\boldsymbol{v}}}{N-n_{p}} \tag{7.86}$$

where

$$\tilde{\boldsymbol{v}}=\tilde{\boldsymbol{z}}-\hat{\tilde{\boldsymbol{y}}} \tag{7.87}$$

In the preceding development, there was no assumption as to whether the parameter vector was real or complex, so it could be either. In practical aircraft parameter estimation problems, the parameter vector is real, so the expressions for the parameter estimate and covariance matrix are simplified to

$$\hat{\boldsymbol{\theta}}=\left[Re\left(\tilde{\boldsymbol{X}}^{\dagger}\tilde{\boldsymbol{X}}\right)\right]^{-1}Re\left(\tilde{\boldsymbol{X}}^{\dagger}\tilde{\boldsymbol{z}}\right) \tag{7.88}$$

$$Cov\left(\hat{\boldsymbol{\theta}}\right)=\sigma^{2}\left[Re\left(\tilde{\boldsymbol{X}}^{\dagger}\tilde{\boldsymbol{X}}\right)\right]^{-1} \tag{7.89}$$

and the fit error variance estimate in Eq. (7.86) remains the same because $\tilde{\boldsymbol{v}}^{\dagger}\tilde{\boldsymbol{v}}$ is always real.

If $\tilde{\boldsymbol{v}}$ is $\mathbb{N}\left(\boldsymbol{0},\sigma^2\boldsymbol{I}\right)$, the likelihood function of $\tilde{\boldsymbol{z}}$ is

$$p(\tilde{\boldsymbol{z}})=\frac{1}{\pi^N\sigma^{2N}}exp\left[-\frac{\left|\tilde{\boldsymbol{z}}-\tilde{\boldsymbol{X}}\hat{\boldsymbol{\theta}}\right|^2}{\sigma^2}\right] \tag{7.90}$$

as developed by Klein [7.1] and Miller [7.14]. The denominator of the exponent in Eq. (7.90) is σ^2, rather than $2\sigma^2$ as in the case of a real random variable. This follows from the assumption

$$p(\tilde{\boldsymbol{z}})=p\left[Re(\tilde{\boldsymbol{z}}),Im(\tilde{\boldsymbol{z}})\right]=p\left[Re(\tilde{\boldsymbol{z}})\right]p\left[Im(\tilde{\boldsymbol{z}})\right]$$

where

$$\tilde{\boldsymbol{z}}=Re(\tilde{\boldsymbol{z}})+j\,Im(\tilde{\boldsymbol{z}})$$

Maximization of $p(\tilde{\boldsymbol{z}})$ with respect to $\boldsymbol{\theta}$ gives the estimator in Eq. (7.88), which means that the maximum likelihood estimator is identical to the least squares estimator. Properties of the least squares estimates are identical to those summarized in Section 5.1.1.

Note that the expressions in Eqs. (7.86), (7.88), and (7.89) for complex data are identical to the expressions developed for real data in Chapter 5, Eqs. (5.24), (5.10), and (5.12), since the conjugate transpose and transpose operations are identical operations when applied to real numbers. Therefore, if the equations in this section are used, the same least squares parameter estimation equations can be used regardless of whether the data are real or complex. In addition, the expressions given in Section 5.5.4 for incorporating prior information into least-squares parameter estimation can be easily extended to the frequency domain.

When the data are complex, the expressions in Eqs. (7.88) and (7.89) are equivalent to using Eqs. (5.10) and (5.12) with the real and imaginary parts of the complex data stacked, so that the resulting data are real. For example, if $\tilde{\boldsymbol{X}}=\boldsymbol{X}_R+j\boldsymbol{X}_I$, where $\boldsymbol{X}_R$ and $\boldsymbol{X}_I$ are $N\times n_p$ matrices of real numbers, then

$$\begin{aligned}Re\left(\tilde{\boldsymbol{X}}^{\dagger}\tilde{\boldsymbol{X}}\right)&=Re\left[\left(\boldsymbol{X}_R-j\boldsymbol{X}_I\right)^T\left(\boldsymbol{X}_R+j\boldsymbol{X}_I\right)\right]=\boldsymbol{X}_R^T\boldsymbol{X}_R+\boldsymbol{X}_I^T\boldsymbol{X}_I\\&=\begin{bmatrix}\boldsymbol{X}_R\\\boldsymbol{X}_I\end{bmatrix}^T\begin{bmatrix}\boldsymbol{X}_R\\\boldsymbol{X}_I\end{bmatrix}\end{aligned} \tag{7.91}$$

and similarly for $Re\left(\tilde{\boldsymbol{X}}^{\dagger}\tilde{\boldsymbol{z}}\right)$. The regression problem for complex data can be therefore be treated as a problem with real data by simply stacking the real and imaginary parts of the complex data. This also shows that the least-squares optimization applies equally to the real and imaginary parts of the complex data, with no relative weighting.

7.7 Computational Aspects

The methods described in Sections 7.3-7.6 use Fourier transform data directly, so there is no need to quantify spectral densities. The data are simply transformed from the time domain to the frequency domain for the modeling. Consequently, there is no need for the averaging or data windowing that must be done to compute accurate frequency responses.

The preceding discussion on the equation-error method for parameter estimation in the frequency domain showed that this method has a relatively simple, noniterative solution, based on linear algebra. The modeling can be applied to the dynamic equations or output equations to estimate dimensional stability and control derivatives, or can be applied using the nonlinear equations of motion to estimate nondimensional stability and control derivatives. The method can be used even when the aircraft is inherently unstable and flying with active automatic feedback control. These features are similar to the features of equation-error in the time domain; however, there are additional advantages to modeling in the frequency domain, which are discussed in this section.

Finite Fourier transform data quantify the projection of time series data onto sinusoids at the transform frequencies. This results in fewer data points in the frequency domain, which improves the speed of modeling computations. For output-error in the frequency domain, the calculations for model outputs and model output sensitivities are algebraic, and do not involve solving sets of differential equations, as in the time domain. This speeds up the calculations considerably in the frequency domain. Because of the linearity of the Fourier transform, output error in the frequency domain can only be applied to linear dynamic systems, whereas in the time domain, the dynamic system can be arbitrarily nonlinear.

7.7.1 Transform Frequencies

It is not necessary to use all of the frequencies defined in Eq. (7.8) for the transformation of time-domain data to the frequency domain. In fact, the high-accuracy finite Fourier transform algorithm explained in Ref. [7.4] and Chapter 11 can be used to concentrate selected transform frequencies in a frequency range corresponding to the dynamic response of the system, independent of the data record length T. Using only these frequencies for the finite Fourier transform is equivalent to smoothing the time domain data with zero lag by using only the frequency-domain data associated with the selected transform frequencies. In effect, wideband noise rejection is achieved by doing fewer transform calculations.

The results of this simple step are fewer calculations in the frequency domain (because of fewer transform frequencies), enhanced signal-to-noise ratio (because of the wideband noise rejection), and unbiased parameter estimates (because of the smoothed regressors, cf. Eq. (5.77) and Ref. [7.18]). As discussed earlier, when the parameter estimation is done in the frequency domain using a limited frequency band, the assumption of a constant value for the spectral density of the residuals in the frequency domain [cf. Eq. (7.70) and Figs. 5.13 and 6.5] matches well with reality, and the estimated parameter error bounds are calculated accurately without any corrections. Furthermore, Ref. [7.18] shows that using least-squares in the frequency domain repairs the adverse effects of data collinearity by effectively removing wideband noise on the regressors.

The frequency vector for the finite Fourier transform can be selected by first transforming the time-domain response data using a wide frequency band, to identify where the dynamic response lies (indicated by relatively large components), then specifying the transform frequencies accordingly. The usable range of frequencies for modeling is limited on the low end to

$$\begin{aligned} f_{min} &= 2/T \ Hz \\ \omega_{min} &= 2\pi f_{min} = 4\pi/T \text{ rad/s} \end{aligned} \tag{7.92}$$

where T is the time length of the maneuver in seconds. This allows at least two full sinusoid waveforms over the length of the maneuver for each frequency ω_k. The frequency limit at the high end is theoretically the Nyquist frequency $f_N = 1/(2\Delta t)$, although the practical limit is lower (see Chapter 8). The upper bound of the transform frequency band should be chosen to include the dynamics of interest, typically around 1.2 Hz for rigid-body modes of full-scale aircraft. For a frequency resolution equal to 0.025 Hz, and $T = 10$ s, a typical transform frequency vector would be

$$f = [0.20, \ \ 0.225, \ \ 0.25, \ \ \ldots, \ \ 1.20]^T \text{ Hz} \tag{7.93}$$

which amounts to 41 data points in the frequency domain.

7.7.2 Detrending

Prior to transforming the time-domain data to the frequency domain, the bias and linear trend with time are removed from the time series. This process is called detrending. The purpose of detrending is to prevent leakage from these relatively large low-frequency components that can pollute the frequency-domain data at the lower end of the frequency range of interest. Chapter 11 explains leakage in detail.

Figure 7.2 shows the effect of leakage using an angle of attack measurement. The plots on the left side show the measured angle of attack and the corresponding finite Fourier transform data. On the right side of Fig. 7.2, the bias and linear trend with time were removed from the angle of attack time series prior to applying the finite Fourier transform. The results show that detrending the time series removes the leakage from the (relatively large) bias and trend, and clearly reveals the frequency-domain data at low frequencies. Detrending is applied for this reason any time the Fourier transform is applied in practice.

Because the time series for the model terms and the response variable are detrended prior to Fourier transformation, the bias term in the model and the linear trend with time are not observable using frequency-domain modeling. Stated another way, frequency-domain data include only dynamic information content, and not static or trend information. Consequently, a separate and subsequent step must be included to update the model bias and linear trend with time. Fortunately, this step is easy to accomplish in the time domain, because these terms are typically among the most well-conditioned of all model terms, and therefore are easy to estimate accurately in the time domain.

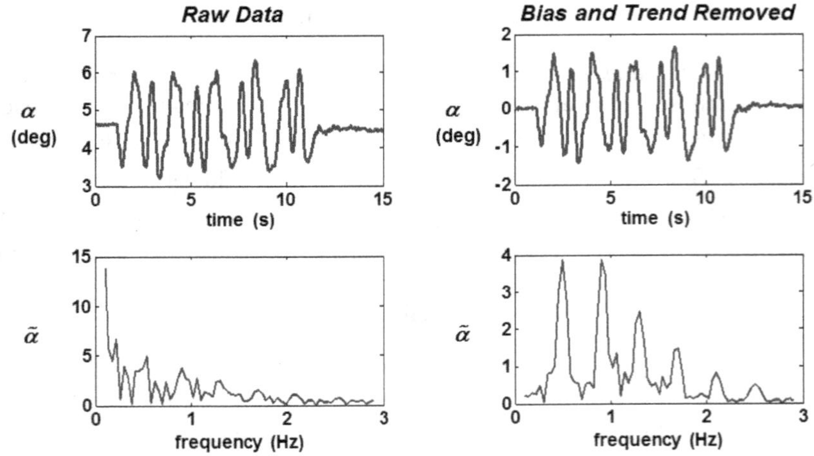

Figure 7.2 **Frequency-domain effect of detrending**

After the frequency-domain modeling is complete, the model parameter vector $\hat{\theta}$ is used to estimate the model bias and trend parameters and associated covariance as follows:

$$\boldsymbol{\theta}_d = \begin{bmatrix} \theta_o \\ \theta_t \end{bmatrix} \qquad \boldsymbol{X}_d = [\mathbf{1} \quad \boldsymbol{t}] \tag{7.94}$$

$$\hat{\boldsymbol{\theta}}_d = \left(\boldsymbol{X}_d^T \boldsymbol{X}_d\right)^{-1} \left[\boldsymbol{X}_d^T \left(\boldsymbol{z} - \boldsymbol{X}\hat{\boldsymbol{\theta}}\right)\right] \tag{7.95}$$

$$\boldsymbol{\Sigma}_{\hat{\boldsymbol{\theta}}_d} = \hat{\sigma}_d^2 \left(\boldsymbol{X}_d^T \boldsymbol{X}_d\right)^{-1} \tag{7.96}$$

$$\hat{\sigma}_d^2 = \frac{1}{(N-2)} \left[\left(\boldsymbol{z} - \boldsymbol{X}\hat{\boldsymbol{\theta}} - \boldsymbol{X}_d\hat{\boldsymbol{\theta}}_d\right)^T \left(\boldsymbol{z} - \boldsymbol{X}\hat{\boldsymbol{\theta}} - \boldsymbol{X}_d\hat{\boldsymbol{\theta}}_d\right)\right] \tag{7.97}$$

where $\boldsymbol{X}$ and $\boldsymbol{z}$ contain detrended time-domain data corresponding to the model regressors and the response variable, respectively. The time-domain bias and trend update is done after the modeling has already been done in the frequency domain, with the associated parameter estimates fixed. This insures that the estimated bias $\hat{\theta}_o$ corresponds to the aerodynamic bias term in a multivariable Taylor series expansion for the aerodynamic coefficient, which is generally of most interest for aerodynamics and flight dynamics. This step can be viewed as restoring the bias and trend information that was removed prior to Fourier transformation, so that the updated model is correct in the time domain.

Note that a simple regression for each time series using the regressors $\boldsymbol{X}_d = [\mathbf{1} \quad \boldsymbol{t}]$ is an easy way to identify the bias and trend that must be removed to detrend the time series.

For output-error modeling in the frequency domain, restoration of bias and trend information is done in the same way as shown earlier in Example 6.1 when evaluating models using prediction data.

7.7.3 Modulating Functions

The finite Fourier transform of time derivatives involves endpoint corrections when the time series endpoints and/or derivatives are nonzero, as shown earlier in Eq. (7.36). These endpoint corrections can be avoided using modulating functions, as first described by Shinbrot [7.30]. The idea is to multiply a differential equation model by a suitably-chosen modulating function, then use integration by parts to transfer the derivatives from the data to the modulating function, incurring endpoint terms that involve the modulating functions and its time derivatives evaluated at the endpoints.

A smooth differentiable function $\phi(t)$ with function value and first $n-1$ time derivatives equal to zero at the endpoints is called an nth order modulating function,

$$\phi^{(k)}(0) = \phi^{(k)}(T) = 0 \qquad k = 0, 1, \ldots, n-1 \tag{7.98}$$

where $\phi^{(1)}(t)$ denotes the first time derivative of $\phi(t)$, and so on.

Pearson [7.12] developed a Fourier-based modulating function. The Fourier-based modulating function for frequency ω is

$$\phi(t) = e^{-j\omega t}\left(1 - e^{-j\omega_o t}\right)^n \qquad \omega_o \equiv 2\pi/T \tag{7.99}$$

The function $\phi(t)$ in Eq. (7.99) satisfies the requirements of Eq. (7.98) for an nth order modulating function. This modulating function can also be written as

$$\phi(t) = \sum_{k=0}^{n} b_k \, e^{-j(\omega + k\omega_o)t} \qquad b_k = (-1)^k \frac{n!}{k!(n-k)!} \tag{7.100}$$

which shows that the modulating function $\phi(t)$ is a summation of complex sinusoidal functions of the same type used in the finite Fourier transform of Eq. (7.3).

The differential equation associated with Eq. (7.75) is

$$\begin{aligned} &z^{(n)}(t) + a_{n-1}z^{(n-1)}(t) + \ldots + a_0 z(t) \\ &= b_m u^{(m)}(t) + b_{m-1}u^{(m-1)}(t) + \ldots + b_0 u(t) + v(t) \end{aligned} \tag{7.101}$$

where $m < n$. Multiplying the equation by $\phi(t)$, and integrating over the time period $[0, T]$,

$$\begin{aligned} &\int_0^T \phi(t) z^{(n)}(t)\,dt + a_{n-1}\int_0^T \phi(t) z^{(n-1)}(t)\,dt + \ldots + a_0 \int_0^T \phi(t) z(t)\,dt \\ &= b_m \int_0^T \phi(t) u^{(m)}(t)\,dt + b_{m-1}\int_0^T \phi(t) u^{(m-1)}(t)\,dt + \ldots + b_0 \int_0^T \phi(t) u(t)\,dt + \tilde{\varepsilon} \end{aligned}$$

where $\tilde{\varepsilon}$ represents the complex-valued equation error for the modulated equation and $\phi(t)$ depends on frequency ω. Using integration by parts, and invoking the properties of an nth order modulating function from Eq. (7.98),

$$(-1)^n \int_0^T \phi^{(n)}(t)z(t)\,dt + a_{n-1}(-1)^{n-1}\int_0^T \phi^{(n-1)}(t)z(t)\,dt + \ldots$$

$$- a_1\int_0^T \phi^{(1)}(t)z(t)\,dt + a_0\int_0^T \phi(t)z(t)\,dt$$

$$= b_m(-1)^m \int_0^T \phi^{(m)}(t)u(t)\,dt + b_{m-1}(-1)^{m-1}\int_0^T \phi^{(m-1)}(t)u(t)\,dt + \ldots$$

$$- b_1\int_0^T \phi^{(1)}(t)u(t)\,dt + b_0\int_0^T \phi(t)u(t)\,dt + \tilde{\varepsilon}$$

or

$$(-1)^n \int_0^T \phi^{(n)}(t)z(t)\,dt = a_{n-1}(-1)^{n}\int_0^T \phi^{(n-1)}(t)z(t)\,dt + \ldots$$

$$+ a_1\int_0^T \phi^{(1)}(t)z(t)\,dt - a_0\int_0^T \phi(t)z(t)\,dt$$

$$+ b_m(-1)^m \int_0^T \phi^{(m)}(t)u(t)\,dt + b_{m-1}(-1)^{m-1}\int_0^T \phi^{(m-1)}(t)u(t)\,dt + \ldots$$

$$- b_1\int_0^T \phi^{(1)}(t)u(t)\,dt + b_0\int_0^T \phi(t)u(t)\,dt + \tilde{\varepsilon} \tag{7.102}$$

Each term in Eq. (7.102) is a complex-valued modulation of the input-output data for frequency ω. Carrying out this process for a vector of frequencies $\boldsymbol{\omega}$ provides frequency-domain data of the same type obtained earlier using the finite Fourier transform. The difference is that the finite Fourier transform required assuming that the endpoint conditions were close to zero, so that the endpoint terms would be negligible, or else n applications of Eq. (7.36) would be required for each nth order derivative. Using the modulating function approach, the endpoint conditions can be arbitrary, because the modulation process removes nonzero endpoint terms analytically. The modulating function approach therefore allows the use of input-output time series data with arbitrary endpoint conditions.

At this point, the parameter estimation proceeds in exactly the same manner as described earlier, using complex linear regression. Because the modulating function in Eq. (7.99) uses complex exponential functions, the required modulation integrals in Eq. (7.102) can be carried out using the same high-accuracy, arbitrary-frequency finite Fourier transform calculation described earlier, cf. Ref. [7.4].

7.7.4 Nonlinear Regressors in the Frequency Domain

Nonlinear regressors can be used in the equation-error method in the frequency domain, by simply assembling the nonlinear regressors in the time domain first, then detrending the resulting time series and applying the finite Fourier transform as usual. This works because the finite Fourier transform is implemented numerically, and therefore it makes no difference how the time series was created. The regression depends on the linearity of the model output with respect to the parameters only, so the regressors can be nonlinear.

In Chapter 5, vector orthogonalization of candidate modeling functions was used to enable the model structure determination in the time domain. The necessary modification for orthogonalizing complex data vectors is simple, and is related to what is called the Euclidean angle between two complex vectors [7.31]. For an arbitrary complex vector $\tilde{\boldsymbol{x}}$ with M generally complex elements, a real isometric vector $\boldsymbol{x}$ with $2M$ elements can be assembled by stacking the real and imaginary parts of the complex elements of $\tilde{\boldsymbol{x}}$

$$\boldsymbol{x} = \begin{bmatrix} Re(\tilde{\boldsymbol{x}}) \\ Im(\tilde{\boldsymbol{x}}) \end{bmatrix} \tag{7.103}$$

Treating the real vector $\boldsymbol{x}$ as the new data vector, all of the orthogonalization techniques and model structure determination techniques developed previously for real data can be applied without modification. The penalty associated with applying the techniques to complex data is simply a doubling of the length of the data vectors involved. References [7.27] and [7.28] demonstrate practical application of nonlinear regressors and model structure determination in the frequency domain to transfer function identification, unsteady aerodynamic modeling, and automated simulation updating based on flight data.

Example 7.1

In this example, the same flight data used in Examples 5.1 and 6.1 to demonstrate equation-error and output-error parameter estimation techniques in the time domain will now be used to demonstrate the use of these same methods in the frequency domain. The flight data come from a lateral maneuver of the NASA Twin Otter (see Figs. 5.3 and 5.4).

All relevant time series were transformed into the frequency domain using the high-accuracy finite Fourier transform described in Chapter 11, for the frequency vector $\boldsymbol{f} = [0.10, 0.12, 0.14, \ldots, 1.5]^T$ Hz. This frequency range includes all the significant spectral components in the measured input-output data for the maneuver. The same model equation (5.78c) was used for least squares estimation of yawing moment coefficient parameters in the frequency domain. Eqs. (7.83) and (7.84) with Eq. (7.86) were used to compute the estimated parameters and covariance matrix.

Table 7.1 contains the parameter estimation results using equation-error linear regression in the frequency domain for the yawing moment coefficient C_n. The results compare favorably with the analogous time-domain results given in Table 5.1, including the indication that the term associated with the $C_{n_{\delta_a}}$ parameter is not significant and probably should be omitted from the model structure. The C_{n_o} parameter was estimated separately in the time domain, after the other parameters were estimated in the frequency domain. Figure 7.3 shows the model fit to measured C_n in the frequency domain. Note that the plot was made using magnitudes of the complex numbers, which includes both real and imaginary parts. The model fit can also be examined by plotting the phase angle of the complex quantities, or by plotting real and imaginary parts separately.

For output-error parameter estimation in the frequency domain, the cost gradient and information matrix were computed from Eqs. (7.59) and (7.60), and the parameter estimate updates were found from Eq. (7.49). Output sensitivities were computed numerically using central finite differences based on Eq. (7.61), although the sensitivities could also have been computed analytically. Note that analytical computation of the output sensitivities involves only algebraic equations in the frequency domain. Similarly, the outputs are computed algebraically in the frequency domain [cf. Eqs. (7.53c) and (7.54a)]. Estimates of the spectral densities of the residuals were obtained from Eq. (7.57).

Table 7.1 **Least squares parameter estimation results, aerodynamic yawing moment coefficient, run 1**

Parameter	$\hat{\theta}$	$s(\hat{\theta})$	$\lvert t_0 \rvert$	$100\left[s(\hat{\theta})/\lvert\hat{\theta}\rvert\right]$
C_{n_β}	8.51×10^{-2}	8.24×10^{-4}	103.2	1.0
C_{n_p}	-5.52×10^{-2}	3.45×10^{-3}	16.0	6.2
C_{n_r}	-2.00×10^{-1}	2.90×10^{-3}	69.2	1.4
$C_{n_{\delta_a}}$	4.06×10^{-4}	1.22×10^{-3}	0.3	301.3
$C_{n_{\delta_r}}$	-1.32×10^{-1}	1.35×10^{-3}	97.5	1.0
C_{n_o}	-1.73×10^{-4}	7.42×10^{-6}	23.3	4.3
$s = \hat{\sigma}$	1.17×10^{-7}			
R^2 (%)	99.9			

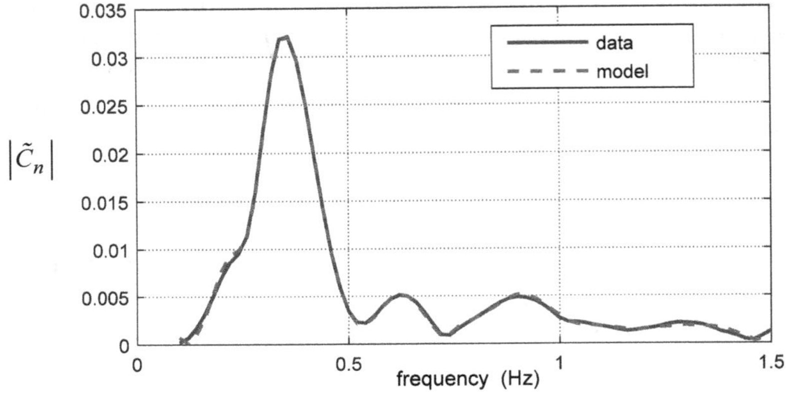

Figure 7.3 **Frequency-domain equation-error model fit to yawing moment coefficient for lateral maneuver, run 1**

In Table 7.2, time-domain output-error results from Example 6.1 are listed along with parameter estimation results using output-error in the frequency domain. The parameter estimation results are in statistical agreement, given the estimated error bounds. The frequency-domain results again indicate that the estimated $C_{n_{\delta_a}}$ parameter has relatively high uncertainty. Figure 7.4 shows the model fit to measured outputs in the frequency domain.

Because of the close agreement between output-error parameter estimates in the time-domain and frequency-domain, the frequency-domain model has excellent prediction capability similar to that seen in Example 6.1 for the time-domain model. However, computing predicted outputs in the frequency domain using a frequency-domain model does not require estimation of bias parameters, as in the time-domain case of Example 6.1. Modeling in the frequency domain omits the biases altogether because the biases and trends are removed prior to Fourier transformation. However, if time-domain matches of predicted model output to measured output are desired, using a model identified in the frequency domain, then the bias terms must be estimated in the time domain using the same method shown in Example 6.1 for the time-domain prediction case. ■

***Table 7.2* Output-error parameter estimation results for lateral maneuver, run 1**

Parameter	Time-domain output-error		Frequency-domain output-error	
	$\hat{\theta}$	$s(\hat{\theta})$	$\hat{\theta}$	$s(\hat{\theta})$
C_{Y_β}	-8.65×10^{-1}	2.83×10^{-3}	-8.65×10^{-1}	7.59×10^{-3}
C_{Y_r}	9.33×10^{-1}	1.67×10^{-2}	9.15×10^{-1}	4.64×10^{-2}
$C_{Y_{\delta_r}}$	3.80×10^{-1}	8.07×10^{-3}	3.59×10^{-1}	2.30×10^{-2}
C_{l_β}	-1.20×10^{-1}	6.07×10^{-4}	-1.18×10^{-1}	1.27×10^{-3}
C_{l_p}	-5.86×10^{-1}	2.61×10^{-3}	-5.84×10^{-1}	5.34×10^{-3}
C_{l_r}	1.89×10^{-1}	1.83×10^{-3}	1.73×10^{-1}	4.02×10^{-3}
$C_{l_{\delta_a}}$	-2.28×10^{-1}	9.21×10^{-4}	-2.30×10^{-1}	1.87×10^{-3}
$C_{l_{\delta_r}}$	4.22×10^{-2}	7.48×10^{-4}	2.75×10^{-2}	1.76×10^{-3}
C_{n_β}	8.64×10^{-2}	2.15×10^{-4}	8.55×10^{-2}	6.02×10^{-4}
C_{n_p}	-6.44×10^{-2}	9.83×10^{-4}	-6.65×10^{-2}	2.76×10^{-3}
C_{n_r}	-1.91×10^{-1}	6.25×10^{-4}	-1.94×10^{-1}	1.78×10^{-3}
$C_{n_{\delta_a}}$	-2.98×10^{-3}	3.62×10^{-4}	-3.42×10^{-3}	1.05×10^{-3}
$C_{n_{\delta_r}}$	-1.36×10^{-1}	2.70×10^{-4}	-1.38×10^{-1}	8.22×10^{-4}

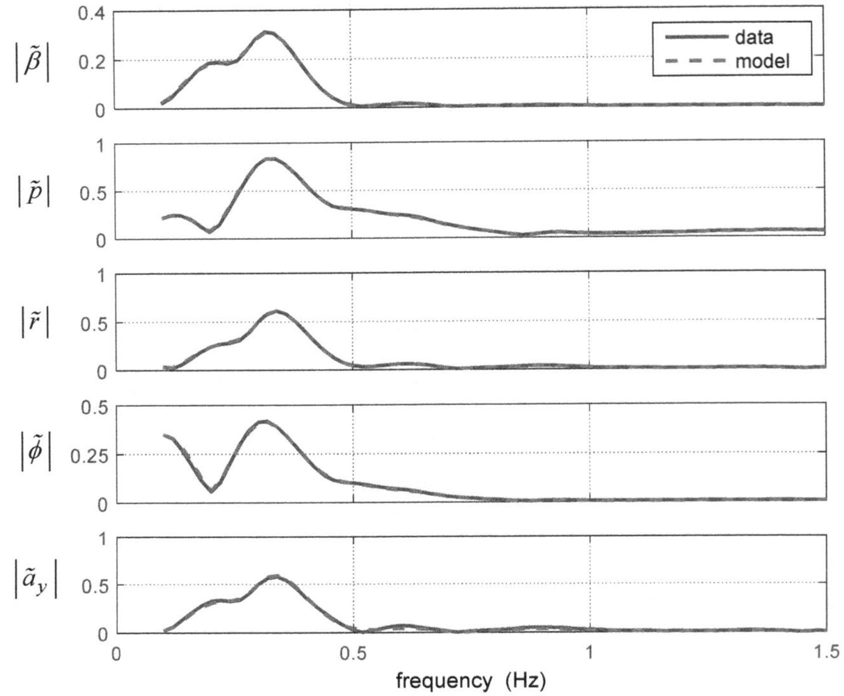

Figure 7.4 **Frequency-domain output-error model fit to measured perturbation outputs for lateral maneuver, run 1**

Example 7.2

Frequency domain methods can also be applied to data from forced oscillation tests conducted in a wind tunnel or water tunnel (see e.g. Refs. [7.21] and [7.24]). This example uses data from Ref. [7.24], where a 2.5 percent model of an F-16XL aircraft was oscillated in pitch in a water tunnel. Nondimensional normal force coefficient was computed from measurements of the hydrodynamic normal force on the model and the dynamic pressure. The pitch angle, which was the same as the angle of attack, was commanded and measured during the test. The configuration of the model was fixed, with control surface deflections at zero. A Schroeder sweep (see Chapter 8) was used to command the pitch angle. This input was selected to excite the dynamic system using a uniform power spectrum for many different frequencies in the range $[0.008, 0.2]$ Hz. Figure 7.5 shows the time series of angle of attack and normal force coefficient for an experimental run. The experiment shown involves oscillations in angle of attack of approximately ±2.5 deg about the mean angle of attack $\alpha_o \approx 35$ deg.

The frequencies were chosen to exhibit unsteady aerodynamic effects, so the postulated model had the form

$$C_N(t) = C_N(\infty) + \int_0^t C_{N_\alpha}(t-\tau)\,\dot{\alpha}(\tau)\,d\tau + \frac{\bar{c}}{2V}\int_0^t C_{N_q}(t-\tau)\,\dot{q}(\tau)\,d\tau$$

where $C_{N_\alpha}(t)$ and $C_{N_q}(t)$ are indicial functions, and $C_N(\infty)$ is the steady-state normal force coefficient at $\alpha = q = 0$. Two assumptions were adopted to simplify the model:

1) The effect of $\dot{q}(t)$ on the normal force coefficient was neglected.

2) The indicial function $C_{N_\alpha}(t)$ was modeled using a simple exponential decay function, $C_{N_\alpha}(t) = C_{N_\alpha}(\infty) - a_1 e^{-b_1 t}$, where $C_{N_\alpha}(\infty)$ is the static α derivative in steady flow conditions, and a_1 and b_1 are unknown constant parameters.

The simplified model then had the form

$$C_N(t) = C_N(\infty) + C_{N_\alpha}(\infty)\alpha(t) - a_1\int_0^t e^{-b_1(t-\tau)}\,\dot{\alpha}(\tau)\,d\tau + \frac{\bar{c}}{2V}C_{N_q}(\infty)\,q(t) \tag{7.104}$$

where the integral characterizes the unsteady aerodynamic effect on the normal force coefficient, and the other terms are the usual linear steady-flow terms. Removing the constant term $C_N(\infty)$, noting that $q(t) = \dot{\alpha}(t)$ for one degree-of-freedom pitching motion, and taking Laplace transforms of both sides results in

$$\tilde{C}_N = C_{N_\alpha}(\infty)\tilde{\alpha} - \frac{a_1 s}{s+b_1}\tilde{\alpha} + \frac{\bar{c}}{2V}C_{N_q}(\infty)\,s\,\tilde{\alpha}$$

Solving for the transfer function of normal force coefficient to angle of attack,

$$\frac{\tilde{C}_N}{\tilde{\alpha}} = \frac{As^2 + Bs + C}{s + b_1}$$

where

$$A = \frac{\bar{c}}{2V} C_{N_q}(\infty)$$

$$B = C_{N_\alpha}(\infty) - a_1 \; + b_1 \frac{\bar{c}}{2V} C_{N_q}(\infty)$$

$$C = b_1 C_{N_\alpha}(\infty)$$

Model parameters $A, B, C,$ and b_1 can be estimated by applying equation-error or output-error in the frequency domain to the transfer function model. If $C_{N_\alpha}(\infty)$ is known from static tests, then a_1 and b_1 can also be computed, resulting in estimates for all of the unknown parameters in the simplified model equation (7.104).

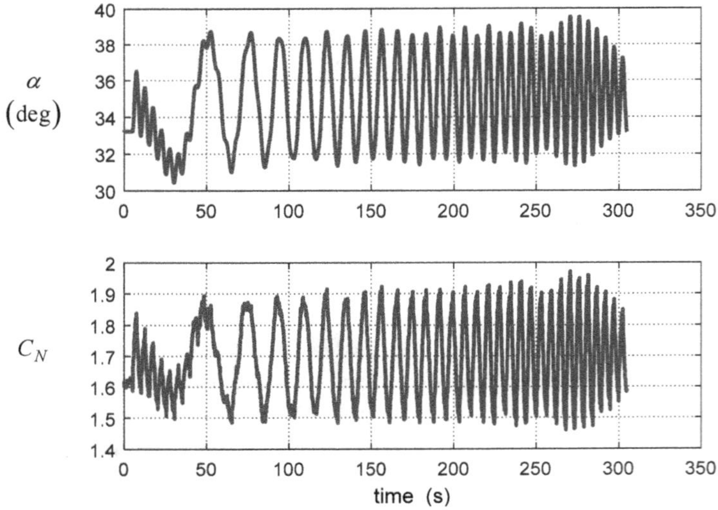

Figure 7.5 **Schroeder sweep in angle of attack, and the normal force coefficient response for an F-16XL 2.5 percent model**

Figure 7.6 shows the model fit to the data in the frequency domain using the equation-error approach with no frequency weighting, cf. Eqs. (7.76) and (7.77). Note the increase in response amplitude with increasing frequency, which is the result of omitting the frequency weighting.

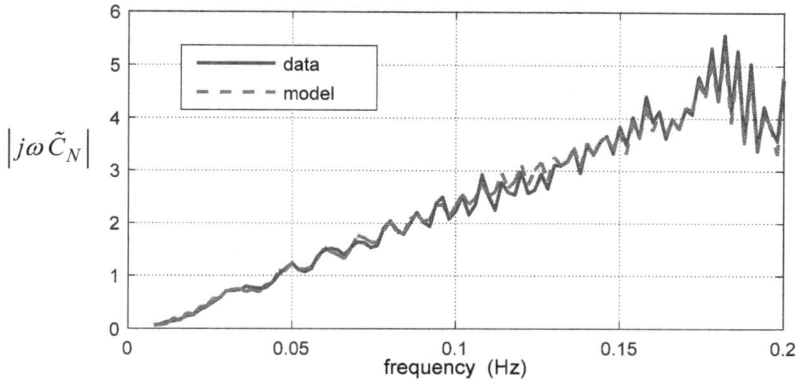

Figure 7.6 **Frequency-domain equation-error model fit to normal force coefficient, F-16XL 2.5 percent model**

Columns 2 and 3 of Table 7.3 give the estimated parameters and standard errors for this case. Columns 4 and 5 of Table 7.3 contain the analogous results using output-error parameter estimation in the frequency domain. Figure 7.7 shows the output-error model fit to the measured data in the frequency domain.

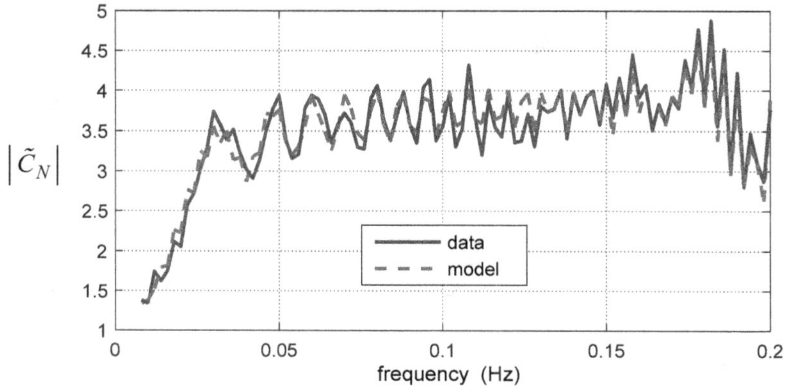

Figure 7.7 **Frequency-domain output-error model fit to normal force coefficient, F-16XL 2.5 percent model**

Table 7.3 **Parameter estimation results for Schroeder sweep forced oscillation on an F-16XL 2.5 percent model**

	Frequency-domain equation-error		Frequency-domain output-error	
Parameter	$\hat{\theta}$	$s(\hat{\theta})$	$\hat{\theta}$	$s(\hat{\theta})$
A	8.58×10^{-1}	3.25×10^{-2}	8.32×10^{-1}	3.44×10^{-2}
B	3.22×10^{0}	3.84×10^{-2}	3.21×10^{0}	3.17×10^{-2}
C	1.82×10^{-1}	7.05×10^{-2}	1.95×10^{-1}	2.80×10^{-2}
b_1	1.46×10^{-1}	2.64×10^{-2}	1.55×10^{-1}	1.19×10^{-2}

The parameter results from the two methods are in statistical agreement, given the estimated error bounds. ■

7.8 Low Order Equivalent System Identification

An important application of parameter estimation in the frequency domain is identifying Low Order Equivalent System (LOES) models from measured data. A LOES model characterizes the closed-loop dynamic response of the airframe and control system, as it appears to the pilot. Consequently, the inputs for LOES modeling are pilot stick and rudder pedal deflections or forces, and the outputs are aircraft response variables, such as angle of attack or roll rate. LOES models identified from flight test data are useful in many applications, including control law design validation, simulation, flying qualities research, aircraft development, and aircraft specification compliance.

The LOES model has the same form as the model for an open-loop unaugmented airplane with classical dynamic modes (see Chapter 3), except the inputs are pilot controls with equivalent time delays, instead of control surface deflections. The equivalent time delay is a pure time delay on the control input, introduced to account for delay resulting from the control system implementation (e.g., sampling delay), and the phase lag at high frequency from control system dynamics and various nonlinearities, such as control surface rate limiting. Control system dynamics can include significant dynamics attributable to the control law implementation, artificial feel systems, sensors, filters, and actuators. The complexity of these control systems results from the desire for improved performance and control over expanded flight envelopes. The LOES model approximates the complete high-order closed-loop aircraft response using a low-order classical linear dynamic model with an equivalent time delay on the input. If this approximation can be done well, the overall closed-loop response can be interpreted more readily in terms of the equivalent low-order model. In addition, military specifications [7.32] relate values of the LOES model parameters to pilot opinions of the aircraft flying qualities [7.33]-[7.35]. This connection can be used to help interpret and evaluate pilot opinions of aircraft flying qualities.

Since the model structure for LOES modeling is fixed *a priori* to correspond to classical linear aircraft dynamic response with an input time delay, the problem reduces to parameter estimation based on measured data. For the short-period longitudinal dynamic mode, the closed-loop pitch rate response to longitudinal stick deflection is modeled in transfer function form as

$$\frac{\tilde{q}}{\tilde{\eta}_e} = \frac{K_{\dot{\theta}}\left(s + 1/T_{\theta_2}\right) e^{-\tau s}}{\left(s^2 + 2\zeta_{sp}\omega_{sp}s + \omega_{sp}^2\right)} = \frac{\left(b_1 s + b_0\right)e^{-\tau s}}{s^2 + a_1 s + a_0} \tag{7.105}$$

where the relationships among the different parameters can be obtained by direct comparison of the coefficients of powers of s. The symbol η_e denotes longitudinal stick deflection, and τ is the equivalent time delay in seconds. The first model parameterization in Eq. (7.105) is used in flying qualities work, and the second is a generic transfer function form.

Based on the dimensional linear equations for short-period motion derived in Chapter 3,

$$\begin{bmatrix} \dot{\alpha} \\ \dot{q} \end{bmatrix} = \begin{bmatrix} -L_\alpha & 1 - L_q \\ M_\alpha & M_q \end{bmatrix} \begin{bmatrix} \alpha \\ q \end{bmatrix} + \begin{bmatrix} L_{\eta_e} \\ M_{\eta_e} \end{bmatrix} \eta_e\left(t - \tau\right) \tag{7.106}$$

where the input is now longitudinal stick deflection, with an equivalent time delay. Assuming $L_q \approx 0$ for low angles of attack, and $L_{\eta_e} \approx 0$, then taking Laplace transforms,

$$\begin{bmatrix} s + L_\alpha & -1 \\ -M_\alpha & s - M_q \end{bmatrix} \begin{bmatrix} \tilde{\alpha} \\ \tilde{q} \end{bmatrix} = \begin{bmatrix} 0 \\ M_{\eta_e} \end{bmatrix} \tilde{\eta}_e\, e^{-s\tau} \tag{7.107}$$

$$\frac{\tilde{q}}{\tilde{\eta}_e} = \frac{M_{\eta_e}\left(s + L_\alpha\right)e^{-\tau s}}{s^2 + \left(L_\alpha - M_q\right)s - \left(M_q L_\alpha + M_\alpha\right)} \tag{7.108}$$

The transfer function in Eq. (7.108) has the same form as Eq. (7.105), but different parameterization.

Other LOES models can be derived in a similar way for other longitudinal and lateral responses to pilot inputs. Note that the dimensional stability and control derivatives in Eq. (7.108) are equivalent derivatives that include both aircraft aerodynamics and closed-loop control effects. It follows that these dimensional derivatives are not, in general, the same as those discussed in Chapter 3, which described the aircraft aerodynamics.

Any of the $\tilde{q}/\tilde{\eta}_e$ models given above could also be expressed in the time domain. For example, the time-domain form of the second version of the transfer function in Eq. (7.105) is

$$\ddot{q}(t)+a_1\dot{q}(t)+a_0q(t)=b_1\eta_e(t-\tau)+b_0\eta_e(t-\tau) \tag{7.109}$$

Estimating the equivalent time delay parameter τ in the time domain is problematic because flight test data are collected at regular sampling intervals Δt, so interpolation of the measured input data is required to implement a value of τ which is not equal to an integer number of sampling intervals. However, interpolation is a smoothing operation, which is a modification of the data. This of course changes the original problem (because the data are changed for every different time lag), causing difficulty in convergence. If values of τ are restricted to integer multiples of Δt, resolution of the τ estimate can be coarse and convergence problems can also occur. These problems can be avoided by analyzing the data in the frequency domain, because the time delay parameter appears as an ordinary real-valued model parameter in the frequency domain.

LOES models of the aircraft dynamic response are usually identified in the frequency domain over a frequency band corresponding to typical pilot inputs, [0.1, 10] rad/s. References [7.6], [7.15]-[7.17] give details of a method that has been used extensively to identify LOES models based on spectral densities estimated from measured data. This method employs output-error in the frequency domain using the cost function formulation in Eq. (7.65) for transfer function parameters.

LOES models can also be identified using Fourier transforms of the measured time-domain data, and applying equation-error or output-error parameter estimation in the frequency domain, as described earlier. This approach avoids estimating spectral densities from measured time-domain data. In fact, the LOES modeling problem is a fairly straightforward application of the techniques described earlier for equation-error or output-error parameter estimation in the frequency domain, as the following example demonstrates.

Example 7.3

The test aircraft for this example is the Tu-144LL supersonic transport aircraft, shown in Fig. 7.8. Measured input-output data for a longitudinal multistep 2-1-1 maneuver (see Chapter 8) are shown in Fig. 7.9. The pilot applied the multistep input to the longitudinal control, η_e, which consisted of fore and aft movements of the control wheel in the cockpit. The goal was to identify an accurate LOES model from the measured data, using the model form given in Eq. (7.105), to characterize the closed-loop pitch rate response for the longitudinal short-period mode. Figure 7.10 shows the single-input, single-output data used for LOES identification.

Figure 7.8 **Tu-144LL supersonic transport aircraft**

The method used to identify the LOES model, described in Ref. [7.22], is a two-step approach using equation-error and output-error formulations in the frequency domain to estimate the parameters. Parameter estimates from the equation-error method are used as starting values for the output-error method.

For the equation-error formulation, the cost function is given by Eqs. (7.76) and (7.77) for the LOES model,

$$J(\boldsymbol{\theta}) = M \sum_{k=0}^{M-1} \tilde{v}^*(k)\, S_{vv}^{-1}\, \tilde{v}(k) \tag{7.110}$$

where

$$\tilde{v}(k) = (j\omega_k)^2\,\tilde{q}(k) - a_1\,(j\omega_k)\,\tilde{q}(k) - a_0\,\tilde{q}(k) + b_1\,(j\omega_k)\,\tilde{\eta}_e(k)e^{-j\omega_k\tau} + b_0\,\tilde{\eta}_e(k)e^{-j\omega_k\tau} \tag{7.111}$$

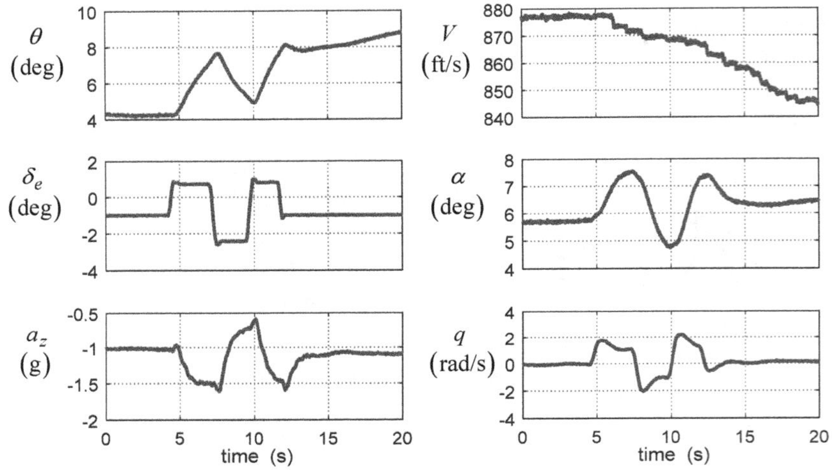

Figure 7.9 **Measured input-output data for a longitudinal maneuver**

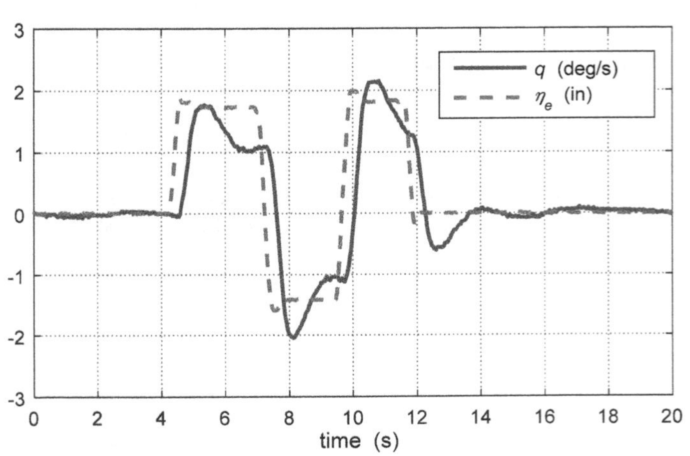

Figure 7.10 **Measured input-output data for longitudinal LOES modeling**

Accurate Fourier transforms were computed using the method of Ref. [7.4], described in Chapter 11, for frequencies evenly spaced at $\Delta\omega = 0.1$ rad/s in the interval $[0.6, 10]$ rad/s. These selected frequencies are the ω_k. The number of data points in the frequency domain was equal to 95, so $M = 95$ in Eq. (7.110). The lower bound of the frequency band for the data analysis corresponds to the 20-second time length of the data, so $\omega_{min} = 4\pi/20 \approx 0.6$ rad/s from Eq. (7.92). The upper bound of the frequency band corresponds to the highest frequency used for LOES modeling, 10 rad/s. The frequency weighting shown in Eq. (7.79) was not included, but could have been. In this problem, the order of the transfer function is low $(n = 2)$, and the selected frequencies ω_k are low, so the frequency weighting can be omitted.

Parameter estimation results from the equation-error (EE) method were used as starting values for the output-error (OE) method, which minimized the cost function given in Eq. (7.58):

$$J_{OE}(\boldsymbol{\theta}) = M \sum_{k=0}^{M-1} \tilde{v}^*(k,\boldsymbol{\theta})\, S_{vv}^{-1}\, \tilde{v}(k,\boldsymbol{\theta}) \tag{7.112}$$

where

$$\tilde{v}(k,\boldsymbol{\theta}) = \tilde{q}(k) - \frac{\left[j\omega_k\, b_1 + b_0\right]}{\left[(j\omega_k)^2 + j\omega_k\, a_1 + a_0\right]}\, \tilde{\eta}_e(k)\, e^{-j\omega_k \tau} \tag{7.113}$$

Final parameter estimates were from the output-error formulation, because of known favorable asymptotic properties (see Chapter 6). The equation-error method is much more robust to starting values of the parameters, because the model output depends linearly on all model parameters except the equivalent time delay τ. The term $\tilde{\eta}_e(k)\, e^{-j\omega\tau}$ makes the equation-error parameter estimation a nonlinear optimization problem, as is the case for the output-error parameter estimation. Because of this, the same optimization routine, implementing the modified Newton-Raphson technique, can be used to solve either formulation of the parameter estimation problem. The difference is only in how the model is formulated. Using an EE/OE sequence gives parameter estimation results with good asymptotic properties, without requiring good starting values for the parameters. Estimated parameter standard errors do not require correction in the frequency domain, as explained previously.

The time-domain data for this example was measured at sampling intervals $\Delta t = 0.03125$ sec (32 Hz), resulting in 641 data points for the 20-second maneuver. The 95 data points in the frequency domain represent a significant reduction in the number of data points to be processed. In addition, the model equations in the frequency domain are always algebraic, so that there is no need for a numerical integration, such as 4th order Runge-Kutta. Both of these factors make the computations go very quickly in the frequency domain, even for the nonlinear optimization that must be done for both equation-error and output-error parameter estimation to identify LOES models. Another advantage of frequency-domain techniques is that the dynamic system can be unstable, and the techniques work just the same. In contrast, time-domain output-error methods must compute model outputs for each iteration of the modified Newton-Raphson technique. This is at best difficult and often not possible for an unstable dynamic system.

Table 7.4 contains the parameter estimation results from the equation-error and output-error parameter estimation in the frequency domain. Figure 7.11 shows that the LOES model fit to the data in the time and frequency domains is very good. The time-domain plot in the lower part of Fig. 7.11 shows that the LOES cannot completely match some high-order response effects, because the low order of the model does not have enough parameters (or model complexity) to capture all the dynamics in the real (high order) physical system. However, the low-order model characterizes the important features of the response very well, and has the advantage of easy interpretation, since the LOES model is only second order.

Table 7.4 **Tu-144LL longitudinal short period LOES modeling results, $\alpha_o = 5.8$ deg, $M_o = 0.88$**

Maneuver 20_4d				
	Equation Error		Output Error	
Parameter	$\hat{\theta}$	$s(\hat{\theta})$	$\hat{\theta}$	$s(\hat{\theta})$
b_1	3.916	0.088	3.788	0.168
b_0	4.921	0.360	5.181	0.249
a_1	3.795	0.097	3.709	0.148
a_0	7.128	0.376	7.396	0.258
τ	0.270	0.003	0.268	0.008
$1/T_{\theta_2}$	1.26		1.37	
ζ_{SP}	0.71		0.68	
ω_{SP}	2.67		2.72	

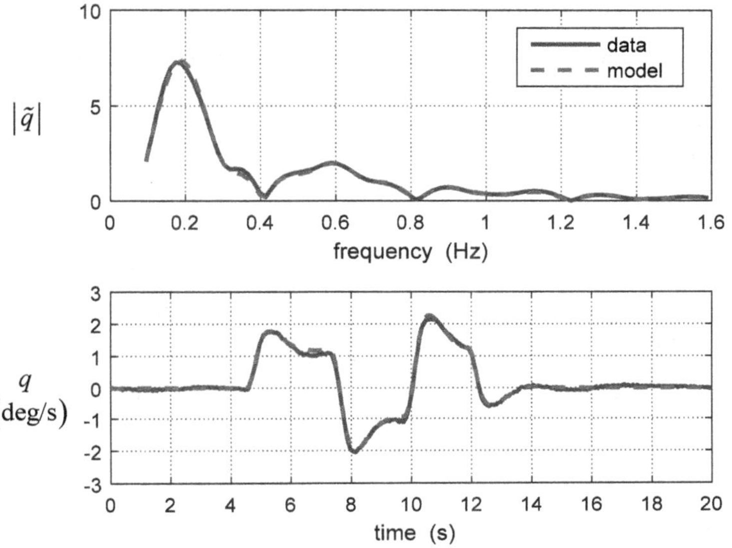

Figure 7.11 **LOES output-error model fit to pitch rate data for a longitudinal maneuver**

Measured data from a similar 2-1-1 maneuver at the same flight condition is shown in Fig. 7.12. The measured longitudinal pilot input from this maneuver was applied to the LOES model identified from the data shown in Fig. 7.10 to compute the pitch rate response for the prediction maneuver. Estimated parameters from the EE/OE sequence (i.e., parameters in column 4 of Table 7.4) were used for the LOES model. Figure 7.12 shows that the match of the model prediction to measured data is very good, and of similar quality to the model fit to the identification data shown in Fig. 7.11. This gives confidence that the identified LOES model has captured the low-order closed-loop short-period pitch rate dynamics of the Tu-144LL at this flight condition. Note that the polarity of the input was reversed for the prediction maneuver in Fig. 7.12, compared to the identification maneuver in Fig. 7.11. This is a check on prediction capability and linearity, since a linear model will respond similarly in either direction and is not sensitive to input form.

The methods described in this example can also be used for multi-input, multi-output models, and can also be used with multiple maneuvers, by combining the frequency-domain data from each maneuver in the manner described at the end of Section 7.1.

The maneuver used for the modeling in this example was a relatively short multistep maneuver. This was possible because Fourier transforms were used directly, and not experimentally-derived frequency responses from spectral estimates. Accurate spectral estimates generally require more data, so methods that use experimentally-derived frequency responses will require more data, such as repeated multistep maneuvers, or, more commonly, frequency sweeps (see Chapter 8). More details on using the EE/OE sequence with Fourier transform data for LOES modeling, as well as applications to the F-18 HARV and Tu-144LL aircraft, can be found in Refs. [7.22] and [7.23]. ■

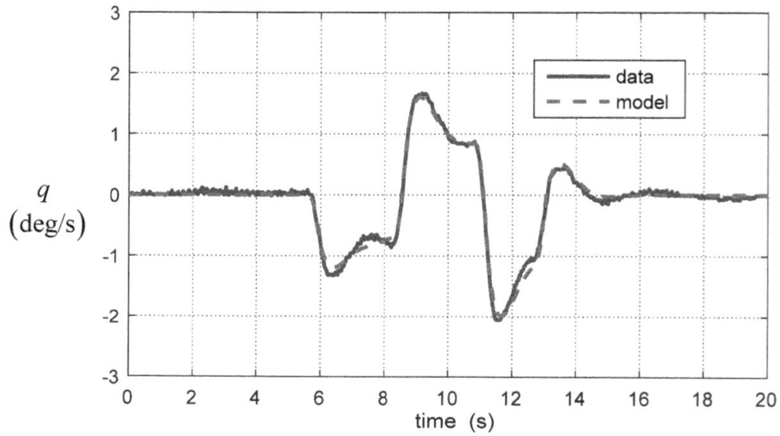

Figure 7.12 **Pitch rate prediction for a longitudinal doublet maneuver**

7.9 Summary and Concluding Remarks

The modeling approaches applied previously to time-domain data were adapted and applied to data transformed into the frequency domain. After a discussion of the data transformation process and the concept of frequency response, the maximum likelihood estimator was developed for frequency-domain data. Following a path similar to Chapter 6 for time-domain data, the output-error and equation-error methods in the frequency domain were developed by making simplifying assumptions in the general problem formulation, then solving the resulting optimization problems. Three aircraft application examples were included to demonstrate the techniques.

Modeling in the frequency domain has several practical advantages, including applicability to unstable dynamic systems, and generally faster computation, as a result of a reduced number of data points in the frequency domain and the fact that the solution to linear differential equations transformed into the frequency domain involves only algebraic operations. It can also be argued that modeling in the frequency domain represents a more robust solution that is more closely aligned with the underlying dynamic system response, because the ordinary or weighted least squares fitting is applied to spectral components, rather than to individual measured data points, as in time-domain methods. When the Fourier transform frequencies are selected in a limited frequency band that includes the dynamic response of interest, signal-to-noise ratio is enhanced, equation-error parameter estimates are unbiased, data collinearity effects are mitigated, and the Cramér-Rao bounds for estimated model parameters do not need correction for colored residuals.

As with any modeling approach, working in the frequency domain has disadvantages as well. Generally, the transformation into the frequency domain omits all biases and linear trends in the time-domain data, so the frequency domain model does not capture the corresponding effects. However, this fact can be considered an advantage when bias and linear trends are not of interest, because the frequency domain modeling then has fewer parameters to estimate. In addition, biases and linear trends can be easily and accurately estimated in the time domain after the frequency-domain analysis. Frequency-domain data are generally more difficult to interpret directly, so there is some disconnection with the physical system when working with frequency-domain data. Using spectral estimates can alleviate this problem, because the Bode plot can be readily interpreted in terms of physical characteristics of the dynamic system, and parametric modeling results can also be presented in the form of Bode plots. The output-error approach in the frequency domain is limited to linear dynamic systems, fundamentally because the finite Fourier transform is a linear operator. On the other hand, the equation-error method based on linear regression can use arbitrarily nonlinear terms in the model, because the nonlinear terms can be generated in the time domain prior to transformation into the frequency domain.

In general, it is advantageous to have both time-domain and frequency-domain techniques available, so that the appropriate tool for the problem at hand can be applied. These two classes of methods are based on different forms of the data, and differ in the formulations of their respective optimization problems. Consequently, successful comparison between time-domain results and frequency-domain results can increase confidence in the identified models.

References

[7.1] Klein, V., "Maximum Likelihood Method for Estimating Airplane Stability and Control Parameters From Flight Data in Frequency Domain," NASA TP-1637, 1980.

[7.2] Bendat, J.S. and Piersol, A.G., *Random Data Analysis and Measurement Procedures*, 2nd Ed., John Wiley & Sons, Inc., New York, NY, 1986.

[7.3] Cooley, J.W. and Tukey, J.W., "An Algorithm for the Machine Calculation of Complex Fourier Series," *Mathematics and Computation*, Vol. 19, No. 90, 1965, pp. 297-301.

[7.4] Morelli, E.A., "High Accuracy Evaluation of the Finite Fourier Transform using Sampled Data," NASA TM 110340, 1997.

[7.5] Bendat, J.S. and Piersol, A.G. *Engineering Applications of Correlation and Spectral Analysis*, 2nd Ed., John Wiley & Sons, Inc., New York, NY, 1993.

[7.6] Tischler, M.B. and Remple, R.K. *Aircraft and Rotorcraft System Identification – Engineering Methods with Flight Test Examples*, 2nd Ed., AIAA Education Series, AIAA, Reston, VA, 2012.

[7.7] Hardin, J.C., "Introduction to Time Series Analysis," NASA RP 1145, November 1990.

[7.8] Press, W.H., S.A. Teukolsky, W.T. Vettering, and B.R. Flannery *Numerical Recipes in FORTRAN: The Art of Scientific Computing*, 2nd Ed., Cambridge University Press, New York, NY, 1992.

[7.9] Otnes, R.K. and Enochson, L., *Applied Time Series Analysis, Volume 1, Basic Techniques*, John Wiley & Sons, Inc., New York, NY, 1978.

[7.10] Graham, D. and McRuer, D., *Analysis of Nonlinear Control Systems*, John Wiley & Sons, New York, NY, 1961.

[7.11] Mehra, R.K., "Identification of Stochastic Linear Dynamic System Using Kalman Filter Representation," *AIAA Journal*, Vol. 9, No. 1, 1971, pp. 28-31.

[7.12] Pearson, A.E., "Aerodynamic Parameter Estimation Via Fourier Modulating Function Techniques," NASA CR 4654, 1995.

[7.13] Åström, K.J., *Introduction to Stochastic Control*, Academic Press, Inc., New York, NY, 1970.

[7.14] Miller, K.S., *Complex Stochastic Processes – An Introduction to Theory and Application*, Addison-Wesley Publishing Co., Inc., New York, NY, 1974.

[7.15] Tischler, M.B., "Frequency-Response Identification of the XV-15 Tilt-Rotor Aircraft Dynamics," NASA TM 89428, 1987.

[7.16] Tischler, M.B. and Cauffman, M.G., "Comprehensive Identification from FrEquency Responses, Vol. 1 – Class Notes," NASA CP 10149, 1994.

[7.17] Tischler, M.B. and Cauffman, M.G., "Comprehensive Identification from FrEquency Responses, Vol. 2 – User's Manual," NASA CP 10150, 1994.

[7.18] Morelli, E.A., "Practical Aspects of the Equation-Error Method for Aircraft Parameter Estimation," AIAA-2006-6144, *AIAA Atmospheric Flight Mechanics Conference*, Keystone, CO, August 2006.

[7.19] Klein, V., "Aircraft Parameter Estimation in Frequency Domain," AIAA Paper 78-1344, *AIAA Atmospheric Flight Mechanics Conference*, Palo Alto, CA, 1978.

[7.20] Klein, V. and Keskar, D.A., "Frequency Domain Identification of a Linear System Using Maximum Likelihood Estimation," *5th IFAC Symposium on Identification and System Parameter Estimation*, Darmstadt, Germany, Pergamon Press, Vol. 2, 1979, pp. 1039-1046.

[7.21] Klein, V., Murphy, P.C., Curry, T.J., and Brandon, J.M., "Analysis of Wind Tunnel Longitudinal Static and Oscillatory Data of the F-16XL Aircraft," NASA/TM-97-206276, 1997.

[7.22] Morelli, E.A., "Identification of Low Order Equivalent System Models from Flight Test Data," NASA TM-2000-210117, 2000.

[7.23] Morelli, E.A., "Low-Order Equivalent System Identification for the Tu-144LL Supersonic Transport Aircraft," *Journal of Guidance, Control, and Dynamics*, Vol. 26, No. 2, 2003, pp. 354-362.

[7.24] Murphy, P.C. and Klein, V., "Validation of Methodology For Estimating Aircraft Unsteady Aerodynamic Parameters from Dynamic Wind Tunnel Tests," AIAA Paper 2003-5397, *AIAA Atmospheric Flight Mechanics Conference*, Austin, TX, 2003.

[7.25] Morelli, E.A., Derry, S.D., and Smith, M.S. "Aerodynamic Parameter Estimation of the X-43A (Hyper-X) from Flight Test Data," AIAA-2005-5921, *AIAA Atmospheric Flight Mechanics Conference*, San Francisco, CA, August 2005.

[7.26] Morelli, E.A. and Cunningham, K. "Aircraft Dynamic Modeling in Turbulence," AIAA-2012-4650, *AIAA Atmospheric Flight Mechanics Conference*, Minneapolis, MN, August 2012.

[7.27] Morelli, E.A. "Transfer Function Identification using Orthogonal Fourier Transform Modeling Functions," AIAA 2013-4749, *AIAA Atmospheric Flight Mechanics Conference*, Boston, MA, August 2013.

[7.28] Morelli, E.A. and Cooper, J., "Frequency-Domain Method for Automated Simulation Updates based on Flight Data," AIAA-2014-0472, *AIAA SciTech 2014 Conference*, National Harbor, MD, January 2014.

[7.29] Morelli, E.A., Rexius, S.L., and Lechniak, J.A. "Flight Test Experiment Design and Aerodynamic Parameter Estimation for the X-51A Waverider," AIAA-2014-2767 *20th AIAA International Space Planes and Hypersonic Systems and Technologies Conference*, Atlanta, GA, June 2014.

[7.30] Shinbrot, M. "On the Analysis of Linear and Nonlinear Dynamical Systems from Transient-Response Data," NACA TN 3288, December 1954.

[7.31] Scharnhorst, K. "Angles in Complex Vector Spaces," *Acta Applicandae Mathematicae*, Vol. 69, 2001, pp. 95-103.

[7.32] *Military Standard – Flying Qualities of Piloted Aircraft*, MIL-STD-1797A, January 1990.

[7.33] Cooper, G.E., and Harper, R.P. Jr., "The Use of Pilot Rating in the Evaluation of Aircraft Handling Qualities," NASA TN D-5153, 1969.

[7.34] Hodgkinson, J., LaManna, W.J., and Heyde, J.L., "Handling Qualities of Aircraft with Stability and Control Augmentation Systems – A Fundamental Approach," *Aeronautical Journal*, February 1976, pp. 75-81.

[7.35] Mitchell, D.G. and Hoh, R.H., "Low-Order Approaches to High-Order Systems: Problems and Promises," *Journal of Guidance, Control, and Dynamics*, Vol. 5, No. 5, 1982, pp. 482-489.

Problems

7.1 Answer the following, using words, graphs, equations, and/or diagrams:

a) Explain why time-series data must be detrended before transforming into the frequency domain.

b) Explain why equation-error and frequency-domain methods can be applied to unstable systems, while time-domain output-error methods generally cannot.

c) Explain the idea of a Low-Order Equivalent System (LOES) model, and describe how these models are used.

d) Explain why output-error in the frequency domain can only be used for linear dynamic systems.

e) Explain how it is possible to avoid correcting estimated parameter errors for colored residuals in the frequency domain.

f) Discuss the important practical considerations for working with measured data transformed into the frequency domain.

7.2 Fly a simple roll doublet maneuver using the F-16 simulation, or use the data in f16_roll_data.mat.

a) Detrend the aileron and roll rate data using zep.m, then identify a transfer function model for roll rate response to aileron input using tfest.m and the roll mode approximation. The tfest.m routine uses equation-error in the frequency domain.

b) Transform the detrended time series for aileron and roll rate into the frequency domain using fint.m. Use fdoe.m with the model file flatss_roll.m (which implements the roll mode approximation) to find output-error parameter estimates in the frequency domain. Use the parameter estimates found in part a) as starting values.

c) Check the model fit in the frequency domain by plotting the measured output and model output in the frequency domain versus frequency, along with the residuals. Use MATLAB® functions abs.m and angle.m to plot the magnitudes and phase angles of these complex quantities.

d) Find model parameter estimates and standard errors using output-error in the time domain.

e) Do a statistical comparison between the results found in parts c) and d). Do the results agree? Why or why not?

7.3 Using the roll mode approximation with dimensional derivatives,

$$\dot{p} = L_p p + L_{\delta_a} \delta_a$$

and using M frequencies, write the model output equation and cost function for:

a) Equation-error parameter estimation in the frequency domain.

b) Output-error parameter estimation in the frequency domain.

7.4 Show analytically that the expression for the least squares solution to a linear parameter estimation problem with complex data (as in the frequency domain) is equivalent to doing the same problem with real numbers, where the real and imaginary parts of the complex data are stacked.

7.5 For the Twin Otter flight data in the file totter_demo_data.mat:

a) Using equation-error and output-error in the time domain, compute estimates of both dimensional and nondimensional stability and control derivatives for the pitching moment.

b) Repeat part a) using equation-error and output-error in the frequency domain.

c) Are the estimates from parts a) and b) consistent, accounting for the estimated uncertainties? Discuss your results, including any reasons why the estimates might differ.

7.6 For the simulated F-16 flight data in f16_05A_msswp_data.mat:

a) Using output-error in the time domain, identify a linear longitudinal short-period dynamic model and a linear lateral dynamic model using nondimensional derivatives.

b) Repeat part a) using equation error in the frequency domain.

c) Compare results from parts a) and b) using any approach that is convincing or illuminating. Compare and contrast the two parameter estimation processes.

7.7 For the data given in ch7_prob7_data.mat, and using a linear dynamic model for a Dutch Roll approximation with dimensional derivatives:

$$\dot{\beta} = Y_\beta \beta + \left(Y_r - \cos\alpha_o\right) r + Y_{\delta_r} \delta_r$$

$$\dot{r} = N'_\beta \beta + N'_r r + N'_{\delta_r} \delta_r$$

$$a_y = \frac{V_o}{g}\left(Y_\beta \beta + Y_r r + Y_{\delta_r} \delta_r\right)$$

a) Find model parameter estimates and standard errors using equation-error in the frequency domain for the frequencies in vector f given in the data file. Use fintd.m to compute the finite Fourier transform of the time derivative of a time series.

b) Find model parameter estimates and standard errors using output-error in the time domain.

c) Do a statistical comparison between the results in parts a) and b). Do the results agree?

d) What additional data would be necessary to do the problem using nondimensional derivatives?

Chapter 8

Experiment Design

Experiment design involves determining which physical quantities will be measured, how those quantities will be measured, test conditions, and how the system being studied will be excited. For aircraft system identification, this translates into specifying the instrumentation and data acquisition system, selecting the aircraft configurations and flight conditions, and designing inputs for the maneuvers. These issues are the topics of this chapter.

The goal of experiment design is to maximize the information content in the data, subject to practical constraints. Some examples of practical constraints are

1) limits on input and/or output amplitudes, e.g., to ensure that a linear model structure can be used to estimate parameters from the measured data;
2) limited resolution or range for the sensors or data acquisition system;
3) hardware or telemetry limitations restricting the rate at which data can be measured or the number of physical quantities that can be measured at an acceptable rate;
4) limited time available for each maneuver and/or for the overall experimental investigation;
5) sensor limitations, characteristics, or availability;
6) limitations on how the aircraft can be excited, e.g., control surface rate or position limits, or the requirement for a continuously-operating feedback control system when the aircraft is open-loop unstable.

As discussed in Chapter 3, inputs for modeling aircraft open-loop or bare-airframe dynamics are the control surface deflections. Outputs are air-relative velocity data (V, α, β), body-axis angular velocities (p, q, r), Euler attitude angles (ϕ, θ, ψ), translational accelerations (a_x, a_y, a_z), and sometimes body-axis angular accelerations $(\dot{p}, \dot{q}, \dot{r})$. Modeling is usually done using some subset of the inputs and outputs listed above, as in the case of linear models for longitudinal and lateral motion described in Chapter 3. For closed-loop modeling, where the modeling includes both the airframe dynamics and the control system, the inputs are one or more of the pilot controls: longitudinal stick deflection, lateral stick deflection, and rudder pedal deflection, or the corresponding forces. Typically, throttle is not moved dynamically, but rather remains constant throughout a maneuver, so the throttle position or power level is treated more like part of the description of the flight test condition. Outputs for closed-loop modeling are again subsets of the aforementioned quantities.

8.1 Data Acquisition System

The data acquisition system records measured time series of input and output variables, along with quantities that define the flight condition and aircraft configuration, such as outside air temperature and static pressure to calculate air density and pressure altitude, throttle position or power level, fuel consumption for estimating aircraft weight and inertia characteristics, landing gear position, flap settings, and other aircraft configuration variables.

Because of high computational requirements, modern system identification methods are implemented on digital computers. Therefore, the measurements of continuous-time signals associated with the airplane must be converted to digital form. Analog signals from the sensors are passed through analog anti-aliasing filters, then possibly scaled to a proper voltage range for digitization using analog-to-digital (A/D) conversion electronics. The output of the A/D conversion is a digital count that is converted to engineering units using laboratory calibrations. Important aspects of the data acquisition system for system identification are discussed below. The discussion here does not include every aspect of aircraft data acquisition system design – only the most critical aspects for system identification. Reference [8.1] provides more information on flight instrumentation and data acquisition.

Sampling Rate

The sampling rate is the rate at which the physical quantities will be measured or sampled, usually expressed in samples per second or Hz. Ideally, all measured signals should be sampled at the same constant rate, but it is not uncommon for measured quantities to be sampled at different rates, for practical reasons. When more than one sampling rate is used for the measured data, interpolation can be used to convert the lower sample rate data to the highest sampling rate. Interpolation is discussed in Chapter 11. Another approach is to thin the high-rate data to match the sample rate of the low-rate data. This is discussed later in this section. In any case, all data should be converted to the same sampling rate before analysis.

The choice of sampling rate is influenced by Shannon's sampling theorem, Ref. [8.2], but also by practical considerations. In theory, the discrete samples of a continuous signal sampled at a frequency f_s capture frequency content up to and including the frequency $f_s/2$, which is known as the Nyquist frequency f_N,

$$f_N = f_s/2 \tag{8.1}$$

Figure 8.1 shows an illustration of this fact, using a continuous sine wave signal. The frequency of the continuous signal is 1 Hz, and the sampling frequency is 2 Hz. Note that if the continuous signal were shifted in time by the equivalent of 90 deg of phase angle, then the samples taken at the times shown would be all zeros. For this reason, sampling at the theoretical minimum rate of twice the Nyquist frequency is not acceptable for capturing frequency content up to and including the Nyquist frequency. In addition, it is clear from Fig. 8.1 that the theoretical minimum sampling rate gives a very rough representation of the continuous signal. To obtain good results in practice, it is necessary to sample at a rate much higher than the theoretical minimum rate. If f_{max} is the maximum frequency of interest in the continuous-time signals from the dynamic system, a good rule of thumb for selecting the sampling rate f_s is:

$$f_s = 25 f_{max} \tag{8.2}$$

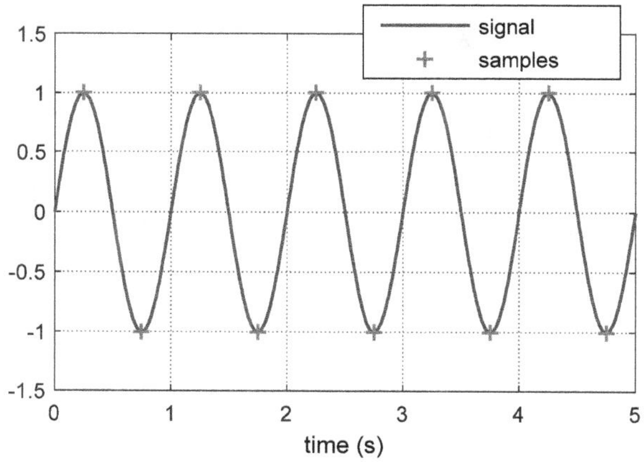

Figure 8.1 **Theoretical minimum sampling rate**

For many aircraft, the frequencies of the rigid-body dynamic modes are below 2 Hz, which would put the sampling rate at 50 Hz.

If the aircraft is a scale model, the frequencies of the dynamic modes scale according to the inverse square root of the model geometric scale,

$$f_{model} = \frac{1}{\sqrt{s}} f_{aircraft} \tag{8.3}$$

where s is the model scale (see Ref. [8.3]). For example, a 1/16 scale model of a full-scale aircraft with rigid-body dynamics at less than 1 Hz would have rigid-body dynamics below 4 Hz, so a sampling rate of 100 Hz would be adequate.

Presampling Data Conditioning

When the continuous signal being sampled has components at frequencies higher than the Nyquist frequency, a phenomena called aliasing occurs. Aliasing is illustrated in Fig. 8.2 for a simple sinusoidal continuous signal. The result of aliasing is that frequency content above the Nyquist frequency f_N is falsely attributed (aliased) to lower frequencies by the sampling process. The frequencies above f_N which are aliased to the frequency f in the range $0 \le f \le f_N$, are

$$2nf_N \pm f \qquad n = 1, 2, \ldots \tag{8.4}$$

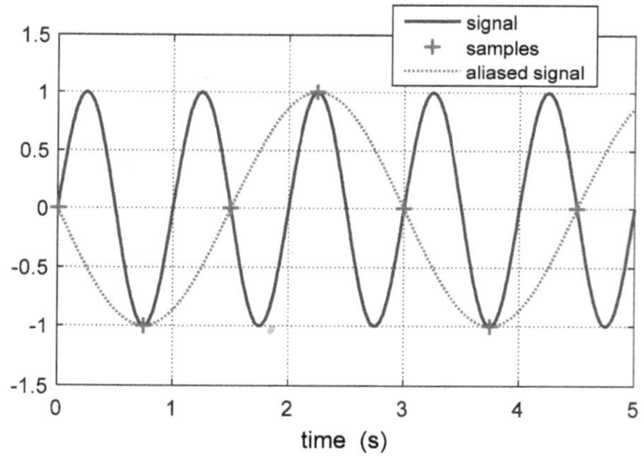

Figure 8.2 **Aliasing**

The manner in which the high frequencies fold downward relative to the Nyquist frequency is illustrated by the diagram in Fig. 8.3. Aliasing pollutes the true frequency content at lower frequencies with frequency content folded down from frequencies above f_N.

Aliasing is a serious problem that must be avoided. The solution is to use analog low-pass filters to remove high frequency components before sampling. Low-pass filters used in this way are called anti-aliasing filters or presampling filters. The filters must be analog to work on the signals before sampling.

Aliasing cannot be repaired with post-flight data analysis or digital filtering, because once the aliasing occurs in the sampling process, the frequencies above the Nyquist frequency have been commingled with the true low frequency content, and there is no way to reliably separate the components.

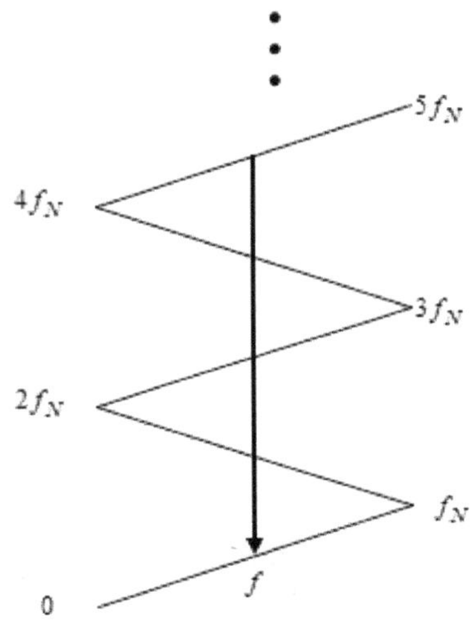

Figure 8.3 **Frequency folding**

Anti-alias filtering can be done with a simple first-order low-pass filter. Figure 8.4 shows a Bode plot of the frequency response for a typical anti-aliasing filter with break frequency at 5 Hz (31.4 rad/s). Low-frequency signal content, which includes the frequency components of interest, is passed through essentially unmodified, and high-frequency content that could be aliased is removed before the sampling occurs. Figure 8.5 shows the amplitude and phase modification for the passband of the same anti-aliasing filter, plotted using a linear frequency scale. The magnitudes are modified only slightly, and the phase change is nearly linear with frequency. This corresponds to a constant time delay τ at all frequencies, since for phase angle ϕ

$$\begin{aligned} \phi &= -\omega\tau \approx \frac{d\phi}{d\omega}\omega \\ \tau &\approx -\frac{d\phi}{d\omega} \end{aligned} \tag{8.5}$$

All sampled signals should have the same anti-aliasing filtering, so that the small time delay from anti-alias filtering prior to sampling is the same for all measured signals at each frequency.

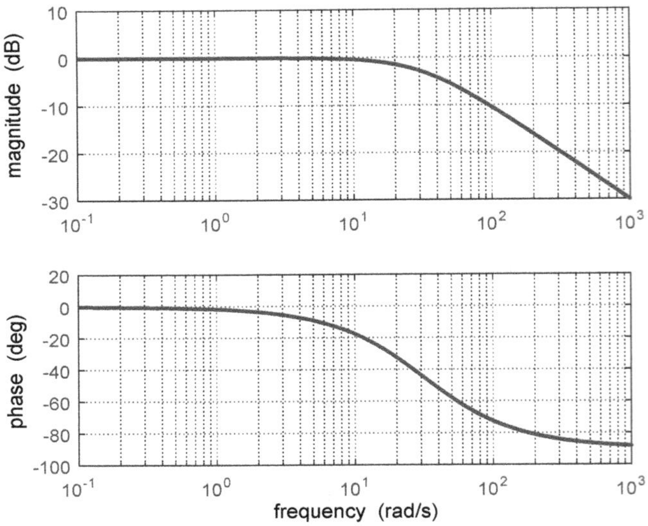

Figure 8.4 **Frequency response for a typical analog anti-aliasing filter:** $\tilde{y}(s)/\tilde{u}(s) = 31.4/(s+31.4)$

The break frequency of the filter must be high enough so that the frequency content of interest is not significantly modified by the anti-alias filtering, but low enough so that high-frequency attenuation is sufficient to remove unwanted high-frequency components. A good rule of thumb is to set the break frequency for the anti-aliasing filters f_a at five times the highest expected frequency of interest,

$$f_a = 5 f_{max} \tag{8.6}$$

which makes the relationships among the important frequencies as follows:

$$f_s = 5 f_a = 25 f_{max} \tag{8.7}$$

If it is unclear what the highest frequency of interest f_{max} might be, it is best to err on the side of a higher sampling rate. A higher sampling rate moves the Nyquist frequency higher, and therefore also raises the level for frequencies that can be aliased. Higher sampling rates also have the advantage of providing more samples than are really needed, and therefore represent protection against problems such as power spikes and data dropouts. In addition, higher sampling rates reduce the possible time skews among the measured signals, or can help to illuminate and quantify existing time skews.

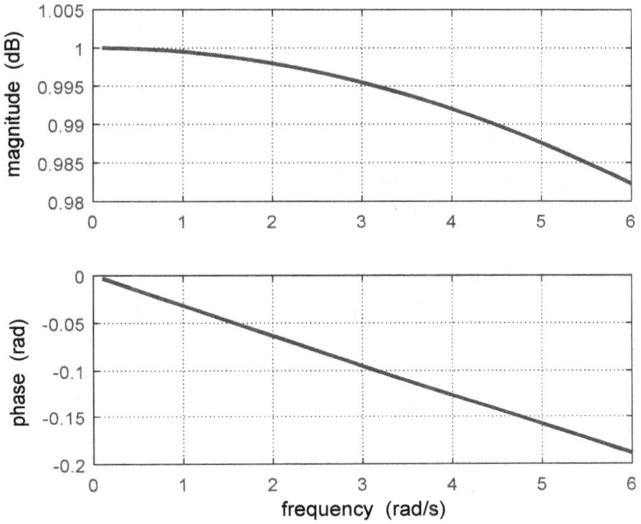

Figure 8.5 **Magnitude and phase modification for the pass band of the analog anti-aliasing filter:** $\tilde{y}(s)/\tilde{u}(s) = 31.4/(s+31.4)$

Thinning high sampling rate data to a lower sampling rate, which is also called downsampling the data, is equivalent to sampling the data at a lower sampling rate. Downsampling therefore lowers the Nyquist frequency, and the same potential for aliasing exists. However, if the original high-rate data collection was done properly, it is only necessary to remove frequency content above the reduced Nyquist frequency associated with the downsampling to avoid aliasing.

For example, downsampling 50 Hz data to 25 Hz by dropping every other data point reduces the effective Nyquist frequency from 25 Hz to 12.5 Hz. To avoid aliasing from the downsampling, there should be no significant components above 12.5 Hz in the original 50 Hz sampled data. Zero phase-shift digital filtering or smoothing, discussed in Chapter 11, can be used prior to downsampling to remove frequency content above 12.5 Hz in the 50 Hz sampled data. The digital filtering or smoothing is done before downsampling, for the same reasons cited above for the analog anti-aliasing filtering.

Implicit in all of the modeling techniques discussed elsewhere in this book is the assumption that all of the measured physical quantities have been sampled simultaneously. It is assumed that there are negligible differences between the instants when the first and last physical quantities are measured for each data sample taken. In practice, there is a small time difference, but the maximum time difference should be negligible (usually less than a millisecond) for modern data acquisition systems intended for collecting flight test data. However, in some cases, such as low-budget flight test programs and accident analysis, the physical quantities may not be sampled at nearly the same time. In that situation, it is necessary to interpolate or time shift the data so that the data are brought to the state of nearly simultaneous sampling of the physical quantities. This is important, because if it is not done, then the time skews that remain can be interpreted as phase lags due to the dynamic response of the physical system, which they are not.

Steers and Iliff [8.4] studied the effects of time skews in the recorded data on output-error parameter estimates, using flight test data from five different aircraft. The results showed that parameters related to high-frequency quantities, such as body-axis angular rates, were more sensitive to time skews in the measured data than parameters associated with lower-frequency quantities, such as angle of attack and sideslip angle. This makes sense, because a given magnitude of time skew represents a larger phase angle change for a high-frequency signal, compared to a lower-frequency signal.

For the full-scale aircraft flight data studied in Ref. [8.4], time skews greater than about 40 msec caused significant inaccuracies in parameter estimation results. Note that if the sampling rate of the data acquisition system is 50 Hz and the system is operating properly, the relative time skews of the measured data can be no greater than the sampling interval, which is 20 msec. This built-in protection against large time skews in the data is another reason to prefer higher sampling rates.

Sensor Range and Resolution

The range of each sensor should be sufficient to include the expected extreme values of the physical quantity to be measured, to allow for unusual motions, and to avoid sensor saturation. For example, physical values greater than the maximum of the sensor range would produce the same maximum sensor reading, resulting in a loss of information.

For typical modern data acquisition systems, the analog-to-digital (A/D) conversion might use a 16-bit binary word. In this case, the sensor resolution can be computed from

$$\text{resolution} = \text{range}/2^{16} = \frac{(\text{max sensor reading} - \text{min sensor reading})}{65{,}536} \tag{8.8}$$

For 16-bit A/D conversion, the resolution is 0.0015 percent of the full range of the sensor. The resolution represents a lower limit on the accuracy of the measurements.

The above equation shows that there is a trade-off between range and resolution, for given A/D conversion hardware. Of course, more bits in the

A/D conversion improves the resolution for a given sensor range, and therefore ameliorates the problem. The range and resolution must be selected carefully for each measured physical quantity. The intent of many flight test programs is to collect data for linear model identification over a relatively large portion of the flight envelope. This requires good resolution to accurately quantify the small changes in physical quantities associated with small perturbation maneuvers, and good range so that the sensors can be used over a large flight envelope. Table 8.1 is an example from Ref. [8.5] of instrumentation characteristics that have worked well for a general aviation airplane.

Table 8.1 **Instrumentation characteristics for a general aviation airplane**

Measured Quantity	Transducer	Working Range	Resolution	rms measurement error
a_x (g)	accelerometer	[–1,1]	0.001	0.0046
a_y (g)	accelerometer	[–1,1]	0.001	0.0050
a_z (g)	accelerometer	[–3,6]	0.001	0.0050
p (deg/s)	rate gyro	[–102,102]	0.12	0.20
q (deg/s)	rate gyro	[–29,29]	0.032	0.19
r (deg/s)	rate gyro	[–29,29]	0.034	0.080
ϕ (deg)	vertical gyro	[–90,90]	0.10	0.077
θ (deg)	vertical gyro	[–87,87]	0.098	0.092
α (deg)	vane	[–12,27]	0.029	0.027
β (deg)	vane	[–29,32]	0.018	0.019
δ_{a_r} (deg)	potentiometer	[–23,10]	0.002	0.019
δ_{a_l} (deg)	potentiometer	[–10,25]	0.002	0.0061
δ_s (deg)	potentiometer	[–16,3]	0.010	0.0037
δ_r (deg)	potentiometer	[–31,28]	0.011	0.0091
V (m/s)	pressure transducer	[0,75]	0.037	0.89
h (m)	altimeter	[–150,2900]	–	–
T (deg C)	thermometer	[–18,38]	–	–

The sensors are also dynamic systems, so their characteristics should be such that the dynamics of the sensor do not interact with the dynamics of the aircraft. This is achieved by building sensors with relatively high natural frequencies, high damping ratios, and small time delays. Table 8.2 gives the dynamic characteristics of the general aviation aircraft instrumentation in Table 8.1.

Table 8.2 **Dynamic characteristics of transducers**

Measured Quantity	Transducer	Natural Frequency (Hz)	Damping Ratio	Time Delay (s)
a_x (g)	accelerometer	402	1.58	0.0012
a_y (g)	accelerometer	216	1.10	0.0016
a_z (g)	accelerometer	921	1.58	0.0005
p (deg/s)	rate gyro	27	0.64	0.0075
q (deg/s)	rate gyro	27	0.64	0.0075
r (deg/s)	rate gyro	27	0.64	0.0075
α (deg)	vane	23*	0.085*	0.0012
β (deg)	vane	23*	0.085*	0.0012

* At $V = 164$ ft/sec

8.2 Instrumentation

Flight test instrumentation is constantly evolving, so any description of current hardware would quickly become outdated. However, important characteristics of the flight instrumentation can be discussed in general, without reference to any particular sensors, and that is the approach taken here. The discussion is organized according to the physical quantities to be measured.

Air-Relative Velocity

Accurate air-relative velocity data, which is composed of measurements of angle of attack, sideslip angle, and airspeed, are perhaps the most difficult to obtain. Part of the reason is that local flows about the airplane influence measured values of these quantities, while the desired quantities for rigid-body dynamic modeling are the values for the aircraft at the c.g. Therefore, no matter where the sensors for these quantities are placed, there will be some corrections involved. Calibration is also more difficult, because air must be flowing over the sensor at the time of the calibrations, and the calibration is really only valid when the sensor is installed on the aircraft. Complicating matters further is the fact that these data are arguably the most important physical quantities in terms of modeling the aerodynamic forces and moments acting on the aircraft. Consequently, the quest for accurate air-relative velocity data is continual.

Sensors for air-relative velocity data can be calibrated very accurately in the wind tunnel, preferably installed on the airplane or on parts of the airplane that are nearby (i.e., the wing or nose cone). The best location for the sensors is on a nose boom, but this location is impractical for aircraft with a propeller mounted on the nose. Sensors are often located on wing tip booms, which should extend 2-3 chord lengths forward of the leading edge of the wing. Local flow at the wingtips in the outboard direction usually biases the sideslip angle measurement, so it is more accurate to use averaged sideslip measurements from sensors on both wing tips. It is possible to use sensors near the wing or fuselage, including sensors based on pressure measurements, but this generally requires careful and extensive calibration for good accuracy. Air flow data from pressure probes or flush air data systems can have pneumatic time lags related to the length and diameter of the tubing.

The position of the air data sensors relative to the aircraft c.g. must be known accurately. This information is necessary to implement corrections for angle of attack and sideslip angle readings resulting from angular motion of the aircraft. Usually, the c.g. and the sensor positions are specified relative to a single fixed reference point, and the sensor position relative to the c.g. is computed from that information. Details of these corrections appear in Chapter 9.

Angular Velocity

Aircraft angular velocity components are usually measured using rate gyros attached to the aircraft and aligned with the body axes. These sensors are among the most reliable and accurate of all aircraft flight instrumentation. Response is very linear, and typical instrumentation errors consist of small biases that are very repeatable among different maneuvers. In theory, the location of these sensors can be anywhere on the aircraft, without the need for any position correction to the c.g., because angular rates are the same at any point on a rigid body. In practice, however, the aircraft is not rigid, and it is necessary to locate the sensors away from the nodes of any significant structural responses, which would cause rotational motion that the sensors will pick up. Often, the rate gyros are packaged with the translational accelerometers.

Translational Acceleration

Translational accelerometers should be located close to the aircraft c.g., and aligned with the body axes. This keeps the corrections small when the measurements are corrected to the aircraft c.g. The corrections come from accelerations due to rotational aircraft motion about the c.g., in conjunction with the location of the accelerometers away from the c.g. Translational accelerometers have excellent linearity and usually only a small bias error. The main negative aspect of these sensors is that their frequency response is excellent, so in addition to rigid-body motion, they also pick up structural response and engine vibrations. This can make the signals quite noisy, and also can cause problems with high-frequency responses folding down to the range of rigid-body frequencies, if the anti-aliasing filters are not designed and implemented properly. The position corrections for these sensors involve body-axis angular accelerations (see Chapter 9), so some additional noise is introduced as a result of the position corrections, since the angular acceleration signals are typically noisy, regardless of whether they are obtained from smoothed numerical derivatives of the angular rates or directly from a sensor. The relatively high noise levels for accelerometer signals also makes it more difficult to filter or smooth the data, because large wideband random noise components can mask the deterministic signal content.

Rotational Acceleration

Sensors for rotational acceleration are not common, but occasionally are included in flight test instrumentation. Most of the sensors available at the time of this writing have relatively high noise levels and/or lags. Because of this, measured angular accelerations are not used often for aircraft system identification. As sensor technologies evolve, this situation may change. Measurements from these sensors can be easily incorporated into the modeling methods. Good angular acceleration measurements would improve parameter estimation results by providing additional information content to the data for output-error modeling, and obviating the need to numerically differentiate the angular rates for instrumentation error corrections and equation-error modeling.

Euler Attitude Angles

Euler angles can be measured using integrating gyros or magnetometers. Often, the Euler angles produced by an inertial measurement unit (IMU) come from a computed attitude solution, which can have time delays because of internal filtering. The Euler angles have secondary importance in system identification for aircraft, because the aerodynamic forces and moments do not depend on aircraft orientation relative to earth axes. Consequently, the Euler angle data is not needed for equation-error methods. However, as discussed in Chapter 3, the Euler angles are needed to include the gravity terms in the equations of motion, which are required for output-error and filter-error methods. The heading angle appears only in the kinematic equations and not in any of the dynamic equations, so it is the least important of the Euler angles for dynamic modeling (but of course very important for guidance). All of the Euler angles are useful in data compatibility analysis, for estimating instrumentation errors on rate gyro measurements, as discussed in Chapter 9.

Control Surface Deflections and Pilot Controls

Sensors that measure control surface deflections and pilot control deflections are typically some type of potentiometer, which produce a voltage proportional to rotational or linear motion. These sensors are very reliable and linear, and have very low noise levels. The latter characteristic is important, because most modeling methods assume that known deterministic inputs can be measured without error. Direct measurements of the input positions are preferred over remote command data, because the latter often include time delays from actuator dynamics and/or telemetry.

8.3 Input Design

When designing inputs for dynamic systems in general, and aircraft in particular, there are two general approaches. The first is to design the input assuming no prior knowledge of how the dynamic system behaves. The goal in that case is to excite the system over a broad frequency range, with nearly constant power for all frequencies. Input designs that employ this approach include frequency sweeps, multisines, and impulse inputs. The second approach is to use prior knowledge about the dynamic system response and tailor the input accordingly. Optimal input designs are in this category, along with square-wave inputs that are designed to excite the dynamic system at or near the *a priori* estimates of the natural frequencies for the dynamic modes.

8.3.1 Maneuver Definition

In Chapter 3, a common aerodynamic model structure was shown to be a locally valid approximation of the global functional dependencies, typically a linear multivariate Taylor series in the aircraft states and controls. The point about which this expansion is made is typically a trimmed flight condition, such as steady, level flight. The coefficients of the Taylor series expansion are the aerodynamic model parameters to be estimated. These parameters therefore depend on the flight condition for the test.

Consequently, one aspect of maneuver definition is selecting the flight conditions where maneuvers will be executed. For the nondimensional aerodynamic force and moment coefficients, the relevant aspects of the flight condition are typically trim angle of attack, Mach number, aircraft configuration, dynamic pressure, and power level. Flight conditions for testing are chosen based on the goals of the particular investigation, weighed against resource limitations and other practical constraints.

An important aspect of the maneuver definition is specifying the excitation inputs. To define the excitation inputs for a maneuver, the following must be specified:

1) maneuver time length;
2) control surfaces or pilot inputs to be moved;
3) input forms (e.g., square wave, frequency sweep, etc.), which includes the input amplitudes.

Input design involves selecting these quantities in a way that provides adequate information content in the data for accurate modeling. The exact meaning of this will be explored in detail.

8.3.2 Input Design Objective

The objective of input design for dynamic model identification is to excite the dynamic system so that the data contain sufficient information for accurate modeling, subject to the practical constraints of the experiment. This section describes data information content and practical constraints mathematically, and provides methods for achieving the desired results in practice.

A general principle of input design for dynamic systems is that the data must include the dynamic response that the model is intended to characterize. If a particular response is not exhibited in the data, it can't be modeled using system identification techniques. For example, if a model characterizing the short period dynamics of an aircraft is desired, then the short period dynamic response must be excited during the maneuver. Similarly, if the effectiveness of a particular control is of interest, it is necessary to move that control during the maneuver, in a manner that is unlike the movement of any of the other explanatory variables that will be included in the model. This allows the system identification techniques to assign dependencies properly, resulting in an accurate identified model that can predict well.

Data Information Content

For a single measured quantity, data information content can be quantified by the signal-to-noise ratio. Specifically, the part of the measured quantity that is called the signal is assumed to be deterministic, whereas the noise is random, with incoherent phase and amplitude variations that can be readily recognized visually in both the time and frequency domains. It is possible to separate the deterministic signal and random noise components of a measured signal using an optimal Fourier smoothing method (see Chapter 11), so that signal-to-noise ratio can be computed using root-mean-square values.

Signal-to-noise ratio can also be estimated visually from plots of the data. For measured aircraft responses, the signal-to-noise ratio should be 10 or more for good modeling results. Usable results can be obtained with signal-to-noise ratios as low as 3.

In the frequency domain, the coherence is a measure of signal-to-noise ratio for the frequency response, assuming a linear relationship between the input and output (see Chapter 7). Coherence values greater than approximately 0.6 are usually required for good linear modeling results.

Both signal-to-noise ratio and coherence could be classified as nonparametric measures of data information content, because they have no connection to any mathematical model with parameters. For most aircraft system identification, the desired end result is an identified parametric model, so mathematical descriptions of data information content for parametric models are important.

For the simple case of a single-input, single-output model with one model parameter, data information content is quantified by the sensitivity of the model output to changes in the parameter, called the output sensitivity, which was discussed in Chapter 6. The best input for a parameter estimation experiment is the input that maximizes the squared output sensitivity over the test time T. This can be expressed as

$$u^* = \max_{u \in U} \sum_{i=1}^{N} \left[\frac{\partial y(i)}{\partial \theta} \right]^2 = \min_{u \in U} \sum_{i=1}^{N} \left\{ \left[\frac{\partial y(i)}{\partial \theta} \right]^2 \right\}^{-1} \tag{8.9}$$

where u^* is the scalar optimal input waveform over the test time $\left[0, T\right]$, U is the set of admissible inputs, and the summation over N time points approximates a time integral.

The optimization in Eq. (8.9) is equivalent to maximizing the signal-to-noise ratio for a parameterized model. The noise does not appear in this scalar case, because the input has no influence on noise, and there is only a single output, so it is not necessary to account for different noise levels on different outputs. High output sensitivity means that small changes in the model parameter will cause large changes in the model output. Consequently, small changes in the model parameter have a large effect on how closely the model output matches the measured output. In this circumstance, parameter estimation routines will be able to accurately determine the parameter value that produces the best match of the model output to the measured output. At the other extreme, if the output sensitivity is low, the parameter estimation routines can adjust the parameter considerably without changing the model output very much, so that a range of values for the parameter may be indistinguishable in terms of the model output match to the measured output. The result is an inaccurate parameter estimate with large uncertainty.

For aircraft dynamic models, there are usually multiple outputs and multiple model parameters. The information content in the data is then quantified by a matrix, called the information matrix, which was discussed in Chapters 5 and 6. For a parameter vector $\boldsymbol{\theta}$ of length n_p, the $n_p \times n_p$ information matrix $\boldsymbol{M}$ is

$$\boldsymbol{M} = \sum_{i=1}^{N} \left[\frac{\partial \boldsymbol{y}(i)}{\partial \boldsymbol{\theta}} \right]^T \boldsymbol{R}^{-1} \left[\frac{\partial \boldsymbol{y}(i)}{\partial \boldsymbol{\theta}} \right] \tag{8.10}$$

which is Eq. (6.35).

When the $\boldsymbol{R}$ matrix is diagonal, as in the usual case, the diagonal elements of $\boldsymbol{R}^{-1}$ introduce a scaling of the output sensitivities according to the inverse of the individual output noise variances. If the sensitivities of the outputs to the parameters are interpreted as the "signal" for parametric modeling purposes, then the information matrix is a discrete-time sum of squared multiple-output, multiple-parameter signal-to-noise ratios. It follows that the information matrix can be loosely interpreted as squared signal-to-noise ratios for multiple-output parameterized models. If the sensitivities are large relative to the noise levels and are uncorrelated with one another, then the output dependence on the parameters is strong and distinct for each parameter. Parameter values can then be estimated with high accuracy when adjusting the parameters so that model outputs match measured outputs.

As discussed in Chapter 6, the partial derivatives in the $n_o \times n_p$ sensitivity matrix $\partial \boldsymbol{y}(i)/\partial \boldsymbol{\theta}$ are the output sensitivities, which can be found using finite differences or by solving the sensitivity equations.

For a linear dynamic system of the form

$$\begin{aligned} \dot{\boldsymbol{x}}(t) &= \boldsymbol{A}(\boldsymbol{\theta})\boldsymbol{x}(t) + \boldsymbol{B}(\boldsymbol{\theta})\boldsymbol{u}(t) \qquad \boldsymbol{x}(0) = \boldsymbol{x}_0 \\ \boldsymbol{y}(t) &= \boldsymbol{C}(\boldsymbol{\theta})\boldsymbol{x}(t) + \boldsymbol{D}(\boldsymbol{\theta})\boldsymbol{u}(t) \end{aligned} \tag{8.11}$$

with discrete measurements

$$\begin{aligned} \boldsymbol{z}(i) &= \boldsymbol{y}(i) + \boldsymbol{v}(i) \qquad i = 1,2,\ldots,N \\ E\left[\boldsymbol{v}(i)\boldsymbol{v}^T(j)\right] &= \boldsymbol{R}\,\delta_{ij} \end{aligned} \tag{8.12}$$

the sensitivity equations for each parameter θ_j, developed in Chapter 6, are

$$\frac{d}{dt}\left(\frac{\partial \boldsymbol{x}}{\partial \theta_j}\right) = \boldsymbol{A}\frac{\partial \boldsymbol{x}}{\partial \theta_j} + \frac{\partial \boldsymbol{A}}{\partial \theta_j}\boldsymbol{x} + \frac{\partial \boldsymbol{B}}{\partial \theta_j}\boldsymbol{u} \quad ; \quad \frac{\partial \boldsymbol{x}(0)}{\partial \theta_j} = \boldsymbol{0}$$

$$\frac{\partial \boldsymbol{y}}{\partial \theta_j} = \boldsymbol{C}\frac{\partial \boldsymbol{x}}{\partial \theta_j} + \frac{\partial \boldsymbol{C}}{\partial \theta_j}\boldsymbol{x} + \frac{\partial \boldsymbol{D}}{\partial \theta_j}\boldsymbol{u} \qquad j = 1,2,\ldots,n_p \tag{8.13}$$

Note that it is necessary to have *a priori* values for the model parameters to solve the dynamic system and sensitivity equations (8.11) and (8.13), respectively. The information matrix depends on the input $\boldsymbol{u}$, which influences the sensitivities both directly as a forcing function in the sensitivity equations and indirectly as an influence on the states, which also force the sensitivity equations.

In Appendix A, it is shown that the inverse of the information matrix is the theoretical lower limit for estimated parameter covariances, computed using an asymptotically unbiased and efficient parameter estimation algorithm. This theoretical lower limit is called the Cramér-Rao lower bound or the dispersion matrix, and is given by

$$\boldsymbol{\Sigma} = \boldsymbol{M}^{-1} = \left\{\sum_{i=1}^{N}\left[\frac{\partial \boldsymbol{y}(i)}{\partial \boldsymbol{\theta}}\right]^T \boldsymbol{R}^{-1}\left[\frac{\partial \boldsymbol{y}(i)}{\partial \boldsymbol{\theta}}\right]\right\}^{-1} \le Cov(\boldsymbol{\theta}) \tag{8.14}$$

Because the information matrix $\boldsymbol{M}$ depends on the input sequence $\boldsymbol{u}$, the theoretical lower bounds on the parameter covariances also depend on the input sequence. The dependence of the Cramér-Rao bounds on the input is nonlinear, regardless of whether or not the dynamic system equations (8.11) are linear, because of the nonlinear character of Eq. (8.14) and the time integrations required to solve Eqs. (8.11) and (8.13). Similarly, the Cramér-Rao bounds depend nonlinearly on the states.

As discussed in Chapter 6, the diagonal elements of the dispersion matrix are the theoretical minimum values of the individual parameter variances. The Cramér-Rao lower bounds for the parameter standard errors are the square root of the diagonal elements of the dispersion matrix,

$$s\left(\hat{\theta}_j\right) = \sqrt{s_{jj}} \qquad j = 1,2,\ldots,n_p \tag{8.15}$$

where s_{jk} are the matrix elements of the dispersion matrix $\boldsymbol{\Sigma}$,

$$\boldsymbol{\Sigma} = \left[s_{jk}\right] \qquad j,k = 1,2,\ldots,n_p \tag{8.16}$$

The jth diagonal element of the dispersion matrix corresponds to the jth parameter in the sensitivity equations (8.13).

The Cramér-Rao bounds are independent of the parameter estimation algorithm used to extract parameter estimates and standard errors from the data, because the Cramér-Rao bounds represent a theoretical lower bound. Thus, the merit of an input design for aircraft parameter estimation can be determined by examining the Cramér-Rao bounds, since these depend only on the information matrix and not on the parameter estimation algorithm. In other words, input designs are evaluated based only on the information content in the data, which is calculable before any parameter estimation is done.

Computation of the Cramér-Rao bounds requires an *a priori* dynamic system and measurement model. So, a model complete with parameter values is necessary in order to design an experiment which will produce data for estimating the model parameters. This has been called the circularity problem [8.6]. For aircraft, the problem is mitigated by use of parameter estimates obtained from aerodynamic calculations, wind tunnel experiments, or previous flight tests. In practice, most input design methods require very little *a priori* information and some require none at all, as discussed later. Only optimal input design techniques require calculation of the information matrix using an *a priori* model. However, the information matrix and the associated calculations are useful in evaluating any input design for parameter estimation, regardless of how that input was designed.

Knowledge of how the information matrix is defined is also helpful for insight into practical input design. For example, calculation of the information matrix and Cramér-Rao bounds implicitly includes the time length of the flight test maneuver. Because the discrete approximation of a time integral is part of the information matrix calculation, it is clear that longer maneuvers will provide more information in the data and lower the Cramér-Rao bounds, assuming that the input continuously excites the system.

Practical Constraints

There are many practical constraints that restrict the manner in which flight testing for aircraft parameter estimation can be conducted, which in turn limits the information content in the data from flight test maneuvers.

Most models identified from flight test data would be called local models, because their range of validity is limited to a relatively small range of important explanatory variables, such as angle of attack and sideslip angle. During the experiment, aircraft response must be limited to perturbations about the initial flight condition, to retain the validity of the local model structure. Under these conditions, the model parameters can be considered constant throughout the maneuver. If the model to be identified is a linear dynamic model, then the output perturbations might also be limited to satisfy the small perturbation requirement necessary for decoupling longitudinal and lateral dynamic equations.

Constraints for model structure validity are best implemented in terms of amplitude constraints on selected aircraft response variables. Typical values for the perturbation amplitude constraints are ±3 deg in angle of attack or sideslip angle, ±20 deg/s in body-axis angular rates and ±0.1g to ±0.3g in translational accelerations. In addition, constraints may be required on aircraft attitude angles for flight test operational considerations, such as flight safety and maintaining the data downlink to the ground. These constraints can be represented mathematically as

$$\left|y_k(t)\right| \le \psi_k \quad \forall t \quad k \in (1,2,\ldots,n_o) \tag{8.17}$$

where ψ_k is the constant amplitude constraint for the *k*th output, and n_o is the number of outputs.

Input amplitudes are limited by mechanical stops, flight control software limiters, or model structure validity. These constraints are specified by

$$\left|u_j(t)\right| \le \mu_j \quad \forall t \quad j \in (1,2,\ldots,n_i) \tag{8.18}$$

where μ_j is the constant amplitude constraint for the *j*th input, and n_i is the number of inputs.

Sometimes the above constraints are implemented indirectly using a constraint on the input energy,

$$\int_0^T \boldsymbol{u}(t)^T \boldsymbol{u}(t)\, dt = E \tag{8.19}$$

where E is a selected input energy. The input energy constraint has been introduced mostly to simplify formulations for input optimization using variational calculus. In practice, it is difficult to choose an appropriate value for the input energy E, since neither the pilot nor control system have any such energy limitations. Use of the input energy constraint as an indirect limit on output amplitudes is imprecise, because the output amplitudes depend on the frequency content of the input, which is not captured by the input energy calculated from Eq. (8.19). Figure 8.6 shows three square wave inputs with very different time lengths and amplitudes, but the same input energy calculated from Eq. (8.19). Further discussion on this issue can be found in Ref. [8.7].

The expression for the information matrix in Eq. (8.10) is a discrete approximation of a time integral. In general, longer maneuvers (or multiple maneuvers) provide more information, as long as the input continues to excite the system within the output amplitude constraints. In many practical situations, such as subscale aircraft flight tests, limited test ranges, or flight testing at conditions that cannot be sustained for a long time (e.g., high angles of attack or hypersonic speeds), the length of individual maneuvers is limited.

In addition, finite resources always put a limit on total flight test time available. Equations (8.13) and (8.14) indicate that large output amplitudes over long time periods will produce the most accurate parameter estimates. However, practical constraints act to limit both the maneuver time and the allowable input and output amplitudes.

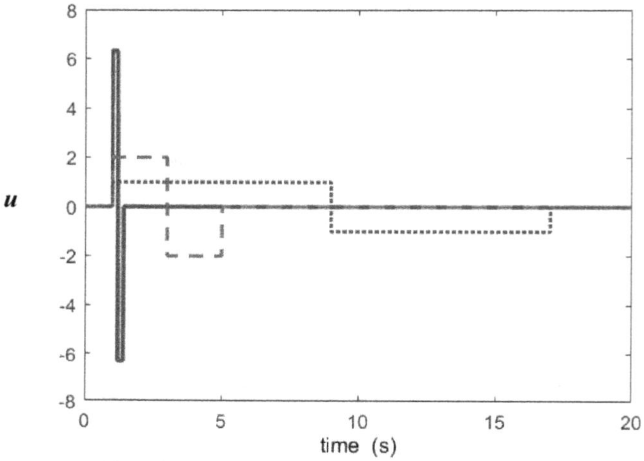

Figure 8.6 **Constant energy square waves with very different forms**

In addition to the foregoing, there are practical considerations related to executing the flight test maneuver. One such consideration is that the allowable frequencies for the input will be limited at the high end, due to limited dynamic response of the sensors, reduced dynamic system response to high frequency inputs, actuator limitations, and the high frequency limitations of the pilot and control system. Inputs must also avoid structural resonance frequencies, for safety, and to maintain the validity of rigid-body modeling. Low-frequency inputs can cause the aircraft to drift away from the flight condition selected for the maneuver.

Another practical limitation on input forms might be called implementation distortion. This refers to the practical fact that a designed input may be distorted when actually implemented, owing to the inherent variability of the pilot, actuator limitations, or actions of the feedback control system. In some cases, the distortion of the input form can actually be helpful. This is discussed in more detail later.

Finally, in addition to the input amplitude limits described earlier, there are limits on the rate that input amplitudes can be changed. For piloted inputs, this is the result of human pilot capability. Useful pilot inputs are generally in the frequency range of [0.1,10] rad/s or [0.016,1.6] Hz. If the input is implemented by a computerized system, the constraint comes from the control surface actuator rate limits. These limits become important if the desired input is a sharp-edged form, such as a square wave.

8.3.3 Single-Input Design

There are many different input forms that can be used for flight test maneuvers. The main types will be discussed below. Single-input design is discussed first, then extended to multiple-input design.

All of the input designs discussed here are for aerodynamic model parameter estimation. They are designed as some type of balanced perturbation about a trim condition, so that the flight condition remains essentially unchanged, and the model parameters can be considered constant throughout the maneuver. The inputs shown are excitations starting from the trim value, which is defined as zero for this discussion. In practice, the input forms would be added to the trim value of the control, which may be nonzero.

Impulse

Perhaps the simplest of the inputs used for aircraft system identification is the impulse input. This is simply a spike or sudden bump, often called a stick rap. Sometimes the impulse input is two-sided, to help return the aircraft to the starting flight condition. Figure 8.7 shows a two-sided impulse input and its power spectrum. The power in the impulse is wideband, but low amplitude. Theoretically, the impulse could be used to collect modeling data when there is no prior information about the aircraft dynamics, but this is often impractical because of low input spectral density. Impulse inputs are best suited for prediction cases.

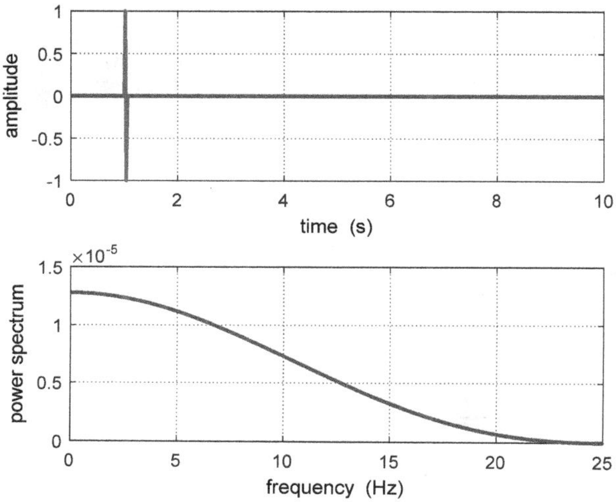

Figure 8.7 **Impulse input**

Frequency Sweeps

A more commonly used input type when there is little or no prior information about the dynamic system is the frequency sweep, shown in Fig. 8.8 (a), along with its power spectrum. The idea behind this input is to apply a sinusoid with frequency increasing slowly and continuously with time, so that the frequency content of the input covers a frequency band of interest.

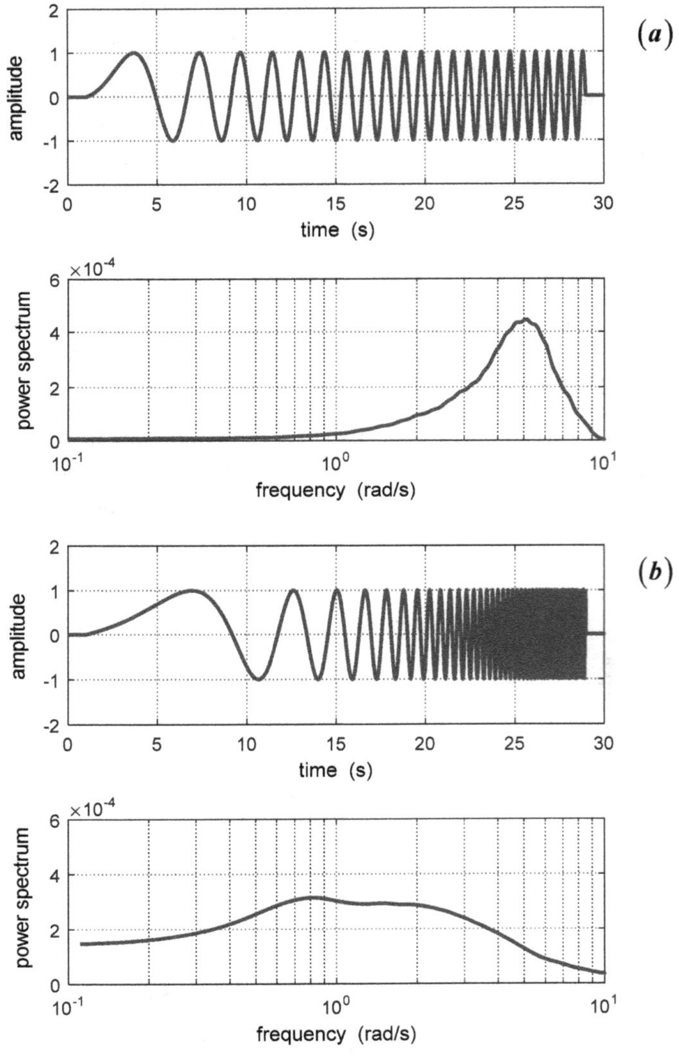

Figure 8.8 **Frequency sweep inputs: (a) linear, (b) logarithmic**

The linear frequency sweep can be described mathematically by

$$u(i) = sin\left[\varphi(i)\right] \qquad i = 0,1,2,\ldots,N-1 \qquad (8.20a)$$

$$\varphi(i) = \omega_0 t(i) + \frac{1}{2}(\omega_1 - \omega_0)\frac{\left[t(i)\right]^2}{T} \qquad i = 0,1,2,\ldots,N-1 \qquad (8.20b)$$

where $t(i) = i\Delta t$, T is the maneuver time, $T = (N-1)\Delta t$, and $[\omega_0, \omega_1]$ rad/s is the frequency band. The sine function is used so that the input is a perturbation that begins and ends at zero. Low frequencies are applied first, to allow time for the dynamic system response to the lower frequency input.

Frequency sweeps are often used to collect data for frequency response estimation, as shown by the Bode plot in Fig. 8.9. The abscissa of the Bode plot is a logarithmic scale, so that using the linear frequency sweep defined in Eq. (8.20) gives sparse frequency content at the lower frequencies. This situation can be remedied by using the logarithmic frequency sweep, which can be implemented using Eq. (8.20a) with

$$\omega(i) = 10^{log_{10}\omega_o + (log_{10}\omega_1 - log_{10}\omega_0)\frac{t(i)}{T}} \qquad i = 0,1,2,\ldots,N-1 \qquad (8.21a)$$

$$\varphi(i) = \omega(i)\,t(i) \qquad i = 0,1,2,\ldots,N-1 \qquad (8.21b)$$

which varies the sweep frequency on a logarithmic scale. Figure 8.8(b) shows a logarithmic frequency sweep and its power spectrum. Other functions can also be used to vary the frequency with time [8.8].

The frequency sweep input should move slowly through the frequency band of interest, and therefore can require a relatively long time, particularly if low frequency excitation is included. For a typical pilot input frequency band between 0.1 rad/s and 10 rad/s, each frequency sweep input might require 60-90 seconds. An example of a piloted frequency sweep applied to the longitudinal control of the Tu-144LL supersonic aircraft (cf. Fig. 7.8) is shown in Fig. 8.10.

Researchers at NASA Ames have extensive experience in frequency sweep flight testing and modeling in the frequency domain based on frequency response estimates, particularly for rotorcraft and V/STOL aircraft. References [8.8]-[8.12] contain detailed design guidelines for frequency sweep inputs, along with many practical examples.

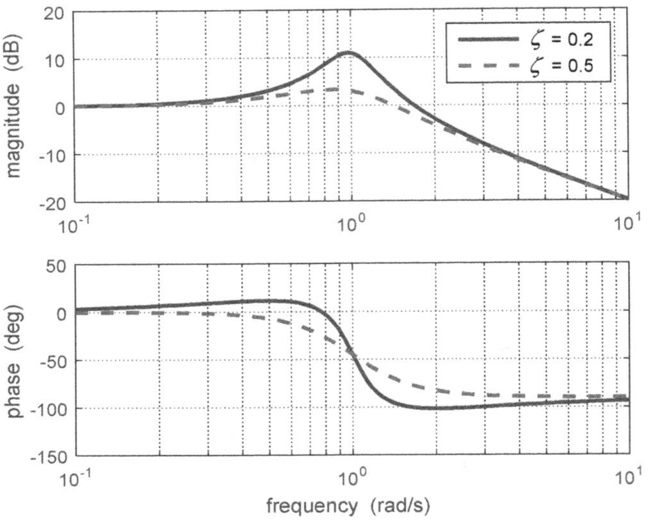

Figure 8.9 **Frequency response for the second order system:**

$$\tilde{y}(s)/\tilde{u}(s) = (s+1)/\left(s^2 + 2\zeta + 1\right)$$

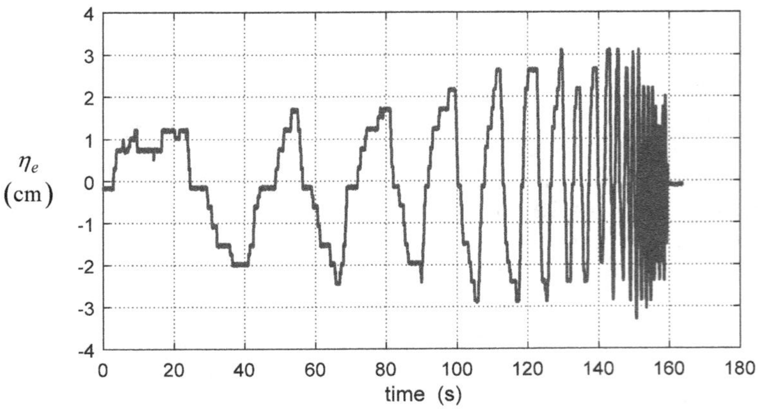

Figure 8.10 **Piloted longitudinal frequency sweep input on the Tu-144LL supersonic aircraft**

It is not important for frequency sweeps to have constant input amplitude, exact input shape, exact frequency progression, exact repeatability, or sustained high frequency inputs. At high frequencies, numerous cycles of excitation can be applied relatively quickly, so the corresponding dwell time required is small.

Modeling results actually improve if the inputs for each run have some variety in shape and frequency, because this enhances the information content in the data. Consequently, computerized frequency sweeps have been found to be inferior to piloted frequency sweeps, because the computer does not introduce variations in the input from run to run. Random noise can be added to a computer implementation of the frequency sweep to simulate human pilot variations, but this represents a good deal of complexity to do something that a pilot does naturally. In any case, additional control inputs are typically applied to suppress off-axis response (e.g., roll response during longitudinal control sweeps) and to maintain the nominal flight condition during the frequency sweep. A good pilot can usually do this.

There are several problems with frequency sweep inputs that sometimes render them unusable for aircraft system identification. One problem is that the sweeps generally take a long time to implement, which rules them out for very short flight test times, such as drop model testing and flight testing at high angles of attack or hypersonic speeds. Often it is difficult to maintain flight condition in the low-frequency part of the sweep at the start of the maneuver. To handle this, the low frequency input amplitudes must be reduced considerably, which lowers signal-to-noise ratio. Frequency sweeps are applied to one input at a time, so they take more time and are less effective for a modeling problem with more than one input. For many aircraft, both the longitudinal and lateral dynamic modeling are multiple input problems. The single-input characteristic also makes frequency sweeps ineffective for investigating control interaction effects. Finally, the sweep input might inadvertently pass through a structural resonance frequency, which can compromise flight safety.

The main advantage of frequency sweep inputs is comprehensive coverage of the frequency band, so that models identified from frequency sweeps typically have very good prediction capability.

Multisines

A multisine input is a sum of sinusoids with various frequencies, amplitudes, and phase angles. The frequencies are chosen to cover a frequency band of interest, similar to frequency sweeps, and the amplitudes are chosen to achieve a specific power distribution over the frequency band. Phase angles can be chosen arbitrarily, and are sometimes set to random values in the interval $[-\pi,\pi]$ rad.

One type of multisine input that has been found very useful in aircraft system identification is composed of summed harmonic sinusoids with individual phase lags. The input takes the form

$$u(i) = \sum_{k=1}^{M} A_k \, sin\left(\frac{2\pi k\, t(i)}{T} + \phi_k\right) \qquad i = 0,1,2,\ldots,N-1 \quad (8.22)$$

where M is the number of harmonically-related frequencies, T is the time length of the excitation, and the ϕ_k are phase angles for each of the harmonic components. The ϕ_k can be chosen to produce a low peak factor PF, defined by

$$PF(\boldsymbol{u}) \equiv \frac{\left[max(\boldsymbol{u}) - min(\boldsymbol{u})\right]/2}{\sqrt{\left(\boldsymbol{u}^T \boldsymbol{u}\right)/N}} \tag{8.23}$$

where $\boldsymbol{u} = \left[u_0, u_1, \ldots, u_{N-1}\right]^T$, or

$$PF(\boldsymbol{u}) = \frac{\left[max(\boldsymbol{u}) - min(\boldsymbol{u})\right]/2}{rms(\boldsymbol{u})} \tag{8.24}$$

A single sinusoidal component from the summation in Eq. (8.22) has $PF = \sqrt{2}$, so the relative peak factor *RPF*, defined by

$$RPF(\boldsymbol{u}) = \frac{\left[max(\boldsymbol{u}) - min(\boldsymbol{u})\right]}{2\sqrt{2}\; rms(\boldsymbol{u})} = \frac{PF(\boldsymbol{u})}{\sqrt{2}} \tag{8.25}$$

quantifies the peak factor of $\boldsymbol{u}$ relative to the peak factor of a single sinusoid. For a single sinusoid, *RPF* equals 1.

The relative peak factor is the input amplitude range divided by a measure of the input energy, which can be interpreted as a measure of the efficiency of an input for parameter estimation purposes. Lower relative peak factors are more desirable for parameter estimation, where the objective is to excite the system with a variety of frequencies without driving it too far away from the nominal operating point. An input with low relative peak factors is efficient in the sense of providing good input energy over a selected frequency band, with low amplitude in the time domain.

The sinusoidal components in Eq. (8.22) can be assigned arbitrary fractions of the total power in the input, to emphasize the excitation at selected frequencies. This is implemented by choosing component amplitudes A_k as

$$A_k = A\sqrt{P_k} \qquad \forall\, k \tag{8.26}$$

where A is the amplitude of the multisine input, and P_k is the power fraction for the *k*th sinusoidal component, so that

$$\sum_{k=1}^{M} P_k = 1 \tag{8.27}$$

This feature can be used to concentrate excitation at frequencies near where the system dynamics are expected to lie. For uniform power distribution,

$$P_k = 1/M \qquad k = 1,2,\ldots,M \tag{8.28}$$

Schroeder [8.13] has shown that a phase-shifted sum of sinusoids, commonly called the Schroeder sweep, provides an input with good frequency content and low peak factor. For a uniform power spectrum, the Schroeder sweep input results from Eq. (8.22) with

$$\begin{aligned} &A_k = A\sqrt{P_k} \qquad P_k = 1/M \qquad k = 1,2,\ldots,M \\ &\phi_1 = 0 \\ &\phi_k = \phi_{k-1} - \frac{\pi k^2}{M} \qquad k = 2,3,\ldots,M \end{aligned} \tag{8.29}$$

where A is the amplitude of the multisine input, and P_k is the power fraction for the kth sinusoidal component.

Comparisons of the Schroeder sweep with linear and logarithmic frequency sweep inputs have indicated that the Schroeder sweep is generally the superior input for dynamic modeling in the frequency domain [8.14]. The Schroeder sweep has been used successfully in practical aircraft system identification problems [8.15]-[8.16].

Using the Schroeder sweep phase angles from Eq. (8.29) results in an input with low relative peak factor, but the input often does not begin and end at zero, as required for a perturbation input. However, the initial phase angle ϕ_1 is arbitrary [set to zero in Eq. (8.29)], so ϕ_1 can be adjusted to achieve an input that begins and ends at zero. This works because the component sinusoids in Eq. (8.22) are harmonics on the time interval $[0, T]$.

The peak factor of a multisine input can be lowered by using a simplex optimization algorithm [8.17] to adjust the component phase angles ϕ_k to minimize the relative peak factor. The optimization problem is nonconvex, as indicated in Fig. 8.11, which shows the variation of relative peak factor with phase angle choices for a multisine input with two component frequencies, i.e., $M = 2$ in Eq. (8.22).

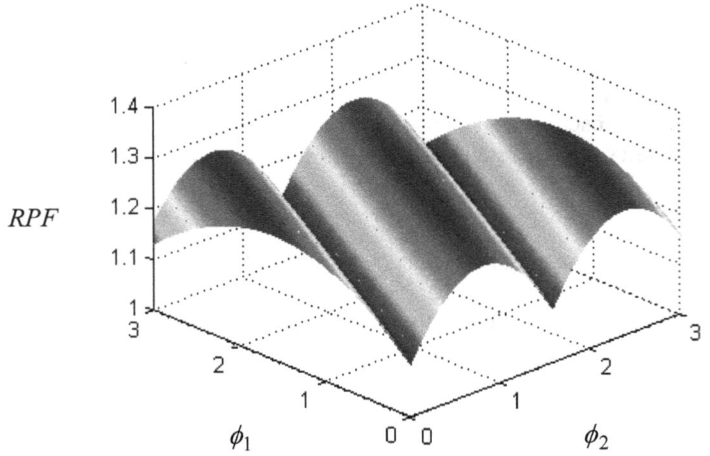

Figure 8.11 **Relative peak factors for a two-component multisine input**

Figure 8.11 also shows that there are several phase angle solutions which are equally good in terms of low *RPF*, or nearly so. This approach to optimizing multisine inputs based on minimum *RPF* is described in Refs. [8.18] and [8.19], for both single-input and multiple-input cases.

To implement the optimized multisine input as a perturbation input, a one-dimensional search is used to find a time offset so that the input begins and ends at zero amplitude. This is equivalent to sliding the input along the time axis until a zero crossing occurs at the origin of the time axis. To implement the same time shift to all of the components, the phase offset for each component must be different, because each component has a different frequency. The phase offset $\Delta\phi_k$ to be added to the optimized phase angle ϕ_k for each sinusoidal component is

$$\Delta\phi_k = \omega_k \, \tau \tag{8.30}$$

where τ is the time shift needed to start the optimized multisine input at zero. Since the sinusoidal components are all harmonics of the base frequency with period T, if the initial value of the input is zero, then the final value of the input at time T will also be zero. The input power spectrum and relative peak factor are unaffected by the phase shifts made to implement a perturbation input.

The process used to produce a perturbation input for the Schroeder sweep is different than for the optimized multisine, because the component phase angles for a Schroeder sweep are predefined, whereas the component phase angles for the optimized multisine are optimized for minimum relative peak factor.

Note that cosine functions can be used in the summation of Eq. (8.22) instead of sine functions, with only an adjustment of the phase angles ϕ_k.

Doublets and Multisteps

Doublet inputs are two-sided pulses, as shown in Fig. 8.12. This input could be viewed as a square wave approximation to a sine wave. For the doublet shown in Fig. 8.12, the dominant frequency would be 0.5 Hz, corresponding to the frequency of a sine wave with the same period. The power spectrum of the doublet shown in Fig. 8.12 is shifted slightly low in frequency, because only a single isolated doublet is used, rather than a train of repeated doublets. The timing of the pulses for a doublet is chosen so that the dominant frequency in the input is at or close to the expected natural frequency of the dynamic system.

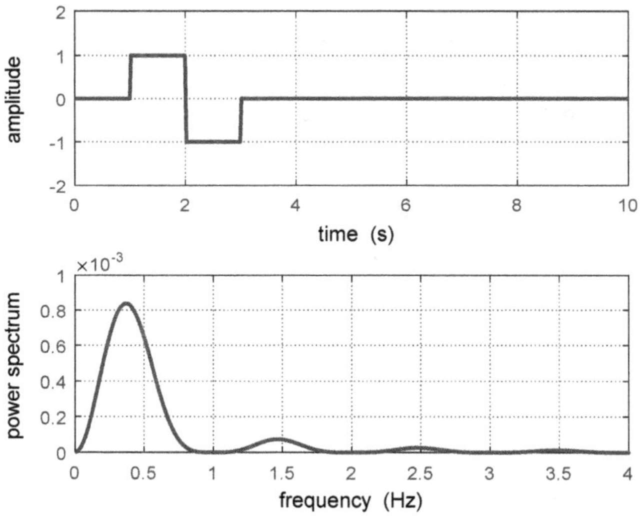

Figure 8.12 **Doublet input**

As an example, assume that the dynamic system can be described by a transfer function with a second-order denominator and a first-order numerator, as shown in the frequency response of Fig. 8.9. This situation exists for the transfer function of pitch rate to elevator deflection using the short-period approximation (see Chapter 3). A frequency sweep for this dynamic system would be designed to contain a wide spectrum of frequencies that includes the expected natural frequency of the system. In contrast, the logic for a doublet input would proceed roughly as follows. The static gain is an easy parameter to estimate, and almost any input will suffice to provide information for that parameter. The system is assumed to be second order, which means that at low and high frequencies relative to the natural frequency, the phase and amplitude characteristics are known, because all second-order systems look similar (except for the static gain) at these frequencies. The natural frequency and damping will be determined most accurately using an input containing frequencies in a band around where the natural frequency of the dynamic system is expected to be. This input does the most good in terms of maximizing information in the data because this is the frequency range where dynamic response of second-order systems differ according to the values of the damping and natural frequency. An input with frequencies in the range of the natural frequency of the system therefore collects the required information in a short length of test time.

The lower plot in Fig. 8.12 shows the power spectrum of the doublet input, with the dominant frequency evident. For square wave inputs such as the doublet, it is straightforward to compute the frequency spectrum analytically, as follows.

Consider the doublet to be composed of two individual pulses, with amplitudes of opposite sign. Each pulse can be considered the sum of two step functions. For a pulse u_k with amplitude A_k on the interval $\left[t_{1_k}, t_{2_k}\right]$, the Fourier transform is:

$$\tilde{u}_k(\omega) = \int_0^\infty A_k \left[\mathbf{1}\left(t - t_{1_k}\right) - \mathbf{1}\left(t - t_{2_k}\right)\right] e^{-j\omega t}\, dt \tag{8.31a}$$

$$= A_k \left(-\frac{1}{j\omega} e^{-j\omega t}\Big|_{t_{1_k}}^{\infty} + \frac{1}{j\omega} e^{-j\omega t}\Big|_{t_{2_k}}^{\infty} \right)$$

$$\tilde{u}_k(\omega) = \frac{A_k}{j\omega}\left(e^{-j\omega\, t_{1_k}} - e^{-j\omega\, t_{2_k}}\right) \tag{8.31b}$$

where $\mathbf{1}\left(t - t_{1_k}\right)$ indicates a unit step function that initiates at time t_{1_k}.

The Fourier transform for a square-wave input is the sum of terms like Eq. (8.31b) for each pulse, at each nonzero frequency ω.

For $\omega = 0$,

$$\tilde{u}_k(0) = A_k\left(t_{2_k} - t_{1_k}\right) \tag{8.31c}$$

If there are m pulses, then the Fourier transform for the square-wave input is

$$\tilde{u}(\omega) = \frac{1}{j\omega} \sum_{k=1}^{m} A_k \left(e^{-j\omega t_{1_k}} - e^{-j\omega t_{2_k}} \right)$$
$$\tilde{u}(0) = \sum_{k=1}^{m} A_k \left(t_{2_k} - t_{1_k} \right) \tag{8.32}$$

The one-sided power spectral density is computed from the squared magnitude of the Fourier transform for the input, as discussed in Chapter 7,

$$G_{uu}(\omega) = 2\, \tilde{u}^*(\omega) \tilde{u}(\omega) \tag{8.33}$$

Figure 8.12 shows that the square edges of the doublet input produce a wider frequency spectrum than the single spike that would be obtained with a pure sine wave at the same basic frequency. The square wave form therefore serves a useful purpose for inputs intended for system identification, in that the spectrum of the input is broadened, which is a hedge against an inaccurate *a priori* estimate of the modal frequency. This consideration is the reason that a single-frequency sinusoid makes a poor input for system identification.

The simplicity and initial sharp edge of the doublet input also makes it very practical for checking polarity and gains, and for identifying time skews in the measured data.

The amplitude of the doublet input is chosen so that the amplitude of the dynamic response output is large enough for good signal-to-noise ratio, but not so large that the model structure assumption is violated or the model parameters can no longer be considered constant. This can generally be achieved by enforcing the output amplitude constraints mentioned earlier.

Taking the preceding concepts further, more pulses of different widths can be added. A common input for aircraft system identification is called the 3-2-1-1 [8.20]-[8.22], which is a multistep input consisting of alternating pulses with widths in the ratio 3-2-1-1. Figure 8.13 shows an example of this input, along with its power spectrum, computed analytically in the manner described earlier. The width of the 2 pulse is selected to correspond to half the period of the expected natural frequency of the dominant dynamic mode. The 3 and 1 pulses then bracket that frequency on either side, resulting in a relatively wideband input. This input has been called a “poor man's frequency sweep” because of the increasing frequency of the square waves that is analogous to the increasing frequency of a sinusoidal frequency sweep. Comparing Figs. 8.12 and 8.13, the 3-2-1-1 is seen to have much richer frequency content than a doublet.

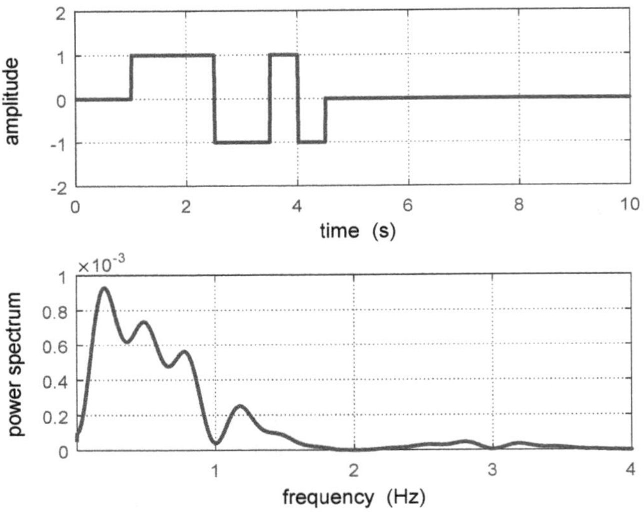

Figure 8.13 **3-2-1-1 input**

The 3-2-1-1 input is sometimes difficult to use because the 3 pulse is relatively long and tends to drive the aircraft off flight condition, in a manner similar to what can happen with a frequency sweep. To address this, the 3 pulse can be reduced in amplitude, or a 2-1-1 input can be used instead.

Figure 8.14 shows an example of the 2-1-1 input and its associated power spectrum. Pulse widths are selected in the ratio 2-1-1. The pulse widths are chosen so that the associated frequencies bracket the expected natural frequency of the dominant dynamic mode to be identified. The following choice for the 1 pulse width has been found to work well:

$$\text{1 pulse width } = \frac{0.7}{2f_n} \tag{8.34}$$

where f_n is the expected natural frequency in Hz for the dynamic mode.

Using Eqs. (8.32) and (8.33), it is possible to investigate the power spectra of arbitrary input square waves with arbitrary amplitudes, or to design a square-wave input that closely approximates an arbitrary input power spectrum, which was done in Ref. [8.22]. Allowing different input amplitudes for each pulse expands the possibilities for input spectra using a multistep input form. The more difficult problem is deciding what the input power spectrum should be. Approximating the chosen input spectra using square wave inputs is relatively straightforward.

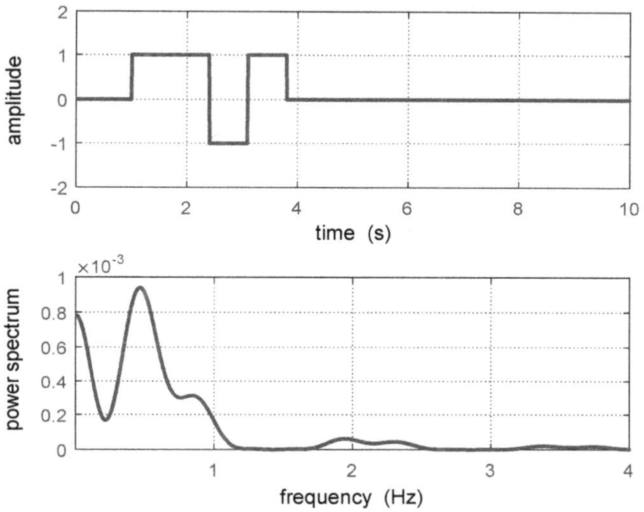

Figure 8.14 **2-1-1 input**

Another practical approach to input design is using a sequence of doublets with different duration. This provides rich frequency content using the simple doublet input form. Figure 8.15 shows a sequence of doublets with different basic frequencies, chosen to bracket the expected natural frequency of the dynamic mode to be modeled. The power spectrum plot in Fig. 8.15 shows that this approach can provide more input power over a widened frequency band, compared to a simple doublet input (cf. Fig. 8.12).

Pilot Inputs

A good test pilot can very efficiently excite the dynamic response of the aircraft. The key to success in this approach is to make sure the pilot fully understands what is desired and why. Once this is achieved, the pilot's ability to fly the aircraft precisely, and to sense and control the aircraft response, can result in excellent inputs for aircraft system identification.

In addition, the adaptability of a human pilot is a great advantage in flight testing for aircraft system identification. For example, if an input is designed prior to flight based on wind tunnel data, it is frequently true that the amplitude and/or frequency content are not quite right for the real aircraft in flight. A pilot can make adjustments for this, and many other eventualities that cannot be predicted prior to flight. This issue is particularly important because most of the time and money necessary to do flight tests are spent getting the aircraft in the air, for tasks such as installing instrumentation, check-out, and safety reviews, and not on the operational costs of the aircraft in flight, such as fuel. Once airborne, the best approach is to do some type of testing, even if it's not optimal or exactly what was planned. A pilot who understands the goals of the experiment can adapt the flight-test plan in flight and collect useful flight data.

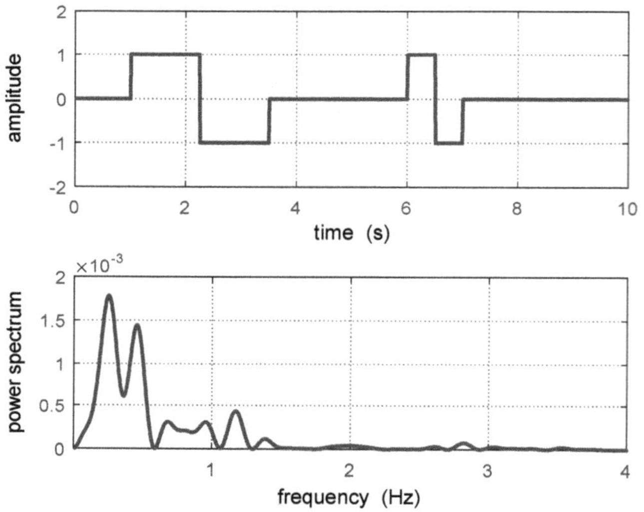

Figure 8.15 **Compound doublet input**

If the goal is closed-loop modeling, then pilot stick and rudder pedal inputs must be used. However, when the model being identified is the open-loop or bare-airframe model, implementing inputs from the cockpit has limitations, in that sometimes closed-loop feedback control systems move control surfaces in proportion to aircraft states, or move more than one surface in a proportional manner, resulting in data collinearity (see Chapter 5). Often the feedback control system cannot be turned off, because the aircraft is open-loop unstable. Approaches to solving this problem include applying computerized movement of individual surfaces (discussed later in this chapter), which addresses the problem by modifying the experiment, and identifying equivalent or combined parameters or using other data sources to augment the flight data (see Chapter 5), which addresses the problem through the modeling process.

Pilots have successfully implemented impulses, frequency sweeps, doublets, and multistep inputs on many different aircraft. As mentioned earlier, distortions in the desired input form by the human pilot implementation usually help the identification by augmenting the range of frequencies and amplitudes applied to the dynamic system. For multisine inputs, a computerized implementation is typically required; however, it has been shown that pilots can implement very effective approximations to optimized multisine inputs [8.23]-[8.26].

Other Input Types

Some specialized inputs that are useful for specific purposes include the longitudinal push-over, pull-up (essentially a low-frequency longitudinal doublet), which is used for lift and drag performance characterization over a relatively large range in angle of attack, and the wind-up turn (a descending spiral with increasing bank angle and aft stick to steadily increase the angle of attack and decrease the turn radius), which is used for data system check-out and data compatibility analysis.

There are other types of inputs for system identification that have been discussed in the literature, such as white noise inputs and pseudo-random binary sequences. The latter input could be described as a long multistep input with constant amplitude and randomly varying pulse widths. Neither of these input types have been widely used in aircraft system identification, mainly because their efficiency is low, and the cost of flight time is very high.

8.3.4 Multiple-Input Design

Many practical aircraft modeling problems involve multiple inputs. The most common is aircraft lateral dynamic modeling, which involves rudder and aileron inputs, or equivalent yawing and rolling moment controls. Modern aircraft have many control effectors for both longitudinal and lateral control, so the multiple-input problem appears often.

In the multiple-input case, there are important considerations that are not present for the single-input case: relative effectiveness, coordination, interaction effects, and correlation. The discussion here is focused on inputs for bare-airframe model identification, which are typically control surface deflections. The material also applies for closed-loop lateral modeling, which involves lateral stick and rudder pedal inputs.

Relative Effectiveness

If an input is moved very little or not at all, there is little or no information on the effect of that input on the aircraft response, and corresponding model parameters cannot be estimated. A control with large effectiveness need not be moved as much as a control with lower effectiveness to produce the same magnitude change in the aerodynamic forces and moments. Generally, the appropriate input amplitudes for a multiple-input maneuver design are estimated based on prior information from wind tunnel tests or previous flight test experience on the same or similar aircraft. Sometimes the input amplitudes are simply set low initially, then increased as needed, based on flight data. For pilot inputs, the test pilot can adjust the input amplitudes in flight to achieve the desired magnitudes of variation in the response variables.

Coordination

Multiple inputs must be coordinated to maximize data information content and to make sure the aircraft responses do not exceed the limits for model structure validity. As a simple example, if two rolling moment control effectors are moved sequentially in the same way, it is easy to exceed roll rate and roll angle limits. In addition, coordination of the inputs can improve the maneuver by taking advantage of dynamic coupling and the natural aircraft motion. The most common example is lateral modeling using a linear model structure and doublet inputs on the rudder and aileron. Applying the rudder doublet first gets the Dutch roll motion of the aircraft started early in the maneuver, so that more data will be collected for this relatively slower dynamic mode by the end of the maneuver. When the aileron doublet is applied subsequently, the roll mode is excited, with some additional excitation of the Dutch roll mode. Since the roll mode is relatively fast, it can be excited later in the maneuver, because less time is required to characterize it. Sequencing the rudder and aileron doublets in this way, with a spacing near the period of the Dutch roll mode, produces a better maneuver than is obtained when applying the aileron doublet first, followed by the rudder doublet.

Interaction Effects

Control effectiveness can be influenced by the position and movement of other controls located in close proximity. Some aircraft have multiple trailing-edge control surfaces on the wing, for example. These controls must be moved simultaneously using uncorrelated multiple-input excitation to provide data for the modeling algorithms to quantify the interaction effects. An approximate alternative is to fix one or more controls at various values while moving the others, but this is much less efficient, and does not include dynamic interaction effects. If the flight data are generated by moving only one control at a time, there is no way to model or quantify the control interaction effects.

Correlation

Input correlation refers to the similarity of the waveforms for multiple inputs. If all of the input waveforms look the same, then any algorithm trying to assign values for the effectiveness of each individual control will fail, because it is impossible to determine which of the multiple inputs, moved in the same manner, was responsible for the changes in the aerodynamic forces and moments. The preceding statement also applies when the input waveforms are scaled versions of one another. Input forms that are completely decorrelated (i.e., orthogonal) will give the most accurate control effectiveness estimates, all other things being equal.

It is common for an automatic control system to move two control surfaces in a proportional manner, bringing about nearly exact linear correlation between the control surface deflections. This is usually done to improve control authority. In this case, the modeling must be done by introducing prior information, such as fixing the effectiveness of all but one of the correlated control surfaces to *a priori* values, or defining a fixed ratio for the control effectiveness of the correlated control surfaces relative to a selected one. Another approach is to estimate a combined control effectiveness, which essentially treats all of the correlated control surfaces as if they were a single control surface. However, if the control system is subsequently changed so that the scheduled movements of the correlated control surfaces is different (not uncommon in development programs), then the results for the combined control effectiveness become useless. However, it is possible to estimate individual control surface effectiveness from a combined set of data, if the ratio of the correlated control surface movements was different for different maneuvers or for different parts of the data.

Mathematically, the correlation between two input vectors $\boldsymbol{u}_1$ and $\boldsymbol{u}_2$ can be quantified by the pairwise correlation coefficient for scaled and centered regressors, introduced in Chapter 5, Eq. (5.162), which is a normalized inner product over the length of the maneuver.

When the experimenter has complete control of the input forms, multiple inputs can be completely decorrelated by skewing the inputs in time or by using orthogonal waveforms. Figure 8.16 shows doublet inputs skewed in time, which makes their inner product zero by Eq. (5.162).

Figure 8.17 shows orthogonal square waves that also have inner products equal to zero by Eq. (5.162). These square wave analogs of harmonic sinusoids are sometimes called Walsh functions.

Multiple multisine inputs have the interesting and very useful property of being mutually orthogonal in the time domain and the frequency domain simultaneously, as described in the next subsection.

It is more efficient in terms of minimizing maneuver time to move multiple control surfaces simultaneously using orthogonal waveforms or inputs with low pairwise correlations. When a feedback control system is operating, desired input forms become distorted by the feedback control. This makes the resulting control surface deflections nonorthogonal, but if the inputs for parameter estimation are moving simultaneously and are mutually orthogonal, it takes an extremely high-gain feedback control system to cause the control surface deflections to be correlated to the point where the modeling results are compromised. If only time skew is being used to implement input orthogonality, high-gain feedback can ruin the orthogonality completely, resulting in correlated aircraft states and/or controls.

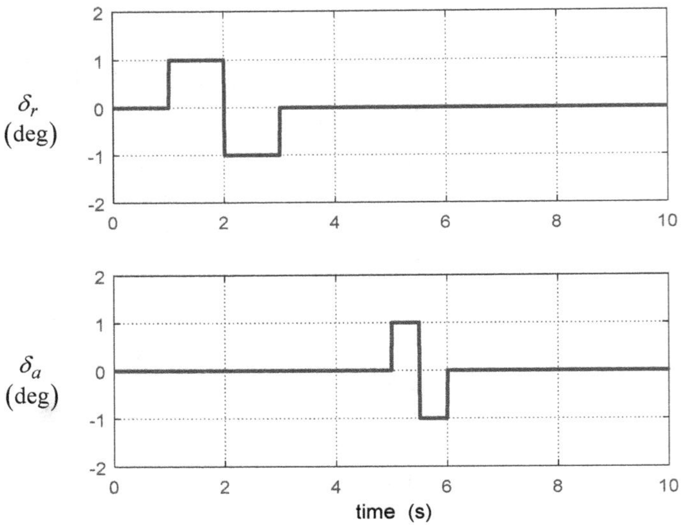

Figure 8.16 **Time-skewed doublet inputs**

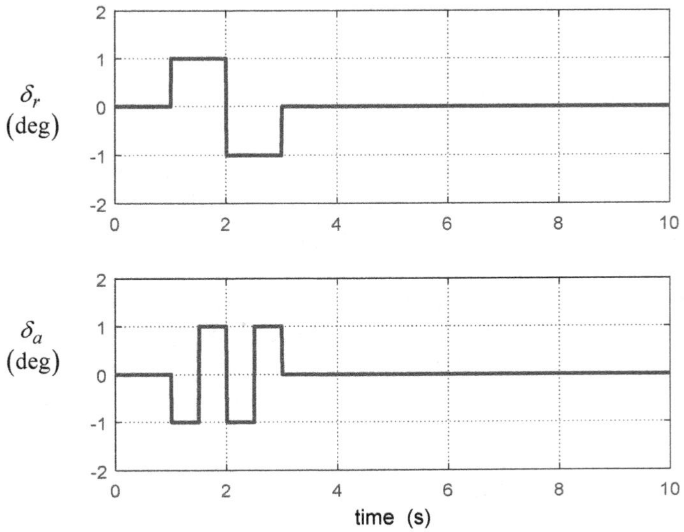

Figure 8.17 **Orthogonal square wave inputs**

Orthogonal Optimized Multisines

Multiple-input design using orthogonal optimized multisines was first developed in Ref. [8.18], and has been applied successfully in many practical aircraft flight tests and wind tunnel tests [8.18]-[8.19], [8.27]-[8.41]. The material in this subsection is based on these references.

The general idea is to excite the aircraft using perturbation inputs with wideband frequency content over a range of frequencies that encompasses the expected modal frequencies of the aircraft dynamic response. The excitations for open-loop or bare-airframe modeling are implemented as perturbations to the control surface deflections by summing designed excitation inputs with the actuator commands from the pilot and feedback control system, just before the actuator command rate and position limiting, as shown in Fig. 8.18.

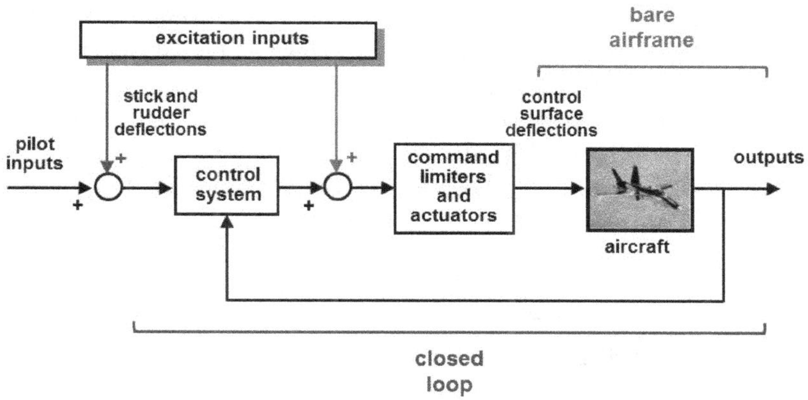

Figure 8.18 **Applying excitation inputs**

Each designed excitation input is a sum of sinusoids with unique frequencies, optimized phase shifts, and specified power distribution. Component frequencies are selected to cover a frequency band of interest, similar to frequency sweeps. The wideband frequency content of the inputs is important because there is naturally some uncertainty in the knowledge of the modal frequencies for the aircraft in flight. Wideband inputs provide robustness to this uncertainty. Phase shifts for the sinusoidal components of each input are optimized to achieve low peak-to-peak amplitude and high input energy content for the sum of sinusoids using the relative peak factor *RPF* from Eq. (8.25). Power fraction for the individual sinusoidal components can be chosen to target specific frequencies or frequency bands.

Multiple inputs are designed to be mutually orthogonal in both the time domain and the frequency domain, and are optimized for maximum data information content in multiple axes over a short time period, while minimizing excursions from the nominal flight condition. These characteristics make the input design very effective for stability and control flight testing. The mutual orthogonality of the inputs allows simultaneous application of multiple inputs, which helps to minimize excitation time, and can provide continuous multi-axis excitation as the aircraft flies through time-varying or precarious flight conditions.

Similarly to the single-input design, each input $\boldsymbol{u}_j$, which is to be applied to the *j*th individual control surface, is composed of a set of summed harmonic sinusoids with individual phase shifts ϕ_k,

$$\boldsymbol{u}_j = \sum_{k \in \{1,2,\ldots,M\}} A_k \sin\left(\frac{2\pi k \boldsymbol{t}}{T} + \phi_k\right) \qquad j = 1,2,\ldots,m \tag{8.35}$$

where M is now the total number of available harmonically-related frequencies, $\boldsymbol{t}$ is the time vector, T is the time length of the excitation, and A_k is the amplitude for the *k*th sinusoidal component. Because there are multiple inputs, each of the m inputs includes selected components from the pool of M harmonic sinusoids with frequencies $\omega_k = 2\pi k/T$, $k = 1,2,\ldots,M$, where $\omega_M = 2\pi M/T$ represents the upper limit of the frequency band for the excitation inputs. The interval $[\omega_1, \omega_M]$ rad/s specifies the range of frequencies where the aircraft dynamics are expected to lie.

If the phase angles ϕ_k in Eq. (8.35) were chosen at random on the interval $[-\pi, \pi]$ rad, then in general, the various harmonic components would add together at some points to produce inputs $\boldsymbol{u}_j$ with relatively large amplitude excursions. This is undesirable, because it can result in the dynamic system being moved too far from the reference condition selected for the experiment. To prevent this, the phase angles ϕ_k for the selected harmonic components are chosen to minimize relative peak factor RPF, defined by Eq. (8.25), for each input $\boldsymbol{u}_j$ individually. As discussed earlier, low relative peak factors are desirable and efficient for estimating dynamic model parameters, because the objective is to excite the dynamic system with good input energy over a range of frequencies while minimizing the input amplitudes in the time domain, to avoid driving the dynamic system too far away from the reference condition.

The integers k specifying the frequencies for the *j*th input $\boldsymbol{u}_j$ are selected to be unique to that input, but are not necessarily consecutive. A good approach for multiple inputs is to assign integers k to each input alternately. This is illustrated in Fig. 8.19 for a flight test maneuver design on the T-2 subscale jet transport aircraft shown in Fig. 8.20. In this case, there were 3 inputs: elevator, rudder, and aileron, and a total of 21 harmonic frequencies

$(M = 21)$ for a maneuver time length $T = 10$ sec. The frequencies were interleaved among the three inputs to achieve wideband frequency content for each input. This provided robustness to uncertainty in how each control excites the dynamic modes of the aircraft. Because each input has wideband frequency content, the same input design can be applied at various flight conditions, which simplifies the flight test and reduces flight computer memory requirements. It is even possible to use the same input design for different aircraft, because of the wideband frequency content of the excitation inputs. Table 8.3 and Fig. 8.19 contain the information needed to assemble the excitation inputs using Eq. (8.35). The resulting excitation inputs are shown in Fig. 8.21.

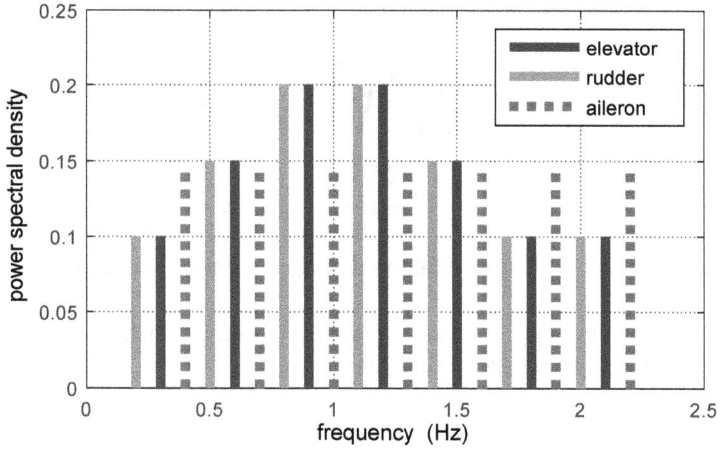

Figure 8.19 **Multiple orthogonal phase-optimized multisine input spectra**

Figure 8.20 **T-2 subscale jet transport aircraft in flight**

Credit: NASA Langley Research Center

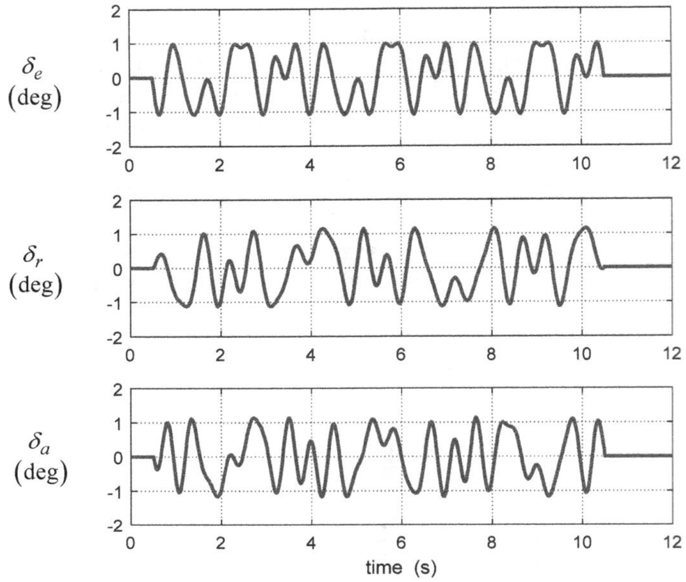

Figure 8.21 **Multiple orthogonal optimized multisine inputs**

In the same manner as for the single-input design, to achieve a uniform power distribution, the A_k are selected as

$$A_k = \frac{A}{\sqrt{n}} \qquad \forall\, k \tag{8.36}$$

where n is the number of sinusoidal components included in the summation of Eq. (8.35) for $\boldsymbol{u}_j$, and A is the amplitude of the composite input $\boldsymbol{u}_j$. Therefore, with uniform power distribution, selection of the A_k reduces to selecting a single value for the input amplitude A. Each input $\boldsymbol{u}_j$ can of course have arbitrary amplitude A, subject to practical flight testing and modeling constraints.

For each input, the power spectrum can be tailored by selecting the A_k in Eq. (8.35) to distribute power over the spectral components. This can focus the excitation on frequencies near where the natural frequencies of the dynamic modes are believed to be, or to avoid exciting structural responses, for example. In Fig. 8.19, the power spectra for elevator and rudder were modified so that more excitation power was applied at middle frequencies where the natural frequencies of the dynamic modes excited by these controls were believed to be. The power spectrum shown in Fig. 8.19 is normalized, so the effects of individual control surface amplitudes are excluded, and the plot shows the power fraction for each sinusoidal component. For each input, the sum of all the spectral line ordinates (sum of the heights of the bars for each input) equals 1.

***Table 8.3* Multiple-input orthogonal optimized multisine design**

Input	A (deg)	A_k (deg)	k	ϕ_k (rad)	RPF
δ_e	2.0	0.6325	3	2.9478	1.03
		0.7746	6	0.6008	
		0.8944	9	−2.6991	
		0.8944	12	−1.6517	
		0.7746	15	2.6902	
		0.6325	18	2.0873	
		0.6325	21	−2.8619	
δ_r	2.0	0.6325	2	2.8435	1.14
		0.7746	5	2.5259	
		0.8944	8	2.7562	
		0.8944	11	−0.5132	
		0.7746	14	−0.7433	
		0.6325	17	2.3959	
		0.6325	20	−0.7581	
δ_a	1.0	0.3780	4	1.5438	1.15
		0.3780	7	−1.6413	
		0.3780	10	1.2011	
		0.3780	13	1.0767	
		0.3780	16	−2.3373	
		0.3780	19	−2.3327	
		0.3780	22	−2.7602	

When the frequency indices k selected for each input $\boldsymbol{u}_j$ in Eq. (8.35) are distinct from those chosen for the other inputs, then the frequency content of each $\boldsymbol{u}_j$ consists of distinct spectral lines in the frequency domain, as can be seen in Fig. 8.19. Therefore, the vectors of Fourier transforms for the inputs as a function of frequency have inner products equal to zero. In this sense, the inputs are mutually orthogonal in the frequency domain, because each input contains frequencies that are not present in the other inputs.

In the time domain, a sum of harmonic sinusoids is orthogonal to any other sum of harmonic sinusoids with different harmonic frequencies, regardless of the constant phase shift of each sinusoidal component. By trigonometric identity,

$$sin\left(\frac{2\pi k \boldsymbol{t}}{T}+\phi_k\right)=sin\left(\frac{2\pi k \boldsymbol{t}}{T}\right)cos\left(\phi_k\right)+cos\left(\frac{2\pi k \boldsymbol{t}}{T}\right)sin\left(\phi_k\right) \quad (8.37)$$

Since ϕ_k is a constant phase shift, $cos(\phi_k)$ and $sin(\phi_k)$ are constants. Harmonic sinusoids with different frequency indices are orthogonal over the base period T. For positive integers k_1 and k_2, where $k_1 \neq k_2$,

$$\begin{aligned} sin\left(\frac{2\pi k_1 \boldsymbol{t}}{T}\right)^T sin\left(\frac{2\pi k_2 \boldsymbol{t}}{T}\right) &= \sum_{i=0}^{N-1} sin\left[\frac{2\pi\, k_1\, i\Delta t}{(N-1)\Delta t}\right] sin\left[\frac{2\pi\, k_2\, i\Delta t}{(N-1)\Delta t}\right] \\ &= \sum_{i=0}^{N-1} sin\left[\frac{2\pi\, k_1\, i}{(N-1)}\right] sin\left[\frac{2\pi\, k_2\, i}{(N-1)}\right] \\ &= \frac{1}{2}\sum_{i=0}^{N-1}\left\{cos\left[\frac{2\pi(k_1-k_2)i}{(N-1)}\right]-cos\left[\frac{2\pi(k_1+k_2)i}{(N-1)}\right]\right\} \\ &= 0 \qquad k_1 \neq k_2 \end{aligned} \quad (8.38)$$

because the sum of a harmonic sinusoid over the base period T is zero. Similar calculations can be done for other multiplications of harmonic sine and cosine functions. It follows that inputs composed of harmonic sinusoids with different frequency indices are also mutually orthogonal in the time domain.

When the $\boldsymbol{u}_j$ applied to the controls are mutually orthogonal, the aircraft dynamics can be excited in all axes simultaneously. This saves expensive flight test time, and is particularly important in situations where the reference flight condition is time-varying or cannot be maintained for very long. Each input is a sum of harmonic sinusoids, so the excitation is double-sided with balanced energy above and below zero, and the aircraft motion remains close to the reference flight condition. Because the excitation inputs are mutually orthogonal and therefore completely uncorrelated, control effectiveness estimates can be computed very accurately. Using the input design method described here, all of the u_j can be made mutually orthogonal in both the time and frequency domains, while also minimizing relative peak factor for each u_j, which keeps the aircraft from departing significantly from the reference flight condition. This gives the analyst the flexibility to use either time-domain or frequency-domain modeling methods, while retaining the desirable feature of mutually orthogonal inputs.

The excitation inputs shown in Fig. 8.21, computed from information in Fig. 8.18 and Table 8.3, are mutually orthogonal in both the time and frequency domains, with phase angles ϕ_k optimized for minimum relative peak factor. Each input is time-shifted individually to implement a perturbation input, as described earlier for single-input optimized multisines.

Because of the various frequencies and phase angles, and the small amplitudes of the excitation inputs, applying these inputs simultaneously to the aircraft produces a dynamic response similar to what might be seen in flight through light to moderate turbulence. Because the inputs are sums of harmonic sinusoids, the excitation inputs have balanced energy above and below zero, and the aircraft stays near the reference condition, but moves energetically around that. Flight test efficiency and data information content are enhanced by the fact that all controls are moving continuously and simultaneously.

In practice, pilot inputs and feedback control (which are summed with the designed excitation inputs) can act to spoil the excitation input orthogonality. However, good modeling results require only low correlations, not zero correlations, so that slightly correlated inputs still work very well. The inputs are designed with perfect zero correlations (mutually orthogonal inputs), so that even with the practical effects of pilot inputs and feedback control, input correlations are still very low.

Multiple orthogonal optimized multisines have been applied successfully in many flight tests. The inputs implement highly efficient, multi-axis, wideband, orthogonal excitation, while keeping the aircraft near the desired test condition. The inputs are particularly useful because they can excite the aircraft dynamics effectively and continuously in all rigid-body degrees of freedom simultaneously, while the pilot or guidance system flies the aircraft through the desired nominal flight conditions. These characteristics make the inputs very effective for hypersonic flight tests [8.19], [8.27], [8.41], real-time dynamic modeling [8.18], [8.29], [8.33], aircraft flight tests at unusual or transient flight conditions, such as stall, post-stall, departure, and high sideslip angle [8.28], [8.32], [8.37], flight envelope expansion and simulator updating for aircraft with many controls [8.29], [8.37], global aerodynamic modeling [8.32], [8.37], flight testing in turbulence [8.34], flight testing to identify unsteady aerodynamics [8.38], stability and control characterization for flight envelope protection in icing conditions [8.30], and real-time frequency response estimation both in flight [8.36], [8.39]-[8.40], and in the wind tunnel

[8.31], [8.35]. Figure 8.22 shows an example of multi-axis orthogonal optimized multisine inputs applied during a slow approach to stall, through departure and recovery. Such maneuvers are very efficient for collecting highly informative flight data over a wide range of uncorrelated variations in many explanatory variables simultaneously. This enables global modeling for a large portion of the flight envelope using flight data from a single maneuver [8.32].

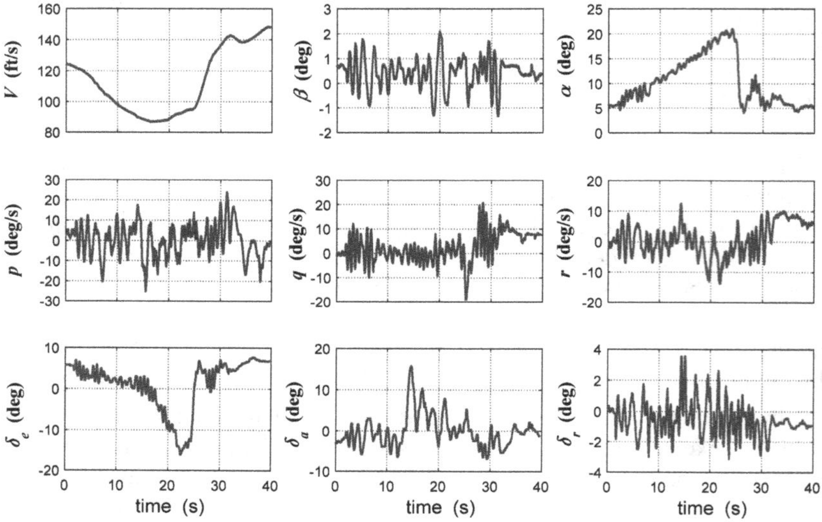

Figure 8.22 **Global modeling maneuver using multiple orthogonal optimized multisine inputs**

More recently, test pilots have implemented similar inputs during flight test maneuvers for global aerodynamic modeling [8.23]-[8.24], [8.26], and frequency response estimation [8.25], demonstrating that the approach can be used on piloted aircraft without onboard computer implementation for the excitation inputs. These inputs, called uncorrelated pilot inputs (UPI), are flown by the pilot moving the cockpit controls in an uncorrelated fashion while varying the frequencies and amplitudes. This emulates the multi-axis orthogonal multisine excitation inputs. With practice, a good test pilot can achieve pairwise control input correlations below 0.2 in absolute value, which is easily sufficient for good modeling results. The UPI can only be used effectively when the aircraft has direct connection from cockpit controls to control surfaces, so that no feedback control is operating to correlate aircraft states and/or controls.

Explanatory variables that are typically changed slowly and/or infrequently can be changed slowly or in a stepwise fashion during the maneuver, to decorrelate their movements from the movements of the other explanatory variables being driven by orthogonal optimized multisines or UPI. This idea has been applied successfully to model flap effects and engine thrust response to throttle inputs [8.37], [8.26].

8.3.5 Optimal Input Design Based on Prior Information

For a single-input, single-output system with one parameter and a fixed maneuver time T, an optimal input for model parameter estimation can be found by optimizing a scalar criterion, as indicated in Eq. (8.9). The optimization is applied to quantities that define the input form. For example, if an optimal square-wave input is desired, then the square wave switching times and amplitudes are quantities that can be adjusted by the optimization. Similarly, a multisine input could be optimized by adjusting amplitudes, frequencies, and phase shifts for the sinusoidal components. This optimal input design problem is the simplest case and is fairly straightforward to solve.

The input optimization problem gets much more complex for multiple outputs and more than one model parameter, which is the typical situation for aircraft problems. This requires the use of the information matrix [cf. Eq. (8.10)] with a prior noise covariance matrix $\boldsymbol{R}$ to account for different noise levels and scaling for multiple outputs, and a prior dynamic model, which is needed to compute the output sensitivities $\partial y(i)/\partial \boldsymbol{\theta}$. The input optimization is also subject to constraints on the input and output, the most significant being the amplitude constraints. The input amplitude constraint is relatively easy to implement, but the output amplitude constraint is not, because the output depends on the amplitude and frequency content of the input acting through the dynamic system, which is characterized only approximately by the prior model. Then a choice must be made as to what function of the information matrix should be used as a scalar optimization criteria. There is not a clear choice – some of the options and their associated implications are discussed in Refs. [8.21], [8.42], and [8.43]. However, it is common to use some scalar function of the information matrix inverse or dispersion matrix [cf. Eq. (8.14)]. Another approach is to minimize one or more selected parameter covariances from the dispersion matrix [8.44].

There have been many approaches to optimal input design for aircraft parameter estimation, including optimal control formulations using variational calculus [8.42]-[8.43], [8.45]-[8.47], a multidimensional optimization approach using multisine input forms [8.6], a suboptimal solution for minimum time to achieve specified parameter error bounds using Walsh functions [8.48], and a globally optimal square-wave solution using dynamic programming [8.7], [8.49]-[8.51]. The last technique balances the theoretical objective of exciting the system as much as possible with the constraints and requirements of the practical flight test situation, such as multiple inputs, constraints on input and output amplitudes, limited flight test time, avoiding structural resonance frequencies, testing with active feedback control, implementation by either a pilot or onboard computer, and robustness to errors in the prior model. This technique has also been demonstrated in flight [8.49]-[8.51].

Optimal input design methods in the frequency domain have been developed by Mehra [8.42], Mehra and Gupta [8.43], and Gupta and Hall [8.47]. These methods have not been popular in practice, because of the relative complexity of the required calculations, dependence on prior models, practical implementation issues, and the fact that simpler inputs such as frequency sweeps, multisteps, and orthogonal optimized multisines provide excellent data for modeling in the frequency domain.

Designing optimal inputs based on prior information for the dynamic model and output noise levels involves a much larger investment of time and resources than alternative methods. However, there are situations when this extra effort is justified. One of these is when the maneuver time is extremely limited for practical reasons. This can occur for unpowered drop model tests, or for specialized tests that are necessarily of short duration because the desired flight condition cannot be maintained for very long, such as flight at high angles of attack and hypersonic flight. The simple heuristic input design methods do not include an effective means for extending the design to multiple inputs applied simultaneously, whereas the most recent and practical optimal input design methods can handle this with only a corresponding increase in computational effort.

Optimal inputs based on prior information represent a different approach to maneuver design, compared to using wideband inputs such as frequency sweeps or multisines. Specifically, optimized inputs based on prior information apply energy in the vicinity of the break frequencies of the dominant dynamic modes, based on the prior model. As discussed earlier in connection with Fig. 8.9, large-scale and important changes in magnitude and phase of a modal response occur near its break frequency, so it is most efficient to concentrate the excitation energy around that frequency when collecting modeling data. The input optimization routine implicitly relies on the prior model to know approximately where the break frequencies are. In a sense, optimized inputs based on prior information are a refinement of designing multistep inputs to target the break frequency of the dominant dynamic mode.

In contrast, wideband inputs like frequency sweeps or multisines require little prior information about the system, because the input power is applied approximately uniformly across the entire frequency range of interest. The trade-off is wideband input maneuver times must be relatively longer and/or maneuvers must be repeated to achieve modeling accuracy equivalent to using an optimal input design, assuming a reasonably accurate prior model is available for the optimal input design. However, wideband inputs are much more robust to deficiencies in prior knowledge of the system being tested and are easier to design.

In practice, inputs optimized based on prior information are most useful in cases with special requirements, such as individual parameter accuracy goals or very short maneuver times.

8.3.6 *Input Comparisons*

It is easy to find works in the literature that compare the effectiveness of various inputs designed for aircraft parameter estimation, particularly for optimal input designs. Unfortunately, making a fair comparison of these inputs is very difficult. This subsection attempts to explain why.

As discussed previously, input designs for parametric modeling of dynamic systems can be evaluated using the Cramér-Rao bounds from Eq. (8.14). The expression for the information matrix is a discrete approximation to a time integral over the maneuver duration. Therefore, when the effectiveness of various input designs are compared using some function of the dispersion matrix Σ as the criterion for comparison, the input designs being compared should have the same maneuver duration.

Practical considerations impose amplitude constraints on the inputs and outputs for flight test maneuvers, and Σ depends nonlinearly on the input and output amplitudes. Consequently, for a fair comparison of input forms, maximum input and output amplitudes should also be the same. This approach contrasts with comparisons presented in many studies in the literature, where input comparisons were based on constant input energy. A wide variety of inputs with varying duration and input amplitudes can have the same constant input energy, as was demonstrated in Fig. 8.6. If the maneuver duration and maximum input and output amplitudes are not the same for the inputs being compared, it is possible to arrange matters so that almost any chosen input form will appear to be the best, using a criterion that depends on Σ. The most fair comparison of input forms is achieved when each input has the same time length, and the same input and output amplitude constraints. Then, the input that produces the best measure of data information content, e.g., minimum $Tr(\Sigma)$, can be considered the best for aircraft parameter estimation.

8.4 Recommendations

In this section, recommendations are given for successfully planning and executing flight experiments for aircraft system identification. The viewpoint encompasses the entire project, rather than specific details such as flight instrumentation, data acquisition system, and specific input design methods, which were covered previously.

8.4.1 Flight Test Planning

The first and most important step in flight test planning is to decide on the objective(s) of the investigation. This decision influences all subsequent decisions, and therefore should be made first, in consultation with management and those providing the funding. The most important issues are the following:

1) Define a successful outcome of the experimentation and modeling. All parties involved should agree on this. Be clear on what the objectives are. For example, is the requirement 10% accuracy on all stability and control derivatives throughout the flight envelope, or is 2% accuracy required only on $\hat{C}_{m_\alpha}$ in approach? The requirements drive how the flight testing and data analysis will be done. In general, higher accuracy requires more information, which means more maneuvers or longer maneuvers will be required to reduce error bounds (via averaging), or else the signal-to-noise ratios must be enhanced with better maneuvers and/or better instrumentation.

2) If possible, build iteration into the plan. If everything that could happen were known *a priori*, there would be no need to conduct the experiment. It is not practical to decide on the entire test program in detail *a priori*. To some, this has the look of making the problem open-ended, but the iterative approach is necessary from a technical standpoint. Here is a fictional (but realistic) exchange that illustrates the problem:

Manager: Tell me how much flight time you will need for the project, so I can enter that number into my spreadsheet and compute the total cost for my budget planning.

Engineer: That depends on what we find.

Manager: Don't be difficult. I want your complete list of maneuvers for the project on my desk by the end of the week.

3) In light of the previous item, spend no more than one quarter of total resources on the initial flight tests. Unfortunately, the time when the most effective experiment can be designed is near the end of the program, when the most is known. The best arrangement is to conduct an initial test flight, then suspend flight test operations and analyze the data all the way through to obtaining model parameter estimates and error bounds. This makes it possible to identify any data recording or instrumentation problems, and to make sure that all the necessary information for successful modeling is being collected in an acceptable form. Any problems discovered can be corrected at a time when most of the flight test time is still in the future. Based on the preliminary results, the flight

test plan (and possibly the goals of the investigation) should be reformulated as necessary before continuing. It is often surprising how much the information from one flight can change the flight test plan. If iteration is impossible for some reason, another approach is to use some form of real-time or global modeling (e.g., see Refs. [8.29], [8.32], [8.52], and Chapters 5 and 10).

4) Make sure the objectives can be achieved with the available resources. Limited resources always have an effect on how flight experiments are conducted, and on the results obtained. When choices must be made, it is usually best to cut back on the number of objectives and get good results for fewer objectives than to try to get everything and obtain mediocre results. Sometimes resource limitations affect how many different flight conditions can be studied. In this case, the flight conditions to be studied must be prioritized by importance to the goals of the investigation. If the goal is flight envelope expansion, then the sequence of flight conditions might be defined by increasing trim angle of attack or Mach number. As with the previous item, real-time and global modeling approaches can be very helpful in this regard.

8.4.2 Data Collection

A general principle of system identification is that the identified model can only capture behavior that is exhibited by the system and embodied in the measured data. For example, the phugoid (long period) dynamics cannot be modeled based on data from a short-duration maneuver. Similarly, if the effect of a particular control surface is to be modeled, that control surface must be moved during the testing, and the movement must be sufficiently different from contemporary variations in other explanatory variables. The same principle applies to all explanatory variables. The fact that the explanatory variables cannot be varied independently for a flying airplane is part of what makes effective experiment design for aircraft system identification flight tests a challenging task.

If it is known that some surfaces will always be moved in a specified way (such as inboard flaps that always move symmetrically with the same amplitude), then these surfaces can be treated as a single control surface for modeling purposes. This simplifies the analysis, since only one combined effectiveness parameter will be estimated, and reduces flight test and instrumentation requirements, since there is only one effective surface being moved. In general, control surfaces should be tested in the same way as they will be used. For example, if surfaces will always be deflected together symmetrically or asymmetrically, then the testing should be done that way, rather than moving each control surface individually. This ensures that any interaction effects are properly characterized. The negative aspect of this approach is that if the scheduled movement of the surfaces is changed, then more testing will be required, where the surfaces will have to be moved according to the new schedule or moved independently, to characterize the control effectiveness properly.

Aircraft system identification can be viewed as processing information from measured data to produce model parameter estimates and associated error bounds. For a fixed amount of information from a given maneuver or set of maneuvers, as more model parameter estimates are requested, the parameter estimates get less accurate. In a sense, the fixed amount of information is spread more thinly over the estimation of more model parameters. Because of this, more complicated model structures with more model parameters require more information to get accurate model parameter estimates. This additional information comes from more or better maneuvers with information content applicable to the model being identified.

At each flight condition, it is advantageous to repeat each maneuver at least once (i.e., 2 maneuvers total), but preferable to repeat each maneuver 3 or 4 times (4 or 5 maneuvers total). Repeats of a maneuver are like insurance from the viewpoint of the data analyst, because of the possibility of data dropouts, bad trims, turbulence, poor maneuver execution, etc., which may render some maneuvers unusable for the intended purpose. Furthermore, significant improvements in accuracy and confidence in the results come with averaging from repeated maneuvers. Variations in the input forms for the repeated maneuvers are desirable, because these differences excite the dynamic system in a more diverse manner, resulting in a more robust identified model. Inputs can be varied by simply reversing input polarity, or using variations of the amplitudes and frequencies of the input waveform. When a pilot implements the inputs, these variations come automatically with the inherent variability of the human pilot.

To set up for a maneuver, the aircraft should be flown in an accurate reference flight condition, with the data collection system turned on, for at least 2 seconds before initiating the maneuver. This enforces a steady initial flight condition and provides data for accurate estimation of the initial condition. When the excitation is complete, the pilot should allow 5 seconds of free response with initial trim controls held, before taking control of the aircraft. This is done to make sure that the entire free response is recorded. Note that these recommendations are the ideal - good modeling results can still be obtained if the starting and ending conditions are unsteady or transient.

For local linear dynamic modeling, unless there are severe practical limits on the time available for each maneuver, a good rule of thumb is to excite the aircraft for a time length of roughly 5 times the period of the dominant dynamic mode. In most cases, it is adequate to use a fixed maneuver time of 10-20 seconds for parameter estimation maneuvers, unless the modeling will include slower dynamics such as the phugoid mode or spiral mode, in which case a much longer maneuver time will be needed.

Table 8.4 contains a list of important considerations for aircraft system identification flight test maneuvers, along with a brief explanation of why each is important. In general, the best approach is to inject orthogonal optimized multisines just upstream of the actuator rate and position limiting, as shown in Fig. 8.18. Referring to Table 8.4, orthogonal optimized multisines have all of the desired features listed. In addition, orthogonal optimized multisines have multi-axis excitation capability, minimized deviation from the desired test condition, orthogonality in both the time and frequency domains simultaneously, an easy design process with no prior information requirements, wideband frequency content, and effectiveness that is practically immune to feedback control and pilot inputs. In practice, a good procedure is to apply orthogonal optimized multisines with relatively low amplitude, then gradually increase the input amplitudes (keeping the input waveforms the same) until sufficient output signal-to-noise ratio (10 or more) is reached. This is a very conservative and efficient process that can be done in flight. Alternatively, the input amplitudes can be estimated prior to flight using flight simulation. This approach has been applied to account for significant dynamic pressure changes in hypersonic flight testing without having to implement many different excitation waveforms in the onboard flight computer [8.27], [8.41].

If the aircraft has a direct connection from the pilot stick and rudder pedal controls to the control surfaces, and there is no feedback control, then the pilot can very effectively implement inputs for flight test maneuvers intended for collecting modeling data. This situation can in fact be preferable, because of the richness associated with the variability in the input implementation by the pilot, and the fact that a pilot can easily change amplitudes, frequencies, and waveforms in real time during the flight. Flight tests and simulator studies have demonstrated that human pilots can implement uncorrelated inputs in multiple axes simultaneously as a good approximation to automated orthogonal optimized multisine inputs [8.23]-[8.26].

It is not always necessary that the aircraft have the full complement of sensors listed in Table 8.1. For example, if the objective is to identify a LOES model to characterize the longitudinal flying qualities associated with the short-period response, only measurements of the longitudinal stick deflection and pitch rate are required. On the other hand, all the values listed in Table 8.1, plus mass/inertia properties and aircraft reference geometry, would be required to estimate longitudinal and lateral nondimensional stability and control derivatives. In general, the quantities that need to be measured are determined by the objectives of the investigation.

Table 8.4 **Summary of important considerations for flight test maneuvers**

Desirable Features	Reasons
Time-skewed inputs or orthogonal inputs	Input decorrelation for accurate control effectiveness estimates, improved excitation
All inputs moving simultaneously	Shortened required flight test time, improved data information content
Low relative peak factors for the inputs	Maintain flight condition, collect data consistent with model structure assumptions, good data information content
High input activity levels	Accurate control effectiveness estimates, persistent excitation for good data information content
Balanced double-sided inputs	Maintain desired flight condition, collect data consistent with model structure assumptions, accurate Fourier transforms
Perturbation inputs that begin and end at zero (2 sec at start, 5 sec at end)	Taylor series modeling, maintain desired flight condition, good initial condition estimate, full dynamic free response at the end, accurate Fourier transforms, useful for data checks and reconstructions
Repeated maneuvers with slightly different inputs	Improves modeling accuracy, tests prediction capability, provides insurance against poorly executed maneuvers

8.5 Open-Loop Parameter Estimation from Closed-Loop Data

Some aircraft are designed so that the airframe is unstable without feedback control. In this case, the aircraft is said to be open-loop unstable. This is often done to improve maneuverability or to reduce trim drag, and means that the aircraft would not be controllable by the pilot if the control surfaces were moved exclusively by pilot stick and rudder pedal inputs. Therefore, an open-loop unstable aircraft must have an automatic feedback control system operating continuously for the aircraft to be flown in a stable and controllable manner.

The impact of automatic feedback control in terms of aircraft system identification is that any input implemented by the pilot or introduced directly at the control surface actuator by an onboard computer system will be distorted by the feedback control. When conducting flight tests for aircraft dynamic modeling, the objective is to excite the natural aircraft motion as much as possible, within the practical constraints of the flight test. The objective of the feedback control is generally stabilization and disturbance rejection. Excitation initiated by the pilot or an onboard system to implement a system identification maneuver is seen by the feedback control system as a disturbance which should be damped and eliminated. To do this, the feedback system moves the control surfaces in a manner that mutes the natural dynamic response of the aircraft. The result is that the excitation input waveform is distorted and the natural dynamic response is subdued. It follows that feedback control works against the objectives of the system identification experiment. Therefore, when flying system identification maneuvers, it is advantageous to reduce feedback control gains as much as possible without risking departure. In practice, however, this is rarely an option.

Note that the bare-airframe dynamic response is not altered by feedback control. The feedback control modifies the aircraft closed-loop response to pilot inputs by altering the control surface deflections, but cannot change the bare-airframe dynamics.

Many modern aircraft, including those that are open-loop unstable, have multiple control surfaces with redundant functions. For example, an aircraft with a canard and all-moving horizontal tail has two control surfaces that can change the aerodynamic pitching moment. Such aircraft normally have an automatic feedback control system. Often, the feedback control system moves the multiple control surfaces so that their motions are highly correlated with one another, or with state variables. To continue the earlier example, the control system on an aircraft with a canard and all-moving horizontal tail will typically move the canard and horizontal tail in a proportional way, with the ratio scheduled as a function of angle of attack. This is done to distribute the aerodynamic loads and reduce trim drag. Consequently, for a small-perturbation flight test maneuver, the ratio of the canard and horizontal tail movements is practically a fixed number, and the data for the two control surfaces are highly collinear, leading to problems in the modeling (cf. Chapter 5). If the same aircraft also has relaxed longitudinal static stability, then both control surfaces may also be highly correlated with the pitch rate, for example, due to the stability augmentation function of the control system. This was seen in Example 5.4 for longitudinal maneuvers on the X-29 aircraft.

This problem can be solved through the modeling process, by estimating equivalent derivatives, or by using biased estimators, possibly with the flight data augmented using information from other sources, as discussed in Chapter 5. However, it is preferable to address the problem by adding orthogonal excitations directly to the control surface actuator commands using an onboard computer system, as depicted in Fig. 8.18. These excitations are sometimes called independent surface inputs or programmed test inputs (PTI). The effect of the added excitations is to decorrelate the resulting control surface motion from other explanatory variables, and thereby allow accurate determination of the control surface effectiveness and other model parameters. For multiple inputs, time-skewed or otherwise uncorrelated excitation inputs must be applied to at least $n_i - 1$ of the n_i inputs to decorrelate all of the inputs. An efficient and effective approach is to additively apply orthogonal optimized multisines simultaneously to all control surfaces of interest, just upstream of the actuator rate and position limiting. When this is done properly, there is practically no pilot input or feedback control that can ruin the data quality for aircraft system identification by correlating the flight data.

Outside of the input distortion and data collinearity caused by the feedback control, data analysis and modeling required to estimate open-loop or bare-airframe model parameters when the maneuvers are performed with a feedback control system operating are exactly the same as described throughout this book. It is therefore possible, and indeed not any more difficult, to estimate open-loop model parameters from data measured with the aircraft operating under closed-loop feedback control. The problem is getting enough uncorrelated excitation past the feedback control system and into the aircraft so that accurate modeling can be done based on measured data from the experiment. This can be done very effectively by adding orthogonal optimized multisines to the actuator commands, even for aircraft with very high-gain feedback control, such as hypersonic vehicles (cf. Refs. [8.19], [8.27], and [8.41]).

An exception to the previous statements occurs when the modeling is done based on frequency responses. Because each frequency response characterizes the linearized response for a single output to a single input, any feedback control will effectively introduce extraneous inputs that adversely affect the modeling (cf. Refs.[8.8]-[8.12]). However, all other modeling techniques described in this book are largely unaffected by this consideration, because they can accommodate moderate multiple correlations among the explanatory variables.

8.6 Summary and Concluding Remarks

An essential part of system identification applied to aircraft is design of the experiment. When this is done well, the data analysis and modeling normally proceed without difficulty. On the other hand, if a significant mistake is made in the experimentation, often there are no post-flight data processing remedies, and the data can be practically useless for modeling purposes.

This chapter laid out important considerations in experiment design for aircraft system identification, including characteristics of the data acquisition system and flight instrumentation, and the design of excitation inputs. Heuristic and optimal design approaches were presented for both single and multiple input design cases. Optimal input design for parameter estimation maneuvers involves optimizing a scalar measure of signal-to-noise ratios for the modeling problem, or an input efficiency metric, subject to practical constraints of the flight test environment. Heuristic input design approaches generally implement some form of wideband input, with frequency content selected to include or be concentrated near the expected modal frequencies of the aircraft. The orthogonal optimized multisine input design method was shown to have attractive theoretical properties, in addition to being very practical, efficient, and easy to design.

The chapter summarized general recommendations for effectively conducting flight test programs for aircraft system identification. Practical problems related to experimentation and data collection for aircraft system identification were discussed, along with suggested solutions. These problems included data collinearity, important considerations for flight test maneuvers, and estimation of bare-airframe model parameters from data collected under closed-loop feedback control.

References

[8.1] Basic Principles of Flight Test Instrumentation Engineering, AGARDograph No. 160, Vol. 1, 1974.

[8.2] Shannon, C.E., "Communication in the Presence of Noise," *Proceedings of the IRE*, Vol. 37, No. 1, 1949, pp. 10-21.

[8.3] Phillips, W.H., "Effects of Model Scale on Flight Characteristics and Design Parameters," *Journal of Aircraft*, Vol. 31, No. 2, 1993, pp. 454-457.

[8.4] Steers, S.T. and Iliff, K.W., "Effects of Time-Shifted Data on Flight-Determined Stability and Control Derivatives," NASA TN D-7830, 1975.

[8.5] Klein, V., "Estimation of Aircraft Aerodynamic Parameters from Flight Data," *Progress in Aerospace Sciences*, Vol. 26, No. 1, 1989, pp. 1-77.

[8.6] Mulder, J.A., "Design and Evaluation of Dynamic Flight Test Manoeuvres," Report LR-497, Delft University of Technology, Department of Aerospace Engineering, Delft, The Netherlands, 1986.

[8.7] Morelli, E.A., "Practical Input Optimization for Aircraft Parameter Estimation Experiments," NASA CR 191462, May 1993.

[8.8] Tischler, M.B., "Frequency-Response Identification of the XV-15 Tilt-Rotor Aircraft Dynamics," NASA TM 89428, 1987.

[8.9] Tischler, M.B., Fletcher, J.W., Diekmann, V.L., Williams, R.A., and Cason, R.W., "Demonstration of Frequency-Sweep Testing Technique Using a Bell 214-ST Helicopter," NASA TM 89422, 1987.

[8.10] Tischler, M.B. and Cauffman, M.G., "Frequency-Response Method for Rotorcraft System Identification with Applications to the BO 105 Helicopter," *American Helicopter Society 46th Annual Forum*, Washington, DC, 1990, pp. 90-137.

[8.11] Williams, J.N., Ham, J.A., and Tischler, M.B., "Flight Test Manual, Rotorcraft Frequency Domain Flight Testing," AQTD Project No. 93–14, U.S. Army Aviation Technical Test Center, Edwards AFB, CA, 1995.

[8.12] Tischler, M.B. and Remple, R.K., Aircraft and Rotorcraft System Identification – Engineering Methods with Flight Test Examples, AIAA, Reston, VA, 2006.

[8.13] Schroeder, M.R., "Synthesis of Low-Peak-Factor Signals and Binary Sequences with Low Autocorrelation," *IEEE Transactions on Information Theory*, Vol. IT-18, No. 1, 1970, pp. 85-89.

[8.14] Young, P. and Patton, R.J., "Comparison of Test Signals for Aircraft Frequency Domain Identification," *Journal of Guidance*, Vol. 13, No. 3, 1990, pp. 430-438.

[8.15] Bosworth, J.T. and Burken, J.J., "Tailored Excitation for Multivariable Stability-Margin Measurement Applied to the X-31A Nonlinear Simulation," NASA TM 113085, August 1997.

[8.16] Klein, V. and Murphy, P.C., "Aerodynamic Parameters of High Performance Aircraft Estimated from Wind Tunnel and Flight Test Data," *System Identification for Integrated Aircraft Development and Flight Testing*, RTO-MP-11, Paper 18, 1999.

[8.17] Press, W.H., Teukolsky, S.A., Vettering, W.T., and Flannery, B.R., *Numerical Recipes in FORTRAN: The Art of Scientific Computing*, 2nd Ed., Cambridge University Press, New York, NY, 1992.

[8.18] Morelli, E.A., "Multiple Input Design for Real-Time Parameter Estimation in the Frequency Domain," Paper REG-360, *13th IFAC Symposium on System Identification*, Rotterdam, The Netherlands, 2003.

[8.19] Morelli, E.A., "Flight-Test Experiment Design for Characterizing Stability and Control of Hypersonic Vehicles," *Journal of Guidance, Control, and Dynamics*, Vol. 32, No. 3, 2009, pp. 949-959.

[8.20] Koehler, R. and Wilhelm, K., "Auslegung von Eingangssignalen für die Kennwertermittlung," ("Design of Input Signals for Identification,") IB 154-77/40, DFVLR Institut für Flugmechanik, Braunschweig, Germany, 1977 (in German).

[8.21] Plaetschke, E. and Schulz, G., "Practical Input Signal Design," *Parameter Identification*, AGARD-LS-104, 1979, Paper 3.

[8.22] Proskawetz, K.-O., "Optimierung stufenförmiger Eingangssignale im Frequenzbereich für die Parameteridentifizierung," ("Optimization of Stepped Input Signals in the Frequency Domain for Parameter Identification,") *Zeitschrift für Flugwissenshcaften und Weltraumforschung*, Vol. 9, No. 6, 1985, pp. 362-370 (in German).

[8.23] Brandon, J.M. and Morelli, E.A., "Nonlinear Aerodynamic Modeling From Flight Data Using Advanced Piloted Maneuvers and Fuzzy Logic," NASA/TM-2012-217778, October 2012.

[8.24] Morelli, E.A., Cunningham, K., and Hill, M.A., "Global Aerodynamic Modeling for Stall/Upset Recovery Training Using Efficient Piloted Flight Test Techniques," AIAA-2013-4976, *AIAA Modeling and Simulation Technologies Conference*, Boston, MA, August 2013.

[8.25] Grauer, J.A. and Martos, B., "Evaluation of Piloted Inputs for Onboard Frequency Response Estimation," AIAA-2013-4921, *AIAA Atmospheric Flight Mechanics Conference*, Boston, MA, August 2013.

[8.26] Brandon, J.M. and Morelli, E.A., "Real-Time Global Nonlinear Aerodynamic Modeling from Flight Data," AIAA-2014-2554, *AIAA Atmospheric Flight Mechanics Conference*, Atlanta, GA, June 2014.

[8.27] Morelli, E.A., Derry, S.D., and Smith, M.S., "Aerodynamic Parameter Estimation of the X-43A (Hyper-X) from Flight Test Data," AIAA-2005-5921, *AIAA Atmospheric Flight Mechanics Conference*, San Francisco, CA, August 2005.

[8.28] Cunningham, K., Foster, J.V., Morelli, E.A., and Murch, A.M., "Practical Application of a Subscale Transport Aircraft for Flight Research in Control Upset and Failure Conditions," AIAA-2008-6200, *AIAA Atmospheric Flight Mechanics Conference*, Honolulu, HI, August 2008.

[8.29] Morelli, E.A. and Smith, M.S., "Real-Time Dynamic Modeling – Data Information Requirements and Flight Test Results," *Journal of Aircraft*, Vol. 46, No. 6, 2009, pp. 1894-1905.

[8.30] Gingras, D.R., Barnhart, B., Ranaudo, R., Ratvasky, T.P., and Morelli, E.A., "Envelope Protection for In-Flight Ice Contamination," NASA TM-2010-216072, February 2010.

[8.31] Heeg, J. and Morelli, E.A., "Evaluation of Simultaneous Multisine Excitation of the Joined Wing SensorCraft Aeroelastic Wind Tunnel Model," AIAA-2011-1959, *52nd AIAA/ASME/ASCE/AHS/ASC Structures, Structural Dynamics and Materials Conference*, Denver, CO, April 2011.

[8.32] Morelli, E.A., "Efficient Global Aerodynamic Modeling from Flight Data," AIAA-2012-1050, *50th AIAA Aerospace Sciences Meeting*, Nashville, TN, January 2012.

[8.33] Morelli, E.A., "Real-Time Aerodynamic Parameter Estimation without Air Flow Angle Measurements," *Journal of Aircraft*, Vol. 49, No. 4, 2012, pp. 1064-1074.

[8.34] Morelli, E.A. and Cunningham, K., "Aircraft Dynamic Modeling in Turbulence," AIAA-2012-4650, *AIAA Atmospheric Flight Mechanics Conference*, Minneapolis, MN, August 2012.

[8.35] Grauer, J.A., Heeg, J., and Morelli, E.A., "Real-Time Frequency Response Estimation of Joined-Wing SensorCraft Aeroelastic Wind Tunnel Data," AIAA-2012-4641, *AIAA Atmospheric Flight Mechanics Conference*, Minneapolis, MN, August 2012.

[8.36] Holzel, M.S. and Morelli, E.A., "Real-Time Frequency Response Estimation from Flight Data," *Journal of Guidance, Control, and Dynamics*, Vol. 35, No. 5, 2012, pp. 1406-1417.

[8.37] Morelli, E.A., "Flight Test Maneuvers for Efficient Aerodynamic Modeling," *Journal of Aircraft*, Vol. 49, No. 6, 2012, pp. 1857-1867.

[8.38] Morelli, E.A., "Transfer Function Identification using Orthogonal Fourier Transform Modeling Functions," AIAA 2013-4749, *AIAA Atmospheric Flight Mechanics Conference*, Boston, MA, August 2013.

[8.39] Grauer, J.A. and Morelli, E.A., "Method for Real-Time Frequency Response and Uncertainty Estimation," *Journal of Guidance, Control, and Dynamics*, Vol. 37, No. 1, 2014, pp. 336-343.

[8.40] Grauer, J.A. and Morelli, E.A., "Reply by the Authors to M. Tischler, C. Ivler, and T. Berger," *Journal of Guidance, Control, and Dynamics*, Vol. 38, No. 3, 2015, pp. 549-550.

[8.41] Morelli, E.A., Rexius, S.L., and Lechniak, J.A., "Flight Test Experiment Design and Aerodynamic Parameter Estimation for the X-51A Waverider," AIAA-2014-2767 *20th AIAA International Space Planes and Hypersonic Systems and Technologies Conference*, Atlanta, GA, June 2014.

[8.42] Mehra, R.K., "Optimal Input Signals for Parameter Estimation in Dynamic Systems - Survey and New Results," *IEEE Transactions on Automatic Control*, Vol. AC-19, No. 6, 1974, pp. 753-768.

[8.43] Mehra, R.K. and Gupta, N.K., "Status of Input Design for Aircraft Parameter Identification," *Methods for Aircraft State and Parameter Identification*, AGARD-CP-172, Paper 12, 1975.

[8.44] Wells, W.R. and Ramachandran, S., "Input Design for Minimum Correlation Between Aerodynamic Parameters," *Fifth Symposium on Nonlinear Estimation*, San Diego, CA, 1974.

[8.45] Reid, D.B., "Optimal Inputs for System Identification," SUDAAR No. 440, Department of Aeronautics and Astronautics, Stanford University, Stanford, CA, 1972.

[8.46] Stepner, D.E. and Mehra, R.K., "Maximum Likelihood Identification and Optimal Input Design for Identifying Aircraft Stability and Control Derivatives," NASA CR-2200, 1973.

[8.47] Gupta, N.K. and Hall, W.E., Jr., "Input Design for Identification of Aircraft Stability and Control Derivatives," NASA CR-2493, 1975.

[8.48] Chen, R.T.N., "Input Design for Aircraft Parameter Identification: Using Time-Optimal Control Formulation," *Methods for Aircraft State and Parameter Identification*, AGARD-CP-172, Paper 13, 1975.

[8.49] Morelli, E.A., "Advances in Experiment Design for High Performance Aircraft," Paper 8, *AGARD System Identification Specialists Meeting*, Madrid, Spain., 1998

[8.50] Morelli, E.A., "Flight Test of Optimal Inputs and Comparison with Conventional Inputs," *Journal of Aircraft*, Vol. 36, No. 2, 1999, pp. 389-397.

[8.51] Cobleigh, B.R., "Design of Optimal Inputs for Parameter Estimation Flight Experiments with Application to the X-31 Drop Model," M.S. Thesis, Joint Institute for the Advancement of Flight Sciences, George Washington University, Hampton, VA, 1991.

[8.52] Morelli, E.A., "In-Flight System Identification," AIAA Paper 98-4261, *AIAA Atmospheric Flight Mechanics Conference*, Boston, MA, 1998.

Problems

8.1 Answer the following, using words, graphs, equations, and/or diagrams:

a) List two considerations that are important for designing experiments to collect data for dynamic modeling. Explain why each is important.

b) Choose an input design type (frequency sweep, multistep, doublet, optimized multisine, etc.) and discuss its advantages and disadvantages for dynamic modeling experiments.

c) Explain the adage: "The best time to design an experiment is after it's over."

d) What are the advantages of running multiple maneuvers at a particular flight condition?

8.2 Create a square wave input using `mksqw.m`. Then use `sqw_psd.m` to find the power spectrum for the input. Experiment with different square waves.

8.3 Create example inputs using mkfswp.m, mksqw.m, and mkmsswp.m with unit amplitudes and excitation time length $T = 10\text{ s}$.

8.4 The T-2 aircraft is a 5.5 percent scale model. The dominant dynamic modes for the full-scale airplane are near 0.25 Hz.

a) At approximately what frequency would the dominant dynamic modes of T-2 be expected to lie?

b) For the T-2 aircraft, what frequencies would be good design choices for the cut-off frequency of a first-order anti-aliasing filter and for the data sampling rate?

8.5 Simulation results for a new aircraft indicate that the dynamic modes of the aircraft lie in the frequency range [0.5, 12] rad/s.

a) Design the data sampling rate for the instrumentation system and the cut-off frequency for the anti-aliasing filters.

b) Design a suitable frequency sweep input for the identifying the system. Set amplitude equal to 1.

c) The aircraft has three control inputs. Design an orthogonal optimized multisine input for identifying the system. Set all amplitudes equal to 1.

d) Discuss the advantages and disadvantages of doing the input design, data analysis, and modeling from experiments using the input designs from parts b) and c).

8.6 Flight data are available for an unmanned aerial vehicle that is flown under closed-loop control because it is open-loop unstable. During a group discussion, a colleague states that it is not possible to do any meaningful dynamic modeling for the aircraft using this data, because of the active feedback control. The boss turns to you and says: "What do you think?" Write your answer.

8.7 Design optimized multisine excitations for three inputs: stabilator, aileron, and rudder. Use a time length equal to 20 seconds, uniform power distribution, and unit amplitudes for all of the excitations, covering a frequency band of [0.1, 2] Hz. The rudder excitation input cannot have any frequency content in the range [1.20, 1.45] Hz, to avoid exciting fuel slosh dynamics. Show that your designed excitation inputs are mutually orthogonal in both the time and frequency domains, and will excite the aircraft over the desired frequency range while satisfying the frequency constraint on the rudder input.

8.8 Flight control designers need an accurate model for a control surface actuator being installed on an aircraft. The dynamic frequency range of interest is [0.1, 25] rad/s, and a test rig is available to apply any desired inputs to the actuator.

a) Design the data sampling rate for the input/output measurements and the cut-off frequency for the analog anti-aliasing filter.

b) Design a suitable frequency sweep input for the identifying the system. Set amplitude equal to 1, and use a time length equal to 30 seconds.

c) Design an optimized multisine input for identifying the system. Set amplitude equal to 1, use a time length equal to 30 seconds, and 20 frequencies.

d) Compare and contrast the inputs in parts b) and c) in terms of how the excitation is applied to the system.

8.9 The longitudinal dynamics of an aircraft flying straight and level at 450 ft/s and 10,000 ft can be approximated by the following linearized equations of motion:

$$\dot{\alpha} = -3\alpha + q - 0.01\delta_e$$
$$\dot{q} = -7.3\alpha - 0.85q - 0.8\delta_e$$
$$a_z = -42\alpha - 0.14\delta_e$$

$$y = \begin{bmatrix}\alpha & q & a_z\end{bmatrix}^T \qquad z = y + v$$

$$v = \mathbb{N}(\mathbf{0}, \boldsymbol{R}) \quad \text{(zero - mean Gaussian, with covariance } \boldsymbol{R}\text{)}$$

$$\boldsymbol{R} = \begin{bmatrix} 2.0e-05 & 0 & 0 \\ 0 & 5.0e-05 & 0 \\ 0 & 0 & 2.5e-05 \end{bmatrix}$$

$$\text{Units: } \alpha \text{ (rad)} \quad q \text{ (rad/s)} \quad \delta_e \text{ (deg)} \quad a_z \text{ (g)}$$

Assume the longitudinal stick commands the elevator directly, with gearing equal to 1.

a) Design a 10 s 3-2-1-1 multistep input to collect data for dynamic modeling purposes. Maneuver load limit on a_z is ± 3g.

b) Design a 10 s optimized multisine input to collect data for dynamic modeling purposes. Maneuver load limit on a_z is ± 3g.

c) Apply the input you designed in part b) to the given dynamic system, add Gaussian white noise with the magnitude specified in $\boldsymbol{R}$ (use randn.m), then apply equation-error parameter estimation in the frequency domain to the resulting data, using the $\dot{q}$ and a_z lines in the dynamic system model. Use the results to assess your input designs.

8.10 Trim and linearize the F-16 in steady, wings-level flight at 10 deg angle of attack and 10,000 ft altitude. Based on the linearized longitudinal short period dynamic model from the simulation at that flight condition, design a long/short doublet sequence input (two doublets in a row, one long, one short), 2-1-1 input, 3-2-1-1 input, linear frequency sweep input, and an optimized multisine input. Apply each designed input to the stabilator of the nonlinear F-16 aircraft simulation, and evaluate each maneuver and the dynamic modeling results from each maneuver, using time-domain output-error parameter estimation in the SIDPAC GUI. Apply 5 percent Gaussian random noise to all measured outputs.

8.11 Show that orthogonal optimized multisine inputs are mutually orthogonal in the frequency domain.

Chapter 9

Data Compatibility

Instrumentation on a research aircraft includes sensors that measure accelerations, rates, and positions associated with the translational motion of the aircraft c.g. and the rotational motion of the aircraft about the c.g., as well as the magnitude and orientation of the air-relative velocity. Kinematic relationships among these quantities can be used to check that the measurements are mutually consistent. An analysis of this type is called a data compatibility analysis or a data consistency check.

If all the measurements were perfect, the data compatibility analysis would show that the kinematic relationships are perfectly satisfied by the sensor measurements. In practice, each sensor measurement has both systematic and random errors. The kinematic relationships are used as a tool to quantify these instrumentation errors and correct the measured data from the sensors for systematic errors. This is an important task, which results in a kinematically consistent data set with improved accuracy. A consistent and accurate data set is a prerequisite for the model structure determination and parameter estimation stages of the system identification process.

Measurements of control surface deflections, power settings, and pilot inputs are notably absent from the foregoing discussion. These measurements cannot be checked against others, because there are no kinematic relationships between these measurements and others in the data set. Consequently, the discussion that follows only applies for measurements associated with the rigid-body motion of the aircraft, which can be checked for compatibility with other measurements.

9.1 Kinematic Equations

The kinematic equations are composed of the translational equations of motion in body axes, the rotational kinematic equations, and the navigation equations in body axes. These equations were derived in Chapter 3. The translational equations of motion in body axes are:

$$\dot{u} = rv - qw + \frac{\bar{q}SC_X}{m} - g\sin\theta + \frac{T}{m} \tag{9.1a}$$

$$\dot{v} = pw - ru + \frac{\bar{q}SC_Y}{m} + g\cos\theta\sin\phi \tag{9.1b}$$

$$\dot{w} = qu - pv + \frac{\bar{q}SC_Z}{m} + g\cos\theta\cos\phi \tag{9.1c}$$

The specific applied forces in g units can be replaced by the translational accelerometer measurements, since (cf. Section 3.7),

$$a_x = \frac{1}{mg}(\bar{q}SC_X + T) \tag{9.2a}$$

$$a_y = \frac{1}{mg}(\bar{q}SC_Y) \tag{9.2b}$$

$$a_z = \frac{1}{mg}(\bar{q}SC_Z) \tag{9.2c}$$

Combining Eqs. (9.1) and (9.2),

$$\dot{u} = rv - qw - g\sin\theta + g a_x \tag{9.3a}$$

$$\dot{v} = pw - ru + g\cos\theta\sin\phi + g a_y \tag{9.3b}$$

$$\dot{w} = qu - pv + g\cos\theta\cos\phi + g a_z \tag{9.3c}$$

In this form, the translational equations of motion relate only kinematic quantities. The rotational kinematic equations and the navigation equations in body axes are used in the forms derived in Chapter 3. The complete set of kinematic equations is given below.

Translational Kinematics

$$\dot{u} = rv - qw - g\sin\theta + g a_x \tag{9.4a}$$

$$\dot{v} = pw - ru + g\cos\theta\sin\phi + g a_y \tag{9.4b}$$

$$\dot{w} = qu - pv + g\cos\theta\cos\phi + g a_z \tag{9.4c}$$

Rotational Kinematics

$$\dot{\phi} = p + \tan\theta\,(q\sin\phi + r\cos\phi) \tag{9.5a}$$

$$\dot{\theta} = q\cos\phi - r\sin\phi \tag{9.5b}$$

$$\dot{\psi} = \frac{q\sin\phi + r\cos\phi}{\cos\theta} \tag{9.5c}$$

<u>*Position Kinematics*</u>

$$\dot{x}_E = u\cos\psi\cos\theta + v\left(\cos\psi\sin\theta\sin\phi - \sin\psi\cos\phi\right) + w\left(\cos\psi\sin\theta\cos\phi + \sin\psi\sin\phi\right) \tag{9.6a}$$

$$\dot{y}_E = u\sin\psi\cos\theta + v\left(\sin\psi\sin\theta\sin\phi + \cos\psi\cos\phi\right) + w\left(\sin\psi\sin\theta\cos\phi - \cos\psi\sin\phi\right) \tag{9.6b}$$

$$\dot{h} = u\sin\theta - v\cos\theta\sin\phi - w\cos\theta\cos\phi \tag{9.6c}$$

The kinematic equations are coupled, nonlinear state differential equations for data compatibility analysis. However, the position kinematic states x_E and y_E do not appear in the other kinematic equations. Altitude is the only position state that has any bearing on the aircraft stability and control (since air density varies with altitude), so the kinematic equations associated with x_E and y_E are generally not used.

Output equations for data compatibility analysis define the relationship between the kinematic states and aircraft responses that are directly measured. For the translational kinematics, measured outputs are related to the states by

$$V = \sqrt{u^2 + v^2 + w^2} \tag{9.7a}$$

$$\alpha = \tan^{-1}\left(\frac{w}{u}\right) \tag{9.7b}$$

$$\beta = \sin^{-1}\left(\frac{v}{\sqrt{u^2 + v^2 + w^2}}\right) \tag{9.7c}$$

The Euler angles and altitude are also measured outputs, so their output equations are simply

$$\phi = \phi \tag{9.7d}$$

$$\theta = \theta \tag{9.7e}$$

$$\psi = \psi \tag{9.7f}$$

$$h = h \tag{9.7g}$$

For the kinematic data consistency check, the measured body-axis angular rates, $p, q,$ and r, and the measured acceleration at the c.g., $a_x, a_y,$ and a_z, are considered inputs in the state equations. The states are $u, v, w, \phi, \theta, \psi,$ and h, and the outputs are $V, \alpha, \beta, \phi, \theta, \psi,$ and h.

The general idea is that a subset of the measurements are used as inputs to the kinematic equations, which are solved and used with the output equations to generate reconstructed values for a different subset of the measurements. If all the measurements are compatible, then the reconstructed outputs will match the measurements of the same output quantities, except for random measurement noise. If the reconstructed outputs do not match the measured outputs, then instrumentation error parameters are introduced into the kinematic equations, and values for these parameters are estimated using techniques such as those discussed in Chapters 4 and 6 in conjunction with state and parameter estimation.

Measured accelerations and angular rates include random measurement noise, so that the kinematic differential equations are in fact stochastic. In some formulations, to be explained in more detail later, this fact is ignored, and the differential equations are solved as if they were deterministic. As long as the measurement noise on the accelerations and angular rates is relatively small and zero mean, the numerical solution of the kinematic equations is not affected significantly by this process noise, because of the smoothing effect of the integration. Filtering or smoothing techniques (see Chapter 11) can be used to remove random noise from the measured acceleration and angular rate signals before using them as inputs to the kinematic equations. However, this is usually not necessary, because the integration removes the random noise, except for any small nonzero mean value of that random noise.

9.2 Data Reconstruction

Solving the kinematic equations numerically using measured values of the inputs a_x, a_y, a_z, p, q, r and initial conditions for the states $u, v, w, \phi, \theta, \psi$, and h, combined with the output equations, results in reconstructed time series for the outputs $V, \alpha, \beta, \phi, \theta, \psi$, and h. The initial conditions can be obtained from smoothed measured values at the initial time. Outputs from this process are called reconstructed outputs, because they have been calculated from other measurements and kinematic relationships. This process is sometimes called kinematic data reconstruction or flight path reconstruction. Note that the reconstruction of air-relative velocity data (V, α, β) does not account for wind.

The reconstructed outputs are sometimes used in lieu of sensor measurements in cases where the sensors are either impractical or simply not available. A good example is the angle of attack on a hypersonic vehicle. Any sort of vane or probe mounted on a vehicle of this type would melt immediately because of aerodynamic heating. Angle of attack can instead be reconstructed from acceleration and angular rate measurements and the kinematic equations, then adjusted for estimated winds aloft.

The main problem with data reconstruction is that the kinematic equations are unstable relative to the inputs a_x, a_y, a_z, p, q, r. For example, a bias in the a_z measurement would cause the w state to drift steadily over time, impacting the reconstructed values of the air data. An example of this is shown in Fig. 9.1 using flight data from a longitudinal maneuver on the NASA F-18 High Alpha Research Vehicle (HARV).

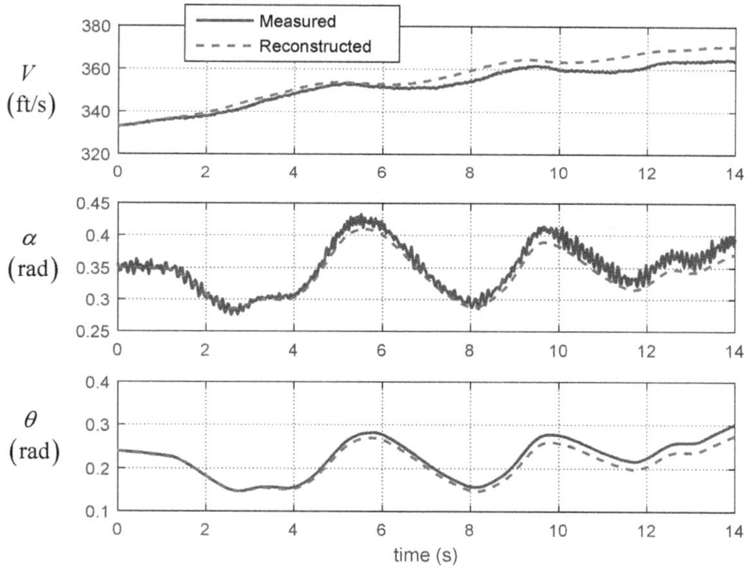

Figure 9.1 **Data compatibility analysis for the NASA F-18 HARV.**

The solid lines in Fig. 9.1 show measured outputs and the dotted lines represent reconstructed outputs. In practice, it is common for the measured kinematic inputs a_x, a_y, a_z, p, q, r to contain bias errors, causing the type of drift in the reconstructed outputs shown in Fig. 9.1. If the bias errors in the measured outputs are ignored, then the drift can be estimated and removed by imposing the condition that the reconstructed outputs must match the measured outputs at the endpoints of the time series.

Another approach to data reconstruction is the complementary filter, where the output is reconstructed as described above for high-frequency response, and the low-frequency response is computed from aerodynamic parameter information. Figure 9.2 shows a block diagram of a complementary filter for an output estimate $\hat{y}$, where the subscript c indicates a value calculated from other measurements. Based on Fig. 9.2, the estimated output $\hat{y}$ can be expressed as

$$\hat{y} = \frac{1}{s+k}\dot{y}_c + \frac{k}{s+k}y_c \tag{9.8a}$$

$$\hat{y} = \frac{s}{s+k}\frac{\dot{y}_c}{s} + \frac{k}{s+k}y_c \tag{9.8b}$$

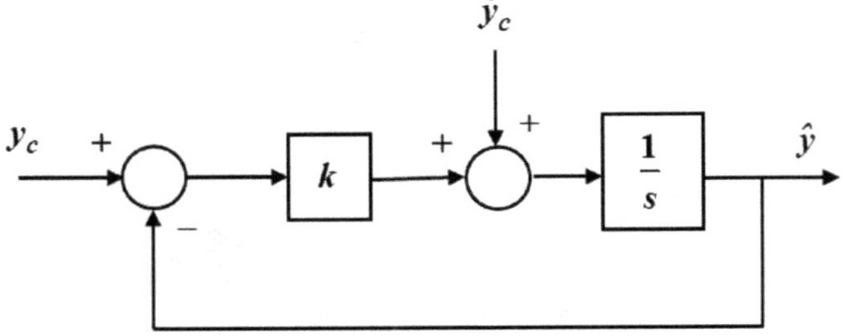

$$\hat{y} = \begin{cases} \text{filtered } y_c & \text{for } \omega < k \\ \text{integrated } \dot{y}_c & \text{for } \omega > k \end{cases}$$

Figure 9.2 **Complementary filter**

The estimated output $\hat{y}$ is computed by integrating $\dot{y}_c$ for frequencies $\omega \gg k$, whereas for $\omega \ll k$, $\hat{y}$ tracks y_c. At frequencies near k, the estimated output $\hat{y}$ is a combination of the two components.

For example, the inputs to a complementary filter for the sideslip angle β could be

$$\dot{\beta}_C \approx \frac{g}{V}\left(a_y + \sin\phi\cos\theta\right) + p\sin\alpha - r\cos\alpha \tag{9.9a}$$

$$\beta_C = \frac{1}{C_{Y_\beta}}\left(\frac{mg}{\bar{q}S}a_y - C_{Y_r}\frac{rb}{2V} - C_{Y_\delta}\Delta\delta\right) \tag{9.9b}$$

where the first equation comes from Eq. (3.39c), assuming the sideslip angle is small, and the second comes from Eqs. (3.54b) and (3.67), with the assumption $C_{Y_p} \approx 0$. Of course, the low-frequency $\hat{\beta}$ will depend on the accuracy of the aerodynamic parameters used in Eq. (9.9b).

Another problem with using inertial measurements a_x, a_y, a_z, p, q, r to reconstruct airspeed, angle of attack, and sideslip angle is that the reconstruction assumes the atmosphere is fixed relative to earth axes. When there are gusts or winds, this assumption is not valid, and the reconstructed data are not accurate. Gusts and winds are usually not measured, or at best are measured at locations and times near the location and time of the flight. If flight testing is done on calm days, this type of error in the reconstructed air-relative velocity data is small, and the instrumentation systematic errors can be accurately estimated.

9.3 Aircraft Instrumentation Errors

Three types of corrections are commonly made to measurements from aircraft instrumentation: ground calibration, sensor alignment and position corrections, and systematic instrumentation error corrections. The first two types of corrections are made as part of the data reduction and do not change unless new sensor hardware is installed on the aircraft. Systematic instrumentation error corrections require analysis and computation that is specific to each maneuver.

9.3.1 Calibration

Calibrations are available from the sensor manufacturer and can also be obtained from laboratory testing on the ground. Calibration for rate gyros can be done using a rate table, where known rates are applied to a surface on which the sensors are mounted. Euler angle sensors and accelerometers can be tested by mounting them on a surface with known angular position. The accelerometers should measure the negative gravity vector, which is the specific force applied by the surface to the sensor. For an extended range, the sensitivity axis of the accelerometer is aligned with the radius of a rate table to experience the normal acceleration of uniform circular motion.

The same general idea can be used to check angle of attack and pitch angle measurements with translational accelerometer measurements a_x and a_z. Using data from steady, wings-level flight:

$$\alpha = \theta = sin^{-1}(a_x) = cos^{-1}(a_z)$$

which follows from Eqs. (9.4a) and (9.4c).

Air flow angle sensors are best calibrated in a wind tunnel, preferably mounted in a manner that mimics the installation on the aircraft. Such calibrations provide local flow corrections, also called upwash corrections for angle of attack and sidewash corrections for sideslip angle. Several flight test methods can be used to calibrate air data sensors [9.1].

9.3.2 Sensor Alignment and Position Errors

The sensitive axes of rate gyros and accelerometers should be aligned with the body axes of the aircraft, so that the measured components of the angular velocity and translational acceleration are assigned to the proper axis. If this alignment cannot be done for some reason, corrections can be applied (see Ref. [9.2]), as long as the angular displacements from the proper alignment are known. The corrections involve expressing a vector in a rotated axis system, which is discussed in Appendix A. For angle of attack, sideslip angle, and the Euler angles, any sensor misalignment is a constant bias error that can be measured on the ground and removed from the raw data as part of the data reduction.

The equations of motion derived in Chapter 3 assume that the aircraft responses are measured at the c.g. However, the c.g. is not a fixed location on the aircraft, because of changes in loading and fuel state. This is not a problem for measured Euler angles $\phi, \theta, \psi,$ angular rates $p, q, r,$ and angular accelerations $\dot{p}, \dot{q}, \dot{r},$ because these quantities are independent of the sensor position on a rigid body.

For air flow angle data, the variation in measured values due to sensor position is the result of the aircraft rotational motion about the c.g. The rotation produces additional air-relative velocity at any sensor location displaced from the c.g. If $\begin{bmatrix} x_\alpha & y_\alpha & z_\alpha \end{bmatrix}^T$ and $\begin{bmatrix} x_\beta & y_\beta & z_\beta \end{bmatrix}^T$ denote the position vectors of the angle of attack and sideslip angle sensors relative to the c.g. in body-axes, then the measured values from the air data sensors are

$$\alpha_E = tan^{-1}\left(\frac{w - q\,x_\alpha + p\,y_\alpha}{u}\right) \tag{9.10a}$$

$$\beta_E = sin^{-1}\left(\frac{v + r\,x_\beta - p\,z_\beta}{\sqrt{u^2 + v^2 + w^2}}\right) \tag{9.10b}$$

where the subscript E denotes the measured value from the experiment. For small angular rates and small aerodynamic angles, Eqs. (9.10) become

$$\alpha \approx \alpha_E + \frac{q\,x_\alpha}{V} - \frac{p\,y_\alpha}{V} \tag{9.11a}$$

$$\beta \approx \beta_E - \frac{r\,x_\beta}{V} + \frac{p\,z_\beta}{V} \tag{9.11b}$$

Similar corrections could be specified for the airspeed measurement, but typically these corrections are small relative to nominal values of airspeed, so the corrections can be omitted.

For translational accelerometers mounted at a location different from the c.g., rotational motion about the c.g. results in tangential and centripetal accelerations that are picked up by accelerometers. For accelerometers located at a location relative to the c.g. defined by $\begin{bmatrix} x_a & y_a & z_a \end{bmatrix}^T$ in body axes, the position corrections are:

$$g\begin{bmatrix} a_x \\ a_y \\ a_z \end{bmatrix}_{c.g.} = g\begin{bmatrix} a_{x_E} \\ a_{y_E} \\ a_{z_E} \end{bmatrix} + \begin{bmatrix} \left(q^2+r^2\right) & -\left(pq-\dot{r}\right) & -\left(pr+\dot{q}\right) \\ -\left(pq+\dot{r}\right) & \left(p^2+r^2\right) & -\left(qr-\dot{p}\right) \\ -\left(pr-\dot{q}\right) & -\left(qr+\dot{p}\right) & \left(p^2+q^2\right) \end{bmatrix}\begin{bmatrix} x_a \\ y_a \\ z_a \end{bmatrix} \tag{9.12}$$

where all of the accelerations are in g units. The position vectors $\begin{bmatrix} x_\alpha & y_\alpha & z_\alpha \end{bmatrix}^T$, $\begin{bmatrix} x_\beta & y_\beta & z_\beta \end{bmatrix}^T$, and $\begin{bmatrix} x_a & y_a & z_a \end{bmatrix}^T$ are found using the current aircraft c.g. location and the known position of the sensors. The units for the position vector elements must be compatible with the correction equations, which usually means the units are feet or meters. Gainer and Hoffman [9.2] is the definitive reference for the instrumentation error corrections described here, along with many other types and variations.

Sensor position corrections in Eqs. (9.10)-(9.12) require values for the body-axis angular rates and angular accelerations. The body-axis angular rates are measured, and the body-axis angular accelerations can be obtained from measurements or from smoothed differentiation of the angular rates (see Chapter 11). All of these quantities can be noisy, which results in increased noise levels on the corrected accelerations and air flow angle data, due to the position corrections. This problem can be solved by applying smoothing or zero-phase filtering techniques to the angular rates and angular accelerations prior to using them for position corrections (see Chapter 11). In addition, the measured angular rates used in the correction equations (9.11) and (9.12) typically have small bias errors (cf. Section 9.4). However, these biases have little impact on the accelerometer sensor position corrections, because the biases do not affect angular accelerations, and the angular rates appear only in second-order terms in Eq. (9.12). In Eq. (9.11), the angular rates are divided by the relatively large value of airspeed, so the air flow angle sensor position correction terms are not significantly affected by small bias errors in the angular rate measurements.

9.3.3 *Time Lags*

Sensors used for air flow angles, airspeed, and altitude sometimes have time lags, particularly when the sensors use pressure measurements and pneumatic tubing, or when the sensors are located close to strong vortical flows near the airframe. Time skews of any kind in the data are detrimental to modeling results, as discussed in Chapter 6. In general, it is difficult to estimate time skews simultaneously with instrumentation error parameters or dynamic model parameters, because of high correlations and consequent identifiability problems. However, time lags in air flow and altitude data can often be estimated and removed separately using data compatibility analysis in the time domain [9.3], and also using frequency domain techniques [9.4]. When other sensor measurements, e.g., control surface position data, could also have relative time skews, a frequency-domain approach can be applied, see Ref. [9.4].

9.3.4 Sensor Dynamics

Dynamic characteristics of sensors are provided by the manufacturer, or can be evaluated from dynamic tests in the laboratory. Because the natural frequencies of sensors in aircraft instrumentation systems are usually very high relative to the frequencies associated with the quantities being measured, the sensor dynamics can be approximated by a small time delay or simply neglected [9.5]. Sensor bandwidth should be high enough so that time delays at the frequencies of interest are small, typically less than 5 ms.

9.4 Model Equations for Data Compatibility Check

All of the errors discussed so far do not vary over the course of a flight test program unless some instrumentation hardware is replaced, modified, or moved. In contrast, systematic instrumentation errors can change with the maneuver type and the flight condition. Consequently, these errors are often estimated for each individual maneuver, or each type of maneuver [9.5].

The measurement equation model for aircraft sensors with typical systematic instrumentation errors is

$$z(i) = (1+\lambda)y(i) + b + v(i) \qquad i = 1,2,\ldots,N \tag{9.13}$$

where $z(i)$ is the ith measured output from the sensor, $y(i)$ is the true output, $v(i)$ is random measurement noise, λ is a constant scale factor error parameter, and b is a constant bias error parameter. If the instrumentation error parameters λ and b are zero, the measured output equals the true output, corrupted only by random measurement noise. Figure 9.3 shows that the instrumentation error model in Eq. (9.13) constitutes a simple model of the relationship between the measured output from the sensor and the true output. Practical experience with flight test instrumentation has shown that this simple model is adequate for most systematic instrumentation errors. More sophisticated measurement equation models could be used instead, if the situation warranted, without any major changes in the methods used to estimate the instrumentation error parameters.

In practice, not all sensors require both the scale factor and bias error parameters. Table 9.1 shows the typical instrumentation error parameters for various sensors. A constant bias term is omitted for altitude h, because the reference values of this variable can be selected arbitrarily.

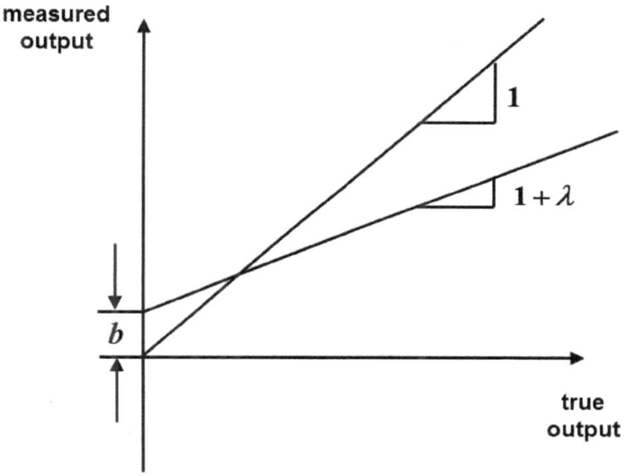

Figure 9.3 **Instrumentation error model**

Table 9.1 **Typical Instrumentation Errors**

Sensor	Variable	Bias Error	Scale Factor Error
translational accelerometer	a_x, a_y, a_z	X	–
rotational accelerometer	$\dot{p}, \dot{q}, \dot{r}$	X	–
rate gyro	p, q, r	X	–
air flow angle vane	α, β	X	X
dynamic pressure sensor	V	X	X
integrating gyro	ϕ, θ, ψ	X	X
pressure altimeter	h	–	X

Combining the information in Table 9.1 with Eqs. (9.4)-(9.6) and (9.13), the state space form of the kinematic equations used for data compatibility analysis is

$$\begin{bmatrix} \dot{u} \\ \dot{v} \\ \dot{w} \\ \dot{h} \end{bmatrix} = \begin{bmatrix} 0 & r_E - b_r & -(q_E - b_q) & 0 \\ -(r_E - b_r) & 0 & p_E - b_p & 0 \\ q_E - b_q & -(p_E - b_p) & 0 & 0 \\ \sin\theta & -\cos\theta\sin\phi & -\cos\theta\cos\phi & 0 \end{bmatrix} \begin{bmatrix} u \\ v \\ w \\ h \end{bmatrix}$$

$$+ \begin{bmatrix} -g\sin\theta + g a_{x_E} - b_{a_x} \\ g\cos\theta\sin\phi + g a_{y_E} - b_{a_y} \\ g\cos\theta\cos\phi + g a_{z_E} - b_{a_z} \\ 0 \end{bmatrix} + \begin{bmatrix} 0 & w & -v & 0 \\ -w & 0 & u & 0 \\ v & -u & 0 & 0 \\ 0 & 0 & 0 & 0 \end{bmatrix} \begin{bmatrix} v_p \\ v_q \\ v_r \\ 0 \end{bmatrix} + \begin{bmatrix} v_{a_x} \\ v_{a_y} \\ v_{a_z} \\ 0 \end{bmatrix} \tag{9.14a}$$

$$\begin{bmatrix} \dot{\phi} \\ \dot{\theta} \\ \dot{\psi} \end{bmatrix} = \begin{bmatrix} 1 & \tan\theta\sin\phi & \tan\theta\cos\phi \\ 0 & \cos\phi & -\sin\phi \\ 0 & \dfrac{\sin\phi}{\cos\theta} & \dfrac{\cos\phi}{\cos\theta} \end{bmatrix} \begin{bmatrix} p_E - b_p + v_p \\ q_E - b_q + v_q \\ r_E - b_r + v_r \end{bmatrix} \tag{9.14b}$$

where the subscript E again indicates measured values from the experiment. For the equations in this section, it is assumed that the measured values have been corrected to the vehicle c.g. using the methods described earlier.

The measured output equations are:

$$V_E(i) = (1+\lambda_V)\sqrt{u^2(i)+v^2(i)+w^2(i)} + b_V + v_V(i) \tag{9.15a}$$

$$\beta_{f_E}(i) = (1+\lambda_{\beta_f})\tan^{-1}\left[\frac{v(i)}{u(i)}\right] + b_{\beta_f} + v_{\beta_f}(i) \tag{9.15b}$$

$$\alpha_E(i) = (1+\lambda_\alpha)\tan^{-1}\left[\frac{w(i)}{u(i)}\right] + b_\alpha + v_\alpha(i) \tag{9.15c}$$

$$\phi_E(i) = (1+\lambda_\phi)\phi(i) + b_\phi + v_\phi(i) \tag{9.15d}$$

$$\theta_E(i) = (1+\lambda_\theta)\theta(i) + b_\theta + v_\theta(i) \tag{9.15e}$$

$$\psi_E(i) = (1+\lambda_\psi)\psi(i) + v_\psi(i) \tag{9.15f}$$

$$h_E(i) = (1+\lambda_h)h(i) + v_h(i) \tag{9.15g}$$

The expression for the measured sideslip angle assumes a vane sensor, so that the angle measured is actually the flank angle, as discussed in Chapter 3. The sideslip angle to be used for aerodynamic modeling is computed from Eq. (3.49),

$$\beta = tan^{-1}\left(tan\,\beta_f\ cos\,\alpha\right) \tag{9.16}$$

State equations (9.14) represent a nonlinear stochastic system where the measurement noise in the acceleration and angular rate signals act as process noise. State and parameter estimation, either simultaneous or separate, would be a difficult task.

A simplification is achieved by assuming that the translational accelerations and angular rates have negligible random measurement errors when used as inputs to the kinematic differential equations. Then Eqs. (9.14) can be written as

$$\begin{bmatrix} \dot{u} \\ \dot{v} \\ \dot{w} \\ \dot{h} \end{bmatrix} = \begin{bmatrix} 0 & r_E - b_r & -(q_E - b_q) & 0 \\ -(r_E - b_r) & 0 & p_E - b_p & 0 \\ q_E - b_q & -(p_E - b_p) & 0 & 0 \\ sin\theta & -cos\theta\, sin\phi & -cos\theta\, cos\phi & 0 \end{bmatrix} \begin{bmatrix} u \\ v \\ w \\ h \end{bmatrix}$$

$$+ \begin{bmatrix} -g\,sin\theta + g\,a_{x_E} - b_{a_x} \\ g\,cos\theta\, sin\phi + g\,a_{y_E} - b_{a_y} \\ g\,cos\theta\, cos\phi + g\,a_{z_E} - b_{a_z} \\ 0 \end{bmatrix} \tag{9.17a}$$

$$\begin{bmatrix} \dot{\phi} \\ \dot{\theta} \\ \dot{\psi} \end{bmatrix} = \begin{bmatrix} 1 & tan\theta\, sin\phi & tan\theta\, cos\phi \\ 0 & cos\phi & -sin\phi \\ 0 & \dfrac{sin\phi}{cos\theta} & \dfrac{cos\phi}{cos\theta} \end{bmatrix} \begin{bmatrix} p_E - b_p \\ q_E - b_q \\ r_E - b_r \end{bmatrix} \tag{9.17b}$$

Further simplification can be achieved by separating the kinematic differential equations and the measured output equations into longitudinal and lateral subsets, when the flight test maneuver contains mainly longitudinal or lateral motion. However, maneuvers used for kinematic consistency checks are often larger amplitude maneuvers such as windup turns and push-over / pull-up maneuvers, and in any case, the kinematic equations should include the full nonlinearity in the equations. Instead, if the maneuver involves mostly longitudinal motion, measured values for the lateral states are substituted into the equations. Similarly, for lateral maneuvers, measured values for the longitudinal states are used in the equations. This technique for decoupling the equations was discussed near the end of Chapter 3, in connection with aerodynamic modeling.

Longitudinal states for the kinematic consistency check are u, w, θ, h, with measured outputs V, α, θ, h. For lateral kinematic consistency check, the states are v, ϕ, ψ, with measured outputs β_f, ϕ, ψ. For maneuvers with combined motion, such as wind-up turns, the full sets of Eqs. (9.14) or (9.17) with (9.15) are used.

Rotational data compatibility analysis can be done separately using Eqs. (9.17b), (9.15d)-(9.15f), because only angular rates and attitude angles are involved.

In some cases, the initial conditions for the states in the kinematic equations are also treated as unknown parameters. However, this must be done carefully to avoid identifiability problems [9.6]-[9.7]. A good approach is to estimate initial conditions separately by applying data smoothing techniques to the measured data (see Chapter 11).

9.5 Instrumentation Error Estimation Methods

Adjusting the unknown instrumentation error parameters in Eqs. (9.14) and (9.15) so that outputs from the model equations match the measured data in a specified way, forms the basis for data compatibility analysis. When combined with the kinematic relationships derived earlier, the measured data from the aircraft instrumentation constitute a redundant set of measurements for the aircraft responses. This redundancy is used to estimate instrumentation error parameters in a postulated model structure for the instrumentation errors. The structure of the instrumentation error model is based largely on experience, but careful residual analysis can be used to help.

Various methods for data compatibility analysis and flight path reconstruction have been developed and applied to simulated and real flight test data. A good general overview is given in Ref. [9.8]. The methods can be divided into two groups:

1) Methods for estimating states and parameters simultaneously, using filtering and/or smoothing methods;
2) Methods for estimating states and parameters separately, using the maximum likelihood method for a stochastic system or the output-error method for a deterministic system.

9.5.1 Filters and Smoothers

The state and output equations for data compatibility analysis, Eqs. (9.14) and (9.15), can also be written as

$$\dot{x}(t) = f\left[x(t), u(t)\right] + B_w\left[x(t)\right] w(t) \tag{9.18a}$$

$$y(t) = g\left[x(t)\right] \tag{9.18b}$$

$$z(i) = g\left[x(i)\right] + v(i) \qquad i = 1, 2, \ldots, N \tag{9.18c}$$

$$\begin{aligned} &E\left[x(0)\right] = \bar{x}_0 && E\left\{\left[x(0) - \bar{x}_0\right]\left[x(0) - \bar{x}_0\right]^T\right\} = P_0 \\ &E\left[w(t)\right] = 0 && E\left[w(t_i) w^T(t_j)\right] = Q(t_i)\delta(t_i - t_j) \\ &E\left[v(i)\right] = 0 && E\left[v(i) v^T(j)\right] = R(i)\delta_{ij} \end{aligned} \tag{9.18d}$$

where $x(0)$, $w(t)$, and $v(i)$ are uncorrelated, and

$$x = \begin{bmatrix} u & v & w & \phi & \theta & \psi & h & b^T & \lambda^T \end{bmatrix}^T \tag{9.18e}$$

$$u = \begin{bmatrix} a_x & a_y & a_z & p & q & r \end{bmatrix}^T \tag{9.18f}$$

$$y = \begin{bmatrix} V & \beta & \alpha & \phi & \theta & \psi & h \end{bmatrix}^T \tag{9.18g}$$

The symbols b and λ denote column vectors containing the bias and scale factor error parameters. The physical state vector has been augmented with the instrumentation error parameters, and the extended Kalman filter can be used to estimate the augmented state vector. For this nonlinear model in continuous-discrete form, the extended Kalman filter can be developed in the same way as in Chapter 4, or can be taken from Ref. [9.9] as

Initial Conditions

$$\begin{aligned}\hat{x}(0\,|\,0) &= \bar{x}_0 \\ P(0\,|\,0) &= P_0\end{aligned} \tag{9.19a}$$

Prediction

$$\frac{d}{dt}\left[\hat{x}(t)\right] = f\left[\hat{x}(t), u(t)\right] \tag{9.19b}$$

$$\frac{d}{dt}\left[P(t)\right] = A\left[\hat{x}(t)\right]P(t) + P(t)A^T\left[\hat{x}(t)\right] + B_w\left[\hat{x}(t)\right]Q(t)\,B_w^T\left[\hat{x}(t)\right]$$

$$(i-1)\Delta t \le t \le i\Delta t \tag{9.19c}$$

Measurement Update

$$\hat{x}(i\,|\,i) = \hat{x}(i\,|\,i-1) + K(i)\left\{z(i) - g\left[\hat{x}(i\,|\,i-1)\right]\right\} \tag{9.19d}$$

$$K(i) = P(i\,|\,i-1)\,C^T(i)\left[C(i)P(i\,|\,i-1)\,C^T(i) + R(i)\right]^{-1} \tag{9.19e}$$

$$P(i\,|\,i) = \left[I - K(i)C(i)\right]P(i\,|\,i-1) \tag{9.19f}$$

The matrices $A\left[\hat{x}(t)\right]$ and $C(i)$ are defined as

$$A\left[\hat{x}(t)\right] = \left.\frac{\partial f\left[x(t), u(t)\right]}{\partial x}\right|_{x(t)=\hat{x}(t)} \tag{9.20a}$$

$$C(i) = \left.\frac{\partial g\left[x(i)\right]}{\partial x}\right|_{x(i)=\hat{x}(i|i-1)} \tag{9.20b}$$

One of the first algorithms for data compatibility analysis was developed by Chen and Eulrich [9.10]. The estimation technique used was an extended Kalman filter with one-stage optimal smoothing. Application of this smoothing can reduce the bias in the estimates, which is inherent to the extended Kalman filter. To further improve the estimates, the algorithm also included a fixed-point smoother, which gives smoothed initial estimates of states and parameters, along with their covariance matrix. The same technique was used in Ref. [9.11], augmented by analysis of residuals, which can help in assessing model adequacy for a given set of measured data.

Mulder [9.12] and Jonkers [9.13] presented four algorithms for data consistency analysis and reconstruction of output variables not available from measurements. These techniques include extended and linearized Kalman filters with and without the fixed-point smoother. Leach and Hui [9.14] describe a novel application of an extended Kalman filter-smoother optimization technique for air data reconstruction. This technique determines the pitot-static position error and calibration curves for the angle of attack and sideslip angle onboard an aircraft.

The algorithm for aircraft state estimation described in Ref. [9.15] is based on a variational solution of a nonlinear, fixed-interval smoothing problem. It is an iterative scheme, providing improved state estimates until the minimum of a squared error measure is achieved. Linearization is about a nominal trajectory. The solution of the fixed-interval smoothing problem consists of determining initial condition estimate $\hat{x}(0)$ and process noise (forcing function) $\boldsymbol{w}(i)$ by minimizing the cost function

$$\begin{aligned} J &= \frac{1}{2}\left(\hat{x}(0)-\bar{\boldsymbol{x}}_0\right)^T \boldsymbol{P}_0^{-1}\left(\hat{x}(0)-\bar{\boldsymbol{x}}_0\right) \\ &+ \frac{1}{2}\sum_{i=1}^{N}\left\{\left[z(i)-y(i)\right]^T \boldsymbol{R}^{-1}\left[z(i)-y(i)\right] + \boldsymbol{w}(i-1)\boldsymbol{Q}^{-1}\boldsymbol{w}(i-1)\right\} \end{aligned} \tag{9.21}$$

subject to the constraints given by Eq. (9.18a) through (9.18c). In the cost function, $\bar{\boldsymbol{x}}_0$ is an *a priori* estimate of $\boldsymbol{x}(0)$, and $\boldsymbol{P}_0$, $\boldsymbol{Q}$, and $\boldsymbol{R}$ are weighting matrices. The estimation algorithm is available as a FORTRAN program called Smoothing for AirCraft Kinematics (SMACK). This program has been used for flight data compatibility analysis at NASA Ames, and for flight path reconstruction to assist the National Transportation Safety Board (NTSB) in investigations of aircraft accidents [9.16].

9.5.2 Output-Error

Data compatibility analysis involves the nonlinear kinematic model and output equations (9.14) and (9.15), with constant unknown instrumentation error parameters to be estimated. In order to simplify the problem, Jonkers [9.13] and Wingrove [9.16] recommended that the output-error method be applied. This is the same type of problem solved in Chapter 6 when the unknowns were the aerodynamic model parameters. Model equations (9.14) are simplified by assuming the translational accelerations and angular rates are measured without random error (no process noise), and output measurement noise matrix is time-invariant. The resulting equations are given above as Eqs. (9.17) and (9.15), which can be written in concise notation as

$$\dot{\boldsymbol{x}}(t) = \boldsymbol{f}\left[\boldsymbol{x}(t), \boldsymbol{u}(t), \boldsymbol{\theta}\right] \tag{9.22a}$$

$$\boldsymbol{y}(t) = \boldsymbol{g}\left[\boldsymbol{x}(t)\right] \tag{9.22b}$$

$$\boldsymbol{z}(i) = \boldsymbol{g}\left[\boldsymbol{x}(i)\right] + \boldsymbol{v}(i) \qquad i = 1, 2, \ldots, N \tag{9.22c}$$

$$E\left[\boldsymbol{v}(i)\right] = \boldsymbol{0} \qquad E\left[\boldsymbol{v}(i)\boldsymbol{v}^T(j)\right] = \boldsymbol{R}\delta_{ij} \tag{9.22d}$$

where

$$\boldsymbol{x} = \left[u \quad v \quad w \quad \phi \quad \theta \quad \psi \quad h\right]^T \tag{9.22e}$$

$$\boldsymbol{u} = \left[a_x \quad a_y \quad a_z \quad p \quad q \quad r\right]^T \tag{9.22f}$$

$$\boldsymbol{y} = \left[V \quad \beta \quad \alpha \quad \phi \quad \theta \quad \psi \quad h\right]^T \tag{9.22g}$$

$$\boldsymbol{\theta} = \left[\boldsymbol{b}^T \quad \boldsymbol{\lambda}^T\right]^T \tag{9.22h}$$

and the symbols $\boldsymbol{b}$ and $\boldsymbol{\lambda}$ denote column vectors containing the bias and scale factor error parameters.

State estimation is reduced to integration of the deterministic dynamic equations (9.22a), which represents Eqs. (9.17a) and (9.17b). The kinematic differential equations are nonlinear, but this does not require any modification of the output-error method, where output sensitivities can be computed from central finite differences applied to numerical solutions of the differential equations. The cost function formulation and optimization method for the parameter estimation are identical to what has been described in Chapter 6 for aerodynamic parameter estimation using the output-error method. In Eqs. (9.22), the model is formulated specifically for data compatibility analysis, and the model parameters characterize instrumentation errors instead of aerodynamic dependencies, but otherwise the estimation problem is conceptually the same.

Initial state values can be estimated from smoothed initial measurements (see Chapter 11), or the initial states can be treated as unknown parameters to be estimated, together with the parameters for instrumentation bias errors and scale factor errors.

After the instrumentation systematic error parameters are estimated, the measured data are corrected by inverting Eq. (9.13), neglecting the random noise,

$$\hat{y}(i) = \left[z(i) - \hat{b}\right] / \left(1 + \hat{\lambda}\right) \qquad i = 1, 2, \ldots, N \tag{9.23}$$

To avoid identifiability problems between scale factor error parameters and initial conditions, which can be particularly severe for maneuvers at high angles of attack, it is advantageous to re-formulate the measurement equations (9.15a)-(9.15c) as follows:

$$V_E(i) = \left(1+\lambda_V\right)\left[\sqrt{u^2(i)+v^2(i)+w^2(i)} - V_0\right] + V_0 + b_V + v_V(i) \tag{9.24a}$$

$$\beta_{f_E}(i) = \left(1+\lambda_{\beta_f}\right)\left\{tan^{-1}\left[\frac{v(i)}{u(i)}\right] - \beta_{f_0}\right\} + \beta_{f_0} + b_{\beta_f} + v_{\beta_f}(i) \tag{9.24b}$$

$$\alpha_E(i) = \left(1+\lambda_\alpha\right)\left\{tan^{-1}\left[\frac{w(i)}{u(i)}\right] - \alpha_0\right\} + \alpha_0 + b_\alpha + v_\alpha(i) \tag{9.24c}$$

In these equations, the values of V_0, β_{f_0}, and α_0 are obtained from smoothed initial measurements. This re-formulation of the measurement equations reduces correlation between the scale factor error parameters, bias error parameters, and the initial conditions, when initial conditions are nonzero. For example, for the airspeed measurement, if the measurement equation (9.15a) is used, then λ_V multiplies both the relatively large and constant part of the airspeed, V_0, as well as the smaller variation due to the maneuver. Since the bias error b_V is usually small, and V_0 is relatively large, small changes in λ_V can approximately account for the bias error in the airspeed measurement with little effect on the model fit related to variations in the airspeed from the maneuver. However, the role of accounting for bias error in airspeed belongs to the bias parameter b_V. The result is a high correlation between λ_V and b_V, which is detrimental to parameter estimation results. Reformulating the measurement equation as in (9.24a) separates the roles of the scale factor and bias error parameters, because λ_V only multiplies the variation in airspeed and not the part represented by V_0. The same issue shows up whenever an initial measured value is not close to zero, which happens for many measured quantities at high angles of attack, for example. Note that role of the scale factor error parameter is different for Eqs. (9.15), compared to Eqs. (9.24), so the estimated values will be different.

More information about the use of the output-error method in data compatibility analysis can be found in Refs. [9.17]-[9.19]. The output-error method has also been applied to the problem of efficient sensor calibration based on flight data [9.20]-[9.21].

9.6 Summary and Concluding Remarks

Successful modeling requires accurate measured data. This chapter is concerned with methods for improving the accuracy of measured data by removing known and estimated systematic errors from the raw measurements.

Every sensor has a static calibration that is done in the laboratory or by the manufacturer. In addition, for sensors that measure aircraft responses, there are errors due to the position of the sensor on the aircraft. Corrections for both of these types of systematic instrumentation errors are relatively straightforward. Equations specific to position corrections for aircraft response sensors are included in this chapter.

Most of the chapter was concerned with the kinematic relationships among measured aircraft response variables, and how these relationships can be used to improve the accuracy of the measured flight data. Typical aircraft response measurements can be checked for compatibility using data from dynamic maneuvers, because there are kinematic relationships among these quantities that would be satisfied if the data were perfectly accurate. The kinematic relationships were developed in this chapter, and it was shown how a subset of the measured aircraft responses can be used with the kinematic equations to reconstruct other aircraft response variables.

If a sensor error model with unknown instrumentation error parameters is introduced, then the kinematic equations can be used in conjunction with a nonlinear optimizer to estimate the most likely values of the instrumentation error parameters. The output-error method of Chapter 6 was shown to apply directly to the solution of this problem. Other methods for data compatibility analysis based on nonlinear filtering and smoothing were also introduced. After the systematic instrumentation errors are successfully estimated, the measured data can be corrected using the identified instrumentation error model, with the result that the final set of data is more accurate and satisfies the kinematic equations as nearly as possible.

Estimating and removing systematic instrumentation errors is important, because if these errors are not removed, then the accuracy of aerodynamic model parameter estimates can be degraded. A good example is if a scale factor error exists for the angle of attack sensor, but the error is ignored and the data are not corrected, then the scale factor error will pollute the estimates of any angle of attack stability derivative, such as C_{m_α}.

References

[9.1] Haering, E.A., Jr., "Airdata Measurement and Calibration," NASA TM 104316, 1995.

[9.2] Gainer, T.G. and Hoffman, S., "Summary of Transformation Equations and Equations of Motion Used in Free-Flight and Wind-Tunnel Data Reduction Analysis," NASA SP-3070, 1972.

[9.3] Blackwell, J. and Feik, R.A. "Identification of Time Delays in Flight Measurements," *Journal of Guidance, Control, and Dynamics,* Vol. 14, No. 1, January-February 1991, pp. 132-139.

[9.4] Morelli, E.A. "Dynamic Modeling from Flight Data with Unknown Time Skews," AIAA-2016-0374, *AIAA SciTech 2016 Conference*, San Diego, CA, January 2016.

[9.5] Klein, V., "Evaluation of the Basic Performance Characteristics of an Instrumentation System," Cranfield Report Aero. No. 22, Cranfield Institute of Technology, Cranfield, UK, 1973.

[9.6] Maine, R.E. and Iliff, K.W., "Identification of Dynamic Systems, Theory and Formulation," NASA RP 1138, 1985.

[9.7] Maine, R.E. and Iliff, K.W., "Application of Parameter Estimation to Aircraft Stability and Control, The Output Error Approach," NASA RP 1168, 1986.

[9.8] Mulder, J.A., Chu, Q.P., Sridhar, J.K., Breeman, J.H., and Laban, M. "Non-linear Aircraft Flight Path Reconstruction Review and New Advances," *Progress in Aerospace Sciences*, Vol. 35, 1999, pp. 673-726.

[9.9] Gelb, A. (editor), *Applied Optimal Estimation,* MIT Press, Cambridge, MA, 1974.

[9.10] Chen, R.T.N. and Eulrich, B.J., "Parameter and Model Identification of Nonlinear System Using a Suboptimal Fixed-Point Smoothing Technique," *Proceedings of the Joint Automatic Control Conference*, 1971.

[9.11] Klein, V. and Schiess, J.R., "Compatibility Check of Measured Aircraft Responses using Kinematic Equations and Extended Kalman Filter," NASA TN D-8514, 1977.

[9.12] Mulder, J.A., "Estimation of Aircraft State in Non-Steady Flight," *Methods for Aircraft State and Parameter Identification*, AGARD-CP-172, Paper 19, 1975.

[9.13] Jonkers, H.L., "Application of the Kalman Filter to Flight Path Reconstruction From Flight Test Data Including Estimation of Instrumental Bias Error Corrections," Report VTH, Delft University of Technology, Delft, The Netherlands, 1976.

[9.14] Leach, B. and Hui, K., "In-Flight Technique For Calibrating Air Data Systems Using Kalman Filtering and Smoothing," AIAA Paper 2001-4260, *AIAA Atmospheric Flight Mechanics Conference*, Montreal, Canada, 2001.

[9.15] Bach, R.E., Jr., "State Estimation Applications in Aircraft Flight Data Analysis," NASA RP 1252, 1991.

[9.16] Wingrove, R.C., "Quasi-Linearization Technique for Estimating Aircraft States From Flight Data," *Journal of Aircraft*, Vol. 10, No. 5, 1973, pp. 303-307.

[9.17] Keskar, D.A. and Klein, V., "Determination of Instrumentation Errors From Measured Data Using Maximum Likelihood Method," AIAA Paper 80-1602, 1980.

[9.18] Klein, V. and Morgan, D.R., "Estimation of Bias Errors in Measured Airplane Responses using Maximum Likelihood Method," NASA TM 89059, 1987.

[9.19] Morelli, E.A., "Optimal Input Design for Aircraft Instrumentation Systematic Error Estimation," AIAA Paper 91-2850, *AIAA Atmospheric Flight Mechanics Conference*, New Orleans, LA, 1991.

[9.20] Martos, B, Kiszely, P., and Foster, J.V., "Flight Test Results Of A GPS-Based Pitot-Static Calibration Method Using Output-Error Optimization For a Light Twin-Engine Airplane," AIAA 2011-6669, *AIAA Atmospheric Flight Mechanics Conference*, Portland, OR, 2011.

[9.21] Siu, M., Martos, B., and Foster, J.V., "Flight Test Results of an Angle of Attack and Angle of Sideslip Calibration Method Using Output-Error Optimization," AIAA 2013-5086, *AIAA Atmospheric Flight Mechanics Conference*, Boston, MA, 2013.

Problems

9.1 Answer the following, using words, graphs, equations, and/or diagrams:

a) Explain why data compatibility analysis can usually only be applied to aircraft response data and not to input measurements.

b) A new angle of attack sensor is installed on an airplane, but there is some uncertainty as to its calibration and the alignment of the sensor as mounted on the airplane. A colleague suggests that the calibrated and trusted accelerometer measurements onboard the aircraft could be used to calibrate the new angle of attack sensor statically, as long as the pilot can fly steady straight-and-level trims so that $\alpha = \theta$, and the flight testing is done on a calm day. Evaluate this idea.

c) Discuss the consequences of not doing data compatibility analysis on modeling results.

9.2 Run dca_lon_demo.m and dca_lat_demo.m, which demonstrate data compatibility analysis using F-18 HARV flight data.

9.3 Repeat the analysis in the previous problem using the SIDPAC GUI, accessed by typing `sid` at the MATLAB® prompt.

9.4 Do a complete system identification for the F-16 at a flight condition of your choice. This includes designing the experiment, flying the maneuver(s) (maneuvers can be flown in batch mode, or piloted in real time, or a combination of these), data compatibility analysis, model structure determination, parameter estimation, and model validation. Discuss how good your identified model is and by what criteria, and state the range of validity for your model. Use of any models derived from the simulation code is not allowed; make conclusions based only on your analysis of simulated noisy flight data. Add 5% Gaussian noise to all simulation outputs (use buzz.m) to generate simulated noisy flight data. Budget constraints dictate that only two maneuvers, each lasting no more than 30 seconds, can be flown at the chosen flight condition. Write a brief report describing your approach and results.

9.5 The longitudinal dynamics of an aircraft flying straight and level at 625 ft/s and 20,000 ft can be approximated by the following longitudinal linearized equations of motion:

$$\dot{\alpha} = -3\alpha + q - 0.035\delta_e$$

$$\dot{q} = -7.3\alpha - 0.85q - 1.2\delta_e$$

$$\dot{\theta} = q$$

$$a_z = 19.4\left(-3\alpha - 0.035\delta_e\right)$$

Assume the longitudinal stick commands the elevator directly, with gearing equal to 1.

a) Design a multistep input to collect data for dynamic modeling purposes.

b) Based on the linearized longitudinal dynamics given above, write the equations for data compatibility analysis. Assume that the usual measurements of $\alpha, q, \delta_e, \theta,$ and a_z are available.

9.6 Derive Eq. (9.12).

Chapter 10

Real-Time Methods

The techniques presented in Chapters 5-7 apply to a complete set of data that is available after the experiment is completed. Such methods are described as batch processing, post-flight modeling, or off-line parameter estimation.

It is also possible to derive parameter estimation algorithms that can be used in real-time, giving interim results as the experiment is being conducted. One of the methods to do this is a recursive formulation of ordinary least squares, where parameter estimates are calculated at each sample time when new measured data are available. This is called recursive least squares. The recursive nature of the calculation avoids reprocessing old data, making the procedure efficient for real-time operation. The extended Kalman filter, which involves augmenting recursive least squares with information about the noise processes and the dynamic system, can also be used for real-time parameter estimation. Another approach is to use batch methods on recent stretches of data to approximate time-varying parameter estimation. This is usually referred to as sequential least squares. Real-time parameter estimation can also be formulated in the frequency domain, using a recursive finite Fourier transform to provide data for a least-squares solution. All of these methods would be classified as on-line processing or real-time parameter estimation. The short list of methods just given is by no means exhaustive, but all of them have been applied to aircraft system identification problems. Each of these methods will be described in this chapter. References [10.1]-[10.5] provide more general information on real-time parameter estimation methods.

An important application of real-time parameter estimation is characterizing changing aircraft dynamics for real-time control law reconfiguration. One approach to this problem is to assume the dynamic model has a linear structure with time-varying parameters. The time variation of the parameters can account for changes in the flight condition, fuel burn, changes in aircraft configuration, or various types of failures, wear, or damage. Allowing the stability and control derivatives in a linear model structure to vary in time can also be used to locally approximate nonlinear aerodynamics. The task is then to estimate time-varying model parameters from measured data in real time, so that adaptive control logic can make the necessary changes to the control law to achieve stability and performance goals. Real-time parameter estimation techniques can also be used for in-flight monitoring of parameter estimates for stability and control testing, flight envelope expansion, or safety monitoring.

The main advantages of real-time parameter estimation can be summarized as follows:

1) Parameter estimates are available in real time without having to process the entire data set as additional measurements are added.
2) Time-varying parameters can be estimated. This feature has proven useful in accident investigations and adaptive/reconfigurable control.
3) Model structure inadequacy and/or identifiability problems are indicated by time variations in the parameter estimates and error bounds.
4) Interim parameter estimation results are available in real time, which is useful for evaluating the effectiveness of various input forms and determining adequate time lengths for excitation. This feature is also important for efficient flight envelope expansion.

The main disadvantages of real-time parameter estimation are:

1) The model structure usually must be fixed. A common approach is to use a fixed linear model structure, but allow the model parameters to vary with time. This implements a time-varying approximating hyperplane model that can be used to account for nonlinearity with local linear approximations that can change with time.
2) Many of the methods have a problem called covariance windup, where the estimated parameter variances continually decrease as time goes on, regardless of whether or not there is excitation to the dynamic system. This leads to inaccurate (i.e., optimistic) error bounds for the parameter estimates.
3) There is a trade-off between rapid response to time variation in the model parameters and smooth time histories for the parameter estimates. This is generally related to data memory, or how the most recent data is treated relative to older data in terms of weighting in the parameter estimation calculations.
4) Periods of low or no excitation combined with a finite data memory can lead to numerical problems for some methods.
5) Data processing tasks such as data compatibility analysis and data smoothing (see Chapters 9 and 11) usually cannot be done in real time.

Because of the aforementioned disadvantages, real-time parameter estimation often involves some engineering judgment or iterative adjustment of values used in the algorithm, sometimes called tuning parameters. A good example is deciding how long into the past the data memory will extend. The adjustments are typically made in simulation or with similar flight test data for which parameter estimates are known from batch methods.

There are many parameter estimation methods, but the requirement of being simple enough to be implemented in real time aboard an aircraft narrows the field. In particular, any method that iterates through the data must be eliminated, which means that equation-error or nonlinear filtering formulations are used. As mentioned earlier, the general approach is to assume a linear model structure with model parameters that are allowed to vary with time.

To estimate dimensional parameters, Eqs. (3.130) and (3.119) can be used:

$$\dot{\alpha} - q = Z_\alpha \Delta\alpha + Z_q q + Z_\delta \Delta\delta \tag{10.1a}$$

$$\dot{q} = M_\alpha \Delta\alpha + M_q q + M_\delta \Delta\delta \tag{10.1b}$$

$$\frac{g}{V_o}\Delta a_z = Z_\alpha \Delta\alpha + Z_q q + Z_\delta \Delta\delta \tag{10.1c}$$

$$\dot{\beta} - p\sin\alpha_o + r\cos\alpha_o - \frac{g\cos\theta_o}{V_o}\phi = Y_\beta \beta + Y_p p + Y_r r + Y_\delta \delta \tag{10.2a}$$

$$\dot{p} - \frac{I_{xz}}{I_x}\dot{r} = L_\beta \beta + L_p p + L_r r + L_\delta \delta \tag{10.2b}$$

$$\dot{r} - \frac{I_{xz}}{I_z}\dot{p} = N_\beta \beta + N_p p + N_r r + N_\delta \delta \tag{10.2c}$$

$$\frac{g}{V_o}\Delta a_y = Y_\beta \beta + Y_p p + Y_r r + Y_\delta \delta \tag{10.2d}$$

Nondimensional parameters can be estimated using the nondimensional forms of the preceding equations,

$$\frac{mV}{\bar{q}S}(\dot{\alpha} - q) = C_{Z_\alpha}\Delta\alpha + C_{Z_q}\frac{q\bar{c}}{2V} + C_{Z_\delta}\Delta\delta \tag{10.3a}$$

$$\frac{I_y}{\bar{q}S\bar{c}}\dot{q} = C'_{m_\alpha}\Delta\alpha + C'_{m_q}\frac{q\bar{c}}{2V} + C'_{m_\delta}\Delta\delta \tag{10.3b}$$

$$-\frac{mg}{\bar{q}S}\Delta a_z = C_{L_\alpha}\Delta\alpha + C_{L_q}\frac{q\bar{c}}{2V} + C_{L_\delta}\Delta\delta \tag{10.3c}$$

$$\frac{mV}{\bar{q}S}\left(\dot{\beta} - p\sin\alpha + r\cos\alpha - \frac{g\cos\theta}{V}\phi\right) = C_{Y_\beta}\beta + C_{Y_p}\frac{pb}{2V} + C_{Y_r}\frac{rb}{2V} + C_{Y_\delta}\delta \tag{10.4a}$$

$$\frac{1}{\bar{q}Sb}\left(I_x\dot{p} - I_{xz}\dot{r}\right) = C_{l_\beta}\beta + C_{l_p}\frac{pb}{2V} + C_{l_r}\frac{rb}{2V} + C_{l_\delta}\delta \tag{10.4b}$$

$$\frac{1}{\bar{q}Sb}\left(I_z\dot{r} - I_{xz}\dot{p}\right) = C_{n_\beta}\beta + C_{n_p}\frac{pb}{2V} + C_{n_r}\frac{rb}{2V} + C_{n_\delta}\delta \tag{10.4c}$$

$$\frac{mg}{\bar{q}S}\Delta a_y = C_{Y_\beta}\beta + C_{Y_p}\frac{pb}{2V} + C_{Y_r}\frac{rb}{2V} + C_{Y_\delta}\delta \tag{10.4d}$$

It is also possible to compute values for the nondimensional aerodynamic force and moment coefficients based on measurements and the nonlinear equations of motion, then set up a linear regression problem, as was done in Chapter 5. The equations are [cf. Eqs. (5.72) and (5.73)],

$$C_X = \frac{1}{\bar{q}S}\left(m\,a_x - T\right) = C_{X_\alpha}\Delta\alpha + C_{X_q}\frac{q\bar{c}}{2V} + C_{X_\delta}\Delta\delta + C_{X_o} \tag{10.5a}$$

$$C_Z = \frac{m\,a_z}{\bar{q}S} = C_{Z_\alpha}\Delta\alpha + C_{Z_q}\frac{q\bar{c}}{2V} + C_{Z_\delta}\Delta\delta + C_{Z_o} \tag{10.5b}$$

$$\begin{aligned} C_D &= -C_X\cos\alpha - C_Z\sin\alpha = -\frac{\left(m\,a_x - T\right)}{\bar{q}S}\cos\alpha - \frac{m\,a_z}{\bar{q}S}\sin\alpha \\ &= C_{D_\alpha}\Delta\alpha + C_{D_q}\frac{q\bar{c}}{2V} + C_{D_\delta}\Delta\delta + C_{D_o} \end{aligned} \tag{10.5c}$$

$$\begin{aligned} C_L &= -C_Z\cos\alpha + C_X\sin\alpha = -\frac{m\,a_z}{\bar{q}S}\cos\alpha + \frac{\left(m\,a_x - T\right)}{\bar{q}S}\sin\alpha \\ &= C_{L_\alpha}\Delta\alpha + C_{L_q}\frac{q\bar{c}}{2V} + C_{L_\delta}\Delta\delta + C_{L_o} \end{aligned} \tag{10.5d}$$

$$\begin{aligned} C_m &= \frac{1}{\bar{q}S\bar{c}}\left[I_y\dot{q} + \left(I_x - I_z\right)pr + I_{xz}\left(p^2 - r^2\right)\right] \\ &= C_{m_\alpha}\Delta\alpha + C_{m_q}\frac{q\bar{c}}{2V} + C_{m_\delta}\Delta\delta + C_{m_o} \end{aligned} \tag{10.5e}$$

$$C_Y = \frac{m\,a_y}{\bar{q}S} = C_{Y_\beta}\beta + C_{Y_p}\frac{pb}{2V} + C_{Y_r}\frac{rb}{2V} + C_{Y_\delta}\delta + C_{Y_o} \tag{10.6a}$$

$$
\begin{aligned}
C_l &= \frac{1}{\bar{q}Sb}\left[I_x\dot{p} - I_{xz}\left(pq+\dot{r}\right)+\left(I_z - I_y\right)qr\right] \\
&= C_{l_\beta}\beta + C_{l_p}\frac{pb}{2V} + C_{l_r}\frac{rb}{2V} + C_{l_\delta}\delta + C_{l_o}
\end{aligned} \tag{10.6b}
$$

$$
\begin{aligned}
C_n &= \frac{1}{\bar{q}Sb}\left[I_z\dot{r} - I_{xz}\left(\dot{p}-qr\right)+\left(I_y - I_x\right)pq\right] \\
&= C_{n_\beta}\beta + C_{n_p}\frac{pb}{2V} + C_{n_r}\frac{rb}{2V} + C_{n_\delta}\delta + C_{n_o}
\end{aligned} \tag{10.6c}
$$

The rest of the chapter is devoted mainly to describing methods for real-time estimation of model parameters in the preceding equations.

10.1 Recursive Least Squares

In Chapter 5, the linear measurement equation was expressed as

$$z = X\theta + v \tag{10.7}$$

where

$$E(v) = 0 \qquad E\left(vv^T\right) = \sigma^2 I \tag{10.8}$$

The ordinary least-squares solution based on n measurements followed from minimization of the cost function

$$J(\theta) = \frac{1}{2}\sum_{i=1}^{n}\left[z(i) - x^T(i)\theta\right]^2 \tag{10.9}$$

as

$$\theta(n) = \mathcal{D}(n)X_n^T Z_n \tag{10.10a}$$

$$Cov\left[\hat{\theta}(n)\right] = \sigma^2\mathcal{D}(n) \tag{10.10b}$$

where

$$\mathcal{D}(n) = \left[X_n^T X_n\right]^{-1} \tag{10.11a}$$

$$X_n^T = \left[x(1) \;\; x(2) \;\; \dots \;\; x(n)\right] \tag{10.11b}$$

$$Z_n = \left[z(1) \;\; z(2) \;\; \dots \;\; z(n)\right]^T \tag{10.11c}$$

When a new measurement $z(n+1)$ is available, the ordinary least-squares solution is

$$\begin{aligned}\boldsymbol{\theta}(n+1) &= \mathcal{D}(n+1)\boldsymbol{X}_{n+1}^T\boldsymbol{Z}_{n+1} \\ &= \mathcal{D}(n+1)\left[\boldsymbol{X}_n^T\boldsymbol{Z}_n + \boldsymbol{x}(n+1)z(n+1)\right]\end{aligned} \tag{10.12}$$

where

$$\begin{aligned}\mathcal{D}(n+1) &= \left[\boldsymbol{X}_{n+1}^T\boldsymbol{X}_{n+1}\right]^{-1} \\ &= \left\{\begin{bmatrix}\boldsymbol{X}_n \\ \boldsymbol{x}^T(n+1)\end{bmatrix}^T \begin{bmatrix}\boldsymbol{X}_n \\ \boldsymbol{x}^T(n+1)\end{bmatrix}\right\}^{-1} \\ &= \left[\boldsymbol{X}_n^T\boldsymbol{X}_n + \boldsymbol{x}(n+1)\,\boldsymbol{x}^T(n+1)\right]^{-1} \\ &= \left[\mathcal{D}^{-1}(n) + \boldsymbol{x}(n+1)\,\boldsymbol{x}^T(n+1)\right]^{-1}\end{aligned} \tag{10.13}$$

Applying the matrix inversion lemma (see Appendix A) to the last expression results in

$$\begin{aligned}\mathcal{D}(n+1) &= \mathcal{D}(n) \\ &- \mathcal{D}(n)\boldsymbol{x}(n+1)\left[1+\boldsymbol{x}^T(n+1)\mathcal{D}(n)\boldsymbol{x}(n+1)\right]^{-1}\boldsymbol{x}^T(n+1)\mathcal{D}(n)\end{aligned} \tag{10.14}$$

Equation (10.14) represents a direct update of the matrix $\mathcal{D}$. Using this expression, the inversion of a $n_p \times n_p$ matrix in Eq. (10.13) has been replaced by the inversion of a scalar in Eq. (10.14), which is a simple division.

Substituting Eq. (10.14) into Eq. (10.12) and using Eq. (10.10),

$$\begin{aligned}\hat{\boldsymbol{\theta}}(n+1) &= \hat{\boldsymbol{\theta}}(n) + \mathcal{D}(n)\boldsymbol{x}(n+1)z(n+1) \\ &- \mathcal{D}(n)\boldsymbol{x}(n+1)\left[1+\boldsymbol{x}^T(n+1)\mathcal{D}(n)\boldsymbol{x}(n+1)\right]^{-1} \\ &\bullet\left[\boldsymbol{x}^T(n+1)\hat{\boldsymbol{\theta}}(n) + \boldsymbol{x}^T(n+1)\mathcal{D}(n)\boldsymbol{x}(n+1)z(n+1)\right]\end{aligned}$$

Rearranging the last two terms

$$\begin{aligned}\hat{\boldsymbol{\theta}}(n+1) &= \hat{\boldsymbol{\theta}}(n) \\ &- \mathcal{D}(n)\boldsymbol{x}(n+1)\left[1+\boldsymbol{x}^T(n+1)\mathcal{D}(n)\boldsymbol{x}(n+1)\right]^{-1}\boldsymbol{x}^T(n+1)\hat{\boldsymbol{\theta}}(n) \\ &+ \mathcal{D}(n)\boldsymbol{x}(n+1)\left[1+\boldsymbol{x}^T(n+1)\mathcal{D}(n)\boldsymbol{x}(n+1)\right]^{-1} \\ &\bullet\left\{\left[1+\boldsymbol{x}^T(n+1)\mathcal{D}(n)\boldsymbol{x}(n+1)-\boldsymbol{x}^T(n+1)\mathcal{D}(n)\boldsymbol{x}(n+1)\right]z(n+1)\right\}\end{aligned}$$

or

$$\begin{aligned}\hat{\boldsymbol{\theta}}(n+1) &= \hat{\boldsymbol{\theta}}(n) \\ &+ \mathcal{D}(n)\boldsymbol{x}(n+1)\left[1+\boldsymbol{x}^T(n+1)\mathcal{D}(n)\boldsymbol{x}(n+1)\right]^{-1} \\ &\bullet\left[z(n+1)-\boldsymbol{x}^T(n+1)\hat{\boldsymbol{\theta}}(n)\right]\end{aligned} \tag{10.15}$$

Therefore, the recursive least-squares estimate can be computed from the following equations:

$$\hat{\boldsymbol{\theta}}(n+1) = \hat{\boldsymbol{\theta}}(n) + \boldsymbol{K}(n+1)\left[z(n+1)-\boldsymbol{x}^T(n+1)\hat{\boldsymbol{\theta}}(n)\right] \tag{10.16a}$$

where $\boldsymbol{K}(n+1)$ is a time-varying matrix computed as

$$\boldsymbol{K}(n+1) = \mathcal{D}(n)\boldsymbol{x}(n+1)\left[1+\boldsymbol{x}^T(n+1)\mathcal{D}(n)\boldsymbol{x}(n+1)\right]^{-1} \tag{10.16b}$$

and $\mathcal{D}(n)$ can be computed recursively from Eq. (10.14),

$$\begin{aligned}\mathcal{D}(n+1) &= \mathcal{D}(n) \\ &- \mathcal{D}(n)\boldsymbol{x}(n+1)\left[1+\boldsymbol{x}^T(n+1)\mathcal{D}(n)\boldsymbol{x}(n+1)\right]^{-1}\boldsymbol{x}^T(n+1)\mathcal{D}(n)\end{aligned} \tag{10.16c}$$

Note that Eqs. (10.16) are the same as the measurement update equations in the continuous-discrete Kalman filter discussed earlier in Chapter 4, cf. Eqs. (4.61d)-(4.61f).

To use the recursive formulas (10.16), starting values $\hat{\boldsymbol{\theta}}(0)$ and $\mathcal{D}(0)$ must be specified. These quantities can be estimated using batch processing in Eqs. (10.9)-(10.11) with an initial data record, or they can be specified using any prior information or parameter estimation results. Recursive least squares can be used to continue a previous analysis by setting $\hat{\boldsymbol{\theta}}(0)$ and $\mathcal{D}(0)$ equal to the parameter vector estimate $\hat{\boldsymbol{\theta}}$ and $\mathcal{D}$ matrix from a previous analysis. When there is no prior information about the parameter estimates, $\hat{\boldsymbol{\theta}}(0)$ can be set to the zero vector, and $\mathcal{D}(0)$ can be chosen as $\mathcal{D}(0)=c\boldsymbol{I}$, where c is a large number, e.g., $c=10^6$. These choices start the algorithm with the statement that the (arbitrary) starting values for the parameters have very large variances, which is equivalent to stating that there is no prior information about the parameter estimates or their accuracy. Another interpretation is that the initial $\boldsymbol{X}^T\boldsymbol{X}$ matrix $\boldsymbol{X}_0^T\boldsymbol{X}_0=\mathcal{D}^{-1}(0)=c^{-1}\boldsymbol{I}$ will be close to the zero matrix, indicating no information content at the outset.

Equations (10.16) constitute the recursive form of the ordinary least-squares solution in Eqs. (10.10). The matrix $\mathcal{D}$ must be converted to the parameter covariance matrix [cf. Eqs. (5.12) and (5.13)],

$$Cov\left[\hat{\boldsymbol{\theta}}(n)\right]=\sigma^2\left[\boldsymbol{X}_n^T\boldsymbol{X}_n\right]^{-1}=\sigma^2\mathcal{D}(n)\equiv\boldsymbol{\Sigma}(n) \tag{10.17}$$

Substituting Eq. (10.17) into (10.16), the recursive least squares algorithm becomes

$$\hat{\boldsymbol{\theta}}(n+1)=\hat{\boldsymbol{\theta}}(n)+\boldsymbol{K}(n+1)\left[z(n+1)-\boldsymbol{x}^T(n+1)\hat{\boldsymbol{\theta}}(n)\right] \tag{10.18a}$$

$$\boldsymbol{K}(n+1)=\boldsymbol{\Sigma}(n)\boldsymbol{x}(n+1)\left[\sigma^2+\boldsymbol{x}^T(n+1)\boldsymbol{\Sigma}(n)\boldsymbol{x}(n+1)\right]^{-1} \tag{10.18b}$$

$$\begin{aligned}\boldsymbol{\Sigma}(n+1)&=\boldsymbol{\Sigma}(n)\\&-\boldsymbol{\Sigma}(n)\boldsymbol{x}(n+1)\left[\sigma^2+\boldsymbol{x}^T(n+1)\boldsymbol{\Sigma}(n)\boldsymbol{x}(n+1)\right]^{-1}\boldsymbol{x}^T(n+1)\boldsymbol{\Sigma}(n)\end{aligned} \tag{10.18c}$$

The recursive computation for the parameter covariance matrix requires the fit error variance σ^2. Usually, σ^2 is not known, so it must be replaced by its estimate $s^2\equiv\hat{\sigma}^2$, computed from prior data analysis. The fit error variance can also be estimated recursively using

$$s^2(n+1)=\frac{1}{(n+1)}\left[ns^2(n)+v^2(n+1)\right] \qquad n<5n_p \tag{10.19a}$$

$$s^2(n+1)=\frac{1}{\left(n+1-n_p\right)}\left[n\,s^2(n)+v^2(n+1)\right] \qquad n=5n_p \tag{10.19b}$$

$$s^2(n+1)=\frac{1}{(n+1-n_p)}\left[(n-n_p)s^2(n)+v^2(n+1)\right] \quad n>5n_p \quad (10.19c)$$

where

$$v(n+1)=z(n+1)-\boldsymbol{x}^T(n+1)\hat{\boldsymbol{\theta}}(n+1) \quad (10.19d)$$

The initial estimates from Eq. (10.19a) are biased because the estimate is not adjusted for the number of parameters in the model, n_p. The transition value $5n_p$ is approximate.

The recursive least-squares algorithm given here is equivalent to the batch algorithm for ordinary least squares presented in Chapter 5. This can be inferred from the derivation shown. Parameter estimation results from recursive least squares at the end of a data record should match ordinary least-squares batch processing results based on the same data.

Example 10.1

In this example, the flight data used in Chapter 5 to demonstrate batch least-squares parameter estimation is used again with the same model structure to demonstrate recursive least squares. The measured data, shown in Fig. 5.4, are from a lateral maneuver of the NASA Twin Otter aircraft (see Fig. 5.3). Recursive least squares implemented by Eqs. (10.16) was used to estimate stability and control derivatives associated with the nondimensional yawing moment coefficient. Figure 10.1 shows time histories of the estimated parameters using recursive least squares. The marks at the right of each plot indicate the parameter estimates from batch least squares, calculated in Chapter 5 (cf. Table 5.1).

The plots in Fig. 10.1 show that the final values of the recursive least-squares parameter estimates match the batch least-squares estimates, as they should. Consequently, using the final parameter estimates shown in Fig. 10.1 produce the same excellent match to the yawing moment coefficient data that was shown in Fig. 5.5.

The recursive least-squares algorithm reveals how the parameter estimates improve and stabilize as information is added from the measurements. Figure 10.1 shows that $\hat{C}_{n_{\delta_r}}$ and $\hat{C}_{n_r}$ stabilize and approach their final accurate values as information comes into the estimator from measurements of rudder movements, and similarly for $\hat{C}_{l_{\delta_a}}$ and $\hat{C}_{l_p}$ in response to aileron movements. ■

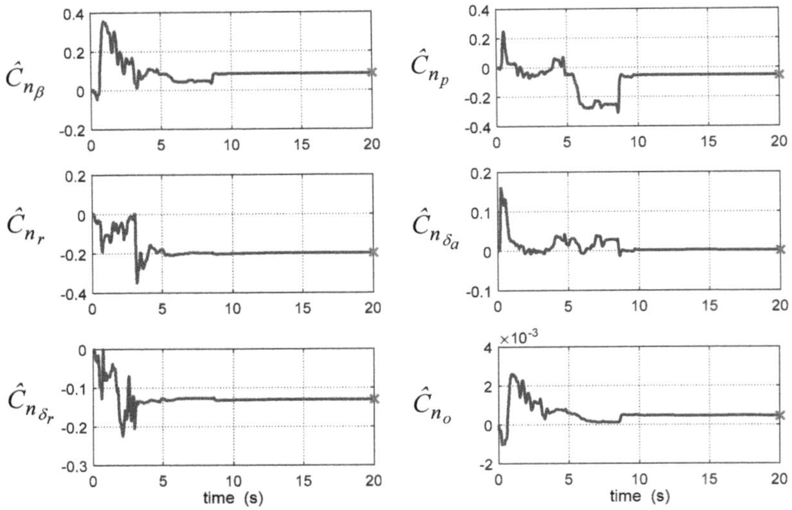

Figure 10.1 **Recursive least squares parameter estimates for lateral maneuver, run 1**

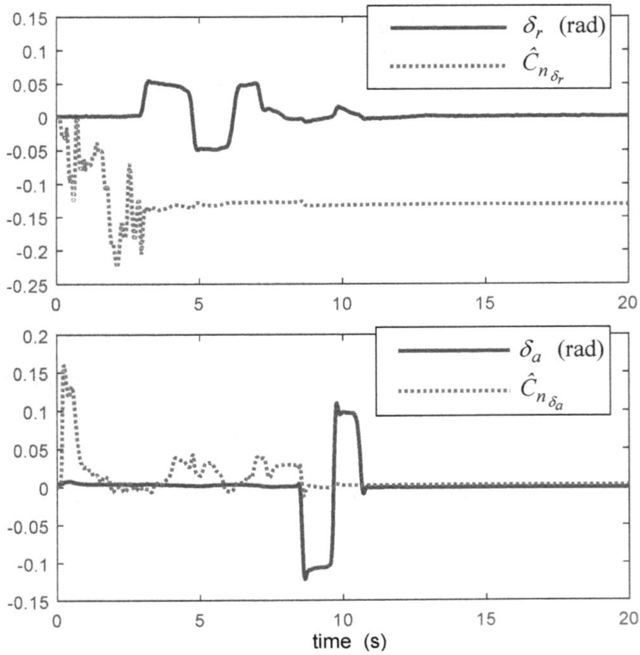

Figure 10.2 **Control derivative recursive estimate convergence**

None of the plots in Figs. 10.1 and 10.2 show standard error estimates for the recursive parameter estimates. The reason is related to the fact that the covariance matrix calculation requires an estimate of the fit error variance $s^2 = \hat{\sigma}^2$ [cf. Eq. (10.17)]. The interim fit error variance estimates from Eqs. (10.19) are inaccurate, because they must use current model parameters to calculate each residual, and the early parameter estimates are inaccurate. Consequently, the fit error estimate varies significantly as time progresses, particularly at the beginning when the estimator has been given little information. Using inaccurate s^2 in the calculation of the estimated parameter covariance matrix leads to inaccurate values for the estimated parameter standard errors. Note that the parameter estimates are unaffected by this issue, because ordinary least squares assumes equal weighting of the equations for each data point (see Chapter 5).

A better approach is to use the estimate $s^2(N)$ at the end of the maneuver to compute the parameter covariance matrix. However, this approach delays the computation of the parameter standard errors until the end of maneuver when the s^2 estimate is improved. Another approach is to use a prior estimate for s^2, but this is difficult because s^2 typically includes both wideband noise and deterministic modeling errors, which are maneuver dependent. There is also the issue of corrections for colored residuals, which has not been addressed.

A much better approach, developed in Refs. [10.6]-[10.7], is to implement a recursive version of the expression developed in Chapter 5 for the estimated parameter covariance matrix accounting for colored residuals. Recalling Eq. (5.89),

$$Cov(\hat{\theta}) = \mathcal{D}\left[\sum_{i=1}^{N} \boldsymbol{x}(i) \sum_{j=1}^{N} \hat{r}_{vv}(i-j)\boldsymbol{x}^T(j)\right]\mathcal{D} \tag{10.20}$$

The right side of Eq. (10.20) involves two sums using regressor data and residual autocorrelation estimates for various time lags. The calculations can be rearranged and made recursive at the nth time step as follows:

$$Cov\left[\hat{\boldsymbol{\theta}}(n)\right] = \mathcal{D}(n)\left[\sum_{k=0}^{m} \hat{r}_{vv}(n,k)\boldsymbol{\Lambda}(n,k)\right]\mathcal{D}(n) \tag{10.21a}$$

where

$$\hat{r}_{vv}(n,k) = \frac{1}{n}\left[(n-1)\,\hat{r}_{vv}(n-1,k) + v(n-k)v(n)\right] \quad k = 0,1,2,\ldots,m \tag{10.21b}$$

$$\Lambda(n,k)=\begin{cases}\Lambda(n-1,k)+\boldsymbol{x}(n)\boldsymbol{x}^T(n) & k=0\\ \Lambda(n-1,k)+\boldsymbol{x}(n-k)\boldsymbol{x}^T(n)+\boldsymbol{x}(n)\boldsymbol{x}^T(n-k) & k>0\end{cases} \tag{10.21c}$$

$$v(n)=z(n)-\boldsymbol{x}^T(n)\hat{\boldsymbol{\theta}}(n) \tag{10.21d}$$

and the quantities $\hat{\boldsymbol{\theta}}(n)$ and $\boldsymbol{\mathcal{D}}(n)$ are computed from Eqs. (10.16),

$$\hat{\boldsymbol{\theta}}(n)=\hat{\boldsymbol{\theta}}(n-1)+\boldsymbol{K}(n)\left[z(n)-\boldsymbol{x}^T(n)\hat{\boldsymbol{\theta}}(n-1)\right] \tag{10.22a}$$

$$\boldsymbol{K}(n)=\boldsymbol{\mathcal{D}}(n-1)\boldsymbol{x}(n)\left[1+\boldsymbol{x}^T(n)\boldsymbol{\mathcal{D}}(n-1)\boldsymbol{x}(n)\right]^{-1} \tag{10.22b}$$

$$\begin{aligned}\boldsymbol{\mathcal{D}}(n)&=\boldsymbol{\mathcal{D}}(n-1)\\&-\boldsymbol{\mathcal{D}}(n-1)\boldsymbol{x}(n)\left[1+\boldsymbol{x}^T(n)\boldsymbol{\mathcal{D}}(n-1)\boldsymbol{x}(n)\right]^{-1}\boldsymbol{x}^T(n)\boldsymbol{\mathcal{D}}(n-1)\end{aligned} \tag{10.22c}$$

Initial values in $\hat{r}_{vv}$ and Λ can be set to zero. Note that the full autocorrelation estimate in Eq. (10.21b) cannot be computed until $n \geq m$. Reference [10.7] recommends 50 lag terms (m=50) for the calculation, although the necessary number of lag terms can depend on the aircraft scale, the dynamic modes being modeled, and the sampling rate. This approach accounts for colored residuals as well as variations in the fit error variance magnitude. Standard errors for the recursive least-squares parameter estimates are found as the square root of the diagonal elements of the covariance matrix calculated from Eq. (10.21a).

10.2 Time-Varying Parameters

The recursive least-squares algorithm described in the previous section assumes that the model parameters are unknown constants, so this algorithm is not directly applicable when the parameters are changing with time. As pointed out earlier, model parameters can vary with time in aerospace applications. In such cases, use of the ordinary least-squares estimator can result in estimates that are much different from the true values, with pronounced variations in the parameter estimates as the algorithm proceeds. Several modifications to the recursive least-squares algorithm have been proposed to estimate time-varying parameters. This section describes several approaches to this problem.

10.2.1 Exponentially-Weighted Least Squares

In this algorithm, old data in the estimation process is gradually devalued and eventually discarded, according to an introduced weighting. The cost function is formulated as

$$J(\boldsymbol{\theta})=\frac{1}{2}\sum_{i=n-m}^{n}\lambda^{n-i}\left[z(i)-\boldsymbol{x}^{T}(i)\boldsymbol{\theta}\right]^{2}\qquad 0<\lambda\leq 1 \tag{10.23}$$

where m is the number of past values included in the data window weighted by powers of λ, which is often called the forgetting factor. The estimation algorithm that minimizes the cost in Eq. (10.23) can be developed in a manner similar to the recursive least squares in the previous section. The resulting equations are:

$$\hat{\boldsymbol{\theta}}(n)=\hat{\boldsymbol{\theta}}(n-1)+\boldsymbol{K}(n)\left[z(n)-\boldsymbol{x}^{T}(n)\hat{\boldsymbol{\theta}}(n-1)\right] \tag{10.24a}$$

$$\boldsymbol{K}(n)=\boldsymbol{\mathcal{D}}(n-1)\boldsymbol{x}(n)\left[\lambda+\boldsymbol{x}^{T}(n)\boldsymbol{\mathcal{D}}(n-1)\boldsymbol{x}(n)\right]^{-1} \tag{10.24b}$$

$$\begin{aligned}\boldsymbol{\mathcal{D}}(n)=\frac{1}{\lambda}\Big\{&\boldsymbol{\mathcal{D}}(n-1)\\ &-\boldsymbol{\mathcal{D}}(n-1)\boldsymbol{x}(n)\left[\lambda+\boldsymbol{x}^{T}(n)\boldsymbol{\mathcal{D}}(n-1)\boldsymbol{x}(n)\right]^{-1}\boldsymbol{x}^{T}(n)\boldsymbol{\mathcal{D}}(n-1)\Big\}\end{aligned} \tag{10.24c}$$

For $\lambda=1$ and m corresponding to the entire data record, Eqs. (10.24) revert to recursive least squares. With $\lambda \ll 1$, a large weighting is placed on recent data by rapidly fading out older data. Thus, the selection of λ is a compromise between fast adaptation to parameter changes and reduced parameter accuracy due to truncation of the data. Typical values of λ are chosen in the range $0.9\leq\lambda<1.0$. Further discussion of the algorithm can be found in Refs. [10.1]-[10.5].

Data forgetting implemented by multiplications of the forgetting factor λ in Eq. (10.23) is called exponential data forgetting. The equivalent time constant for exponential data forgetting depends on the time step Δt and the forgetting factor λ. The time evolution for a quantity ψ subject to multiplicative data forgetting applied at each time step can be computed from

$$\frac{d\psi}{dt}=\frac{(\lambda\psi-\psi)}{\Delta t}=\frac{(\lambda-1)}{\Delta t}\psi \tag{10.25a}$$

which has the exponential solution

$$\psi=\psi_{o}\,exp\left[\frac{(\lambda-1)}{\Delta t}t\right] \tag{10.25b}$$

with time constant

$$\tau = \frac{\Delta t}{(1-\lambda)} \tag{10.25c}$$

Figure 10.3 shows that for $\Delta t = 0.02$ seconds, a forgetting factor $\lambda = 0.99$ corresponds to an exponential forgetting time constant $\tau = 2$ seconds.

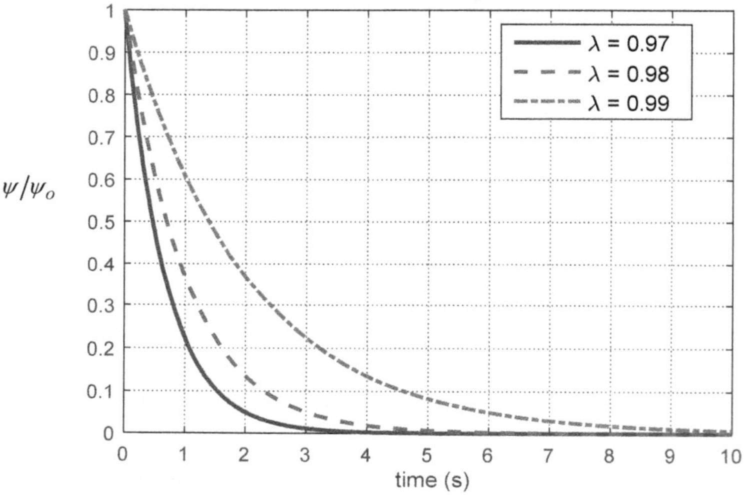

Figure 10.3 **Exponential data forgetting**

10.2.2 Kalman Filter

In this approach, it is assumed that the parameter variations with time can be modeled by a stochastic linear difference equation,

$$\boldsymbol{\theta}(i) = \boldsymbol{\Phi}(i-1)\,\boldsymbol{\theta}(i-1) + \boldsymbol{w}(i-1) \tag{10.26a}$$

with measurement equation

$$z(i) = \boldsymbol{x}^T(i)\,\boldsymbol{\theta}(i) + v(i) \tag{10.26b}$$

where $\boldsymbol{\Phi}(i-1)$ is assumed to be known for all i, and

$$\begin{aligned} E\left[\boldsymbol{w}(i)\boldsymbol{w}^T(j)\right] &= \boldsymbol{Q}(i)\,\delta_{ij} \\ E\left[v(i)v(j)\right] &= \sigma^2(i)\,\delta_{ij} \end{aligned} \tag{10.26c}$$

For the model specified by Eqs. (10.26), the best linear unbiased estimate of $\boldsymbol{\theta}$ based on past measurements can be obtained from Kalman filter equations (4.51), realizing that in Eqs. (10.26), $\boldsymbol{\Phi}(i-1)$ is the state transition matrix and $\boldsymbol{x}^T(i)$ relates the state $\boldsymbol{\theta}(i)$ to the output. Starting with Eqs. (4.51), and replacing $\boldsymbol{x}$ with $\boldsymbol{\theta}$, $\boldsymbol{C}(i)$ with $\boldsymbol{x}^T(i)$, and $\boldsymbol{\Gamma}_w(i-1)$ with $\boldsymbol{I}$, the new set of equations takes the form

$$\hat{\boldsymbol{\theta}}(i|i-1)=\boldsymbol{\Phi}(i-1)\hat{\boldsymbol{\theta}}(i-1|i-1) \tag{10.27a}$$

$$\boldsymbol{P}(i|i-1)=\boldsymbol{\Phi}(i-1)\boldsymbol{P}(i-1|i-1)\boldsymbol{\Phi}^T(i-1)+\boldsymbol{Q}(i-1) \tag{10.27b}$$

$$\hat{\boldsymbol{\theta}}(i|i)=\hat{\boldsymbol{\theta}}(i|i-1)+\boldsymbol{K}(i)\left[z(i)-\boldsymbol{x}^T(i)\hat{\boldsymbol{\theta}}(i|i-1)\right] \tag{10.27c}$$

$$\boldsymbol{P}(i|i)=\left[\boldsymbol{I}-\boldsymbol{K}(i)\boldsymbol{x}^T(i)\right]\boldsymbol{P}(i|i-1) \tag{10.27d}$$

$$\boldsymbol{K}(i)=\boldsymbol{P}(i|i-1)\boldsymbol{x}(i)\left[\boldsymbol{x}^T(i)\boldsymbol{P}(i|i-1)\boldsymbol{x}(i)+\sigma^2(i)\right]^{-1} \tag{10.27e}$$

The state covariance matrix $\boldsymbol{P}$ in the above equations is the same as the parameter covariance matrix $\boldsymbol{\Sigma}$ in Eqs. (10.18), since the state vector of the Kalman filter in this case is the parameter vector $\boldsymbol{\theta}$. However, the covariance matrix produced by this algorithm is not accurate, as explained in Ref. [10.8].

Since the resulting algorithm is rather complicated for practical application, simplified equations have been suggested [10.4]. Simplification is achieved by setting $\boldsymbol{\Phi}(i)=\boldsymbol{I}$ in Eq. (10.26a), so that the state equation is

$$\hat{\boldsymbol{\theta}}(i)=\hat{\boldsymbol{\theta}}(i-1)+\boldsymbol{w}(i-1) \tag{10.28}$$

This state equation is equivalent to specifying that the dynamics of the estimated parameters exhibit a random walk behavior. Furthermore, it is assumed that $\boldsymbol{w}(i)$ and $v(i)$ are stationary random sequences, so that the process noise covariance matrix $\boldsymbol{Q}$ and the measurement noise variance σ^2 are constant. The new set of Kalman filter equations is then

$$\hat{\boldsymbol{\theta}}(i|i-1)=\hat{\boldsymbol{\theta}}(i-1|i-1)\equiv\hat{\boldsymbol{\theta}}(i-1) \tag{10.29a}$$

$$\boldsymbol{P}(i|i-1)=\boldsymbol{P}(i-1|i-1)+\boldsymbol{Q} \tag{10.29b}$$

$$\hat{\boldsymbol{\theta}}(i)=\hat{\boldsymbol{\theta}}(i-1)+\boldsymbol{K}(i)\left[z(i)-\boldsymbol{x}^T(i)\hat{\boldsymbol{\theta}}(i-1)\right] \tag{10.29c}$$

$$\boldsymbol{P}(i|i)=\left[\boldsymbol{I}-\boldsymbol{K}(i)\boldsymbol{x}^T(i)\right]\boldsymbol{P}(i|i-1) \tag{10.29d}$$

$$K(i)=P(i\,|\,i-1)\,x(i)\left[x^T(i)\,P(i\,|\,i-1)\,x(i)+\sigma^2\right]^{-1} \tag{10.29e}$$

When $w(i)=\mathbf{0}$ for all i, $Q=\mathbf{0}$ and the state model is

$$\theta(i)=\theta(i-1) \tag{10.30}$$

which indicates that the parameters are constants. In that case, Eq. (10.29b) becomes $P(i\,|\,i-1)=P(i-1\,|\,i-1)\equiv P(i-1)$, and the algorithm becomes identical to the recursive least squares algorithm of Eqs. (10.18). This demonstrates that the Kalman filter applied in this way differs from recursive least squares only in that a stochastic dynamic model is assumed for the time variation of the parameters [10.9].

10.2.3 Extended Kalman Filter

In Chapter 4, the extended Kalman filter was introduced as a state estimation algorithm for nonlinear dynamic systems. The same algorithm can also be used to estimate both the states and parameters in a dynamic system model. To demonstrate this, the following model equations are considered:

$$\dot{x}=A(\theta)x+B(\theta)u+w \tag{10.31a}$$

$$z(i)=C(\theta)x(i)+v(i) \qquad i=1,2,\ldots,N \tag{10.31b}$$

$$E\left[x(0)\right]=\bar{x}_0 \qquad E\left\{\left[x(0)-\bar{x}_0\right]\left[x(0)-\bar{x}_0\right]^T\right\}=P_{x_0} \tag{10.31c}$$

The noise sequences w and v are zero-mean white Gaussian noise sequences with covariance matrices Q and R, respectively, i.e.,

$$v \text{ is } \mathbb{N}(\mathbf{0},R) \qquad Cov\left[v(i)\right]=E\left[v(i)v^T(j)\right]=R\,\delta_{ij} \tag{10.31d}$$

$$w \text{ is } \mathbb{N}(\mathbf{0},Q) \qquad Cov\left[w(t)\right]=E\left[w(t_i)w^T(t_j)\right]=Q\,\delta(t_i-t_j) \tag{10.31e}$$

If the state vector is augmented with the parameters, the resulting augmented state vector is given by

$$x_a=\begin{bmatrix} x \\ \theta \end{bmatrix} \tag{10.32}$$

The augmented state vector with both the original states and the unknown model parameters will be considered as the state vector for the extended Kalman filter. The parameters θ appear in the dynamic system matrices, and are assumed to be unknown constants. Therefore, the dynamics of the estimated parameters are governed by the equations

$$\dot{\boldsymbol{\theta}} = \mathbf{0} \tag{10.33}$$

The system equations for the augmented state are then

$$\dot{\boldsymbol{x}}_a = \boldsymbol{A}_a \boldsymbol{x}_a + \boldsymbol{B}_a \boldsymbol{u} + \boldsymbol{w}_a \tag{10.34a}$$

$$\boldsymbol{z}(i) = \boldsymbol{C}_a\ \boldsymbol{x}_a(i) + \boldsymbol{v}(i) \qquad i = 1, 2, \ldots, N \tag{10.34b}$$

$$E\left[\boldsymbol{x}_a(0)\right] = \overline{\boldsymbol{x}}_{a_0} \qquad E\left\{\left[\boldsymbol{x}_a(0) - \overline{\boldsymbol{x}}_{a_0}\right]\left[\boldsymbol{x}_a(0) - \overline{\boldsymbol{x}}_{a_0}\right]^T\right\} = \boldsymbol{P}_{a_0} \tag{10.34c}$$

where

$$\boldsymbol{A}_a \equiv \begin{bmatrix} \boldsymbol{A}(\boldsymbol{\theta}) & \mathbf{0} \\ \mathbf{0} & \mathbf{0} \end{bmatrix} \quad \boldsymbol{B}_a \equiv \begin{bmatrix} \boldsymbol{B}(\boldsymbol{\theta}) \\ \mathbf{0} \end{bmatrix} \quad \boldsymbol{w}_a \equiv \begin{bmatrix} \boldsymbol{w} \\ \mathbf{0} \end{bmatrix} \quad \overline{\boldsymbol{x}}_{a_0} \equiv \begin{bmatrix} \overline{\boldsymbol{x}}_0 \\ \boldsymbol{\theta}_0 \end{bmatrix}$$

$$\boldsymbol{C}_a \equiv \begin{bmatrix} \boldsymbol{C}(\boldsymbol{\theta}) & \mathbf{0} \end{bmatrix} \tag{10.34d}$$

The dynamic system for the augmented state is no longer linear, because the system matrices depend on elements of the augmented state. This nonlinearity is present regardless of the linearity of the original dynamic system. Thus, the system described by Eqs (10.34) can be presented in a more concise form as

$$\dot{\boldsymbol{x}}_a = \boldsymbol{f}(\boldsymbol{x}_a, \boldsymbol{u}) + \boldsymbol{w}_a \tag{10.35a}$$

$$\boldsymbol{z}(i) = \boldsymbol{h}\left[\boldsymbol{x}_a(i)\right] + \boldsymbol{v}(i) \qquad i = 1, 2, \ldots, N \tag{10.35b}$$

with the initial conditions given by Eq. (10.34c).

An extended Kalman filter algorithm for the system given in Eqs. (10.35) is obtained from the general equations developed in Chapter 4 for a continuous dynamic system with discrete measurements. The equations for state propagation and measurement update are as follows:

Initial Conditions

$$\begin{aligned} \hat{\boldsymbol{x}}_a(0) &= \overline{\boldsymbol{x}}_{a_0} \\ \boldsymbol{P}_a(0) &= \boldsymbol{P}_{a_0} \end{aligned} \tag{10.36a}$$

$$\boldsymbol{P}_{a_0} \equiv \begin{bmatrix} \boldsymbol{P}_{x_0} & \mathbf{0} \\ \mathbf{0} & \boldsymbol{P}_{\theta_0} \end{bmatrix} \qquad \overline{\boldsymbol{x}}_{a_0} \equiv \begin{bmatrix} \overline{\boldsymbol{x}}_0 \\ \boldsymbol{\theta}_0 \end{bmatrix} \tag{10.36b}$$

$$E\left[\boldsymbol{x}(0)\right]=\bar{\boldsymbol{x}}_0 \qquad E\left\{\left[\boldsymbol{x}(0)-\bar{\boldsymbol{x}}_0\right]\left[\boldsymbol{x}(0)-\bar{\boldsymbol{x}}_0\right]^T\right\}=\boldsymbol{P}_{x_0} \tag{10.36c}$$

$$E\left[\boldsymbol{\theta}(0)\right]=\boldsymbol{\theta}_0 \qquad E\left\{\left[\boldsymbol{\theta}(0)-\boldsymbol{\theta}_0\right]\left[\boldsymbol{\theta}(0)-\boldsymbol{\theta}_0\right]^T\right\}=\boldsymbol{P}_{\theta_0} \tag{10.36d}$$

Prediction

$$\frac{d}{dt}\left[\hat{\boldsymbol{x}}_a(t\,|\,k-1)\right]=\boldsymbol{A}_a\hat{\boldsymbol{x}}_a(t\,|\,k-1)+\boldsymbol{B}_a\boldsymbol{u}(t) \tag{10.36e}$$

$$\frac{d}{dt}\left[\boldsymbol{P}_a(t\,|\,k-1)\right]=\boldsymbol{A}_a\boldsymbol{P}_a(t\,|\,k-1)\boldsymbol{A}_a^T+\boldsymbol{Q}_a \tag{10.36f}$$

for $(k-1)\Delta t \le t \le k\Delta t$

$$\boldsymbol{A}_a \equiv \begin{bmatrix} \boldsymbol{A}(\hat{\theta}) & \mathbf{0} \\ \mathbf{0} & \mathbf{0} \end{bmatrix} = \left.\frac{\partial \boldsymbol{f}}{\partial \boldsymbol{x}_a}\right|_{x_a=\hat{x}_a} \qquad \boldsymbol{B}_a \equiv \begin{bmatrix} \boldsymbol{B}(\hat{\theta}) \\ \mathbf{0} \end{bmatrix} = \left.\frac{\partial \boldsymbol{f}}{\partial \boldsymbol{u}}\right|_{x_a=\hat{x}_a}$$

$$\boldsymbol{Q}_a \equiv \begin{bmatrix} \boldsymbol{Q} & \mathbf{0} \\ \mathbf{0} & \mathbf{0} \end{bmatrix} \tag{10.36g}$$

Measurement Update

$$\hat{\boldsymbol{x}}_a(k\,|\,k)=\hat{\boldsymbol{x}}_a(k\,|\,k-1)+\boldsymbol{K}_a(k)\left\{\boldsymbol{z}(k)-\boldsymbol{h}\left[\hat{\boldsymbol{x}}_a(k\,|\,k-1)\right]\right\} \tag{10.36h}$$

$$\boldsymbol{P}_a(k\,|\,k)=\left[\boldsymbol{I}-\boldsymbol{K}_a(k)\boldsymbol{C}_a\right]\boldsymbol{P}_a(k\,|\,k-1) \tag{10.36i}$$

$$\boldsymbol{K}_a(k)=\boldsymbol{P}_a(k\,|\,k-1)\,\boldsymbol{C}_a^T\left[\boldsymbol{C}_a\boldsymbol{P}_a(k\,|\,k-1)\boldsymbol{C}_a^T+\boldsymbol{R}\right]^{-1} \tag{10.36j}$$

$$\boldsymbol{C}_a \equiv \left[\boldsymbol{C}(\hat{\theta}) \;\; \mathbf{0}\right] = \left.\frac{\partial \boldsymbol{h}}{\partial \boldsymbol{x}_a}\right|_{x_a=\hat{x}_a(k|k-1)} \qquad \boldsymbol{K}_a \equiv \begin{bmatrix} \boldsymbol{K}_x \\ \boldsymbol{K}_\theta \end{bmatrix} \tag{10.36k}$$

The augmented state covariance matrix has four partitions corresponding to the original state error covariance, the estimated parameter error covariance, and their cross variance,

$$\boldsymbol{P}_a \equiv \begin{bmatrix} \boldsymbol{P}_x & \boldsymbol{P}_{x\theta} \\ \boldsymbol{P}_{\theta x} & \boldsymbol{P}_\theta \end{bmatrix} \tag{10.37}$$

The expressions given here are the standard extension of the Kalman filter to a nonlinear dynamic system, which is called the extended Kalman filter, as discussed in Chapter 4. The only difference between the equations shown here and the Kalman filter equations for a linear dynamic system are in the definitions of the augmented matrices given by Eqs. (10.32), (10.36b), (10.36g), (10.36k), and (10.37).

The extended Kalman filter produces time histories of the parameter estimates, since the parameters are elements of the augmented state vector. As a result, the extended Kalman filter is a recursive algorithm which is applicable to real-time parameter estimation. Despite the dynamic equation in Eq. (10.33), the parameter estimates can exhibit time variations, due to the measurement updates. Equation (10.33) simply specifies that the estimated parameters cannot change during the propagation of the augmented state from one sample time to the next. Therefore, the extended Kalman filter can track parameter variations along with the states of the dynamic system. Because of the inherent feedback in the algorithm (cf. Eq. (10.36h), the extended Kalman filter can be applied to an unstable system. The measurement updates continually correct the state estimates and prevent divergence. An arbitrary nonlinear dynamic model can be used, because the system matrices are computed by partial differentiation of the nonlinear functions $\boldsymbol{f}(\boldsymbol{x}_a, \boldsymbol{u})$ and $\boldsymbol{h}(\boldsymbol{x}_a)$, cf. Eqs. (10.36g) and (10.36k).

The extended Kalman filter has some disadvantages. As pointed out in Chapter 4, the Kalman gain matrix cannot be computed in advance, as for a linear filter. In addition, the parameter estimates are generally correlated with the state estimates. This correlation can decrease the accuracy of the parameter estimates. Values for $\boldsymbol{Q}$ and $\boldsymbol{R}$ must be chosen, and it is difficult to find a method to justify these choices. Consequently, the values are usually treated as tuning parameters, and are selected using simulation cases or flight test data for which batch parameter estimates are known. Estimates of the initial states and their variances can be obtained from initial measurements. However, the initial values $\boldsymbol{x}_{a_0}$ and $\boldsymbol{P}_{a_0}$ must also include initial estimates of the parameters and their variances. Parameter estimation results from the extended Kalman filter can be sensitive to the choices made for these quantities. Further problems can appear in convergence of the parameter estimates, due to Eq. (10.33), cf. Ref. [10.10]. This problem can be ameliorated by assuming that the parameter dynamics are driven by random noise $\boldsymbol{w}_\theta$, as in Section 10.2.2, so that Eq. (10.33) would be replaced by

$$\dot{\boldsymbol{\theta}} = \boldsymbol{w}_\theta \tag{10.38}$$

Applications of the extended Kalman filter to aircraft state and parameter estimation can be found, e.g., in Refs. [10.10]-[10.11].

10.3 Sequential Least Squares

Ordinary least squares can be applied repeatedly to recent time segments of measured data to generate a sequence of parameter estimation results. This is the equivalent of approximating time-varying parameters with parameter estimates that are piecewise constant with respect to time. The ordinary least-squares solution applied to the cost function in Eq. (10.23) results in

$$\hat{\boldsymbol{\theta}}(n) = \left[\boldsymbol{M}_\lambda(n)\right]^{-1} \boldsymbol{S}_\lambda(n) \tag{10.39a}$$

$$Cov\left[\hat{\boldsymbol{\theta}}(n)\right] = \sigma^2 \left[\boldsymbol{M}_\lambda(n)\right]^{-1} \tag{10.39b}$$

where

$$\boldsymbol{M}_\lambda(n) = \sum_{i=n-m}^{n} \lambda^{n-i}\, \boldsymbol{x}(i)\boldsymbol{x}^T(i) \tag{10.40a}$$

$$\boldsymbol{S}_\lambda(n) = \sum_{i=n-m}^{n} \lambda^{n-i}\, \boldsymbol{x}(i)\, z(i) \tag{10.40b}$$

The fit error variance estimate $\hat{\sigma}^2 = s^2$ can be computed using Eqs. (10.19), or from prior data.

The quantities $\boldsymbol{M}_\lambda(n)$ and $\boldsymbol{S}_\lambda(n)$ can be updated recursively using

$$\boldsymbol{M}_\lambda(n) = \lambda \boldsymbol{M}_\lambda(n-1) + \boldsymbol{x}(n)\boldsymbol{x}^T(n) \tag{10.41a}$$

$$\boldsymbol{S}_\lambda(n) = \lambda \boldsymbol{S}_\lambda(n-1) + \boldsymbol{x}(n)\, z(n) \tag{10.41b}$$

In the preceding expressions, the data window is not limited, but repeated multiplications by the forgetting factor $\lambda < 1$ make the influence of the older data approach zero.

The sequential estimates and covariance matrix can be computed from Eqs. (10.39) at any sample time n, using the latest updated values of $\boldsymbol{M}_\lambda(n)$ and $\boldsymbol{S}_\lambda(n)$ from Eqs. (10.41) and $\hat{\sigma}^2 = s^2(n)$ from Eqs. (10.19). However, the parameter estimate update calculations are typically done at a slower rate, such as 1 or 2 Hz. This saves computation time, while also providing a good approximation to the time variation in the parameters. The matrix inversion $\left[\boldsymbol{M}_\lambda(n)\right]^{-1}$ in Eqs. (10.39) can be done efficiently using Cholesky factorization or singular value decomposition, cf. Ref. [10.12].

10.4 Regularization

Aircraft flight often includes extended periods of steady conditions. If such periods are at least as long as the data window implemented by the forgetting factor λ in Eq. (10.23), then the information in the regressors is mainly noise. This obviously has very detrimental effects on the accuracy of the estimated parameters. To address the problem, prior information about the parameters can be introduced to regularize the information matrix, and thereby numerically stabilize the parameter estimation. This is the same singular information matrix issue discussed in Chapter 5. The discussion here is modified slightly for the real-time parameter estimation problem, and is based on the work presented by Ward et al. [10.13].

To include prior information in the parameter estimation process, the cost function in Eq. (10.23) can be augmented with constraint equations of the form

$$\boldsymbol{C}(t) = \boldsymbol{L}\boldsymbol{\theta}(t) \tag{10.42}$$

where $\boldsymbol{C}(t)$ is a time-varying vector of constraint values, and $\boldsymbol{L}$ is a constant matrix. If the number of constraint equations is n_c, then $\boldsymbol{L}$ is $n_c \times n_p$, and $\boldsymbol{C}(t)$ is an $n_c \times 1$ vector. This general form of the constraint equations can be used to incorporate prior parameter estimates (spatial constraints), or to limit time variations in the real-time parameter estimates (temporal constraints), or to enforce relationships among the parameters (physical constraints). The augmented cost function takes the form

$$\begin{aligned} J[\boldsymbol{\theta}(n)] &= \frac{1}{2} \sum_{i=n-m}^{n} \lambda^{n-i} \left[z(i) - \boldsymbol{x}^T(i)\boldsymbol{\theta}(n) \right]^2 \\ &\quad + \frac{\gamma}{2} \left[\boldsymbol{C}(n) - \boldsymbol{L}\boldsymbol{\theta}(n) \right]^T \boldsymbol{W} \left[\boldsymbol{C}(n) - \boldsymbol{L}\boldsymbol{\theta}(n) \right] \end{aligned} \tag{10.43}$$

where $\boldsymbol{W}$ is a diagonal weighting matrix and γ is proportional to the area under the windowing function applied to the residuals in the cost function,

$$\gamma = \sum_{i=n-m}^{n} \lambda^{n-i} \tag{10.44}$$

The constant γ is included in the constraint term of the cost function so that the weighting implemented by the matrix $\boldsymbol{W}$ will not be affected by the choice of λ. The diagonal elements of the weighting matrix $\boldsymbol{W}$ quantify the relative influence of the prior information in each constraint equation, compared to recent measured data. Weighting matrix elements are usually chosen using simulation data or flight test data for which batch estimates of the parameters are known.

Taking the partial derivative of the cost function with respect to $\boldsymbol{\theta}(n)$, and setting the result equal to zero gives the normal equations,

$$\sum_{i=n-m}^{n} \lambda^{n-i} \boldsymbol{x}(i)\boldsymbol{x}^T(i)\boldsymbol{\theta}(n) + \gamma \boldsymbol{L}^T \boldsymbol{W} \boldsymbol{L} \boldsymbol{\theta}(n)$$
$$- \sum_{i=n-m}^{n} \lambda^{n-i} \boldsymbol{x}(i) z(i) - \gamma \boldsymbol{L}^T \boldsymbol{W} \boldsymbol{C}(n) = \boldsymbol{0} \tag{10.45}$$

Solving these equations for $\boldsymbol{\theta}(n)$ gives the constrained least-squares estimate,

$$\hat{\boldsymbol{\theta}}(n) = \left[\boldsymbol{M}_\lambda(n) + \gamma \boldsymbol{L}^T \boldsymbol{W} \boldsymbol{L}\right]^{-1} \left[\boldsymbol{S}_\lambda(n) + \gamma \boldsymbol{L}^T \boldsymbol{W} \boldsymbol{C}(n)\right] \tag{10.46}$$

where

$$\boldsymbol{M}_\lambda(n) = \sum_{i=n-m}^{n} \lambda^{n-i} \boldsymbol{x}(i)\boldsymbol{x}^T(i) \tag{10.47a}$$

$$\boldsymbol{S}_\lambda(n) = \sum_{i=n-m}^{n} \lambda^{n-i} \boldsymbol{x}(i) z(i) \tag{10.47b}$$

In Eq. (10.46), the prior information in the constraints leads to a regularization term for the matrix being inverted. The term due to the prior information in the rightmost brackets of Eq. (10.46) also influences the parameter estimates, unless there is significant information in the recent data, in which case the $\boldsymbol{S}_\lambda(n)$ term dominates. The selection of weighting matrix $\boldsymbol{W}$ determines the information level for transition between the estimates being influenced by the prior information as opposed to recent measured data.

The quantities $\boldsymbol{M}_\lambda(n)$ and $\boldsymbol{S}_\lambda(n)$ can be updated recursively as shown previously in Eqs. (10.40). With these recursive updates, real-time parameter estimation can be done using Eq. (10.46) to estimate the parameter vector at selected times, which is just the sequential least squares approach described earlier. This use of sequential least squares with prior information to regularize the parameter estimation has been called modified sequential least squares (cf. Ref. [10.13]). The approach is analogous to the mixed estimator described in Chapter 5, so that the parameter covariance matrix can be calculated from

$$Cov\left[\hat{\boldsymbol{\theta}}(n)\right] = \left[\boldsymbol{M}_\lambda(n)/\sigma^2 + \gamma \boldsymbol{L}^T \boldsymbol{W} \boldsymbol{L}\right]^{-1} \tag{10.48}$$

where σ^2 can be estimated using data smoothing methods from Chapter 11 or prior information.

10.5 Recursive Orthogonalization

In Chapter 5, multivariate orthogonal functions were generated from ordinary multivariate functions in the explanatory variables using a Gram-Schmidt orthogonalization procedure or a $\boldsymbol{QR}$ decomposition, applied to all of the data at once. For real-time operation, recursive orthogonalization can be implemented starting from a standard $\boldsymbol{QR}$ decomposition of the matrix of candidate regressors,

$$X = QR \tag{10.49}$$

where the columns of X contain the candidate modeling functions, Q is an orthonormal matrix with the same dimensions as X, and R is a square upper-triangular matrix. The recursive $\boldsymbol{QR}$ decomposition process is initialized by applying a $\boldsymbol{QR}$ decomposition algorithm to the X matrix built from the first n_c data points, where n_c is the number of candidate modeling functions. Implementations of $\boldsymbol{QR}$ decomposition algorithms are available in many numerical analysis software packages, including MATLAB®. Substituting the decomposition in Eq. (10.49) into Eq. (5.9b),

$$R^T R\hat{\theta} = R^T Q^T z \tag{10.50}$$

where

$$Q^T Q = I \tag{10.51}$$

for the orthonormal matrix Q. Assuming R is nonsingular,

$$R\hat{\theta} = Q^T z \tag{10.52}$$

From Eq. (10.52), the elements of $\hat{\theta}$ can be found by simple back substitution, because R is an upper-triangular matrix. Note that Eq. (10.52) is just an alternate form of Eq. (5.9b), and therefore is a form of the least-squares solution. However, Eq. (10.52) is convenient for recursion, because the R matrix must be an upper-diagonal $n_c \times n_c$ matrix, and only the inner products of the orthonormal columns of Q with the dependent variable vector z appear in the equation, and not the Q matrix itself. Consequently, the dimension of both sides of Eq. (10.52) is always $n_c \times 1$, regardless of the number of data points N.

Writing Eq. (10.52) in component form,

$$\begin{bmatrix} r_{11} & r_{12} & \cdots & r_{1n_c} \\ 0 & r_{22} & \cdots & r_{2n_c} \\ \vdots & & \ddots & \vdots \\ 0 & \cdots & 0 & r_{n_c n_c} \end{bmatrix} \begin{bmatrix} \hat{\theta}_1 \\ \hat{\theta}_2 \\ \vdots \\ \hat{\theta}_{n_c} \end{bmatrix} = \begin{bmatrix} \boldsymbol{q}_1^T z \\ \boldsymbol{q}_2^T z \\ \vdots \\ \boldsymbol{q}_{n_c}^T z \end{bmatrix} \tag{10.53}$$

where $\boldsymbol{q}_j$, is the jth column of the matrix $\boldsymbol{Q}$. The right side of Eq. (10.53) is a vector of projections of the dependent variable vector z onto the orthonormal functions in the columns of $\boldsymbol{Q}$. The absolute values of these quantities indicate the degree of correlation of the orthonormal functions in the columns of $\boldsymbol{Q}$ with z, and consequently, the effectiveness of each orthonormal function in modeling the dependent variable data.

When new data arrive, Eq. (10.53) is augmented by appending the new data in the bottom row,

$$\begin{bmatrix} r_{11} & r_{12} & \cdots & r_{1n_c} \\ 0 & r_{22} & \cdots & r_{2n_c} \\ \vdots & & \ddots & \vdots \\ 0 & \cdots & 0 & r_{n_c n_c} \\ \xi_1 & \xi_2 & \cdots & \xi_{n_c} \end{bmatrix} \begin{bmatrix} \hat{\theta}_1 \\ \hat{\theta}_2 \\ \vdots \\ \hat{\theta}_{n_c} \end{bmatrix} = \begin{bmatrix} \boldsymbol{q}_1^T z \\ \boldsymbol{q}_2^T z \\ \vdots \\ \boldsymbol{q}_{n_c}^T z \\ \zeta \end{bmatrix} \tag{10.54}$$

where $\begin{bmatrix} \xi_1 & \xi_2 & \cdots & \xi_{n_c} \end{bmatrix}$ is the new row of data for the $\boldsymbol{X}$ matrix, and ζ is the new dependent variable data. To maintain the $\boldsymbol{QR}$ decomposition including the appended data, the matrix multiplying the parameter vector must be transformed so that the last row contains all zeros. This can be done by applying Givens rotation matrices, as described in Refs. [10.14]-[10.15]. For example, to remove the value of ξ_1 from the bottom row, the Givens rotation matrix is

$$\boldsymbol{G}_1 = \begin{bmatrix} c & 0 & \cdots & 0 & s \\ 0 & 1 & \cdots & & 0 \\ \vdots & & \ddots & & \vdots \\ 0 & \cdots & 0 & 1 & 0 \\ -s & 0 & \cdots & 0 & c \end{bmatrix} \tag{10.55a}$$

where

$$c = r_{11} \Big/ \sqrt{r_{11}^2 + \xi_1^2} \qquad s = \xi_1 \Big/ \sqrt{r_{11}^2 + \xi_1^2} \tag{10.55b}$$

Applying the rotation matrix $\boldsymbol{G}_1$ to Eq. (10.54) gives

$$\begin{bmatrix} cr_{11}+s\xi_1 & cr_{12}+s\xi_2 & \cdots & cr_{1n_c}+s\xi_{n_c} \\ 0 & r_{22} & \cdots & r_{2n_c} \\ \vdots & & \ddots & \vdots \\ 0 & \cdots & 0 & r_{n_c n_c} \\ 0 & -sr_{12}+c\xi_2 & \cdots & -sr_{1n_c}+c\xi_{n_c} \end{bmatrix} \begin{bmatrix} \hat{\theta}_1 \\ \hat{\theta}_2 \\ \vdots \\ \hat{\theta}_{n_c} \end{bmatrix} = \boldsymbol{G}_1 \begin{bmatrix} \boldsymbol{q}_1^T \boldsymbol{z} \\ \boldsymbol{q}_2^T \boldsymbol{z} \\ \vdots \\ \boldsymbol{q}_{n_c}^T \boldsymbol{z} \\ \zeta \end{bmatrix} \tag{10.56}$$

This makes the lower left element in the matrix equal to zero. This process is repeated until the last row of the matrix on the left side contains all zeros, which indicates that the $\boldsymbol{QR}$ decomposition has been updated. In general, this will involve n_c rotation matrices, which are applied to both sides of Eq. (10.54), resulting in

$$\begin{bmatrix} r'_{11} & r'_{12} & \cdots & r'_{1n_c} \\ 0 & r'_{22} & \cdots & r'_{2n_c} \\ \vdots & & \ddots & \vdots \\ 0 & \cdots & 0 & r'_{n_c n_c} \\ 0 & 0 & \cdots & 0 \end{bmatrix} \begin{bmatrix} \hat{\theta}_1 \\ \hat{\theta}_2 \\ \vdots \\ \hat{\theta}_{n_c} \end{bmatrix} = \begin{bmatrix} \boldsymbol{q}_1'^T \boldsymbol{z} \\ \boldsymbol{q}_2'^T \boldsymbol{z} \\ \vdots \\ \boldsymbol{q}_{n_c}'^T \boldsymbol{z} \\ \varepsilon \end{bmatrix} \tag{10.57}$$

where the primed notation indicates the updated $\boldsymbol{QR}$ decomposition, including the appended data. The value ε remaining in the bottom row on the right side of Eq. (10.57) is the residual for the appended data point, assuming a model that includes all the $\boldsymbol{q}_j$, $j=1,2,\ldots,n_c$. Equivalently, this is the portion of the appended dependent variable data that cannot be projected onto the updated orthonormalized candidate modeling functions. The combined rotation matrix for the orthonormalization update process is

$$\boldsymbol{G} = \boldsymbol{G}_{n_c} \ldots \boldsymbol{G}_2 \boldsymbol{G}_1 \tag{10.58}$$

After all of the rotations have been applied to arrive at Eq. (10.57), the bottom row of zeros in the matrix on the left side is discarded. The residual ε can be used for real-time residual variance estimation or stored for colored residual calculations required for accurate parameter covariance estimation. The process is then repeated as each new data sample arrives. The result is a recursive algorithm that efficiently updates the orthonormalization of the n_c candidate modeling functions using simple matrix multiplications after each new data point is appended to the $\boldsymbol{QR}$ decomposition from the previous time step. The recursive orthonormalization is implemented with n_c simple matrix multiplications, which can be done very efficiently. Note that the model parameters $\hat{\theta}_j$, $j = 1, 2, \ldots, n_c$, are associated with the original multivariate modeling functions in the columns of $\boldsymbol{X}$, and not with the orthonormal functions in the columns of $\boldsymbol{Q}$.

As before, the ordinary multivariate functions can be arbitrary functions of the explanatory variables, and the orthogonalization is dependent on the order of the ordinary multivariate functions in the columns of $\boldsymbol{X}$. However, this order can be modified in the recursive $\boldsymbol{QR}$ decomposition, by reordering the columns of $\boldsymbol{R}$, then applying the appropriate Givens rotations to produce an upper-triangular $\boldsymbol{R}$. This is convenient because the reordering effectively changes the columns of the $\boldsymbol{Q}$ matrix, but the columns of $\boldsymbol{Q}$ are never directly manipulated. The reordered $\boldsymbol{QR}$ decomposition corresponds to an $\boldsymbol{X}$ matrix modified with the same column permutations.

The multivariate orthogonal function modeling approach described in Section 5.4.4 can be applied without modification to the recursive $\boldsymbol{QR}$ decomposition to implement real-time model structure determination and model parameter estimation. When using the $\boldsymbol{QR}$ decomposition, the relevant expressions become even simpler, because the orthogonalized candidate modeling functions in the columns of $\boldsymbol{Q}$ are orthonormal, cf. Eq. (10.51). When the multivariate orthogonal functions are orthonormal, Eqs. (5.143) and (5.145) are simplified to

$$\hat{a}_j = \boldsymbol{q}_j^T z \qquad j = 1, 2, \ldots, n_c \tag{10.59}$$

$$J(\hat{\boldsymbol{a}}) = \frac{1}{2}\left[z^T z - \sum_{j=1}^{n_c} \left(\boldsymbol{q}_j^T z \right)^2 \right] \tag{10.60}$$

The quantities $\boldsymbol{q}_j^T z$, $j = 1, 2, \ldots, n_c$, appearing in Eq. (10.60) are the parameter estimates for the orthonormal multivariate functions from Eq. (10.59), which are obtained directly from the right side of the recursive $\boldsymbol{QR}$ decomposition in Eq. (10.57). The squared elements on the right side of the $\boldsymbol{QR}$ decomposition quantify the reduction in the least-squares cost function associated with adding the corresponding orthonormal function to the model individually, regardless of the terms already in the model.

This approach is described and applied to real-time global aerodynamic modeling from flight data in Ref. [10.16].

10.6 Sequential Least Squares in the Frequency Domain

Sequential least squares can also be implemented in the frequency domain [10.17]-[10.25]. This approach has important practical advantages, as will be discussed later. As in time-domain sequential least squares, parameter estimation is repeated at short intervals to produce piecewise-constant estimates for time-varying model parameters in a linear model structure. The identified models involve perturbations of aircraft states, controls, and outputs from values associated with a reference flight condition.

10.6.1 Equation-Error in the Frequency Domain

The finite Fourier transform of a signal $x(t)$ is defined by (cf. Chapter 7)

$$\tilde{x}(\omega) \equiv \int_0^T x(t)\, e^{-j\omega t}\, dt \tag{10.61}$$

which can be approximated by

$$\tilde{x}(\omega) \approx \Delta t \sum_{i=0}^{N-1} x(i)\, e^{-j\omega i \Delta t} \tag{10.62}$$

where $x(i) \equiv x(i\Delta t)$, and Δt is the sampling interval. The summation in Eq. (10.62) is the discrete Fourier transform discussed in Chapter 7,

$$X(\omega) \equiv \sum_{i=0}^{N-1} x(i)\; e^{-j\omega\, i\Delta t} \tag{10.63}$$

The finite Fourier transform approximation in Eq. (10.62) can be written as

$$\tilde{x}(\omega) \approx \Delta t\, X(\omega) \tag{10.64}$$

Chapter 11 details some fairly straightforward corrections that can be made to remove the inaccuracy resulting from the fact that Eq. (10.64) is a simple Euler approximation to the finite Fourier transform of Eq. (10.61). However, if the sampling rate is much higher than the frequencies of interest (as is often the case for flight data analysis), then the corrections are relatively small and can be safely ignored for real-time parameter estimation, to reduce computations.

Applying the Fourier transform to Eq. (10.1b), for example, gives

$$j\omega_k\, \tilde{q}(k) = M_\alpha\, \tilde{\alpha}(k) + M_q\, \tilde{q}(k) + M_\delta\, \tilde{\delta}(k) \qquad k = 1,2,\ldots,M \tag{10.65}$$

where the Δ notation indicating perturbation quantities has been dropped. The Fourier transforms are done for M frequencies of interest ω_k, $k = 1,2,\ldots,M$.

The index k denotes the dependence of the Fourier transform values on the frequencies ω_k.

The least-squares cost function to be minimized is

$$J(\boldsymbol{\theta}) = \frac{1}{2}\sum_{k=1}^{M} \left| j\omega_k \, \tilde{q}(k) - M_\alpha \, \tilde{\alpha}(k) - M_q \, \tilde{q}(k) - M_\delta \, \tilde{\delta}(k) \right|^2 \quad (10.66)$$

where the vertical lines denote the modulus of the complex number enclosed. The symbol $\tilde{q}(k)$ denotes the Fourier transform of the measured pitch rate for frequency ω_k, and similarly for the other terms. This is the equation-error method in the frequency domain, described in Section 7.5. Similar cost expressions can be written for other individual state and output equations given earlier in this chapter, using either dimensional or nondimensional model parameters.

As in Chapter 7, estimation of the unknown parameters can be formulated as a standard least-squares regression problem with complex data. For this example,

$$\tilde{\boldsymbol{z}} = \tilde{\boldsymbol{X}}\boldsymbol{\theta} + \tilde{\boldsymbol{\nu}} \quad (10.67)$$

where

$$\tilde{\boldsymbol{z}} \equiv \begin{bmatrix} j\omega_1 \, \tilde{q}(1) \\ j\omega_2 \, \tilde{q}(2) \\ \vdots \\ j\omega_M \, \tilde{q}(M) \end{bmatrix} \quad (10.68)$$

$$\tilde{\boldsymbol{X}} \equiv \begin{bmatrix} \tilde{\alpha}(1) & \tilde{q}(1) & \tilde{\delta}(1) \\ \tilde{\alpha}(2) & \tilde{q}(2) & \tilde{\delta}(2) \\ \vdots & \vdots & \vdots \\ \tilde{\alpha}(M) & \tilde{q}(M) & \tilde{\delta}(M) \end{bmatrix} \quad (10.69)$$

$$\boldsymbol{\theta} = \begin{bmatrix} M_\alpha \\ M_q \\ M_\delta \end{bmatrix} \quad (10.70)$$

and $\tilde{\boldsymbol{\nu}}$ represents the complex equation error in the frequency domain. The least-squares cost function is

$$J(\boldsymbol{\theta})=\frac{1}{2}\left(\tilde{z}-\tilde{\boldsymbol{X}}\boldsymbol{\theta}\right)^{\dagger}\left(\tilde{z}-\tilde{\boldsymbol{X}}\boldsymbol{\theta}\right) \tag{10.71}$$

which is the same as Eq. (10.66) when Eqs. (10.68)-(10.70) are used. The parameter vector estimate that minimizes this cost function is computed from (cf. Section 7.6),

$$\hat{\boldsymbol{\theta}}=\left[Re\left(\tilde{\boldsymbol{X}}^{\dagger}\tilde{\boldsymbol{X}}\right)\right]^{-1}Re\left(\tilde{\boldsymbol{X}}^{\dagger}\tilde{z}\right) \tag{10.72}$$

with estimated parameter covariance matrix

$$Cov\left(\hat{\boldsymbol{\theta}}\right)\equiv E\left[\left(\hat{\boldsymbol{\theta}}-\boldsymbol{\theta}\right)\left(\hat{\boldsymbol{\theta}}-\boldsymbol{\theta}\right)^{T}\right]=\sigma^{2}\left[Re\left(\tilde{\boldsymbol{X}}^{\dagger}\tilde{\boldsymbol{X}}\right)\right]^{-1} \tag{10.73}$$

The equation-error variance σ^2 can be estimated from the residuals,

$$\hat{\sigma}^{2}=\frac{1}{\left(M-n_{p}\right)}\left[\left(\tilde{z}-\tilde{\boldsymbol{X}}\hat{\boldsymbol{\theta}}\right)^{\dagger}\left(\tilde{z}-\tilde{\boldsymbol{X}}\hat{\boldsymbol{\theta}}\right)\right] \tag{10.74}$$

where n_p is the number of elements in parameter vector $\boldsymbol{\theta}$, so $n_p = 3$ for this example. Parameter standard errors are computed as the square root of the diagonal elements of the $Cov\left(\hat{\boldsymbol{\theta}}\right)$ matrix from Eq. (10.73), using $\hat{\sigma}^2$ from Eq. (10.74).

To implement sequential least-squares parameter estimation in the frequency domain, the preceding parameter estimation equations are applied to frequency-domain data at selected time intervals. The frequency-domain data must be available at any time, so the Fourier transforms are computed using a recursive Fourier transform, which is described next.

10.6.2 Recursive Fourier Transform

For a given frequency ω, the discrete Fourier transform in Eq. (10.63) at sample time $i\Delta t$ is related to the discrete Fourier transform at time $(i-1)\Delta t$ by

$$X_i(\omega)=X_{i-1}(\omega)+x(i)\,e^{-j\omega i\Delta t} \tag{10.75}$$

where

$$e^{-j\omega i\Delta t}=e^{-j\omega\Delta t}e^{-j\omega(i-1)\Delta t} \tag{10.76}$$

The quantity $e^{-j\omega\Delta t}$ is constant for a given frequency ω and constant sampling interval Δt. It follows that the discrete Fourier transform can be computed for a given frequency at each time step using one addition in Eq. (10.75) and two multiplications – one in Eq. (10.76) using the stored constant $e^{-j\omega\Delta t}$ for frequency ω, and one in Eq. (10.75). There is no need to store the time-domain data in memory when computing the discrete Fourier transform in this way, because each sampled data point is processed immediately. Time-domain data from all preceding maneuvers can be used in all subsequent analysis by simply continuing the recursive calculation of the Fourier transform. In this sense, the recursive Fourier transform acts as memory for the information in the data. More data from more maneuvers improves the quality of the data in the frequency domain without increasing memory requirements to store it. Furthermore, the Fourier transform is available at any time $i\,\Delta t$. The approximation to the finite Fourier transform is completed using Eq. (10.64).

The recursive computation of the Fourier transform does not use a Fast Fourier Transform (FFT) algorithm, and therefore would be comparatively slow, if the entire frequency band up to the Nyquist frequency $1/(2\Delta t)$ (see Chapter 8) were of interest. However, rigid-body dynamics of aircraft lie in a rather narrow frequency band of approximately 0.01-2.0 Hz. Since the frequency band is limited, it is efficient to compute the discrete Fourier transform using Eqs. (10.75) and (10.76), which are a recursive formulation of Eq. (10.63), for only the selected frequencies ω_k, $k = 1,2,\ldots,M$. With this approach, it is possible to select closely spaced fixed frequencies for the Fourier transform and still do the calculations efficiently.

Excluding zero frequency removes trim values and measurement biases, so it is not necessary to estimate bias parameters. Using a limited frequency band for the Fourier transformation confines the data analysis to the frequency band where the dynamics lie, and automatically filters wideband measurement noise or structural response outside the frequency band of interest. These automatic filtering features are important for real-time applications, where instrumentation error corrections (see Chapter 9) and noise filtering (see Chapter 11) would require additional computational resources that may not be available.

In past work on fighter aircraft longitudinal short-period modeling, frequency spacing of 0.04 Hz on the interval [0.1-1.5] Hz was found to be adequate, giving 36 evenly-spaced frequencies for each transformed time-domain signal [10.17]. Finer frequency spacing requires slightly more computation, but was found to have little effect on the results. When the frequency spacing is very coarse, there is a danger of omitting important frequency components, and this can lead to inaccurate parameter estimates. In general, a good rule of thumb is to use frequencies evenly spaced at 0.04 Hz over the bandwidth for the dynamic system. For good results, the bandwidth should be limited to the frequency range where the signal components in the frequency domain are at least twice the amplitude of the wideband noise level. However, the algorithm is robust to these design choices, so the selections to be made are not difficult.

For airplane dynamic modeling, the number of time-domain signals to be transformed is usually low (typically 10 – more if there are additional control surfaces), so this approach requires a small amount of computer memory. In the frequency domain, the memory required is fixed and independent of the time length of the flight maneuvers. Each state, control, or output requires memory for M complex numbers to hold the current values of its Fourier transform for M frequencies.

The recursive Fourier transform update need not be done for every sample time point. Systematically skipping time points effectively lowers the sampling rate of the data prior to Fourier transformation. This saves computation, and provided that aliasing is avoided (see Chapter 8), does not adversely impact the parameter estimation results until the Fourier transform update rate gets below approximately 5 times the highest frequency of interest for the dynamic system. This lower limit on the update rate for the recursive Fourier transform is simply to provide sufficient sampling to accurately characterize the frequency content in the measured signals. The parameter estimation and covariance calculations in Eqs. (10.72)-(10.74) can be done at any time, but are usually done at 1 or 2 Hz, to save computations. Linearized aerodynamic characteristics rarely change faster than this, except in cases of strong nonlinearity, damage, failure, or rapid maneuvering. For these cases, the update rate can be increased, at the cost of more computations.

The states, controls, and outputs in the linear equations are perturbation quantities, not the measured quantities. Therefore, it is first necessary to remove the reference values, which would more generally be called the constant offsets or biases in cases where there is no distinct steady reference condition. The bias and low-frequency drift or trend with time can be removed using a high-pass filter for each signal before the recursive Fourier transform is applied. If the bias and trend in each signal are not removed before the recursive Fourier transform, then the relatively large spectral components at low frequency due to the bias and trend will spill over to neighboring frequencies through a process called leakage (cf. Chapter 11). This pollutes the frequency content at the lower frequencies, which is detrimental to the modeling results. This is the same detrending issue discussed earlier in Section 7.7.2.

As discussed in Chapter 7, the estimated parameter standard errors computed from the covariance matrix in the frequency domain do not require correction for colored residuals. Standard errors computed using Eq. (10.73) are therefore a good representation of the error in the estimated parameters. High-quality error measures are important for problems such as failure detection and control law reconfiguration. The fit error variance estimate from Eq. (10.74) is well-conditioned and appropriate for each covariance matrix calculation. This means that the error bounds computed at any time are an appropriate representation of the parameter accuracy, based on the information available.

The recursive Fourier transform in Eqs. (10.75) and (10.76) represents a data information memory for as long as the running sum is incremented. When the aircraft dynamics change, the older data should be discounted in some way, as was done in the time domain using the forgetting factor. If this is not done, then the speed of response for the real-time parameter estimator is progressively degraded, as new information has to overwhelm an increasingly longer memory. Consequently, there is a trade-off between the desired rapid response of the parameter estimator to changes in the aircraft dynamics, versus retaining enough information from past data for sufficiently accurate model parameter estimates.

It is possible to trigger a reduction in the magnitudes of the Fourier transforms computed from past data, based on an event such as a detected failure or a significant increment in the information content of the measured data. In the latter case, the detected increase in recent data information content would presumably support an update to the parameter estimates. In practice, the criteria and discounting schemes for data forgetting depend on the problem and the aircraft.

If past values of the Fourier transform $X_i(\omega)$ computed from Eq. (10.75) are saved in computer memory, then it is possible to implement selective amnesia by simply subtracting past values of the running sum corresponding to the Fourier transform, or differences between past values of the running sum. For example, forgetting all data information content older than 10 seconds (i.e., removing that data information content from the complex regression problem) could be implemented by subtracting the value of the running sums for the Fourier transforms at 10 seconds ago from the current running sums. Similarly, to forget data information content collected between 5 and 7 seconds ago, the difference between the running sums at 5 and 7 seconds ago would be subtracted from the current running sum. The price to pay for this capability is the computer memory required to store past values of the running sums associated with the Fourier transforms for each signal at each frequency. The memory requirements could be reduced by perhaps only saving the running sums at intervals of 2 seconds, for example.

The simplicity of Eq. (10.75) makes it easy to see how exponential data forgetting could be implemented for the recursive Fourier transform. As discussed earlier, for exponential data forgetting, each past value of the time-domain signal is multiplied by a forgetting factor λ, $0 < \lambda < 1$, at each time step. In this way, old data are gradually devalued and eventually discarded. To implement this, Eq. (10.75) is modified slightly to

$$X_i(\omega) = \lambda X_{i-1}(\omega) + x(i)\, e^{-j\omega i \Delta t} \tag{10.77}$$

Everything else remains the same as before. This simple approach is possible because the Fourier transform is linear with respect to the measured data $x(i)$.

The challenge with data forgetting is not in the implementation, but rather in deciding how much data information content to forget, and when. There are currently no concrete guidelines for choosing a value for λ, nor for determining when and how much to forget using selective amnesia. Consequently, these design choices are typically made based on simulation runs and analysis of previous flight data. Alternatively, the problem can be simply restarted by setting the recursive Fourier transforms to zero, based on an event, such as a flight condition change, the start of a new maneuver, a change to the aircraft configuration, or some type of damage or failure.

Using the recursive Fourier transform with a limited frequency band also provides the important advantage of automatic data smoothing, because wideband noise components are automatically removed, since the high-frequency components are never transformed into the frequency domain. That same process also enhances the signal-to-noise ratio for the frequency-domain data and makes the technique robust to high-frequency noise and structural dynamic response, as well as occasional data dropouts. Furthermore, any periods of steady flight have no adverse impact on the model parameter estimates or uncertainty measures, because any low-information flight data dominated by noise will have no effect on the recursive Fourier transforms applied using a limited frequency band. This avoids both parameter estimate and uncertainty measure inaccuracies that plague many other real-time modeling methods. Because the technique is both a frequency-domain technique and an equation-error technique, the approach works well and without modification for open-loop unstable aircraft.

Because sequential least squares in the frequency domain employs linear regression with complex numbers, incorporating prior information can be done in the same manner as explained in Chapter 5. In the case of real-time dynamic modeling, including prior information of this kind can reduce variations in the real-time parameter estimates, and improve convergence speed.

By analogy to the Bayesian cost function in Eq. (4.28), or equivalently, the mixed estimator formulation of the cost function in Eq. (5.188),

$$J(\boldsymbol{\theta}) = \frac{1}{2\sigma}\left(\tilde{z} - \tilde{X}\boldsymbol{\theta}\right)^{\dagger}\left(\tilde{z} - \tilde{X}\boldsymbol{\theta}\right) + \frac{1}{2}\left(\boldsymbol{\theta} - \boldsymbol{\theta}_p\right)^T \boldsymbol{\Sigma}_p^{-1}\left(\boldsymbol{\theta} - \boldsymbol{\theta}_p\right) \tag{10.78}$$

where $\boldsymbol{\theta}_p$ is a vector of parameter estimates from a prior analysis, with associated covariance matrix $\boldsymbol{\Sigma}_p$. The vector of parameter estimates that minimize this modified least squares cost function is

$$\hat{\boldsymbol{\theta}} = \left[Re\left(\tilde{\boldsymbol{X}}^{\dagger} \tilde{\boldsymbol{X}} \right) + \boldsymbol{\Sigma}_p^{-1} \right]^{-1} \left[Re\left(\tilde{\boldsymbol{X}}^{\dagger} \tilde{\boldsymbol{z}} \right) + \boldsymbol{\Sigma}_p^{-1} \boldsymbol{\theta}_p \right] \tag{10.79}$$

with covariance matrix

$$Cov\left(\hat{\boldsymbol{\theta}}\right) = \sigma^2 \left[Re\left(\tilde{\boldsymbol{X}}^{\dagger} \tilde{\boldsymbol{X}} \right) + \boldsymbol{\Sigma}_p^{-1} \right]^{-1} \tag{10.80}$$

where σ^2 is estimated from Eq. (10.74), as before.

Sequential least squares parameter estimation in the frequency domain is very practical, computationally simple, and produces highly accurate modeling results with valid error measures. Two independent evaluations by different groups [10.26]-[10.27] have identified sequential least squares parameter estimation in the frequency domain as the best method available for real-time dynamic modeling. The method has been used for real-time dynamic modeling in flight for many different situations, such as normal and unusual flight conditions [10.17], [10.19]-[10.20], [10.24], icing conditions [10.21]-[10.22], open-loop instability [10.19]-[10.20], no air flow angle (α, β) sensors [10.23], and turbulence [10.25].

Example 10.2

Returning again to the data from the lateral maneuver on the NASA Twin Otter aircraft, the yawing moment parameters will now be estimated using sequential least squares in the frequency domain. The measured data were plotted in Fig. 5.4. The same model equation (5.78c) was used, but the data were transformed into the frequency domain using the recursive Fourier transform of Eqs. (10.74) and (10.75) for frequencies spaced evenly at 0.04 Hz over the frequency range [0.1, 1.5] Hz, which corresponds to a spacing of 0.25 rad/sec over the range [0.63, 9.42] rad/sec.

Figure 10.4 shows the sequential parameter estimates, including the estimated 95 percent confidence interval (±2 standard errors), shown as a vertical bar on each estimate. Parameter estimation calculations in Eqs. (10.72)-(10.74) were done at 1 Hz, and the recursive Fourier transforms were updated with each measurement, at 50 Hz. The marker at the right of each plot indicates the batch least-squares parameter estimate using time-domain data. After 9 seconds, the sequential parameter estimates change very little, and the confidence intervals are so small that the symbols representing the parameter estimates obscure them.

The parameter estimates shown were based only on the measured data – the algorithm was started with no prior information about the parameters or their uncertainties. Final values of the sequential parameter estimates in the frequency domain matched the parameter estimates from batch least squares in the time domain, as expected.

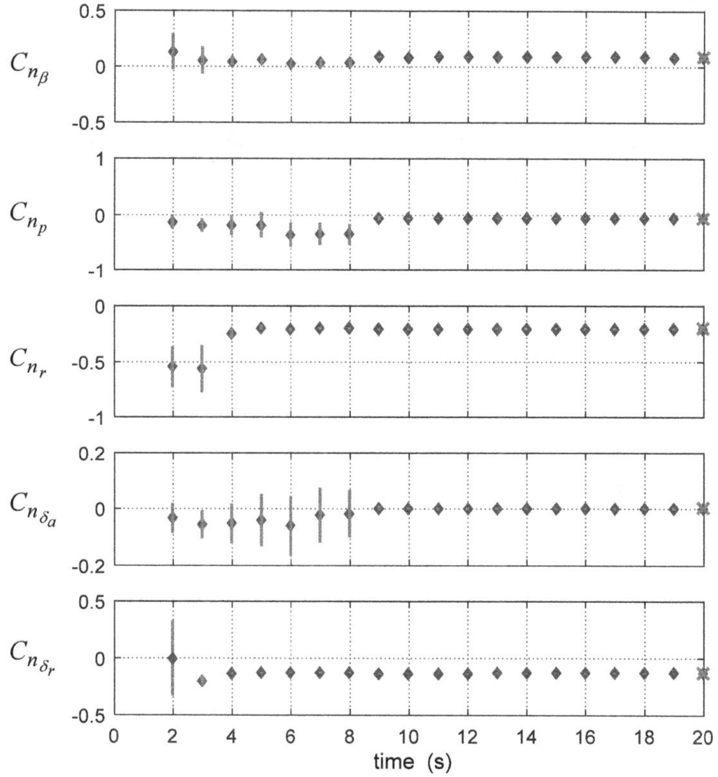

***Figure 10.4* Sequential least squares parameter estimates in the frequency domain for lateral maneuver, run 1**

Estimated standard errors shown in Fig. 10.4 accurately reflect the quality of each sequential estimate, including the effect of colored residuals. This comes about because the modeling is done in the frequency domain, where the fit error variance estimate is well-defined and the residual coloring is properly incorporated in the parameter covariance matrix calculation. The parameter estimates have larger uncertainties at the beginning of the maneuver, due to low information. The uncertainties decrease as information becomes available to the estimator, and the estimated uncertainties do not change when there is no excitation (i.e., no additional information) at the end of the maneuver. This is an accurate characterization of the quality of the parameter estimates.

Note that the C_{n_p} parameter has some uncertainty until the aileron begins to move. This happens because the initial rudder movement excites the roll rate somewhat, but not as much as the aileron. Once the aileron moves, the roll rate is excited, and the estimator can accurately determine both C_{n_p} and $C_{n_{\delta_a}}$. The other parameters are already accurately determined by this time, because of excitation from the rudder.

Figure 10.5 for sequential least squares in the frequency domain shows behavior similar to that shown in the lower plot of Fig. 10.2 for recursive least squares in the time domain. Again, the aileron derivative estimate does not converge to an accurate and steady value until the aileron is moved, providing the necessary information. ■

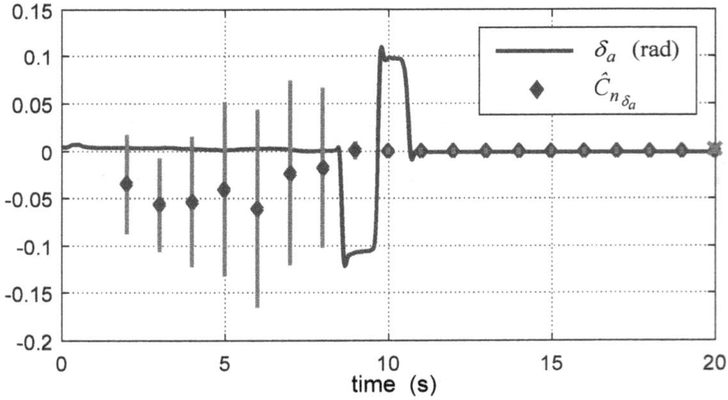

Figure 10.5 **Aileron control derivative estimate convergence using sequential least squares in the frequency domain**

10.7 Summary and Concluding Remarks

This chapter presented methods that can be used in real time to estimate aircraft dynamic model parameters, based on measured flight data. The methods were recursive least squares, Kalman filter, extended Kalman filter, sequential least squares, modified sequential least squares, and sequential least squares in the frequency domain. The first technique is a recursive formulation of batch least-squares linear regression. The Kalman filter approach modifies this slightly by introducing a stochastic dynamic model for the time evolution of the model parameters. The extended Kalman filter provides simultaneous estimates of state variables and model parameters. This technique involves a nonlinear estimation problem, even for a linear dynamic system model. Sequential least squares applies batch methods to a recent measured data record. Regularization methods were introduced to address information deficiency problems. Sequential least squares in the frequency domain applies least squares parameter estimation to frequency-domain data generated by a recursive Fourier transform in a limited frequency band. This technique has important practical advantages, including robustness to measurement noise, biases, and data dropouts, accurate parameter estimates and uncertainty measures, computational simplicity, and efficient storage of data information content.

The success of real-time parameter estimation depends mainly on data information content issues and practical considerations.

All real-time methods must contend with the issue of how far back in time the data record should extend to form the basis for parameter estimation. This issue can be addressed using a forgetting factor to decrease the influence of older data, or by including various constraints in the cost function for parameter estimation.

For airplane flight, lack of information content in the data can be problematic, because normal flight operations include extended periods where the state, control, and output variables are fairly constant. During these times, the dynamic content of the signals are at or below the (relatively constant) noise level. In this circumstance, a time-domain regression method will generally give very inaccurate parameter estimates unless the estimation is regularized by including a term in the cost function that penalizes movement of the parameters away from *a priori* values (e.g., values from wind tunnel tests or prior modeling based on flight data), and/or a term that penalizes time variation of the parameter estimates. Tuning parameters must be adjusted for this approach, because the magnitude of the penalty term(s) must be balanced relative to the least-squares part of the cost function associated with the measured data. For sequential least squares in the frequency domain, this problem does not exist, because the recursive Fourier transform applied in a limited frequency band only changes when there is data information content in the frequency band of interest.

The data information content problem can also be addressed by implementing a very long data memory. But this has the disadvantage that new data are combined with old data, resulting in parameter estimates that are some weighted average over the entire data record, which may include several disparate flight conditions or configurations. Adaptation to new situations then becomes progressively slower, as new data information must overcome an increasingly long memory with past information. The recursive Fourier transform used in sequential least squares parameter estimation in the frequency domain efficiently implements a long data memory. To make this method, and others, responsive to sudden changes in the aircraft dynamics, there must be some method for deciding how much of the past data information content should be forgotten. Actually implementing the data memory loss in any of the methods is straightforward – the difficult question is how much to forget and when.

In the Kalman filter or extended Kalman filtering approaches, weighting matrices that represent assumed measurement and process noise covariances are used to discriminate signal from noise. These matrices are usually treated as tuning parameters, which are adjusted in simulation, often in an *ad hoc* fashion. These methods also sometimes exhibit convergence problems.

Standard errors for the model parameter estimates, which are important in all practical applications, generally cannot be computed accurately using recursive time-domain methods. An exception is the recursive colored residual correction for the parameter covariance matrix in recursive least squares parameter estimation, described in Section 5.1. Sequential least squares in the frequency domain computes accurate parameter estimates and uncertainty measures because the analysis is done in the frequency domain using a limited frequency band.

In real-time operation, there is no time for typical data conditioning to remove problems such as noise, biases, and infrequent data dropouts. Time-domain methods are sensitive to these problems, and will produce inaccurate parameter estimates as a consequence, whereas applying the recursive Fourier transform in a limited frequency band largely removes these errors without further processing. This makes sequential least squares in the frequency domain robust to data anomalies, as well as computationally simple.

A problem that relates to both data information content and practical considerations is data collinearity due to the control system. Many control laws move more than one control surface at the same time in a nearly proportional way, or move control surfaces nearly in proportion to state variables. When states and controls are nearly proportional to one another, data collinearity exists, and it is difficult to identify individual parameters from the measured data alone, as discussed in Chapter 5. When the proportionality is perfect, the task becomes impossible without some regularization or assumptions to make the parameter estimation tractable. The data collinearity problem appears often when real-time parameter estimation is attempted on aircraft operating normally, as opposed to being flight tested specifically for parameter estimation. Chapter 8 covered input design methods that can be used to excite the aircraft response so that real-time parameter estimation methods will provide useful and accurate results. Some data information augmentation of this type is required for accurate parameter estimates at times when the aircraft is in a steady flight condition and/or has a feedback control system operating.

Real-time parameter estimation can improve efficiency and effectiveness of stability and control flight testing and flight envelope expansion. The quality of flight test maneuver excitations and the required duration of those excitations can be determined readily using real-time parameter estimation results. In addition, many types of indirect reconfigurable control and failure detection require real-time parameter estimation. Other uses for real-time parameter estimation include real-time safety monitoring, flight envelope protection, and accident investigation.

References

[10.1] Goodwin, G.C. and Payne, R.L., *Dynamic System Identification: Experiment Design and Data Analysis*, Academic Press, New York, NY, 1977.

[10.2] Norton, J.P., *An Introduction to Identification*, Academic Press, London, UK, 1986.

[10.3] Young, P.C., *Recursive Estimation and Time-Series Analysis*, Springer-Verlag, New York, NY, 1984.

[10.4] Hsia, T.C., *System Identification*, Lexington Books, D.C. Heath and Company, Lexington, MA, 1977.

[10.5] Ljung, L., System Identification – Theory for the User, 2nd Ed., Prentice Hall, Upper Saddle River, NJ, 1999.

[10.6] Holzel, M.S. and Morelli, E.A. "Real-Time Frequency Response Estimation from Flight Data," *Journal of Guidance, Control, and Dynamics*, Vol. 35, No. 5, September-October 2012, pp. 1406-1417.

[10.7] Grauer, J.A. and Morelli, E.A. "Parameter Covariance for Aircraft Aerodynamic Modeling using Recursive Least Squares," AIAA-2016-2009, *AIAA SciTech 2016 Conference*, San Diego, CA, January 2016.

[10.8] Cao, L. and Schwartz, H.M., "The Kalman Filter Based Recursive Algorithm: Windup and Its Avoidance," *Proceedings of the Automatic Control Conference*, Arlington, VA, 2001, pp. 3606-3611.

[10.9] Sorenson, H.W. "Least-squares estimation: from Gauss to Kalman," *IEEE Spectrum*, July 1970, pp. 63-68.

[10.10] Jategaonkar, R. and Plaetschke, E., "Estimation of Aircraft Parameters Using Filter Error Methods and Extended Kalman Filter," DFVLR Forschungsbericht 88-15, 1988.

[10.11] Garcia-Velo, J. and Walker, B., "Aerodynamic Parameter Estimation for High-Performance Aircraft Using Extended Kalman Filter," AIAA Paper 95-3500, *AIAA Atmospheric Flight Mechanics Conference*, Baltimore, MD, 1995.

[10.12] Press, W.H., Teukolsky, S.A., Vettering, W.T., and Flannery, B.R., *Numerical Recipes in FORTRAN: The Art of Scientific Computing*, 2nd Ed., Cambridge University Press, New York, NY, 1992.

[10.13] Ward, D.G., Monaco, J.F., and Bodson, M., "Development and Flight Testing of a Parameter Identification Algorithm for Reconfigurable Control," *Journal of Guidance, Control, and Dynamics*, Vol. 21, No. 6, 1998, pp. 948-956.

[10.14] Gentleman, W.M. "Least Squares Computation by Givens Transformation Without Square Roots," *Journal of the Institute for Mathematics Applications*, Vol. 12, 1973, pp. 329-336.

[10.15] Gentleman, W.M. "Regression Problems and the QR Decomposition," *Bulletin of the Institute for Mathematics Applications*, Vol. 10, 1974, pp. 195-197.

[10.16] Morelli, E.A. "Real-Time Global Nonlinear Aerodynamic Modeling for Learn-To-Fly," AIAA-2016-2010, *AIAA SciTech 2016 Conference*, San Diego, CA, January 2016.

[10.17] Morelli, E.A., "Real-Time Parameter Estimation in the Frequency Domain," *Journal of Guidance, Control, and Dynamics*, Vol. 23, No. 5, 2000, pp. 812-818.

[10.18] Morelli, E.A., "In-Flight System Identification," AIAA paper 98-4261, *AIAA Atmospheric Flight Mechanics Conference*, Boston, MA, 1998.

[10.19] Morelli, E.A. "Multiple Input Design for Real-Time Parameter Estimation in the Frequency Domain," Paper REG-360, *13th IFAC Symposium on System Identification*, Rotterdam, The Netherlands, August 2003.

[10.20] Morelli, E.A. and Smith, M.S. "Real-Time Dynamic Modeling – Data Information Requirements and Flight Test Results," *Journal of Aircraft*, Vol. 46, No. 6, November-December 2009, pp. 1894-1905.

[10.21] Ranaudo, R., Martos, B., Norton, B., Gingras, D.R., Barnhart, B. Ratvasky, T.P., and Morelli, E.A. "Piloted Simulation to Evaluate the Utility of a Real Time Envelope Protection System for Mitigating In-Flight Icing Hazards," NASA TM-2011-216951, October 2011.

[10.22] Gingras, D.R., Barnhart, B., Ranaudo, R., Martos, B., Ratvasky, T.P., and Morelli, E.A. "Development and Implementation of a Model-Driven Envelope Protection System for In-Flight Ice Contamination," NASA TM-2011-216960, October 2011.

[10.23] Morelli, E.A. "Real-Time Aerodynamic Parameter Estimation without Air Flow Angle Measurements," *Journal of Aircraft*, Vol. 49, No. 4, July-August 2012, pp. 1064-1074.

[10.24] Morelli, E.A. "Flight Test Maneuvers for Efficient Aerodynamic Modeling," *Journal of Aircraft*, Vol. 49, No. 6, November-December 2012, pp. 1857-1867.

[10.25] Morelli, E.A. and Cunningham, K., "Aircraft Dynamic Modeling in Turbulence," AIAA-2012-4650, *AIAA Atmospheric Flight Mechanics Conference*, Minneapolis, MN, 2012.

[10.26] Basappa, K. and Jategaonkar, R., "Evaluation of Recursive Methods for Aircraft Parameter Estimation," AIAA-2004-5063, *AIAA Atmospheric Flight Mechanics Conference and Exhibit*, Providence, RI, 2004.

[10.27] Song, Y., Campa, G., Napolitano, M.R., Seanor, B., and Perhinschi, M.G., "Online Parameter Estimation Techniques Comparison Within a Fault Tolerant Flight Control System," *Journal of Guidance, Control, and Dynamics*, Vol. 25, No. 3, 2002, pp. 528-537.

Problems

10.1 Answer the following, using words, graphs, equations, and/or diagrams:

a) Explain why real-time parameter estimation methods generally use an equation-error approach, rather than an output-error approach.

b) What is the difference between a recursive parameter estimation method and a sequential parameter estimation method? If there is a difference, is it important? Why or why not?

c) Explain one practical use of real-time dynamic modeling.

d) Explain why the computed error bounds for the estimated model parameters are generally wrong when using conventional recursive least squares in the time domain.

10.2 Using a pure sine wave with unit amplitude and frequency 1 Hz over a 10 s time length, use rft.m to investigate when the recursive transform is exactly correct and when it is not. Is the error acceptable?

10.3 Repeat Example 10.2 using prior information $C_{n_{\delta_a}} = -0.01 \pm 0.002$ for the aileron derivative. How does the prior information change the parameter estimate history plot in Fig. 10.5?

10.4 Use the real-time parameter estimation demonstration called rtpid_demo.m to investigate different piloted inputs for the F-16 longitudinal short-period dynamics. What are the characteristics of a good or bad input? Provide an explanation for each characteristic you identify.

10.5 Using Twin Otter data from Example 10.2 in totter_f1_017_data.mat, compare the conventional recursive least squares covariance matrix calculation in the time domain to the recursive calculation accounting for colored residuals in the time domain. Compare the results to sequential least squares in the frequency domain.

Chapter 11

Data Analysis

This chapter describes some common operations performed on measured data to condition or transform the data for system identification purposes. Most of the techniques have very general applicability, but were selected because they have been found useful for aircraft system identification. The selected techniques certainly do not represent the full spectrum of techniques available. Reference [11.1] is an excellent resource for techniques of the type described here.

The first two sections of the chapter discuss filtering and smoothing, which are concerned with extracting deterministic signal from noisy measured time series. This is an important operation in system identification, because the general aim is to identify a mathematical model based on the deterministic parts of the measurements. The next section is concerned with interpolation, which provides a means to reconstruct missing or bad data points, and has links to ideas used for smoothing. Specialized methods for smoothed numerical differentiation are described next. These methods find use in equation-error modeling and sensor position error correction. Practical methods for accurately computing the finite Fourier transform and power spectral density estimates, which form the basis for the frequency-domain methods of Chapter 7, are explained in detail. Finally, the chapter concludes with discussions of methods for comparing signal waveforms and visualizing aircraft motion during a flight test maneuver using animated computer graphics.

11.1 Filtering

Filters intended for removing random noise or other unwanted signal components can be implemented in hardware using analog electronic components, or by a digital computer implementing transfer functions. Filters behave like dynamic systems, with associated amplitude and phase changes as a function of input frequency. Consequently, a filter represents additional system dynamics introduced between the physical system and the signal used to identify dynamic models. Since the objective of system identification is often to characterize the amplitude and phase angle changes from input to output for the physical system, filtering can distort the modeling results by modifying the amplitude and phase of the measured signals.

The danger occurs when the break frequency of the filter is near the frequencies of interest, because amplitude and phase angle changes due to the filter are greatest near that frequency. This can be seen from the frequency response plot for a typical low-pass filter shown in Figure 8.4. The anti-aliasing filter, with break frequency set according to Eq. (8.6), does not introduce significant amplitude or phase angle modification to the frequencies of interest, because the frequencies of interest are much lower than the break frequency of the anti-aliasing filter.

Once the data are sampled, digital filtering can be applied using the same principles discussed in Chapter 8 for anti-aliasing filters. Some analysts apply identical digital filtering to all the measured signals. This approach is fine as long as only linear relationships are being investigated, but is not acceptable for nonlinear modeling. The reason is that inputs to a linear dynamic system produce outputs at the same frequencies, with possibly altered amplitude and phase angle, as described by the transfer function of the system. For a nonlinear system, this is not true, so amplitude changes and phase shifts from the filtering are not applied uniformly when the same filtering is used for all measured signals. Since the signal amplitudes and phase angle relationships are at the core of system identification, it is preferable to separate signal from noise using zero-phase-shift filtering or smoothing, with very low amplitude distortion.

All filtering methods use only current and past data, as discussed in Chapter 4. While this approach must be taken for real-time applications, post-flight data processing can use subsequent data as well. The associated additional information gives significant advantages to smoothing techniques. Furthermore, filtering methods assume that the analyst knows what the frequency range for the filtering should be. In system identification applications, it is not always clear what this frequency band should be, and a good choice for the frequency band is often not the same for different measured time series. This is important, because the modeling will be adversely affected by either admitting noise into the signals or by discarding parts of the deterministic signals. A global smoothing method based on Fourier analysis, described in the next section, can provide information on appropriate cut-off frequencies to be used for system identification.

For post-flight analysis, an ordinary digital filter can be run forward and backward in time, which effectively cancels the phase effects of the filter and squares the amplitude. This makes any amplitude distortion worse, but also doubles the filter order, so that the frequency cut-off is sharper.

11.2 Smoothing

Smoothing uses data points both before and after the data point to be smoothed. Since subsequent data are used, smoothing can only be applied post-flight, or in real time up to a prior time, a few time steps into the past. However, these situations constitute the majority of cases for system identification applied to aircraft.

11.2.1 Frequency-Domain Filter Implemented in the Time-Domain

A filter defined in the frequency domain can be implemented in terms of smoothing weights in the time domain, without introducing phase shifts in the smoothed data. Graham [11.2] describes a procedure for doing this, which is summarized here.

A low-pass filter can be defined in the frequency domain according to

$$\tilde{g}(\omega) = \begin{cases} 1 & |\omega| \le \omega_c \\ \dfrac{1}{2}\left\{ cos\left[\dfrac{\pi(\omega+\omega_c)}{(\omega_t-\omega_c)}\right]+1\right\} & -\omega_t < \omega < -\omega_c \\ \dfrac{1}{2}\left\{ cos\left[\dfrac{\pi(\omega-\omega_c)}{(\omega_t-\omega_c)}\right]+1\right\} & \omega_c < \omega < \omega_t \\ 0 & |\omega| \ge \omega_t \end{cases} \tag{11.1}$$

where ω_c is the selected cut-off frequency for the low-pass filter, and ω_t is the end of the frequency roll-off (see Fig. 11.1). Note that this filter definition is symmetric about the zero frequency axis. Using this definition of the low-pass filter, the inverse Fourier transform

$$g(t) = \frac{1}{2\pi}\int_{-\infty}^{\infty} \tilde{g}(\omega)\, e^{j\omega t}\, d\omega \tag{11.2}$$

can be done analytically, resulting in

$$g(t) = \frac{\pi}{2t}\left[\frac{sin\,\omega_t t + sin\,\omega_c t}{\pi^2 - (\omega_t-\omega_c)^2 t^2}\right] \tag{11.3}$$

The value of $g(t)$ when $t=0$ can be obtained using L'Hôpital's rule,

$$g(0)=\frac{(\omega_t+\omega_c)}{2\pi}$$

In discrete-time form, for $t_k=k\Delta t$,

$$g(k)=\left\{\begin{array}{ll}\dfrac{\pi}{2k\Delta t}\left[\dfrac{\sin\omega_t k\Delta t+\sin\omega_c k\Delta t}{\pi^2-(\omega_t-\omega_c)^2(k\Delta t)^2}\right] & k\neq 0\\[2ex] \dfrac{(\omega_t+\omega_c)}{2\pi} & k=0\end{array}\right\} \tag{11.4}$$

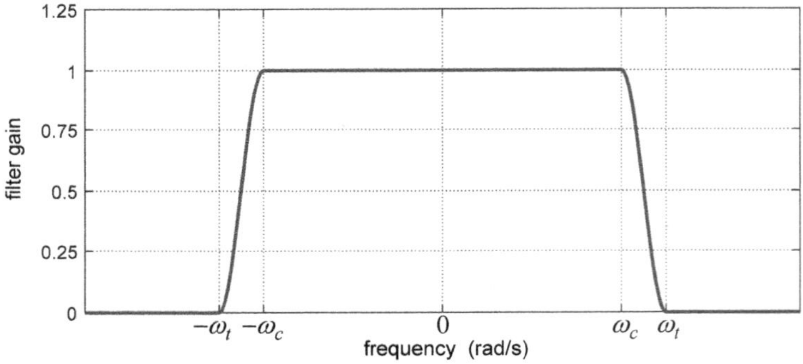

Figure 11.1 **Low-pass filter gain**

The smoothing is brought about by convolving the measured data with the discrete-time smoothing weights $g(k)$, which is equivalent to a multiplication by $\tilde{g}(\omega)$ in the frequency domain (see Appendix A). The values of $g(k)$ are normalized so that

$$\sum_{k=-N_s}^{N_s} g(k)=1 \tag{11.5}$$

where N_s is the number of smoothing weights used on either side of the measured data point to be smoothed. Figure 11.2 shows an example of the smoothing weights $g(k)$ corresponding to the filter definition in Fig. 11.1. The smoothed value of the ith data point is computed by applying the smoothing weights shown in Fig. 11.2 to the measured data, with the weighting centered at the ith point. For N discrete measured values $z(i)$, the smoothed values $z_s(i)$ are computed from

$$z_s(i) = \sum_{k=-N_s}^{N_s} g(k)z(i+k) \qquad i = 1,2,\ldots,N \tag{11.6}$$

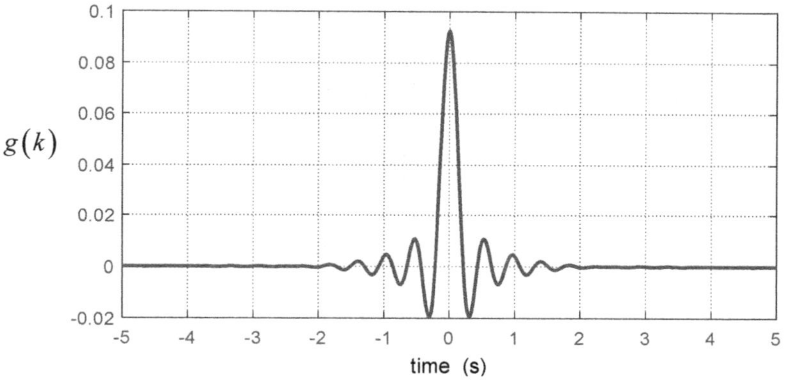

Figure 11.2 **Time-domain fixed smoothing weights**

Near the endpoints, measured values on the interior must be used twice. Since the smoothing weights are symmetric about the smoothed point, this can be done by augmenting the measured data with reflections about the endpoints. Then the convolution can be done in a straightforward manner for all points in the data record, using the augmented data.

To use this method, the frequency cut-off ω_c must be chosen. There is a trade-off between values for ω_t and N_s, with larger N_s giving a sharper roll-off in the frequency domain, and vice versa. The relationship among these quantities that is consistent with achieving less than 0.5 percent gain error in the pass band is

$$N_s \geq \frac{2\pi^2}{\Delta t(\omega_t - \omega_c)} \tag{11.7}$$

In practice, it is easiest to choose N_s fairly large for a given ω_c, e.g., set N_s equal to half the original data length, $N_s = N/2$, then compute ω_t from Eq. (11.7).

The actual frequency characteristic realized by the implemented smoothing weights can be calculated from

$$\hat{\hat{g}}(\omega) = \sum_{k=-N_s}^{N_s} g(k)\cos(\omega k \Delta t) \tag{11.8}$$

since $g(k)$ is an even function that is nonzero only for $-N_s \le k \le N_s$. The implemented filter computed from Eq. (11.8) can be compared to the ideal filter defined in Eq. (11.1).

11.2.2 Local Smoothing in the Time Domain

Another method for data smoothing is local smoothing in the time domain. This involves fitting a local polynomial in time to measured data points near the point to be smoothed. Using a second-order polynomial model implies that the local data points lie on a parabola, except for random errors. This is equivalent to assuming the second derivative is a constant over the same time period. Defining time relative to the data point to be smoothed, the local model to be identified can be written as

$$y = a_0 + a_1 t + \frac{1}{2} a_2 t^2 \tag{11.9}$$

If $z(i)$ denotes the ith measured data point sampled at time $i\Delta t$, then the equations used for the local fit to the data are:

$$\begin{aligned}
a_0 + a_1(-2\Delta t) + a_2(-2\Delta t)^2 &= z(i-2) \\
a_0 + a_1(-\Delta t) + a_2(-\Delta t)^2 &= z(i-1) \\
a_0 &= z(i) \\
a_0 + a_1(\Delta t) + a_2(\Delta t)^2 &= z(i+1) \\
a_0 + a_1(2\Delta t) + a_2(2\Delta t)^2 &= z(i+2)
\end{aligned} \tag{11.10}$$

There are five equations for three unknowns, so the least squares solution of Chapter 5 applies. Once the solution is obtained, the smoothed value for $z(i)$ is simply equal to the estimate of a_0, since the time was defined to be zero at the ith data point for the local modeling. The equations can be written in matrix form as

$$\begin{bmatrix} 1 & -2\Delta t & 4(\Delta t)^2 \\ 1 & -\Delta t & (\Delta t)^2 \\ 1 & 0 & 0 \\ 1 & \Delta t & (\Delta t)^2 \\ 1 & 2\Delta t & 4(\Delta t)^2 \end{bmatrix} \begin{bmatrix} a_0 \\ a_1 \\ a_2 \end{bmatrix} = \begin{bmatrix} z(i-2) \\ z(i-1) \\ z(i) \\ z(i+1) \\ z(i+2) \end{bmatrix} \tag{11.11}$$

The resulting normal equations are

$$\begin{aligned} 5a_0 + 10(\Delta t)^2 a_2 &= \sum_{k=i-2}^{i+2} z(k) \\ 10\Delta t\, a_1 &= \sum_{k=i-2}^{i+2} (k-i) z(k) \\ 10a_0 + 34(\Delta t)^2 a_2 &= \sum_{k=i-2}^{i+2} (k-i)^2 z(k) \end{aligned} \tag{11.12}$$

so the least squares estimate of a_0 is

$$\hat{y}(i) = \hat{a}_0 = \frac{34}{70} \sum_{k=i-2}^{i+2} z(k) - \frac{1}{7} \sum_{k=i-2}^{i+2} (k-i)^2 z(k) \tag{11.13}$$

or

$$\hat{y}(i) = z_s(i) = \hat{a}_0 = \frac{1}{70}\left[-6z(i-2) + 24z(i-1) + 34z(i) + 24z(i+1) - 6z(i+2)\right] \tag{11.14}$$

Equation (11.14) can be used for local smoothing in the time domain at each interior time point. When $z(i)$ is at or near an endpoint, the nearest local regression can be used for local smoothing. For the example used in Eqs. (11.9)-(11.14), the smoothed value of the first $k = 2$ data points can be computed from Eq. (11.9),

$$\hat{y}(1) = \hat{a}_0 + \hat{a}_1(-2\Delta t) + \frac{1}{2}\hat{a}_2(-2\Delta t)^2 \tag{11.15a}$$

$$\hat{y}(2) = \hat{a}_0 + \hat{a}_1(-\Delta t) + \frac{1}{2}\hat{a}_2(-\Delta t)^2 \tag{11.15b}$$

where the parameter estimates $\hat{a}_0$, $\hat{a}_1$, and $\hat{a}_2$ come from the local regression at $i = k + 1 = 3$. A similar approach can be used at the end of the time series.

The same procedure can be used to implement local smoothing for different numbers of neighboring points k and various approximating polynomial orders n. Since the method involves repeated local smoothing solutions moved along the time series, it is difficult to determine an exact cut-off frequency for smoothing over the entire time series. In general, larger values of k for constant n give lower cut-off frequencies. Increasing n improves the capability of the local model to match higher frequency variations, which moves the cut-off frequency higher.

For $n = 2$, the cut-off frequency in Hz for this smoother can be approximated by

$$f_c = \frac{1}{2k\,\Delta t} \tag{11.16}$$

For $\Delta t = 0.02$ sec, using k=10 and n=2 gives good smoothing results with cut-off frequency near 2.5 Hz. Figure 11.3 shows results from applying local smoothing with k=10 and n=2 to a noisy time series. The lower plot in Fig. 11.3 shows that the local smoothing effectively removed noise without introducing phase shift.

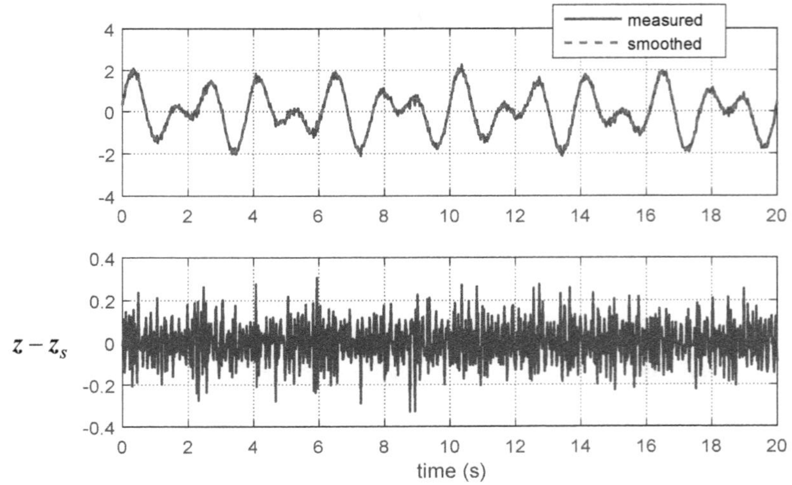

Figure 11.3 **Time-domain local smoothing**

11.2.3 Global Smoothing in the Frequency Domain

Both of the smoothing techniques described earlier compute smoothed values based on local measured data in the time domain. A different approach can be formulated by treating the smoothing problem in a global sense. This approach uses Fourier analysis, and is discussed in Refs. [11.3] and [11.4]. The latter reference is the source of the material presented here.

Consider a noisy measured scalar time history $z(t)$ sampled at constant intervals Δt,

$$z(i) = z(i\Delta t) \qquad i = 0,1,2,\ldots,N-1 \tag{11.17}$$

The goal is to separate signal and noise, or equivalently to find $y(i)$ and $v(i)$ such that

$$z(i) = y(i) + v(i) \qquad i = 0,1,2,\ldots,N-1 \tag{11.18}$$

The strategy for determining the noise sequence $v(i)$, $i = 0,1,2,\ldots,N-1$, is to find an accurate description for the signal in the frequency domain, invert the Fourier transform to get $y(i)$, then find $v(i)$ from Eq. (11.18) in the form

$$v(i) = z(i) - y(i) \qquad i = 0,1,2,\ldots,N-1 \tag{11.19}$$

Fourier series expansion implicitly assumes that the time series under consideration is periodic. For most measured time series from flight tests, making such an assumption implies discontinuities in the amplitude and first time derivative at the endpoints. Figure 11.4 shows a typical measured time series for sideslip angle from a flight test maneuver. Lanczos [11.3] shows that Fourier series for functions with these discontinuities have much slower convergence than the Fourier series for a function with no discontinuities in either the amplitude or first derivative. In the former case, the magnitudes of the Fourier coefficients decrease asymptotically with k^{-1}, where k is the number of terms in the Fourier series expansion, whereas in the latter case, the asymptotic decrease goes according to k^{-3}. This seems reasonable, because the sinusoids in the Fourier series have no discontinuities in amplitude or first derivative anywhere, and therefore would be expected to have difficulty representing a time series with those discontinuities.

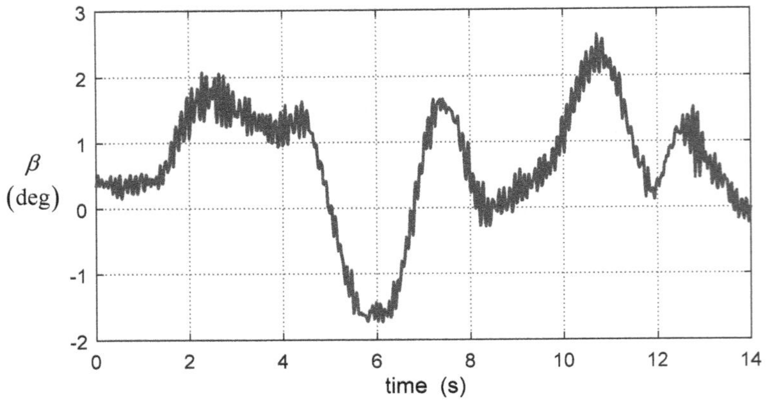

Figure 11.4 **Sideslip angle measured time series**

It is possible to remove the endpoint discontinuities from any arbitrary time series, and thereby achieve the more abrupt decrease in the magnitude of the Fourier coefficients, which corresponds to a faster convergence of the Fourier series expansion. The importance of the higher rate of convergence will become clear in the subsequent discussion. To remove the discontinuities, subtract a linear trend from the original measured time series, in order to make the amplitudes at the endpoints equal to zero, and then reflect the result about the origin to remove the slope discontinuities at the endpoints. Define the new time series as $g(i)$, with $g(-N+1) = g(0) = g(N-1) = 0$. The values of $g(i)$ are computed from

$$g(i) = z(i) - z(0) - i\left[\frac{z(N-1) - z(0)}{N-1}\right] \qquad i = 0,1,2,\ldots,N-1 \qquad (11.20a)$$

$$g(-i) = -g(i) \qquad i = 1,2,\ldots,N-1 \qquad (11.20b)$$

Figure 11.5 shows the result of performing these operations on the time series of Fig. 11.4. The vector

$$\boldsymbol{g} = \left[g(-N+1) \;\; g(-N+2) \;\; \ldots \;\; g(-1) \;\; g(0) \;\; g(1) \;\; \ldots \;\; g(N-1)\right]^T \qquad (11.21)$$

is an odd function of time, and therefore can be expanded using a Fourier sine series:

$$\hat{g}(i) = \sum_{k=1}^{N-1} b(k) \sin\left[k\pi\left(\frac{i}{N-1}\right)\right] \qquad i = 0,1,2,\ldots,N-1 \qquad (11.22)$$

where $\hat{g}(i)$ denotes the approximation to $g(i)$ using the Fourier sine series and the $b(k)$ are Fourier sine series coefficients. The summation over frequency index k omits $k=0$ (zero frequency), since this is a pure sine series for an odd function. Only positive values of i are included in Eq. (11.22), since those values correspond to the original function.

The abrupt k^{-3} decrease in the magnitudes of the Fourier sine series coefficients $b(k)$ can now be expected because the discontinuities in the amplitude and first time derivative at the endpoints have been removed.

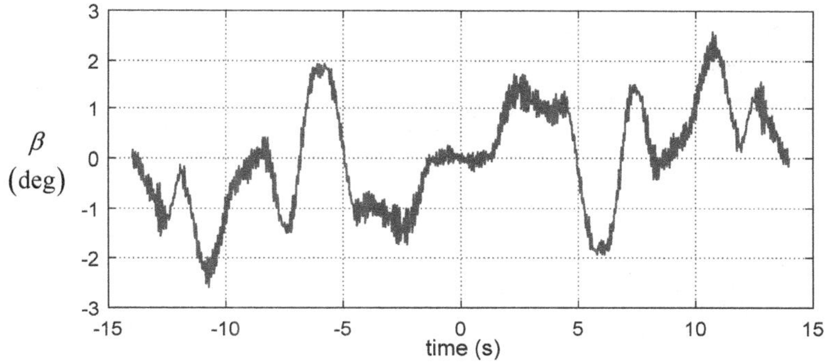

Figure 11.5 **Measured sideslip angle with endpoint discontinuities removed**

The Fourier sine series coefficients for $\boldsymbol{g}$ are computed as [11.3]

$$b(k)=\frac{2}{N-1}\sum_{i=1}^{N-2} g(i)\,sin\left[k\pi\left(\frac{i}{N-1}\right)\right] \qquad k=1,2,\ldots,N-1 \quad (11.23)$$

where the index i runs from 1 to $(N-2)$, because $g(0)$ and $g(N-1)$ are zero. The upper limit for k is $N-1$, which corresponds to the Nyquist frequency. From Eq. (11.22), the period T_k of the kth Fourier sine series term is given by

$$T_k=\frac{2(N-1)\Delta t}{k} \qquad (11.24a)$$

so the kth frequency f_k is related to the frequency index k by

$$f_k=\frac{k}{2(N-1)\Delta t} \qquad (11.24b)$$

To effectively smooth the data, it is necessary that the Nyquist frequency $f_N = 1/(2\Delta t)$ be much higher than the highest frequency of the deterministic components in the signal. This consideration is rarely a problem with modern data acquisition systems on aircraft, when proper attention is paid to analog anti-aliasing filtering before sampling (cf. Chapter 8).

Using Eqs. (11.20) to construct $\boldsymbol{g}$ guarantees that any endpoint discontinuities are in the second-order and higher time derivatives of the original time series. Convergence of the Fourier sine series approximation would be further accelerated if endpoint discontinuities in higher derivatives were also removed. However, since the measured time histories are from a dynamic system where application of forces and moments can produce discontinuities in second-order and higher time derivatives, any attempt to remove additional endpoint discontinuities runs the risk of inadvertently modifying the signal. Therefore, removing amplitude and first-order time derivative endpoint discontinuities is accepted as the best that can be done without corrupting the desired signal.

The Fourier sine series expansion in Eq. (11.22) contains all spectral components that can be computed from the given finite time series, including both signal and noise. The Fourier series for a coherent signal is fundamentally different than the Fourier series for noise. As mentioned earlier, a coherent signal with discontinuities only in the second-order time or higher time derivatives at the endpoints has Fourier series coefficient amplitudes that rapidly decrease asymptotically to zero with increasing k, i.e., $|b(k)|$ is proportional to k^{-3}. On the other hand, since noise is incoherent and theoretically has constant power over the frequency range up to the Nyquist frequency, its Fourier series coefficients do not decrease asymptotically to zero, but instead have a relatively small and constant magnitude for all frequencies. This reflects the fact that the Fourier series expansion for an incoherent time series is divergent, due mainly to the inconsistent phase-amplitude relationships. In the case of a Fourier sine series, the Fourier series coefficients of the noise appear as a relatively constant amplitude oscillation about zero, representing random phase changes in the spectral components. The abrupt decrease in the magnitudes of the Fourier sine series coefficients for the coherent signal contrasts sharply with the relatively constant magnitude of the Fourier sine series coefficients for the noise. The use of Eqs. (11.20) to remove endpoint discontinuities prior to performing the Fourier transform was done specifically to enhance this contrast.

The contrast is so stark, that it is often easy to find the separation point visually. Figure 11.6 shows the situation for the measured time series of Fig. 11.4. For low-pass smoothing typical of aircraft applications, a simple method for implementing the smoothing would be to visually determine the highest frequency in the cubic roll-off of the Fourier sine series coefficients

$b(k)$, set all the sine series coefficients at higher frequencies equal to zero, and then compute the smoothed time series using the inverse Fourier sine transform,

$$\hat{g}_s(i) = \sum_{k=1}^{k_{max}} b(k) sin\left[k\pi\left(\frac{i}{N-1}\right)\right] \qquad i = 0,1,2,\ldots,N-1 \qquad (11.25)$$

where k_{max} is the maximum frequency index determined visually. The end of the cubic roll-off for the coherent signal is defined as the frequency where the Fourier coefficient magnitudes for the coherent signal reach the noise floor, defined by the relatively constant magnitude of the Fourier coefficients at high frequencies. Figure 11.7 shows the simple frequency domain filter that results, based on the frequency domain data in Fig. 11.6.

This simple form of the filter in the frequency domain can be improved upon, in a way that provides a hedge against an inaccurate visual selection of the frequency cut-off. Consider the Fourier transform of the measured time series, which is composed of signal plus noise,

$$\tilde{z}(f) = \tilde{y}(f) + \tilde{v}(f) \qquad (11.26)$$

The goal is to design a frequency domain filter that is optimal in the sense of minimizing the squared difference between the true signal $\tilde{y}(f)$ and the estimated signal $\hat{\tilde{y}}(f)$ over the entire frequency range up to the Nyquist frequency, i.e., the integral

$$\int_0^{f_N} \left[\tilde{y}(f) - \hat{\tilde{y}}(f)\right]^2 df \qquad (11.27)$$

is to be minimized. The estimated signal in the frequency domain will be obtained by multiplying a filter $\Phi(f)$ with $\tilde{z}(f)$,

$$\hat{\tilde{y}}(f) = \Phi(f)\tilde{z}(f) = \Phi(f)\left[\tilde{y}(f) + \tilde{v}(f)\right] \qquad (11.28)$$

Substituting for $\hat{\tilde{y}}(f)$ from Eq. (11.28) and recognizing that $\tilde{y}(f)\tilde{v}(f)$ integrated over all frequencies will be approximately zero due to the incoherence of the noise, Eq. (11.27) becomes

$$\int_0^{f_N} \left\{\tilde{y}^2(f) - 2\Phi(f)\tilde{y}^2(f) + \Phi^2(f)\left[\tilde{y}^2(f) + \tilde{v}^2(f)\right]\right\} df \qquad (11.29)$$

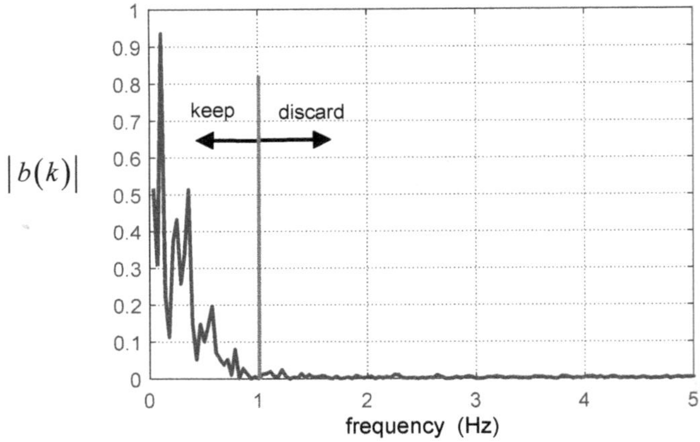

Figure 11.6 **Fourier sine series coefficients for measured sideslip angle with endpoint discontinuities removed**

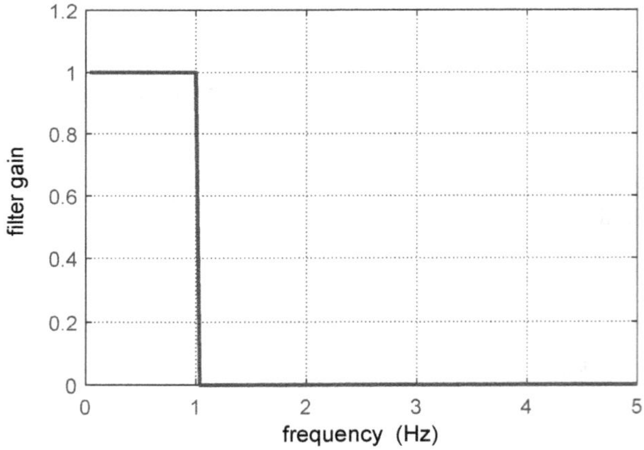

Figure 11.7 **Simple frequency-domain smoothing filter for $|b(k)|$ in Fig. 11.6**

Taking the derivative of the integrand in Eq. (11.29) with respect to $\Phi(f)$, setting the result equal to zero, and solving for $\Phi(f)$ gives

$$\Phi(f) = \frac{\tilde{y}^2(f)}{\tilde{y}^2(f) + \tilde{v}^2(f)} \qquad 0 \le f \le f_N \tag{11.30}$$

or, in terms of the discrete frequency index k,

$$\Phi(k)=\frac{\tilde{y}^2(k)}{\tilde{y}^2(k)+\tilde{v}^2(k)} \qquad 0 \le k \le N-1 \qquad (11.31)$$

Equation (11.31) gives the form of the optimal filter in the frequency domain, also called the Wiener filter [11.1]. For the problem at hand, the Wiener filter can be constructed by assuming that the Fourier coefficients for the signal are those below the frequency cut-off determined visually, and the constant noise level can be estimated using an average of the Fourier coefficient magnitudes for frequencies higher than the selected cut-off. Figure 11.8 shows the Wiener filter constructed in this manner, using Eq. (11.31) and the same frequency cut-off determined for Fig. 11.7, with frequency domain data from Fig. 11.6. The optimal filter from Eq. (11.31) then multiplies the $b(k)$ in the inverse Fourier transform, so that the smoothed signal (with endpoint discontinuities still removed) is computed from

$$\hat{g}_s(i)=\sum_{k=1}^{N-1}\Phi(k)\,b(k)\,sin\left[k\pi\left(\frac{i}{N-1}\right)\right] \qquad i=0,1,2,\ldots,N-1 \quad (11.32)$$

The summation over the frequency index k can be truncated when values of $\Phi(k)\,b(k)$ approach zero.

The Wiener filter computed from Eq. (11.31) is near unity at low frequencies, passing the Fourier sine series components for the coherent signal, then transitions smoothly to near zero at high frequencies, removing Fourier sine series components associated with the noise. The optimality of the Wiener filter gives some room for error in the visual selection of the cut-off frequency. Furthermore, the magnitudes of the $b(k)$ values are small in the region of the cut-off frequency, so including or excluding a few components improperly will have minimal adverse effect on the final smoothed time series.

Naturally, some components of the noise lie in the low-frequency range, but there is no way to distinguish this noise from the signal, which also resides in the low-frequency range. Typically, the large majority of the noise power resides at high frequency relative to the frequencies of the signal, and this noise can be removed very effectively. When the signal-to-noise ratio is high, the power of the noise relative to that of the signal is small in an overall sense, but this situation is improved by the fact that the noise power is spread over a wide frequency range, whereas the signal power resides in a smaller frequency range. Once the high-frequency noise has been removed, the remaining noise components in the frequency range of the signal have very low power, and therefore have little impact on the smoothed time series.

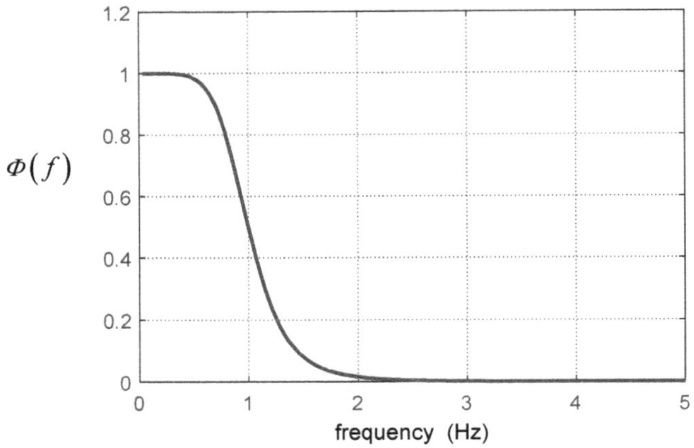

Figure 11.8 **Wiener frequency-domain smoothing filter for $|b(k)|$ in Fig. 11.6**

As a final step, the linear trend removed from the original time series using Eqs. (11.20) must be restored to the smoothed values $\hat{g}_s(i),\ i=0,1,2,\ldots,N-1$ from Eq. (11.32). The final smoothed time series is computed from

$$\hat{y}(i)=z_s(i)=\hat{g}_s(i)+z(0)+i\left[\frac{z(N-1)-z(0)}{N-1}\right] \qquad i=0,1,2,\ldots,N-1 \quad (11.33)$$

In the preceding development, it is clear that the endpoints of the measured time history, $z(0)$ and $z(N-1)$, are completely excluded from all smoothing operations. For very noisy data, this can produce significant error in the smoothed time history. To correct this, the endpoints can be smoothed locally using one of the methods described previously. The cut-off frequency for the local endpoint smoothing should be chosen at roughly the same value selected for the global Fourier smoothing, and the endpoints should be smoothed before the global Fourier smoothing is begun. Using this procedure, all the data points are smoothed using a consistent cut-off frequency. Figure 11.9 shows the smoothed coherent signal found by applying the global Fourier smoother to the measured time history in Fig. 11.4.

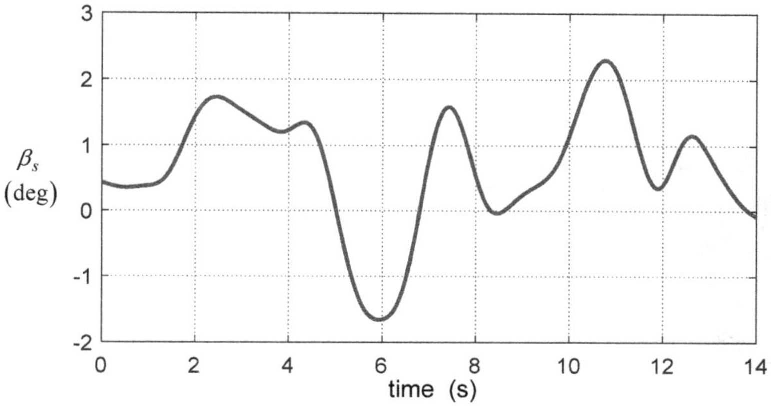

Figure 11.9 **Smoothed sideslip angle time series**

The noise sequence $\nu(i)$, $i = 0,1,2,\ldots,N-1$, can be obtained from Eq. (11.19) using $\hat{y}(i)$ computed from Eqs. (11.32) and (11.33). The noise sequence from the global Fourier smoother has only wideband random components, i.e., the noise is not colored by deterministic modeling error. In practice, this noise can be assumed Gaussian and stationary, so that noise characteristics can be estimated from

$$\bar{\nu} = \frac{1}{N}\sum_{i=1}^{N} \nu(i) \tag{11.34}$$

$$\hat{\sigma}_\nu^2 = \frac{1}{N-1}\sum_{i=1}^{N}\left[\nu(i) - \bar{\nu}\right]^2 \tag{11.35}$$

Since the noise sequence is comprised of relatively wideband and high-frequency random components, the computed estimate for the mean value is close to zero. Figure 11.10 shows the estimated noise sequence associated with Figs. 11.4 and 11.9.

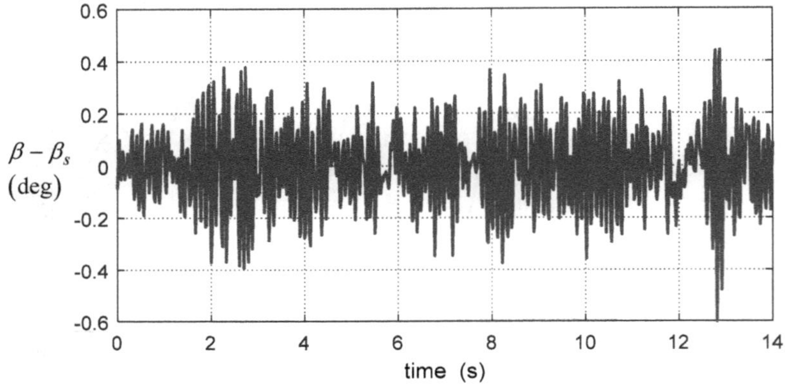

Figure 11.10 **Estimated noise sequence for the sideslip angle time series**

For an experiment with n_o measured outputs, the measured time series can be arranged in an $N \times n_o$ matrix,

$$\boldsymbol{Z} = \begin{bmatrix} z_1 & z_2 & \cdots & z_{n_o} \end{bmatrix} \tag{11.36}$$

where z_j, $j = 1, 2, \ldots, n_o$, are $N \times 1$ vectors of measured outputs. Similarly, define

$$\hat{\boldsymbol{Y}} = \begin{bmatrix} \hat{y}_1 & \hat{y}_2 & \cdots & \hat{y}_{n_o} \end{bmatrix} \tag{11.37}$$

where $\hat{y}_j$, $j = 1, 2, \ldots, n_o$, are $N \times 1$ vectors of the respective smoothed signals. An estimate of the measurement noise covariance matrix $\boldsymbol{R}$ can then be obtained from

$$\hat{\boldsymbol{R}} = \frac{\left(\boldsymbol{Z} - \hat{\boldsymbol{Y}}\right)^T \left(\boldsymbol{Z} - \hat{\boldsymbol{Y}}\right)}{N - 1} \tag{11.38}$$

Noise covariance estimated from measured data in this way can be very useful for experiment design and instrumentation evaluation (Chapter 8), for model structure determination (Chapter 5), and for evaluation of modeling success. The effective separation of signal and noise also makes it easy to compute signal-to-noise ratio for the time series, e.g., using root-mean-square values for the signal and noise time series. This is important for preliminary data evaluation, where the signal-to-noise ratios for the measured outputs should be at least 10 to achieve good modeling results, but useful results can still be obtained with signal-to-noise ratios as low as 3.

Figure 11.11 shows the frequency response data of Fig. 11.6 for the entire frequency range, up to the Nyquist frequency at 25 Hz. The data in Fig. 11.11 illustrate the fact that the global smoothing technique can be used to identify other deterministic components in the data, in addition to the rigid-body deterministic components at low frequency. These other deterministic components show up with the same characteristic cubic decrease in the Fourier sine coefficients.

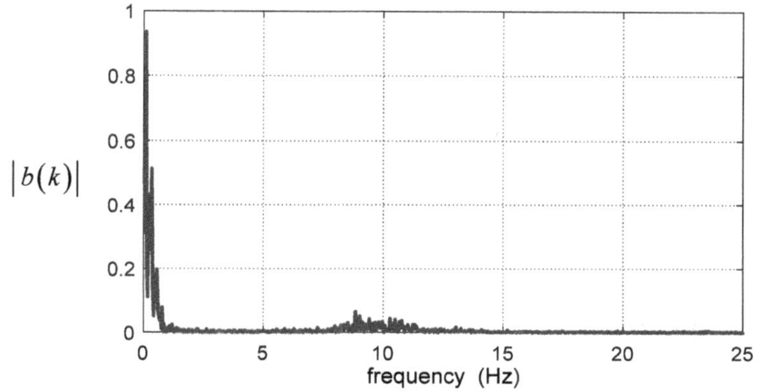

Figure 11.11 **Fourier sine series coefficients for measured sideslip angle with endpoint discontinuities removed for the full frequency range**

In Fig. 11.11, the Fourier coefficients near 9 and 11 Hz are from structural responses of the wingtip boom on which the sideslip angle sensor is mounted and from wing bending. The deterministic structural response can also be isolated and smoothed by simply discarding all the Fourier sine coefficients at frequencies above and below the relatively large components around 9-11 Hz, then using the inverse Fourier transform, as before. This can also be achieved easily with two applications of the global Fourier smoother, using cut-off frequencies near 11 Hz and 9 Hz, then subtracting the resulting smoothed time series. Figure 11.12 shows the deterministic structural response in the frequency range 8-12 Hz, extracted in this way from the sideslip angle measurements shown in Fig. 11.4.

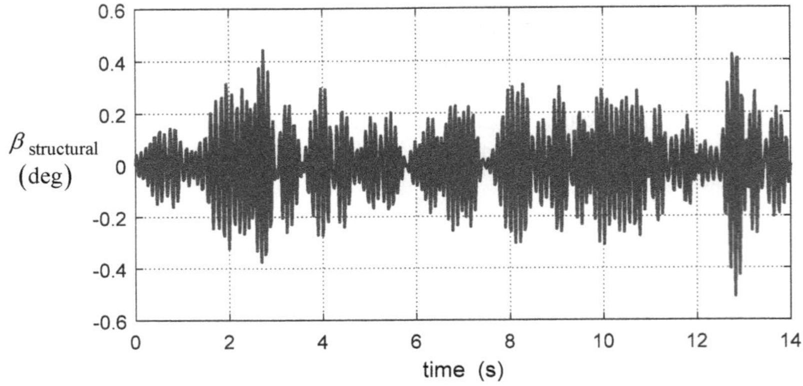

Figure 11.12 **Smoothed structural response from the sideslip angle data**

11.3 Interpolation

Local interpolation in the time domain is similar to local time-domain smoothing described earlier. The difference is that the measurement for the center point in the local smoothing is now missing. Otherwise, the principle of fitting a local polynomial to the nearby data is exactly the same. The interpolation is done by evaluating the local fitted polynomial at the location of the missing data point, which is normally the center location. This gives a locally smoothed value for the interpolated data point. The simplest form of this type of interpolation is the use of a linear function fitted to data points adjacent to the missing point.

The same principle can be used when interpolating data from a lower to a higher sampling rate. When the noise level is low, the simple approach of using linear interpolation between neighboring points can be used; however, a local polynomial model for the deterministic signal gives better results.

A more rigorous approach to generating values at a high sampling rate based on measurements at a lower sampling rate is given by the sampling theorem [11.1]. This theorem states that, for a time series $z(i)$, where the underlying continuous function of time is bandwidth-limited so that all components lie at frequencies below the Nyquist frequency $f_N = 1/(2\Delta t)$, the value of the underlying continuous function can be reconstructed at any time using the measured samples, by the following expression:

$$z(t)=\sum_{k=-\infty}^{\infty} z(k)\frac{sin\left[\pi\left(\frac{t}{\Delta t}-k\right)\right]}{\left[\pi\left(\frac{t}{\Delta t}-k\right)\right]} \tag{11.39}$$

Unfortunately, this expression requires an infinite amount of data. However, a useful approximation can be made using only the available measured data, $z(i),\ i=0,1,2,\ldots,N-1$,

$$z(t)\approx\sum_{k=0}^{N-1} z(k)\frac{sin\left[\pi\left(\frac{t}{\Delta t}-k\right)\right]}{\left[\pi\left(\frac{t}{\Delta t}-k\right)\right]} \tag{11.40}$$

The interpolation results from this method are very good overall, but degrade slightly near the endpoints, because the interpolated points near the endpoints have less measured data nearby. This approach is sensitive to noise, because there is no mechanism to distinguish between deterministic signal and noise in the interpolation.

The global Fourier smoothing technique described above can also be used for interpolation, by simply using a time vector with a higher sampling rate in the reconstruction of the time history via the inverse Fourier transform in Eq. (11.32). A time vector with a higher sampling rate can be implemented by increasing the value of N in Eq. (11.32). This approach has the advantage of smoothing the data in addition to the interpolation, resulting in a more accurate result for noisy measured data.

11.4 Numerical Differentiation

A common problem in data preparation for system identification is computing a numerical time derivative based on noisy measurements. For example, a numerical time derivative is required when computing values of nondimensional aerodynamic moment coefficients based on measured data, using the dynamic equations for rotational motion. This was discussed in conjunction with equation-error methods using linear regression in Chapter 5. Flight test instrumentation on an aircraft often does not include sensors for body-axis angular acceleration, so the values of $\dot{p}$, $\dot{q}$, and $\dot{r}$ in the equations must be computed by numerically differentiating body-axis angular rate measurements p, q, and r. Other uses for numerical differentiation include conducting a rough check on data polarity by comparing $\dot{\alpha}$ and q or $\dot{\beta}$ and $-r$ (see Eqs. (3.34b) and (3.34c)), incorporating time-derivative terms in a model, such as $C_{L_{\dot{\alpha}}} \dfrac{\dot{\alpha}\bar{c}}{2V}$, and correcting sensor measurements to the c.g., discussed in Chapter 9.

The most straightforward method for computing numerical time derivatives is to use finite differences applied to the measured data. However, measured data are usually noisy, and subtracting nearly equal quantities (i.e., neighboring measured values) and dividing by the time step to compute the derivative magnifies the noise.

A better approach is to differentiate the local smoothing solution discussed earlier [11.5]. The local fitted polynomial was identified with sample times defined relative to the current point. If the local polynomial model is differentiated with respect to time, then evaluated at the current time, defined as $t = 0$, the value of the local smoothed derivative is equal to the coefficient of the linear term in the polynomial model identified for local smoothing. Using the local model of Eq. (11.9),

$$\dot{y} = a_1 + a_2 t \tag{11.41}$$

The local smoothed derivative is simply equal to the value of a_1 at the current point, where $t = 0$. This local smoothed derivative calculation is applied at each data point, as was done for local time-domain smoothing. Using this insight, it is possible to compute both smoothed values and smoothed time derivatives for a noisy time series with a single pass through the measured data.

Smoothed numerical differentiation can also be done using the global Fourier smoothing technique described earlier. In this case, the deterministic part of the measured signal is represented as a weighted sine series with an added linear trend [Eqs. (11.33) and (11.32)],

$$\hat{y}(i) = z_s(i) = \hat{g}_s(i) + z(0) + i\left[\frac{z(N-1) - z(0)}{N-1}\right] \qquad i = 0,1,2,\ldots,N-1 \tag{11.42}$$

$$\hat{g}_s(i) = \sum_{k=1}^{N-1} \Phi(k)\, b(k) \sin\left[k\pi\left(\frac{i}{N-1}\right)\right] \qquad i = 0,1,2,\ldots,N-1 \tag{11.43}$$

The smoothed derivative can be found by differentiating the sine series and the linear trend before transforming back to the time domain. The expression for the smoothed derivative is:

$$\frac{d\hat{y}(i)}{dt} = \left[\frac{z(N-1) - z(0)}{N-1}\right] + \sum_{k=1}^{N-1} \Phi(k)\, b(k)\left(\frac{k\pi}{N-1}\right) \cos\left[k\pi\left(\frac{i}{N-1}\right)\right]$$

$$i = 0,1,2,\ldots,N-1 \tag{11.44}$$

11.5 Finite Fourier Transform

In Chapter 7, it was noted that the finite Fourier transform

$$\tilde{x}(f) \equiv \int_0^T x(t)\, e^{-j2\pi f t}\, dt \tag{11.45}$$

can be approximated by

$$\tilde{x}(f) \approx \Delta t \sum_{i=0}^{N-1} x(i)\, e^{-j2\pi f i \Delta t} \tag{11.46}$$

which is a simple Euler approximation for the finite Fourier transform using N discrete samples of the continuous time function $x(t)$. For a conventional finite Fourier transform, the frequencies are chosen as

$$f_k = \frac{k}{N\Delta t} \qquad k = 0,1,2,\ldots,N-1 \tag{11.47a}$$

or

$$\omega_k = 2\pi f_k = 2\pi \frac{k}{N\Delta t} \qquad k = 0,1,2,\ldots,N-1 \tag{11.47b}$$

Using the discrete frequencies defined in Eqs. (11.47), the approximation to the finite Fourier transform in Eq. (11.46) becomes

$$\tilde{x}(k) \approx \Delta t \sum_{i=0}^{N-1} x(i)\; e^{-j(2\pi k/N)\, i} \qquad k = 0,1,2,\ldots,N-1 \tag{11.48}$$

The summation in Eq. (11.48) is the discrete Fourier transform $X(k)$ from Chapter 7, so the finite Fourier transform can be approximated by multiplying the discrete Fourier transform (DFT) by the sampling interval Δt,

$$\tilde{x}(k) \approx \Delta t\, X(k) \tag{11.49a}$$

where

$$X(k) \equiv \sum_{i=0}^{N-1} x(i)\; e^{-j(2\pi k/N)\, i} \qquad k = 0,1,2,\ldots,N-1 \tag{11.49b}$$

For flight data analysis, there are two main disadvantages to using the preceding approximation:

1) The spacing of the discrete frequencies for the *DFT* in Hz is roughly equal to the reciprocal of the data record length, so that the frequencies for the transformed data are fixed at values defined by N and Δt in

Eqs. (11.47). Efficient fast Fourier transform (*FFT*) algorithms for computing the *DFT* apply to this particular selection of frequencies, which means that the frequencies cannot be selected arbitrarily for a given data record length.

2) The approximation of the finite Fourier transform using the *DFT* is a zeroth-order Euler approximation of the integrand $x(t)\,e^{-j\omega t} = x(t)\,cos(\omega t) - j\,x(t)\,sin(\omega t)$. Since the integrand oscillates as the integration variable t changes, the Euler approximation can be inaccurate. The integrand oscillates more rapidly as ω increases, which means that the Euler approximation gets worse with increasing frequency. The same problem occurs as the selected Δt for the discrete approximation gets larger.

Using the *FFT* implies use of the frequencies in Eqs. (11.47). In this case, the frequency resolution becomes more coarse as the data record length decreases, leading to a loss of detail in the frequency domain, with consequent degraded data analysis and modeling results. Adding zeros to the measured time-domain data to artificially increase the data record length, known as zero padding, results in interpolation of the frequency-domain data obtained from the original time series, rather than increased resolution in the frequency domain. Zero padding is also done to make the number of data points in the time domain equal to an integer power of 2, which is a requirement for most *FFT* algorithms.

Modal frequencies for rigid-body dynamics of full-scale aircraft are fairly low, usually below 2 Hz. For a small frequency band like this, most of the processing involved in the usual calculation of the discrete Fourier transform is for frequencies outside the range of interest. Finally, typical post-flight data analysis does not involve tight limitations on computation time, so the speed of *FFT* algorithms is not needed.

To overcome these problems, the finite Fourier transform can be computed using an improved approximation to the defining integral in Eq. (11.45), for frequencies that can be chosen arbitrarily. The idea is to evaluate the finite Fourier transform for selected frequencies, using values of x that are interpolated based on the sampled values $x(i)$, to improve the accuracy of the transform. Because the interpolation approximates the continuous function $x(t)$, rather than $x(t)\,e^{-j\omega t}$, the accuracy of the transform is not dependent on the selected frequencies. The work by Filon [11.6] is the original source for this approach, using a quadratic interpolation scheme for the time-domain data in the integrand of the finite Fourier transform. The approach was extended in Ref. [11.1] to use cubic interpolation and in Ref. [11.7] to use the chirp *z*-transform to achieve arbitrary frequency resolution in the Fourier transform, regardless of the length of the data record. The chirp z-transform is equivalent to the discrete Fourier transform with frequencies chosen arbitrarily. By combining these techniques, the frequency-domain data points can be concentrated in the frequency band of interest and computed very accurately, resulting in high-quality modeling and data analysis results with excellent computational efficiency. Equations for the finite Fourier transform approach described in Refs. [11.1] and [11.7] are included here.

The *DFT* in Eq. (11.49b) can be re-written as

$$X(k) \equiv \sum_{i=0}^{N-1} x(i)\ AW^{-ki} \tag{11.50a}$$

where

$$A = 1 \qquad W = exp\left(j\frac{2\pi}{N}\right) \tag{11.50b}$$

The quantity AW^k for $k = 0,1,\ldots,N-1$ represents equally-spaced points on the unit circle in the complex plane, separated by the angle $2\pi/N$. Interpreting AW^k as a complex transform variable z, the selections in Eqs. (11.50b) correspond to a *z*-transform of the time-domain data $x(i),\ i = 0,1,\ldots,N-1$ for values of z placed uniformly around the unit circle in the complex plane.

The values of A and W in Eq. (11.50a) can be changed to implement the *z*-transform for a different contour in the complex plane. This is called a chirp *z*-transform. Specifically, the angular steps implemented by W can be chosen, while the value of A can be selected to start the contour at an arbitrary location in the complex plane. For example, A can be chosen to remain on the unit circle in the complex plane and start the contour at an arbitrary frequency, corresponding to the lower limit of a frequency band of interest. The value of W can be selected to implement the desired frequency resolution for the transform. The chirp *z*-transform is therefore a discrete Fourier transform with selectable frequency range and resolution.

Assuming the contour in the complex plane is along the unit circle, the general forms for A and W are

$$A = exp(j\theta_o) \qquad W = exp(j\Delta\theta) \tag{11.50c}$$

where $\omega_o = \theta_o/\Delta t$ represents the lower limit of the frequency band for the chirp *z*-transform in rad/sec, and $\Delta\omega = \Delta\theta/\Delta t$ is the frequency resolution. The values of θ_o and $\Delta\theta$ can be chosen arbitrarily, with the limitation that both $\theta_o/\Delta t$ and $\Delta\theta/\Delta t$ must be in the range $[0,\pi/\Delta t]$, where $\pi/\Delta t$ is the

Nyquist frequency in rad/sec. Figure 11.13 shows the values of the complex transform variable z for the conventional discrete Fourier transform [cf. Eq. (11.49b)] and a chirp z-transform defined for a selected frequency range and resolution.

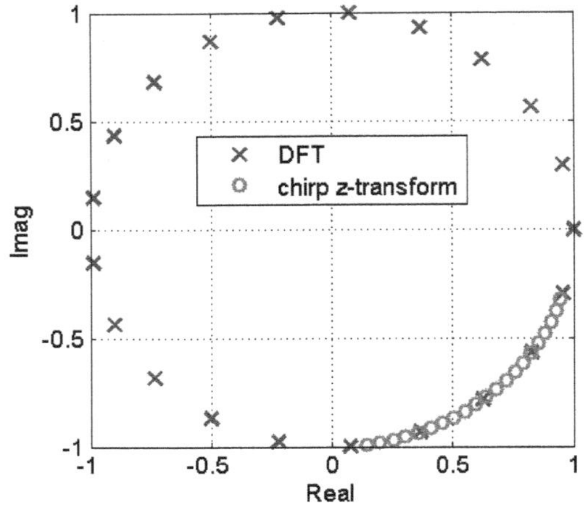

Figure 11.13 **Values of the transform variable *z* for the discrete Fourier transform and the chirp *z*-transform**

When the selected frequencies are regularly spaced in a frequency band of interest, Rabiner et al. [11.8] explain how the chirp z-transform can be computed efficiently using the *FFT* to implement a high-speed convolution. Uniform frequency spacing is the preferred approach for aircraft system identification, to make sure that all important features in the frequency domain are captured. Interpolation of the time-domain data samples to achieve high accuracy for the finite Fourier transform can be implemented by weightings applied to the values of the chirp z-transform, as shown in Refs. [11.1] and [11.7].

The chirp z-transform values are just the discrete Fourier transform evaluated at M arbitrary frequencies f_k in Hz, selected in the range $0 \le f_k \le f_N$,

$$X(k) = \sum_{i=0}^{N-1} x(i)\; e^{-j2\pi f_k i} \qquad 0 \le f_k < f_N\,, \quad k = 0,1,2,\ldots,M-1 \qquad (11.51)$$

The number of frequencies M can be chosen, but M should not be greater than approximately $2N$, or else numerical problems can occur.

Defining

$$\theta \equiv \omega_k \Delta t = 2\pi f_k \Delta t \qquad k = 0,1,2,\ldots,M-1 \qquad (11.52)$$

The expression for high accuracy calculation of the finite Fourier transform is

$$
\begin{aligned}
\tilde{x}(\theta) \approx \Delta t \{ & W(\theta) X(k) \\
& + \gamma_0(\theta) x(0) + \gamma_1(\theta) x(1) + \gamma_2(\theta) x(2) + \gamma_3(\theta) x(3) \\
& + e^{j\theta T/\Delta t} \left[\gamma_0^*(\theta) x(N-1) + \gamma_1^*(\theta) x(N-2) \right. \\
& \left. \left. + \gamma_2^*(\theta) x(N-3) + \gamma_3^*(\theta) x(N-4) \right] \right\}
\end{aligned}
\tag{11.53}
$$

where the weights $W(\theta), \gamma_0(\theta), \gamma_1(\theta), \gamma_2(\theta)$, and $\gamma_3(\theta)$ are found by analytically evaluating the finite Fourier transform integral in Eq. (11.45), using cubic Lagrange interpolation applied to the sampled time-domain data $x(i),\ i = 0,1,2,\ldots,N-1$. The resulting expressions for the weights are

$$
W(\theta) = \left(\frac{6+\theta^2}{3\theta^4} \right) (3 - 4\cos\theta + \cos 2\theta) \approx 1 - \frac{11}{720}\theta^4 + \frac{23}{15120}\theta^6 \tag{11.54a}
$$

$$
\begin{aligned}
\gamma_0(\theta) = & \frac{(-42 + 5\theta^2) + (6+\theta^2)(8\cos\theta - \cos 2\theta)}{6\theta^4} \\
& - j \frac{(-12\theta + 6\theta^3) + (6+\theta^2)\sin 2\theta}{6\theta^4} \\
& \approx -\frac{2}{3} + \frac{1}{45}\theta^2 + \frac{103}{15120}\theta^4 - \frac{169}{226800}\theta^6 \\
& - j\theta \left(\frac{2}{45} + \frac{2}{105}\theta^2 - \frac{8}{2835}\theta^4 + \frac{86}{467775}\theta^6 \right)
\end{aligned}
\tag{11.54b}
$$

$$
\begin{aligned}
\gamma_1(\theta) = & \frac{14(3-\theta^2) - 7(6+\theta^2)\cos\theta}{6\theta^4} \\
& - j \frac{30\theta - 5(6+\theta^2)\sin\theta}{6\theta^4} \\
& \approx \frac{7}{24} - \frac{7}{180}\theta^2 + \frac{5}{3456}\theta^4 - \frac{7}{259200}\theta^6 \\
& - j\theta \left(\frac{7}{72} - \frac{1}{168}\theta^2 + \frac{11}{72576}\theta^4 - \frac{13}{5987520}\theta^6 \right)
\end{aligned}
\tag{11.54c}
$$

$$\gamma_2(\theta) = \frac{-4\left(3-\theta^2\right)+2\left(6+\theta^2\right)cos\,\theta}{3\theta^4}$$

$$-j\frac{-12\theta+2\left(6+\theta^2\right)sin\theta}{3\theta^4} \tag{11.54d}$$

$$\approx -\frac{1}{6}+\frac{1}{45}\theta^2-\frac{5}{6048}\theta^4+\frac{1}{64800}\theta^6$$

$$-j\theta\left(-\frac{7}{90}+\frac{1}{210}\theta^2-\frac{11}{90720}\theta^4+\frac{13}{7484400}\theta^6\right)$$

$$\gamma_3(\theta) = \frac{2\left(3-\theta^2\right)-\left(6+\theta^2\right)cos\,\theta}{6\theta^4}$$

$$-j\frac{6\theta-\left(6+\theta^2\right)sin\theta}{6\theta^4} \tag{11.54e}$$

$$\approx \frac{1}{24}-\frac{1}{180}\theta^2+\frac{5}{24192}\theta^4-\frac{1}{259200}\theta^6$$

$$-j\theta\left(\frac{7}{360}-\frac{1}{840}\theta^2+\frac{11}{362880}\theta^4-\frac{13}{29937600}\theta^6\right)$$

In the preceding equations, θ takes M different values, corresponding to the M selected frequencies f_k, resulting in M values of the finite Fourier transform. Detailed derivation of the above expressions can be found in Ref. [11.7]. The truncated series expansions for $W(\theta)$, $\gamma_0(\theta)$, $\gamma_1(\theta)$, $\gamma_2(\theta)$, and $\gamma_3(\theta)$ given in Eqs. (11.54) are used for small θ, where θ is defined to be small when it is less than the largest value of θ for which identical results are obtained to machine precision using the analytic expression and its truncated series expansion. The series expansions are necessary because of high-order cancellations that make the analytic expressions inaccurate when θ is small.

It is advantageous to select the frequencies for the finite Fourier transform to be closely spaced within the frequency band of interest, to ensure that details of the frequency spectrum are accurately captured. The chirp *z*-transform can be used with arbitrary frequency resolution, independent of the length of the time record, and without the restriction that N be a power of 2. The chirp *z*-transform therefore decouples the frequency resolution from the length of the time record and can place all calculated frequency points within the frequency band of interest. The price for this flexibility is more computation time, but the magnitude of this extra computation time is relatively small.

For example, a flight test data analyst might select M discrete frequencies in a frequency band $[f_0, f_1]$ Hz, so that the selected frequencies would be

$$f_k = f_0 + k\,\Delta f \qquad k = 0,1,2,\ldots,M-1 \tag{11.55}$$

where

$$\Delta f = \frac{(f_1 - f_0)}{(M-1)} \tag{11.56}$$

For rigid-body dynamics of a full-scale aircraft, typical values might be $f_0 = 0.1$ Hz, $\Delta f = 0.02$ Hz, and $f_1 = 2$ Hz, so that the vector of selected frequencies is $\boldsymbol{f} = [0.10, 0.12, 0.14, \ldots, 2]$ Hz and $M = 96$. The result is 96 data points in the frequency domain, evenly spaced in the interval [0.1,2] Hz, regardless of the number of data points in the time domain. This is a significant advantage, particularly for iterative methods, because the data analysis and modeling computations will be faster for fewer data points.

Note that the chirp z-transform will compute the conventional *DFT* in Eq. (11.49b) for the selections $M = N$ and f_k from Eq. (11.47a).

In summary, the procedure for computing the high accuracy finite Fourier transform is:

1) Choose three of the four quantities in Eq. (11.56), and compute the remaining value from Eq. (11.56). This defines the frequencies for the analysis, using Eq. (11.55).
2) Compute the M values of θ from Eq. (11.52).
3) Use the chirp z-transform to compute the *DFT* for the selected frequencies, with Eqs. (11.50).
4) Use Eqs. (11.53)-(11.54) to compute the high-accuracy finite Fourier transform for each θ.

Although other quadrature methods can be used to accurately compute the finite Fourier transform, the technique presented here has the advantages of convenient and efficient calculation, while at the same time achieving high accuracy and arbitrary frequency resolution that is independent of the data record length.

11.6 Power Spectrum Estimation

Spectral densities can be used in frequency-domain modeling methods, as discussed in Chapter 7, as well as for the general purpose of examining frequency content of time series. Spectral densities are also called power spectral densities or the power spectrum. Computing accurate power spectral density estimates from measured time series involves several important practical considerations that will be discussed in this section.

Intuitively, the total power of a signal $x(t)$ should be the same when expressed in either the time domain or the frequency domain. This idea is the basis of Parseval's theorem,

$$\text{Total power } = \int_{-\infty}^{\infty} x^2(t)\,dt = \int_{-\infty}^{\infty} \tilde{x}^*(f)\,\tilde{x}(f)\,df \tag{11.57}$$

Assuming that $x(t) = 0$ for $t < 0$,

$$\text{Total power } = \int_{0}^{\infty} x^2(t)\,dt = \int_{-\infty}^{\infty} \tilde{x}^*(f)\,\tilde{x}(f)\,df \tag{11.58}$$

In the literature, many different definitions of power spectral densities and total power are used to enforce this concept [11.1]. In this book, and in the accompanying software, the power spectral density is normalized so that the sum of the squared spectral density estimates equals the mean squared value of the time series. The discrete form of Parseval's theorem is

$$\frac{1}{N}\sum_{i=0}^{N-1} x^2(i)\,\Delta t = \frac{1}{N}\sum_{k=0}^{N-1}\left[X^*(k)\ \Delta t\right]\left[X(k)\ \Delta t\right]\Delta f \tag{11.59}$$

where Eqs. (11.49) were used. For the frequencies specified in Eq. (11.47a), the discrete form of Parseval's theorem reduces to

$$\frac{1}{N}\sum_{i=0}^{N-1} x^2(i) = \frac{1}{N^2}\sum_{k=0}^{N-1} X^*(k)\,X(k) \tag{11.60}$$

or

$$\boldsymbol{x}^T\boldsymbol{x} = \frac{1}{N}\boldsymbol{X}^{\dagger}\boldsymbol{X} \tag{11.61}$$

where

$$\begin{aligned}\boldsymbol{x} &= \left[x(0)\ \ x(1)\ \ \ldots\ \ x(N-1)\right]^T \\ \boldsymbol{X} &= \left[X(0)\ \ X(1)\ \ \ldots\ \ X(N-1)\right]^T\end{aligned} \tag{11.62}$$

When the frequency spacing is arbitrarily chosen, as in Eqs. (11.55) and (11.56), Parceval's theorem only holds when all of the signal power lies within the frequency band selected for the analysis. Assuming that is the case, the discrete form of Parseval's theorem follows from Eq. (11.59) as

$$\frac{1}{N}\sum_{i=0}^{N-1} x^2(i)\Delta t = \frac{2}{N}\sum_{k=0}^{M-1} \tilde{x}^*(k)\,\tilde{x}(k)\,\Delta f \tag{11.63}$$

The factor of 2 on the right side of Eq. (11.63) accounts for the terms associated with negative frequencies in Eq. (11.59), which are not present for the chirp z-transform. Because the *DFT* has conjugate symmetry [cf. Eq. (7.14)], the values in the summation of the right side of Eq. (11.63) must be doubled, except for the endpoint frequencies of 0 and $f_N = 1/(2\Delta t)$ Hz. If the f_k of the chirp z-transform include 0 or f_N, the corresponding term in the summation has a factor of 1 instead of 2. Power spectra that use the factor of 2 and only positive frequencies are called one-sided power spectra. If negative frequencies are included, there is no factor of 2 and the power spectra are called two-sided.

There are three important practical issues associated with the use of the discrete Fourier transform to compute power spectral density. The first is that the number of frequencies used for the Fourier transformation is finite, so that in practice the underlying continuous spectral density function is characterized by discrete values for small frequency bands centered at each frequency f_k, namely $[f_k - \Delta f/2, f_k + \Delta f/2]$. The power spectral density estimates are therefore equivalent values for the continuous frequencies contained in each frequency bin centered on f_k.

The second issue arises because a finite length of data is used to estimate the power spectrum. A finite length of data on the time interval $[0,T]$ can be obtained by multiplying an infinite length of data by a function that equals 1 on the time interval $[0,T]$ and is zero otherwise. This is called the boxcar function, shown in the upper plot of Fig. 11.14. From Chapter 7 and Appendix A, it is known that the Fourier transform of the product of two time-domain functions is equivalent to convolution of their Fourier transforms in the frequency domain. The Fourier transform of the boxcar function is fairly wide, with a poor rate of decrease for the side lobes as a function of frequency, as shown in the lower plot of Fig. 11.14. This behavior is related to the fact that it is difficult for a Fourier series to represent a function with discontinuities, such as the boxcar function. Therefore, the Fourier transform of a finite length of data is equivalent to convolution in the frequency domain of the Fourier transform of the time function of interest (assumed infinite in length) with the Fourier transform of the boxcar function. This causes a smearing or leakage of frequency components from each frequency bin into adjacent ones, resulting in reduced accuracy of the power spectral estimates. Note that there is no leakage if the only frequencies present in the data are those associated with the frequency bins, see the lower plot of Fig. 11.14.

To mitigate leakage, the finite data can be multiplied by a windowing function that changes gradually from zero to one, then returns to zero. A windowing function without the discontinuity of the boxcar function has fewer and smaller side lobes, and therefore reduces leakage. There are many windowing functions that can be used, see Refs. [11.1], [11.9], and [11.10]. A good practical choice is the Bartlett window, defined by

$$w(i) = 1 - \left| \frac{i - N/2}{N/2} \right| \qquad i = 0, 1, \ldots, N-1 \tag{11.64}$$

which is a simple ramp from 0 to 1 and back to 0. When data windowing is implemented, the time series $x(i)$ in Eq. (11.60), for example, is modified to $w(i)x(i)$, where $w(i)$ is the value of the windowing function at the ith data point. To maintain the validity of Parseval's theorem, the power spectral density estimates must be divided by the mean sum of squares of the weighting function,

$$\frac{1}{N}\sum_{i=0}^{N-1} x^2(i) = \frac{1}{N\sum_{i=0}^{N-1} w^2(i)} \sum_{k=0}^{N-1} X_w(k)\, X_w^*(k) \tag{11.65}$$

where $X_w(k)$ is computed from the windowed time series $w(i)x(i)$, $i = 0, 1, \ldots, N-1$.

The third issue relates to the accuracy of power spectral estimates computed from measured time series data. In Chapter 7, the expression for computing the one-sided spectral density of a time function $x(t)$ was given as

$$\hat{G}_{xx}(\omega) = \frac{2}{T}\tilde{x}^*(\omega, T)\,\tilde{x}(\omega, T) \qquad \omega > 0 \tag{11.66}$$

where $\tilde{x}(\omega, T)$ is the Fourier transform of $x(t)$ for a data record T seconds long. Unfortunately, spectral density estimates computed directly from Eq. (11.66) have standard errors equal to 100% of the computed values [11.1], [11.9]. In practice, this difficulty can be overcome using some form of averaging. One common method is to partition the measured time series into n segments of data. The standard calculation of the spectral density in Eq. (11.66) is applied to each data segment. Then the results at each frequency are averaged, resulting in a spectral density estimate with random error variance reduced by a factor of $1/n$, due to the averaging. To achieve more averaging, the data segments are normally overlapped by 50%, meaning that adjacent segments share 50% of their data points. This increases the number of averages n, but decreases the variance reduction factor to $9/(11n)$, because some data are reused in adjacent data segments. However, the increased number of averages from overlapping data segments more than compensates for the decrease in the variance reduction factor.

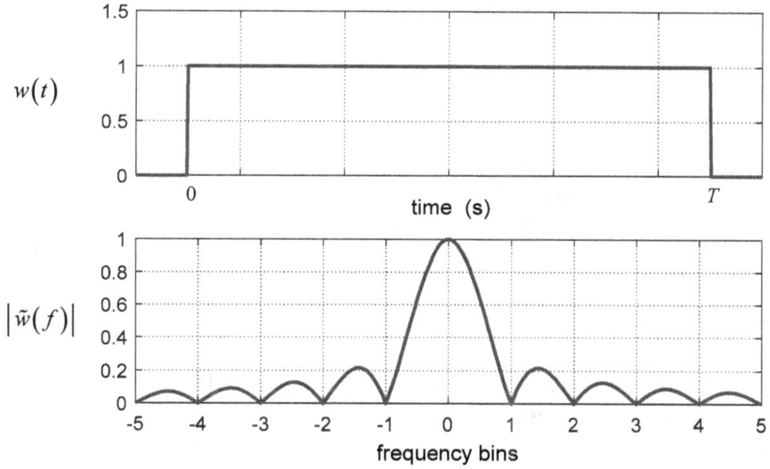

Figure 11.14 **Boxcar function and its Fourier transform**

It is not easy to choose a single best window length for a given finite time length of data, because longer window lengths are needed to capture low-frequency content, but shorter windows allow more averaging, and consequently produce results with higher accuracy. A compromise solution can be achieved by combining results from different window lengths using an optimized weighting based on coherence [11.11].

Another method for reducing the variance in the power spectral density estimates, which is nearly identical mathematically to the method just described, is to compute the Fourier transform at a frequency resolution n times finer than is desired for the end result. Eq. (11.66) is used to compute the spectral densities on the fine frequency mesh, then the results are summed for adjacent n frequencies to produce results for the desired (coarser) frequency mesh. This results in spectral density estimates with random error variance reduced by a factor of $1/n$. The spectral density values are summed rather than averaged, to maintain the validity of Parseval's theorem. The summing operation is also consistent with the earlier discussion concerning discrete frequency bins.

This second method is easy to implement with the chirp z-transform described earlier, because the chirp z-transform has the capability for arbitrary frequency resolution. The approach is simply to determine the number of values n required to achieve the desired accuracy for the spectral density estimates, and select a frequency mesh fine enough to provide the required number of adjacent values. Then implement Eq. (11.66) and sum the adjacent results to compute an accurate spectral estimate at each desired frequency.

11.7 Frequency Response Estimation using Multisine Inputs

The spectral density estimation expression in Eq. (11.66) comes from random data analysis, where the spectral density as a function of frequency is unknown for the finite time series. This knowledge deficiency leads to the large uncertainty in the spectral estimates, and drives the need to use averaging techniques to obtain spectral estimates with acceptable accuracy. Such techniques are often used when the input design has unknown frequency content, such as when piloted frequency sweeps are used for aircraft flight testing, as shown in Fig. 8.10. Furthermore, frequency responses cannot be calculated accurately at any analysis frequency where input power is low or nonexistent, cf. Eq. (7.20).

An alternate approach to frequency response estimation is to use orthogonal optimized multisine inputs, which have frequency content that is discrete and known. This obviates the need for spectral estimation, because the spectral content in the output is known to lie at the discrete input design frequencies, assuming linear system dynamics. This greatly simplifies the calculation of frequency responses, because Eq. (7.20) can be applied directly at the discrete frequencies where input power is known to exist. Error can be introduced because of the nonlinearity inherent in all practical aircraft, which produces spectral content in the output at frequencies other than the input frequencies. However, this error is present in any case when using frequency responses, which characterize only linear relationships.

Optimized orthogonal multisine inputs (cf. Chapter 8) can be applied at the same time to different inputs, so that many frequency responses can be extracted simultaneously, rather than only the frequency responses for one input at a time, as must be done when using frequency sweeps and spectral estimation. Careful design of the multisines and excitation time length or some windowing of the data record can be used to minimize or eliminate leakage.

A frequency response describes the steady-state response of a single input to a single output, as discussed in Chapter 7. Orthogonal optimized multisine inputs apply excitation at discrete frequencies in parallel for many inputs simultaneously, as opposed to a frequency sweep, which applies excitation at sequential frequencies for one input at a time. Consequently, using orthogonal optimized inputs provides advantages in reducing flight test time and improving data quality, particularly at lower frequencies, because the dynamic system reaches steady state for all input frequencies relatively close to the beginning of the maneuver. In contrast, a frequency sweep must be lengthy so that the input frequencies can be traversed slowly, to approach steady state for the sequential frequencies. Using discrete frequencies in the inputs also avoids the spectral estimation problems that require windowing, averaging, and coherence weighting, so that the data analysis can be done simply with Eq. (7.20).

The combination of orthogonal optimized multisines and Eq. (7.20) has been applied successfully in wind tunnel tests with many inputs and outputs [11.12], and in flight tests [11.13]-[11.14], to efficiently compute frequency responses from time series data. This general approach is also the basis of effective methods for real-time frequency response estimation, both in the time domain [11.13], and in the frequency domain [11.12], [11.14]-[11.16].

11.8 Signal Comparisons

In many cases, system identification techniques are used to find linear relationships between input and output variables. A simple example would be estimating stability and control derivatives for the nondimensional pitching moment coefficient in an equation-error formulation (cf. Chapter 5). In that case, the model equation might be

$$C_m = C_{m_o} + C_{m_\alpha}\alpha + C_{m_q}\frac{q\bar{c}}{2V} + C_{m_\delta}\delta + v_m \tag{11.67}$$

In matching the measured C_m with the expansion on the right, the C_{m_o} parameter takes care of the bias, and the terms involving stability and control derivatives model the variations in C_m over the time period for which the data is collected. The stability and control derivative estimates result in the best match with the measured nondimensional pitching moment coefficient in a least-squares sense.

The stability and control derivatives are therefore linear scaling parameters relating regressors formed from explanatory variables (states and controls) to the dependent variable (nondimensional coefficient). Linear scaling parameters can modify the magnitude of the regressors, but cannot modify the time variation in the regressors. The variation of the dependent variable over time must be matched by the terms in the model equation (11.67), or equivalently, the model structure. When the model structure is inadequate, the error term v_m will be different from a white noise sequence. In that case, the task is to find a regressor that varies in the same manner as the remaining deterministic part of v_m, and then add that regressor to the model structure, possibly improving the model. The scaling of this added regressor is irrelevant for model structure determination, because the associated model parameter estimate will be optimized to provide the necessary scaling.

It is therefore useful to have a method to compare the time variations of potential regressors and the current model residual, with all linear scaling and biases removed. This gives some insight as to which regressors have a time variation that can be used to model the remaining variation in the dependent variable. Such a comparison can be made by removing a linear trend from each time series, then scaling the second signal to have the same root-mean-square amplitude as the first signal.

Figure 11.15 shows a model residual time series and a potential model regressor in their original form (upper plot), and after removing biases and scaling (lower plot). The figure shows how this simple data processing can significantly clarify the question of whether or not a postulated model regressor would be useful in the model equation. An alternate graphical method is to cross-plot the time series. If the time series are perfectly correlated, the result will be a straight line. Figure 5.17 shows examples of this type of plot. Either plot type can be considered a graphical analog of the correlation coefficient defined in Eq. (5.120).

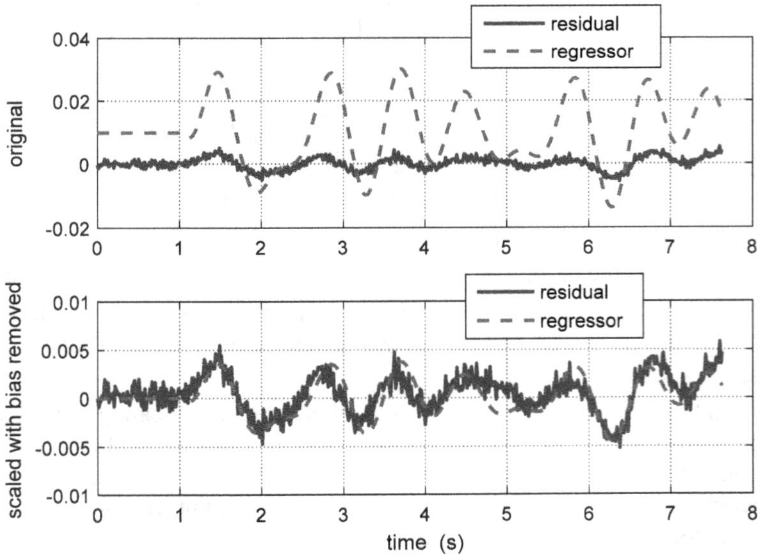

Figure 11.15 **Comparison of model residuals and postulated regressor**

11.9 Maneuver Visualization

Measured data from a flight test maneuver includes time series for many different physical quantities. It is extremely difficult to get a good overall mental picture of what the aircraft is doing by looking sequentially at two-dimensional plots, such as angle of attack or elevator deflection versus time. This is particularly true for maneuvers with combined longitudinal and lateral motion, or for maneuvers at high angles of attack, among others.

Modern three-dimensional graphics software and hardware make it possible to create real-time visualizations of the aircraft in flight. In effect, the flight test maneuver can be replayed for the analyst, as often as desired, using whatever vantage point is useful. This capability gives the analyst insight that simply is not available from studying two-dimensional plots or streams of numbers.

Figure 11.16 shows what the display looks like when replaying a maneuver using measured flight data with a three-dimensional solid model of the F-16. The software used is called Aviator Visual Design Simulator (AVDS), available from RasSimTech, Ltd. at http://www.rassimtech.com. There are other software packages available with similar capability. Note that a head-up display (HUD) is included so that the analyst has access to the numerical values of key quantities while watching the recreated aircraft motion. The AVDS software can also show the air relative velocity vector and movement of the individual control surfaces on the aircraft. The real-time display can be driven by a nonlinear aircraft simulation running in MATLAB® in real time, using pilot inputs from a joystick or computer mouse (see Appendix D).

Figure 11.16 **Maneuver visualization**

11.10 Summary and Concluding Remarks

This chapter includes a selection of data analysis techniques that have been found useful in practice for aircraft system identification. The list is by no means exhaustive, but instead represents some proven approaches to common problems. The techniques described here are often among the first to be applied to measured data. It is therefore important that the procedures are well-understood and properly implemented, so that subsequent modeling results are not compromised.

Most of the data analysis techniques described provide some alternate view of the data. For example, filtering and smoothing can be used to strip away high-frequency noise, revealing underlying deterministic signal, Fourier transformation and power spectral estimates indicate how power in the time-domain signal is distributed as a function of frequency, and maneuver visualization shows a realistic playback of the flight data using a three-dimensional depiction of the aircraft in flight.

Several of the methods described in this chapter are required for implementing modeling methods described earlier in the book. Examples are smoothed numerical differentiation, which is required for equation-error parameter estimation methods discussed in Chapters 5 and 6, as well as sensor position corrections in Chapter 10. The high-accuracy Fourier transform for arbitrary frequencies and the power spectral estimation methods are important for frequency-domain methods described in Chapter 7.

This chapter marks the end of the presentation of methods for aircraft system identification. The next and final chapter provides details of the accompanying software package and how to use it effectively.

References

[11.1] Press, W.H., Teukolsky, S.A., Vettering, W.T., and Flannery, B.R., *Numerical Recipes in FORTRAN: The Art of Scientific Computing*, 2nd Ed., Cambridge University Press, New York, NY, 1992.

[11.2] Graham, R.J., "Determination and Analysis of Numerical Smoothing Weights," NASA TR R-179, 1963.

[11.3] Lanczos, C., *Applied Analysis*, Dover Publications, Inc., New York, NY, 1988.

[11.4] Morelli, E.A., "Estimating Noise Characteristics from Flight Test Data using Optimal Fourier Smoothing," *Journal of Aircraft*, Vol. 32, No. 4, 1995, pp. 689-695.

[11.5] Morelli, E.A., "Practical Aspects of the Equation-Error Method for Aircraft Parameter Estimation," AIAA-2006-6144, *AIAA Atmospheric Flight Mechanics Conference*, Keystone, CO, 2006.

[11.6] Filon, L.N.G., "On a Quadrature Formula for Trigonometric Integrals," *Proceedings of the Royal Society of Edinburgh*, Vol. 49, 1928, pp. 38-47.

[11.7] Morelli, E.A., "High Accuracy Evaluation of the Finite Fourier Transform using Sampled Data," NASA TM 110340, 1997.

[11.8] Rabiner, L.R., Schafer, R.W., and Rader, C.M., "The Chirp *z*-Transform Algorithm and Its Application," *The Bell System Technical Journal*, Vol. 48, Issue 5, 1969, pp. 1249-1292.

[11.9] Bendat, J.S. and Piersol, A.G., *Random Data Analysis and Measurement Procedures*, 2nd Ed., John Wiley & Sons, Inc., New York, NY, 1986.

[11.10] Otnes, R.K. and Enochson, L., *Applied Time Series Analysis, Volume 1, Basic Techniques*, John Wiley & Sons, New York, NY, 1978.

[11.11] Tischler, M.B. and Remple, R.K. *Aircraft and Rotorcraft System Identification – Engineering Methods with Flight Test Examples*, AIAA, Reston, VA, 2006.

[11.12] Grauer, J.A., Heeg, J., and Morelli, E.A. "Real-Time Frequency Response Estimation of Joined-Wing SensorCraft Aeroelastic Wind Tunnel Data," AIAA-2012-4641, *AIAA Atmospheric Flight Mechanics Conference*, Minneapolis, MN, August 2012.

[11.13] Holzel, M.S. and Morelli, E.A., "Real-Time Frequency Response Estimation from Flight Data," *Journal of Guidance, Control, and Dynamics*, Vol. 35, No. 5, 2012, pp. 1406-1417.

[11.14] Grauer, J.A. and Morelli, E.A. "Method for Real-Time Frequency Response and Uncertainty Estimation," *Journal of Guidance, Control, and Dynamics*, Vol. 37, No. 1, January-February 2014, pp. 336-343.

[11.15] Tischler, M., Ivler, C., and Berger, T., "Comment on paper 'Method for Real-Time Frequency Response and Uncertainty Estimation'," *Journal of Guidance, Control, and Dynamics*, Vol. 38, No. 3, 2015, pp. 547-548.

[11.16] Grauer, J.A. and Morelli, E.A. "Reply by the Authors to M. Tischler, C. Ivler, and T. Berger," *Journal of Guidance, Control, and Dynamics*, Vol. 38, No. 3, 2015, pp. 549-550.

Problems

11.1 Answer the following, using words, graphs, equations, and/or diagrams:

a) Explain one practical problem encountered when computing frequency response estimates from measured time series data.

b) Explain why it is better to smooth a noisy time series first, then differentiate, rather than the reverse.

11.2 Using smoo.m, isolate the structural dynamic response of the wing and boom in the frequency range [8, 12] Hz from the sideslip angle measurement in the data file flt_beta_250k2.mat.

11.3 Using the transfer function:

$$\tilde{q}(s)/\tilde{\delta}_s(s) = -(0.1s + 0.08)/(s^2 + 1.8s + 4)$$

a) Apply a 30 s frequency sweep input to the dynamic system. Use buzz.m to add 5 percent noise to the simulated output from tfsim.m.

b) Use fresp.m and frespc.m to estimate the frequency response.

c) Discuss the results from both methods and compare to the true frequency response computed from the transfer function using cmpbode.m.

11.4 Using the transfer function from problem **11.3**,

a) Apply a 30 s optimized multisine input to the dynamic system. Use buzz.m to add 5 percent noise to the simulated output from tfsim.m.

b) Use fint.m applied at the optimized multisine frequencies only, then use Eq. (7.20) to estimate the frequency response.

c) Discuss the results and compare to the true frequency response computed from the transfer function using cmpbode.m.

d) Compare with the results from problem **11.3**, including the relative difficulty in obtaining good results.

11.5 Create a time series using mkmsin.m and compare the high-accuracy Fourier transform obtained using fint.m with the recursive Fourier transform obtained using rft.m. Remember that rft.m computes the discrete Fourier transform recursively, so that Eq. (11.49a) will be needed for comparison with the results from fint.m. Experiment with different frequency content and data lengths.

a) What are the differences as a function of time and frequency?

b) Why do the differences depend on time and frequency?

Chapter 12

MATLAB® Software

This chapter explains the practical use of MATLAB® software that implements the aircraft system identification methods described in the preceding chapters. The software was written in the MATLAB® language, but is not a product of The MathWorks, Inc., which produces and sells MATLAB®. MATLAB® is a high-level programming language with extensive capabilities for numerical computation and graphics. Familiarity with basic operations in MATLAB® is needed to use the software associated with this book. Introductory material in Ref. [12.1] provides the necessary background information. Excellent help files and tutorials are also readily available within the MATLAB® computing environment.

In the following descriptions, commands typed by the user in the MATLAB® command window at the >> prompt will appear in `this font`. All such commands are followed by pressing the enter or return key. Variable names will appear in **this font**. Names of m-files will appear in this font.

12.1 Overview

The software associated with this textbook is a collection of m-files called System IDentification Programs for AirCraft, or SIDPAC. An m-file in MATLAB® is the equivalent of a subroutine in FORTRAN or a function subprogram in C. SIDPAC was developed at NASA Langley Research Center, more or less continuously since 1990, and has been applied to flight data, wind tunnel data, and simulation data from many different projects. SIDPAC is available by request from NASA Langley at the following web site: https://software.nasa.gov/software/LAR-16100-1. Full contact information and organizational affiliation will be collected from all requestors, and all requestors must pass NASA security screening before receiving SIDPAC.

SIDPAC was developed using several versions of MATLAB® over the years. Proper operation of the SIDPAC software has been validated for MATLAB® version 8.6, R2015b. Software validation and the examples in the book were done on a personal laptop computer with Microsoft Windows 7 Enterprise operating system. However, MATLAB® is a platform-independent programming language, so the software should work properly on any computer running any version of MATLAB®. One exception is that the graphical user interface (GUI) in SIDPAC will not run properly in MATLAB® versions earlier than version 8.6, R2015b.

Each m-file in SIDPAC performs one task or implements one method. There are SIDPAC routines that implement the techniques discussed in the book, as well as other techniques that were not discussed. Comments in each SIDPAC m-file include a list of other SIDPAC m-files called. Any standard MATLAB® routine called by a SIDPAC m-file is not listed as a called routine. SIDPAC does not require any specialized add-on MATLAB® toolboxes, just the standard MATLAB® software.

Some m-files in SIDPAC also have a C mex-file equivalent. The reason is that C mex-files run much faster, so the m-files with C mex-file equivalents are computationally intensive and/or called frequently. The source code for each C mex-file has the same name as its equivalent m-file, but has a .c file extension instead of .m. Compiled mex-files have the same name, with a file extension that includes mex, and is specific to the computer and operating system being used. MATLAB® treats mex-files in the same manner as m-files, in terms of execution; however, MATLAB® will always use a mex-file before using an m-file with the same name. Because of this, SIDPAC will always try to use the faster mex-file first.

The SIDPAC .c files should be recompiled for each particular computer to produce appropriate mex-files. To do this, first type `mex -setup` at the MATLAB® command prompt, which implements a one-time set up process. Then type `mex filename.c`, where filename.c is any of the SIDPAC files with .c file extension, to produce the mex-file. If the mex-file is deleted or placed in a location that is not in the MATLAB® path, then SIDPAC will use the .m file equivalent, which is slower.

The folder containing SIDPAC m-files can be placed anywhere in the MATLAB® path. Drop-down menus or the path command in MATLAB® can be used to include the SIDPAC folder in the MATLAB® path. The startup.m file included with SIDPAC will set the MATLAB® path to automatically include a subdirectory called SIDPAC. Commands in the startup.m file are executed automatically when MATLAB® is started. Make sure this startup.m file is the first one encountered in the MATLAB® path, which will be true if startup.m is in the current working directory.

The calling syntax and descriptive material for the m-files in SIDPAC appear in the header of each m-file. The header is displayed in the MATLAB® command window in response to typing `help filename`, where `filename` is the name of a SIDPAC m-file, without its .m file extension. To execute any m-file, the correct calling syntax can be copied from the header directly to the MATLAB® command window and executed by pressing the enter or return key.

The SIDPAC m-files were originally written for research purposes and not for public release, so error handling, user interface, and the like, are Spartan. SIDPAC includes a GUI, which allows the user to call individual m-files using point-and-click, rather than typing the m-file command line syntax in the MATLAB® command window or using m-file scripts to execute a sequence of commands. To use the SIDPAC GUI, type `sid` at the MATLAB® command prompt. The GUI displays information and options in drop-down menus. SIDPAC m-files can also be used individually from the MATLAB® command line, even after the GUI has been started. Note that the GUI is not necessary for using SIDPAC. The GUI simply makes invoking SIDPAC m-files and displaying results easier for the user. Only a subset of the m-files in SIDPAC can be invoked via the GUI. SIDPAC was designed for use from the MATLAB® command line or inside m-file scripts, so all SIDPAC m-files can be used in these ways.

SIDPAC m-files will sometimes make plots or ask the user for information via the MATLAB® command window. Because of this, it is best to arrange the MATLAB® command and figure windows so that they do not overlap and obscure one another.

Examples in the book have been implemented in an m-file called sidpac_text_examples.m. By typing `sidpac_text_examples`, then entering an example number from the text, the user is taken through all the steps necessary to duplicate the selected textbook example. The MATLAB® code inside sidpac_text_examples.m provides examples of how SIDPAC routines can be used to solve aircraft system identification problems.

SIDPAC also includes demonstration programs for some of the more commonly used routines in SIDPAC. All of these programs have names that include _demo.m, preceded by the name of the SIDPAC m-file being demonstrated, e.g., swr_demo.m. is a demonstration program for the SIDPAC m-file swr.m. The demonstration programs are executed by simply typing their name (without the .m file extension) at the MATLAB® command prompt, followed by the enter or return key.

Reference [12.2] was used as a source of theoretical and practical information in the development of SIDPAC. The software associated with this book is SIDPAC version 3.0. SIDPAC version 2.0 was released with the first edition of this book. An overview of SIDPAC version 1.1 is given in Ref. [12.3]. SIDPAC version 1.0 was released in January 2001 as the first version of SIDPAC available to the public. More information about SIDPAC is available at http://sunflyte.com/SIDBook_SIDPAC.htm.

Much of SIDPAC is based on the concept of a standardized arrangement of the measured data. For typical uses of SIDPAC, the measured data are arranged in a standard data matrix called **fdata**, which is shorthand for "flight data". Each row of **fdata** corresponds to a data point, and each column represents a particular measurement with specified units. Measured data from a flight test maneuver or other experiment are arranged by column in a standard format, defined in the files SIDPAC_Data_Channels.doc or

SIDPAC_Data_Channels.pdf. This approach has been taken because each experimental program has its own nomenclature and units for measured signals. By standardizing the form of the measured data, validated SIDPAC routines can be used for many different projects, with little chance of errors in data handling, such as different sign conventions or unit conversion errors. Programming SIDPAC routines is also more standardized and straightforward, reducing the possibility of errors.

The conversion of measured data into standard SIDPAC data format is done first, before any analysis. Normally, the user creates an m-file for each experimental program to convert measured data to the standard units, and assign each measured signal to the proper column in the **fdata** matrix. There are spare (unused) columns in **fdata** for special cases, and the size of **fdata** is limited only by computer memory. SIDPAC m-file f16_fltdatsel.m is an example of an m-file created to convert data from the F-16 nonlinear simulation described in Appendix D into the standard **fdata** format expected by SIDPAC. Output from the F-16 nonlinear simulation is defined in the header of ac_sim.m, which is the program used for batch nonlinear simulation of the F-16. All F-16 nonlinear simulation programs are written in MATLAB®, and are included with SIDPAC.

SIDPAC includes some utilities for loading data into the MATLAB® workspace; however, the native MATLAB® tools for this purpose are very capable. The initial screen of the SIDPAC GUI (which appears in response to typing `sid` at the MATLAB® command prompt) includes a point-and-click utility for assembling the standard **fdata** matrix from variables loaded into the MATLAB® workspace. A data structure named **fds** (for "flight data structure") is initialized when the SIDPAC GUI is invoked, unless a variable named **fds** already exists in the MATLAB® workspace. The SIDPAC GUI assumes that the user has already loaded data into the MATLAB® workspace, and the data have been assembled into the standard **fdata** format.

Brief descriptions of all of the programs in SIDPAC can be displayed in the MATLAB® command window by typing `help sidpac` at the MATLAB® command prompt. This list is grouped approximately by functionality, and is intended to familiarize the user with SIDPAC capabilities and help in selecting appropriate tools. The list of SIDPAC programs can also be found in the SIDPAC file named Contents.m, in the SIDPAC directory.

The remainder of this chapter lists and describes individual m-files in SIDPAC that implement the main techniques. Table 12.1 gives a convenient reference for the main SIDPAC routines, including a brief description of what each one does, along with references to earlier chapters in the book and other sources for explanations of the algorithms and their implementation. A description of the purpose and operation of each m-file listed in Table 12.1 is given next, along with practical hints for effective application.

Table 12.1 SIDPAC main programs

Regression Methods, Chapter 5	
Filename	Description
lesq.m	linear regression
r_colores.m	colored residual corrections for Cramér-Rao bounds of linear regression parameter estimates
swr.m	stepwise regression
mof.m	orthogonal function modeling using Gram-Schmidt orthogonalization
Maximum Likelihood Methods, Chapter 6	
Filename	Description
oe.m	output-error parameter estimation in the time domain
colores.m	colored residual corrections for Cramér-Rao bounds of output-error parameter estimates
m_colores.m	vectorized (faster) version of colores.m
Frequency Domain Methods, Chapter 7	
Filename	Description
fint.m	high-accuracy finite Fourier transform for arbitrary frequencies
fdoe.m	output-error parameter estimation in the frequency domain
tfest.m	equation-error parameter estimation for transfer function models
Experiment Design, Chapter 8	
Filename	Description
mksqw.m	generates arbitrary multistep square wave inputs
mkfswp.m	generates frequency sweep inputs
mkmsswp.m	generates multiple orthogonal optimized multisine sweep inputs

Table 12.1 **SIDPAC main programs (completed)**

Data Compatibility, Chapter 9	
Filename	Description
dcmp.m	kinematic equations and instrumentation error model for data compatibility analysis
rotchk.m	kinematic check on data for rotational motion
airchk.m	kinematic check on data for translational motion
Chapter 10, Real-Time Methods	
Filename	Description
rlesq.m	recursive least squares parameter estimation
rtpid.m	sequential least squares parameter estimation in the frequency domain
Chapter 11, Data Analysis	
Filename	Description
smoo.m	global Fourier smoothing for noisy time series
deriv.m	compute locally smoothed derivatives for noisy time series
lsmep.m	smooth endpoints for noisy time series
compfc.m	compute nondimensional aerodynamic force coefficients
compmc.m	compute nondimensional aerodynamic moment coefficients
cmpsigs.m	graphically compare signals on a common scale with biases removed

12.2 Regression Methods

12.2.1 lesq.m

Program lesq.m estimates model parameters for linear least-squares regression problems. The routine can be used with either real or complex data, which means that the same m-file can be used for time-domain data or frequency-domain data. The matrix inversion is done using singular value decomposition, with automatic modification of the singular values when the matrix of regressors is ill-conditioned. Model fit error variance and parameter covariance matrix estimates are based on calculations that use the estimated parameter values. Since MATLAB® uses double precision arithmetic by default, it is not necessary to normalize the regressors to account for differences in the scale of the regressors. Accordingly, lesq.m does not normalize the regressors.

Program lesq.m does not automatically include a constant regressor, so it is necessary to include a column of ones in the regressor matrix **x** to estimate a bias term. Estimated parameter standard errors can be computed from the estimated covariance matrix **crb** by typing `serr=sqrt(diag(crb))`. In general, omit the **svlim** input and let the program compute and use the default value. The input **svlim** can be used to address collinearity problems in some cases. Optional inputs **p0** and **crb0** can be used to include prior information from a previous analysis or another source.

The m-file named totter_lesq_demo.m is a longitudinal example application of lesq.m, using flight test data from the NASA Twin Otter aircraft. Type `totter_lesq_demo` at the MATLAB® prompt to run the demonstration.

12.2.2 r_colores.m

Program r_colores.m estimates model parameters for a linear least-squares regression problem and computes Cramér-Rao bounds for the covariances of the estimated parameters, both conventionally and accounting for the frequency content of the residuals. This routine is for time-domain data – the correction is not necessary for frequency-domain data. Program r_colores.m should be run once at the end of the analysis, when the parameter estimates from least-squares parameter estimation (lesq.m, swr.m, or mof.m) are judged to be acceptable.

Output of r_colores.m includes the conventional Cramér-Rao bounds `crbo` and the corrected Cramer-Rao bounds **crb**. The **crbo** output should match the Cramér-Rao bounds from conventional least-squares parameter estimation (lesq.m, swr.m, or mof.m), within round-off error. Input **x** to r_colores.m should be the regressor matrix for the final selected model. When using swr.m or mof.m for least-squares parameter estimation, the regressor matrix **x** will in general be an output from the routine that is not the same as the input regressor matrix. In general, omit the **svlim** input and let the program compute and use the default value.

The m-file totter_crb_demo.m is an example application using r_colores.m, with comments and explanations output to the command window.

12.2.3 swr.m

Program swr.m identifies general least-squares models from measured input-output data. All regressors to be considered for inclusion in the model are input as columns of the input matrix **x**. Candidate regressors can be swapped in and out of the model manually by the analyst. Least-squares parameter estimation for each candidate model is done using lesq.m. Several statistical diagnostics are computed at each step to help in deciding which regressors should be retained in the model. The swr.m routine can be used with either real or complex data, which means the same m-file can be used for time-domain data or frequency-domain data.

The swr.m program is interactive, and requires direction from the analyst as to which regressor to move in or out of the candidate model. Generally, if adding a particular regressor (column of **x**) to the model decreases the fit error, increases R^2, decreases predicted squared error (*PSE*), and has a partial F ratio greater than the given cut-off value, the regressor should be retained in the model. The regressor to be added to the model at any point is the one with the highest squared partial correlation. Sometimes the addition of a regressor will render a previously selected regressor superfluous. This is indicated by the partial F value of the previously selected regressor becoming small. In that case, the two regressors (the newly added one and the one whose partial F became small) are probably highly correlated. This can be checked using corx.m or regcor.m.

Estimated parameter standard errors can be computed from the estimated covariance matrix **crb** by typing `serr=sqrt(diag(crb))`. In general, omit the **svlim** input and let the program compute and use the default value.

Program swr.m automatically includes a constant regressor, so it is not necessary to include a column of ones in the **x** regressor matrix. The regressor matrix **x** must be assembled in MATLAB®, outside of swr.m. For example, if spline regressors are desired, each candidate spline regressor must be included as a column in **x** before calling swr.m. Splines of various orders can be generated using splgen.m. Polynomial terms of arbitrary order can be generated using reggen.m. If it is not clear whether or not a particular regressor influences the dependent variable, include that regressor in the input matrix **x**, and swr.m will help to figure this out using statistical metrics, as described in Chapter 5. The *PSE* computed for the final model is conservative. Therefore, the squared error for prediction cases will generally be less than the value of *PSE* computed for the final model.

Program swr.m implements stepwise regression, not modified stepwise regression. This simply means that no regressors (linear or otherwise) are automatically included in the model, except for the constant term, which is always included. The estimated parameter vector **p** always has length equal to the number of candidate regressors (i.e., the number of columns of x), plus one (for the constant term). Output **pindx** is a vector of indices that specify which columns of **x** were selected for the final model.

The m-file swr_demo.m is an example with comments and explanations output to the command window. The m-file totter_swr_demo.m is an example using NASA Twin Otter flight data.

12.2.4 mof.m

Program mof.m identifies general multivariate models from measured input-output time-domain data. Important explanatory variables and combinations are determined automatically by generating orthogonal modeling functions directly from the measured data, using Gram-Schmidt orthogonalization, as described in Chapter 5. Program mof.m then uses these multivariate orthogonal functions in a least-squares formulation to determine model structure and associated model parameter values. Because of the method used to generate the multivariate orthogonal basis functions, it is possible to decompose each retained basis function into ordinary functions in the explanatory variables. Once this is done and the results are combined, the final model form is an ordinary multivariate function in the explanatory variables. The covariance matrix estimate for the final model parameters is computed based on the error bounds computed for the parameters associated with the retained orthogonal modeling functions, using the fact that the decomposition from orthogonal modeling functions to ordinary modeling functions is exact.

The mof.m program is an upgrade of the original offit.m program in SIDPAC version 1.1. The main difference in these programs is the method for generating and ordering the orthogonal functions, which is improved in mof.m. Generally, mof.m produces more accurate models with fewer terms than offit.m. Both offit.m and mof.m functions are included in SIDPAC.

Each column of input matrix **x** should contain values for an explanatory variable that the analyst believes might influence the dependent variable **z**. Program mof.m automatically includes a constant regressor, so it is not necessary to include a column of ones in the **x** matrix of explanatory variables.

The input matrix **x** must be assembled in MATLAB® outside of mof.m. Input **nord** is a vector of integers with length equal to the number of explanatory variables (columns of **x**), where each integer indicates the maximum order of the corresponding explanatory variable in any model term. Input **maxord** is an integer indicating the maximum order for each model term. For example, if model terms of order 4 or less were desired for explanatory variables vectors $\boldsymbol{\alpha}$ and $\boldsymbol{\beta}$, with maximum order of $\boldsymbol{\alpha}$ in any model term equal to 2, and maximum order of $\boldsymbol{\beta}$ in any model term equal to 3, then **x**=[$\boldsymbol{\alpha}$, $\boldsymbol{\beta}$], **nord**=[2,3], and **maxord**=4. These choices will allow orthogonalized modeling functions based on ordinary multivariate functions such as $\boldsymbol{\alpha}^2\boldsymbol{\beta}^2$, $\boldsymbol{\alpha}^2$, or $\boldsymbol{\alpha}\boldsymbol{\beta}^3$, for example, but not $\boldsymbol{\alpha}^2\boldsymbol{\beta}^3$ or $\boldsymbol{\beta}^4$.

Input values for **nord** and **maxord** are not critical. A good general choice is to use 4 for all values. Use higher values if you suspect a highly nonlinear relationship, and lower values to force a simpler model. If the values chosen are too high, the algorithm will generate more orthogonal functions, which will take more execution time, but the best model in terms of minimum predicted square error (*PSE*) will be identified just the same.

If the column of **x** containing explanatory variable 2 has only 4 distinct values (even if they are repeated many times), then the highest order for that explanatory variable (corresponding to the second element of input **nord**) cannot be higher than 3. This is a fundamental information limitation, similar to the fact that 4 distinct points are fit exactly with a third-order polynomial.

If it is not clear whether or not a particular explanatory variable influences the dependent variable, then include that explanatory variable in the input matrix **x**, and mof.m will figure it out automatically. The price for this is more execution time.

The number of orthogonal functions chosen should generally correspond to the minimum *PSE*, as shown on the display. Occasionally, with real data, two or three local minima will occur. In such cases, it is generally best to choose the lowest number of orthogonal functions (i.e., the first local minima), which results in a simpler model. Resist the temptation to include more orthogonal function for a better model fit to the data. Although the model fit for the given data points may be better with more orthogonal functions, the prediction capability deteriorates as higher numbers of orthogonal functions are chosen. In general, stick with the number of orthogonal functions that gives minimum *PSE*. Program mof.m will use the orthogonal function model corresponding to minimum *PSE* if the program is run in automatic mode (**auto**=1), or if the user just hits enter or return when asked how many orthogonal functions to keep in manual mode (**auto**=0).

The orthogonal functions are computed sequentially, and when the later orthogonal functions are no longer orthogonal to those already generated, the information in the explanatory variables has been exhausted. The practical result is that the orthogonal functions generated beyond when the orthogonality remains intact cannot be properly decomposed into ordinary functions. This shows up as a mismatch in the model fit error using orthogonal functions compared to using ordinary functions. Consequently, the program will not allow the analyst to choose orthogonal functions beyond the point where the orthogonalization process cannot produce additional mutually orthogonal functions, which is noted in the display output as the number of orthogonal functions for orthogonality intact.

The quantities **rmsd** and **drmsd** displayed by mof.m are the root-mean-square deviation of measured dependent variable from the model output, and the change in this quantity as each orthogonal function is added to the model, respectively. The generated orthogonal functions are added to the model in order of most effective to least effective in reducing **rmsd**. This can be seen from the **rmsd** and **drmsd** data.

The most compact models are generated when the explanatory variables in the matrix of explanatory variables **x** are arranged as most important to least important left to right by column. For example, most aerodynamic modeling would have angle of attack in the first column, followed by perhaps a control surface deflection in the second column, etc. Input parameter **ivar** should usually be omitted. This input is used when it is desired that every term in the model include at least one power of a particular explanatory variable. For example, if sideslip angle is explanatory variable number 1 (i.e., in the first column of input matrix **x**), then setting **ivar**=1 will force every term in the resulting model to contain sideslip angle to at least the first power. This is useful if the dependent variable is always zero when the sideslip angle is zero, because the model will then always be exactly correct for zero sideslip angle.

Using the default value for the dependent variable noise variance **sig2** is conservative, so that the squared error for prediction cases will generally be less than the output value **pse** for the final model. The value of **sig2** affects the model term selection. In general, allow the program to use the default conservative value for **sig2**, unless an accurate estimate of the dependent variable noise variance is available from other data. Estimated parameter standard errors can be computed from the estimated parameter covariance matrix **crb** by typing `serr=sqrt(diag(crb))`.

The m-file mof_demo.m is an example with comments and explanations output to the command window. The m-file of_mdoe_demo.m is an example using wind tunnel data.

12.3 Maximum Likelihood Methods

12.3.1 oe.m

Program oe.m estimates model parameters from measured input-output time-domain data using the output-error method. The nonlinear optimization used by oe.m is modified Newton-Raphson, as described in Chapter 6. If this method fails to produce a decrease in the cost function, the optimizer automatically switches to the simplex method for 50 iterations, then goes back to modified Newton-Raphson. This procedure continues until convergence criteria for parameter estimates, cost function, and cost gradients specified in Chapter 6 are satisfied. After each such convergence, the noise covariance matrix estimate is updated, and the entire process is repeated until all of the convergence criteria given in Chapter 6 are satisfied, including parameter estimates, cost function, cost function gradients, and the estimated noise covariance matrix elements.

The modified Newton-Raphson approach has fast convergence, but sometimes diverges when initiated far from the solution, because of inaccurate cost gradient information. The simplex method always moves toward the solution, but slowly. The optimization approach used in oe.m combines the advantages of both, and has been found to work well in practice. The optimization can also be controlled manually by the user by setting input **auto**=0. In this case, the user can specify the number of modified Newton-Raphson steps to be taken, and when to update the noise covariance matrix estimate.

The first input to oe.m, is a user-defined model file that can be of arbitrary complexity, but must have the following call syntax: `y=dsname(p,u,t,x0,c)`. The user-defined function dsname.m contains the model definition and the MATLAB® code to compute outputs **y** based on a candidate parameter vector **p**, and the other inputs. The input **dsname** to oe.m should be a string containing the name of the user-defined model function, without the .m file extension. For example, if the user-defined model file is tlonss.m, with the required calling syntax (i.e., function definition statement) `[y,x]=tlonss(p,u,t,x0,c)`, then the first input to oe.m should be `'tlonss'`, including the single quotes. Note that extra variables can be output from the model file (e.g., **x** for tlonss.m), but the first variable must be the computed output of the model. The output **y** must correspond to the measured outputs in the columns of **z**, so **y** and **z** have the same dimensions. There are several examples of the user-defined model files included with SIDPAC, including tlonss.m (longitudinal linear state-space dynamic model with dimensional parameters), tlontf.m (longitudinal transfer function model), nldyn.m (nonlinear dynamic model with nondimensional parameters) and tlatss.m (lateral linear state-space dynamic model with dimensional parameters), among others, for time-domain analysis. In general, it is good practice to test the user-defined model file separately in the command window, e.g., using `[y,x]=tlonss(p,u,t,x0,c)`, prior to using oe.m, to make sure the user-defined model file is programmed as intended.

The parameters in the initial parameter vector **p0**, which is the starting point for the nonlinear optimization, are also treated as prior parameter estimates with covariance matrix **crb0**, but only if the input **crb0** is provided. This amounts to using the same input **p0** for two different roles. However, using parameter estimates based on prior information as starting values for the nonlinear optimization usually works well. If some of the parameters in **p0** should not be treated as prior estimates, then their associated diagonal elements in **crb0** can be set to a large value, e.g., 10^6, which indicates that those elements in **p0** contain no prior information for parameter estimation. If all of the elements in **p0** are simply starting values, then input **crb0** should be omitted.

As with any nonlinear optimization, the starting values of the parameters in **p0** are important. Program `oe.m` will always try to take a modified Newton-Raphson step first, but if that step results in an increased cost, the program automatically switches to a slower, but more robust simplex method. This assures that the program will converge, as long as the problem has been set up properly (e.g., all model parameters can be estimated from the data, parameters have independent roles in modeling the output, etc.). The closer the initial parameter values are to the final answer, the faster that answer will be found, because more of the optimization steps will be modified Newton-Raphson steps. The price for poor starting values is a longer run time for oe.m. For most practical aircraft system identification problems, it is not difficult to choose starting values that will lead to convergence to the global minimum of the cost function. However, it is also possible to choose starting values so poor that the optimizer will not converge at all. This is simply the nature of nonlinear optimization problems.

In general, let the program compute **del**, **svlim**, and **auto** by omitting these inputs from the oe.m command line. The vector **del** controls the size of finite central difference perturbations for sensitivity calculations, and the quantity **svlim** can be used to address collinearity problems in some cases. Input parameter **auto** can be used to manually control the optimization process, as mentioned earlier.

Measured inputs **u** and outputs **z** must be assembled in MATLAB® outside of oe.m. The model structure to be used in oe.m is specified in the user-defined model file, as described earlier. Normally, this model structure is chosen from analysis using swr.m or mof.m, or by analysis of residuals, experience, or judgment.

The m-file totter_demo.m includes a demonstration of the use of nldyn.m, which implements a nonlinear dynamic model with nondimensional aerodynamic parameters. The m-files oe_lon_demo.m and oe_lat_demo.m are examples for longitudinal and lateral cases, respectively, with comments and explanations output to the command window. These examples use flight data from the NASA Twin Otter aircraft and demonstrate the use of linear state-space models with dimensional derivatives.

12.3.2 colores.m and m_colores.m

Programs colores.m and m_colores.m compute corrected parameter covariance matrices by post-processing results from output-error parameter estimation routine oe.m. The correction, discussed in Chapter 6, accounts for the practical fact that the output-error residuals are colored, not white, as assumed in the output-error maximum likelihood formulation. Corrected parameter standard errors from colores.m or m_colores.m are consistent with the scatter in parameter estimates from repeated flight test maneuvers, and therefore accurately represent estimated parameter uncertainty.

Program m_colores.m is a vectorized version of colores.m, so m_colores.m runs faster than colores.m. The results from both programs are identical within numerical round-off error. The colored residual correction code should be run once at the end of the analysis, when the parameter estimates from the output-error method are judged to be acceptable. Inputs to colores.m and m_colores.m are the same as for oe.m, except that input **p** to colores.m and m_colores.m is the final parameter estimate **p** from oe.m.

In general, let the program compute **del** by omitting this input. The vector **del** controls the size of finite central difference perturbations for sensitivity calculations. Output of m_colores.m or colores.m includes the conventional Cramér-Rao bounds **crbo** and the corrected Cramér-Rao bounds **crb**. The **crbo** output should match the Cramér-Rao bounds from conventional output-error parameter estimation (oe.m), within round-off error.

The model structure to be used in m_colores.m or colores.m must be the same as that used for the parameter estimation in oe.m, and is defined by the user in a separate m-file or mex-file.

The m-files oe_lon_demo.m, oe_lat_demo.m, and totter_demo.m include the use of m_colores.m, with comments and explanations output to the command window.

12.4 Frequency Domain Methods

12.4.1 fint.m

Program fint.m computes the finite Fourier integral for arbitrary frequencies. The finite Fourier integral is computed based on the chirp

z-transform, with cubic interpolation applied to the integrand for high accuracy.

The fint.m program expects input data **x** to be sampled at regular intervals corresponding to the time points in **t**. The code is vectorized, so the time-domain data **x** can be either a single column vector or a matrix of column vectors. The frequencies selected for the Fourier integral in the input vector **w** can have arbitrary resolution on the interval $\left[0, \pi/\Delta t\right)$ rad/sec.

Use of fint.m is demonstrated inside tfest.m in the m-files tfest_demo.m and totter_tf_demo.m.

12.4.2 fdoe.m

Program fdoe.m estimates model parameters from measured input-output data using output error in the frequency domain. This routine is the frequency-domain analog of oe.m, so the inputs, outputs, optimization algorithm, and printed display for fdoe.m are similar to those of oe.m. Fourier transformed inputs **U** and outputs **Z** are supplied to fdoe.m as frequency-domain data, so the Fourier transformation must be done outside of program fdoe.m. Program fint.m should be used for this purpose. The vector of frequencies **w** in rad/sec used for the Fourier transformation is an additional input to fdoe.m, and the model file for fdoe.m must have the following call syntax: `y=dsname(p,U,w,x0,c)`. In general, SIDPAC programs use capital letters for frequency-domain data. The model structure to be used in fdoe.m is defined by the user in a separate m-file or mex-file (dsname), similarly to oe.m. Some example m-files for frequency-domain parameter estimation included in SIDPAC are flonss.m (longitudinal linear state-space model with dimensional parameters), flontf.m (longitudinal transfer function model), flatss.m (lateral state-space model with dimensional parameters), and flattf.m (lateral transfer function model).

Although fdoe.m was developed for output-error parameter estimation using frequency-domain data, the routine can also be used for equation-error parameter estimation, by modifying the user-defined m-file that specifies the model (dsname). The example model file called flonss.m includes code for both equation-error and output-error parameter estimation in the frequency domain using a state-space model for longitudinal short-period dynamics. The optimizer used in fdoe.m is the same one used in oe.m, but modified for frequency-domain data.

In general, let the program compute **del**, **svlim**, **crb0**, and **auto** by omitting these inputs from the fdoe.m call. Matrix **crb0** is the covariance matrix associated with the initial parameter vector **p0**. Inputs **p0** and **crb0** are treated by fdoe.m in the same way that oe.m treats these inputs. Input parameter **auto** can be used to manually control the optimization process, in the same manner as for oe.m.

The m-file fdoe_demo.m runs an example with comments and explanations output to the command window. The example uses simulated data for the closed-loop pitch rate response of a supersonic transport aircraft.

12.4.3 tfest.m

Program tfest.m estimates parameters in a transfer function model using equation-error in the frequency domain and fint.m to compute high-accuracy finite Fourier integrals for arbitrary frequencies. Numerator order **nord** and denominator order **dord** must be supplied as input. The routine works only for single-input, single-output (SISO) transfer function parameter estimation.

Parameter estimation results from program tfest.m can be examined using Bode plots with program cmpbode.m. Program tfest.m uses linear regression with complex numbers to match the highest output derivative in the frequency domain. Frequency scaling is introduced so that the parameter estimation will not be weighted toward the higher frequencies. This can be modified by changing the regression problem formulation at the end of program tfregr.m. Program tfest.m expects the input data **u** and **z** to be sampled sequentially at regular intervals corresponding to the time points in **t**. The selected frequencies in the input frequency vector **w** can have arbitrary resolution on the interval $[0, \pi/\Delta t)$ rad/sec.

The m-file tfest_demo.m shows an example with comments and explanations output to the command window. The example uses roll rate response data from the F-16 fighter aircraft simulation documented in Appendix D.

12.5 Experiment Design

12.5.1 mksqw.m

Program mksqw.m generates arbitrary square waves of length **T** with sampling interval **dt**. Input **npulse** is a row vector which represents the multiples of **tpulse** seconds to be used for each pulse of the square wave. For example, a doublet input with the basic pulse width equal to one second can be generated with `tpulse=1` and `npulse=[1,1]`. To generate a 3-2-1-1 input with the basic pulse width equal to one second, use `tpulse=1` and `npulse=[3,2,1,1]`. The pulses in the square wave input are adjacent and alternate in sign by default, but this can be changed with the **amp** input. For example, two pulses with alternating sign, amplitude 2, and 1 sec duration, separated by 4 sec, can be generated with `tpulse=1`, `npulse=[1,4,1]`, and `amp=[2,0,-2]`. The second pulse, with zero amplitude, is automatically assigned a negative sign, because of the automatic alternating sign, so the third element of **amp** is negative to get a pulse with amplitude equal to -2. If input **amp** is a scalar, the amplitudes of all pulses are set equal to **amp**. The input **tdelay** specifies the time delay before the square wave begins.

After generating a desired square wave with mksqw.m, program ratelim.m can be used to implement rate limiting for a practical input.

12.5.2 mkfswp.m

Program mkfswp.m generates linear or logarithmic frequency sweeps of length **T** with sampling interval **dt**, covering the frequency range [**wmin, wmax**] in rad/sec. The logarithmic frequency sweep dwells longer on the lower frequencies than the linear frequency sweep. The result is more power at the lower frequencies for the logarithmic frequency sweep. This is generally advantageous when the dynamic modes of interest are in the lower part of the specified frequency band. Each frequency sweep ends automatically at the zero crossing closest to the final time **T**, so the input starts and ends at zero. This implements a perturbation input.

After generating a desired frequency sweep with mkfswp.m, program ratelim.m can be used to implement rate limiting for a practical input.

12.5.3 mkmsswp.m

Program mkmsswp.m generates multiple orthogonal phase-optimized multisine sweep inputs. Each sweep has minimized relative peak factor **pf** for a selected power spectrum **pwr** for selected frequencies **fu** in Hz. The relative peak factor **pf** for each designed input (column of output **u**) is computed from Eq. (8.25).

If the frequencies **fu** of the sinusoidal components are not specified, mksswp.m uses the maximum frequency resolution possible for the sinusoidal components that compose the sweep. The only way to get finer frequency resolution for the sweep is to increase the time length **T**. Finer frequency resolution produces a richer and smoother power spectrum in the selected frequency band.

Program mksswp.m can generate multiple inputs that are mutually orthogonal in both the time domain and the frequency domain simultaneously. This is advantageous for practical accurate modeling. The program uses flat, uniform power spectra if the power spectra are not specified in **pwr**. Elements in each column of **pwr** indicate the fraction of the total signal power for the corresponding component sinusoids whose frequencies are specified in **fu**. The *k*th column of **fu** and **pwr** correspond to the *k*th column of the output matrix **u**. Consequently, the sum of each column of **pwr** must equal 1. This provides the capability to create multiple orthogonal multisine inputs with arbitrary power spectra and minimum peak factors.

Phase angles of the sweeps with minimized peak factor are adjusted so that each input begins and ends at zero. This implements a perturbation input.

After generating multisine sweep inputs with mksswp.m, program ratelim.m can be used to implement rate limiting for practical inputs.

12.6 Data Compatibility

As discussed in Chapter 9, data compatibility analysis can be done using output-error parameter estimation in the time domain. Consequently, the output-error parameter estimation routine oe.m, described earlier, is applicable to this problem. For data compatibility analysis, the user-defined model file integrates the kinematic equations, with instrumentation errors introduced as the unknown parameters. The model equations, inputs, outputs, and parameters are different for data compatibility than for aerodynamic parameter estimation, but these two output-error parameter estimation problems are otherwise conceptually identical. A general model file called dcmp.m, described next, has been developed for data compatibility analysis using oe.m.

12.6.1 dcmp.m

Program dcmp.m is a model file that computes data compatibility model outputs for output-error parameter estimation in the time domain. Program dcmp.m is a user-defined model file for oe.m, so that the first input to oe.m should be `'dcmp'`, including the single quotes. The syntax of the user-defined m-file or mex-file (like dcmp.m) must be the same as defined above for other model files used with oe.m.

Output-error parameter estimation goes much faster if a mex-file based on C code is used for the user-defined model file. The dcmp.m program can use C mex-files for the kinematic equations (dcmp_eqs.c) and the numerical integration (adamb3.c). The command issued at the MATLAB® prompt for compiling dcmp_eqs.c into a mex-file is `mex dcmp_eqs.c`, and similarly for adamb3.c. Compiling this or any other C mex-file requires a C compiler. Once the user-defined mex-files are compiled, the operation and call syntax of a user-defined mex-file is exactly the same as for a user-defined m-file. Maximum execution speed is achieved when both dcmp_eqs.c and adamb3.c are first compiled into mex-files.

The model structure for data compatibility analysis (i.e., which instrumentation error parameters should be estimated) can be determined by the analyst based on deficiencies in the comparison of measured quantities (airspeed, angle of attack, sideslip angle, Euler roll, pitch, and heading angles, and altitude) with reconstructed values obtained by integrating derivative-type measurements (linear accelerations and body-axis angular rates), setting all instrumentation error parameters equal to zero. The m-files dcmp_demo.m, dca_lon_demo.m, and dca_lat_demo.m show plots that can be used for this preliminary step. These demonstration examples use dcmp.m with oe.m for data compatibility analysis of flight data from the NASA F-18 High Alpha Research Vehicle (HARV).

It is not uncommon for some trial-and-error to be necessary to arrive at an appropriate instrumentation error parameter set. Once found, though, this model structure usually does not change over the course of a flight test program, unless the instrumentation is changed. The instrumentation error parameters and the manner in which they are used is specified in the user-defined model file.

The outputs to be matched and instrumentation error parameters to be estimated in dcmp.m are specified in dcmp_psel.m. Set-up for the data compatibility problem can be done by setting flags inside dcmp_psel.m and then typing `dcmp_psel` at the MATLAB® prompt to implement the settings, following by running oe.m with dcmp.m as the user-defined model file. The entire procedure can also be done using the SIDPAC GUI.

The m-files dca_lon_demo.m and dca_lat_demo.m contain longitudinal and lateral data compatibility examples, with comments and explanations shown in the command window. For the longitudinal case, measured values are substituted for the lateral states. This is done because the longitudinal maneuver has very little excitation in the lateral states, so any instrumentation error parameters associated with the lateral motion cannot be estimated from this data. Similarly, for the lateral case, measured values are substituted for the longitudinal state variables. If the maneuver has both longitudinal and lateral excitation (e.g., a windup turn), then it is possible to estimate values for all of the instrumentation error parameters from a single maneuver.

12.6.2 rotchk.m

Program rotchk.m checks the compatibility of measured data for body-axis angular rates and Euler angles. With some modification, this routine could be used as a user-defined model file (dsname) for oe.m, to estimate instrumentation error parameters for the body-axis angular rates and Euler angle sensors. Program rotchk.m computes the Euler angle time series based on the rotational kinematic differential equations with body-axis angular rates as inputs and initial conditions for the Euler angles. Input data are provided using standard **fdata** format. The relationships included in rotchk.m are a subset of the entire set of equations used for data compatibility.

A good way to evaluate the output from rotchk.m is to use cmpplot.m to compare reconstructed Euler angle time series computed from rotchk.m with the measured Euler angle data. This comparison can be used to get an idea of which instrumentation error parameters should be estimated using oe.m. For example, if Euler roll angle **phi** computed from rotchk.m drifts linearly with time, compared to measured **phi**, that would indicate that one or more of the measured body-axis angular rates has a bias error. A bias integrated over time produces a drift with time. If **phi** computed from rotchk.m differed from measured **phi**, such that the size of the difference correlated with the absolute value of measured **phi**, then measured **phi** may have a scale factor error.

12.6.3 airchk.m

Program airchk.m checks the compatibility of measured air data, translational acceleration, and body-axis angular rates. The code computes reconstructed time series of airspeed, angle of attack, and sideslip angle, based on the translational kinematic differential equations with translational accelerations and body-axis angular rates as inputs, and initial conditions for the body-axis velocity components **u**, **v**, and **w**. Input data are provided using standard **fdata** format. The relationships included in airchk.m are a subset of the entire set of equations used for data compatibility.

A good way to evaluate the output from airchk.m is to use cmpplot.m to compare reconstructed time series of airspeed, angle of attack, and sideslip angle computed from airchk.m with measured air-relative velocity data. The comparisons can indicate which instrumentation error parameters should be estimated using oe.m, as described earlier for rotchk.m.

12.7 Real-Time Methods

12.7.1 rlesq.m

Program rlesq.m computes least-squares linear regression model parameter estimates and covariance matrix estimates recursively. Although rlesq.m implements recursive least squares, the inputs to the routine include all of the measurements collected during a particular experiment or flight test maneuver. Consequently, the input regressor matrix **x** includes all of the data points for each regressor, where each row is one data point. The measured output **z** is a vector containing all of the measured output values.

The recursive identification loop is inside rlesq.m. To use the technique in a real-time application, the statements within the loop inside rlesq.m would be called at each time step when new measurements are available.

A typical range of possible values for the forgetting factor **ff** is $0.95 \leq \mathbf{ff} \leq 1.00$. Input covariance matrix **crb0** quantifies the confidence in the initial parameter estimates **p0**. For arbitrary starting values **p0** or low confidence in the **p0** values, corresponding diagonal elements of **crb0** should be set to a large positive number, e.g. 10^6.

Since **rlesq.m** computes the fit error estimate **s2** recursively, and the parameter estimates are usually of low quality early in the maneuver, the final estimated covariance matrix **crb** computed recursively in **rlesq.m** will not agree exactly with the batch estimate from **lesq.m**. However, the computed inverse information matrix $\left(X^T X\right)^{-1}$ is the same, whether computed by the batch method in **lesq.m** or recursively in **rlesq.m**. In addition, the final parameter estimates **p** from **rlesq.m** with **ff** = `1` should agree exactly with the batch parameter estimates from **lesq.m**, using the same data.

The sequence of parameter estimates and standard error estimates are contained in outputs **ph** and **seh**, respectively, stored by column. To examine the results, type `plot(t,ph(:,k))` or `plot(t,seh(:,k))` for sequences associated with the *k*th parameter. Program **rlesq.m** uses the final value of the estimated parameter vector **p** to compute the model output **y**.

12.7.2 rtpid.m

Program **rtpid.m** computes real-time linear regression model parameter and covariance matrix estimates using sequential least squares in the frequency domain.

Although program **rtpid.m** implements real-time parameter estimation, the inputs to the routine include all of the measurements collected during a particular experiment or flight test maneuver. The recursive Fourier transform and sequential least-squares parameter estimation calculations are inside program **rtpid.m**. To use the technique in a real-time application, the recursive Fourier transform called inside **rtpid.m** would be called at each time step when new measurements are available. The sequential least squares would be executed at desired times to produce real-time estimates of the parameters and covariance matrix. The settings that determine the update rate for the recursive Fourier transform and the sequential least squares parameter estimation are defined inside **rtpid.m**.

Program **rtpid.m** implements real-time parameter estimation for a single linear regression model, using all of the regressors in the input regressor matrix **x**. The regressor matrix **x** should not include a vector of ones, because the bias terms are omitted in the frequency domain. The model will be a least-squares fit to the measured output **z** when `lder=0`, or to the first time derivative of **z** when `lder=1`. In either case, the model output **y** corresponds to the time-domain measured output **z**. If `lder=1`, the time derivative of **z** is computed in the frequency domain inside **rtpid.m**.

The sequence of parameter estimates, model fit error variance estimates, and standard error estimates are contained in outputs **ph**, **s2h**, and **seh**, respectively, stored by column. The output vector **th** contains the times when each sequential estimate was made. To examine the results, type `plot(th,ph(:,k))` or `plot(th,seh(:,k))` for sequences associated with the kth parameter, and `plot(th,s2h)` for the model fit error variance estimate sequence.

Program rtpid.m uses the final value of the estimated parameter vector to compute the model output **y**. All biases and slow trends are taken out of the problem when the analysis is done in the frequency domain, because these components correspond to very low frequencies which are omitted from the analysis. Consequently, there may be a bias and/or drift mismatch between measured output **z** and model output **y** in the time domain. In practice, a high-pass filter can be used to detrend the data in real time prior to applying the recursive Fourier transform. Example code for that process is included inside rtpid.m, but it is commented out. Consequently, data should be detrended prior to applying rtpid.m, unless the internal high-pass filter code is uncommented.

The m-file rtpid_demo.m contains a demonstration of sequential least-squares parameter estimation in the frequency domain using the longitudinal short-period dynamics of the F-16 aircraft, derived from the F-16 nonlinear simulation described in Appendix D. Fore and aft mouse movements in the figure window command the stabilator for the F-16 linear simulation, and the real-time parameter estimation is executed using the resulting noisy data. Plots of the real-time parameter estimation results appear during the simulation run.

12.8 Data Analysis

12.8.1 smoo.m

Program smoo.m implements global Fourier smoothing with an optimal Wiener filter, as described in Chapter 11. The routine makes a plot of Fourier sine series coefficient magnitudes and allows the user to manually choose the frequency cut-off for global smoothing. It is also possible to run smoo.m in automatic mode, where the software identifies frequency domain models for signal and noise, using k^{-3} and a constant, respectively, where k is the frequency index. These identified models are then used to find the frequency cut-off and construct the Wiener filter.

Program smoo.m is vectorized, so the measured **z** can be either a single column vector or a matrix of column vectors. Filter cut-off frequencies **fco** can be determined automatically based on the data, or chosen manually by the analyst based on frequency-domain data plots. As currently implemented, the filters are low pass, so the deterministic signal is assumed to be in a frequency range $\left[0,\mathbf{fco}\right]$ Hz.

Program smoo.m can be used to find smoothed time series from noisy measured data, and to estimate noise variances. The measurements **z** must be made sequentially at regular intervals, because of the use of Fourier series.

Input **fc** applies only to the endpoints. The value chosen therefore is generally not critical, and the default value suffices in most cases. The value of **fc** should be set higher for high-frequency content in the deterministic signal near the endpoints. To avoid problems with endpoints, the best approach is to include extra data points at the beginning and end of **z**, run smoo.m, then discard the extra data points at the beginning and end from the result. Setting input `auto`=0 allows the analyst to make the decision on where the filter cut-off frequency should be. Because of the formulation of the Fourier smoothing problem, and also because of the use of the Wiener filter, the algorithm is robust to the choice of cut-off frequency for the filter. This means that virtually the same filtering will occur for a range of cut-off frequencies near the best one. Using the frequency-domain data plots, it is only necessary to place the filter cut-off frequency between the large Fourier sine series components and the small components. Since the decay of the Fourier sine series for the deterministic signal goes with the inverse cube of the number of terms in the series, this separation is easy to make visually.

In practice, the data occasionally will show more than one group of relatively large Fourier sine series components. Each of these corresponds to deterministic signal components. For flight data, the lowest frequency group of large magnitude Fourier coefficients corresponds to the deterministic rigid-body response, while the higher frequency groups of large magnitude Fourier coefficients are often from a structural response.

Manual operation of smoo.m, (`auto` = 0), allows the analyst to get a better idea of the frequency content of the measured data, and more control over the final results. The automatic method is faster and less trouble to use, but does not provide the same insight.

12.8.2 deriv.m

Program deriv.m computes a local smoothed numerical time derivative based on measured time series data. Program deriv.m is vectorized, so the input measured **z** can be either a single column vector or a matrix of column vectors. Program deriv.m assumes the measurements **z** are made sequentially at regular time intervals of size **dt**.

For very noisy data, the smoothed derivative at the endpoints of **z** can be degraded. To avoid problems with endpoints, the best approach is to include 5-10 extra data points at the beginning and end of **z**, run deriv.m, then discard the extra data points at the beginning and end from the result. If deriv.m is being used with smoo.m (or any other smoothing or filtering routine), it is best to apply the smoother or filter first, then deriv.m, cf. Ref. [12.4]. The final result is a very smooth and accurate numerical time derivative.

Program nderiv.m is a generalization of the algorithm in deriv.m, where the order of the local least-squares polynomial fit to the measured data, and the number of neighboring data points used in that fit, are integer inputs to the nderiv.m code. The default values inside program nderiv.m have been found to give good results in many flight data analysis cases. Using more neighboring points or a lower-order polynomial fit smoothes the local fit more and therefore removes high frequencies, and vice versa.

A related program called lsmoo.m does the same local smoothing as nderiv.m, but returns the local smoothed value instead of the time derivative. The result is a locally-smoothed time series.

12.8.3 *lsmep.m*

Program lsmep.m is a local smoother for endpoints of a time series. This program is useful for estimating initial conditions from measured states and controls. Program lsmep.m uses a local polynomial smoother similar to lsmoo.m.

Other routines that perform a similar function are: smep.m, which uses a time convolution implementation of a low-pass filter, then extrapolates the smoothed points near the endpoint to obtain smoothed endpoint data; xsmep.m, which is the same as smep.m, except the endpoint data are excluded from the initial smoothing; csmep.m, which uses a local cubic polynomial fit over a selected time period near the endpoints to compute the smoothed endpoints; zep.m, which sets the endpoints to zero by removing the bias and a linear trend; and szep.m, which is the same as zep.m, except that smoothed endpoints from lsmep.m are used to remove the bias and linear trend. All of the programs mentioned can be used on a single column vector of data or on matrices of column vector data.

12.8.4 *compfc.m*

Program compfc.m computes the nondimensional aerodynamic force coefficients based on aircraft geometry and measured data arranged in standard **fdata** format. Corrections for instrument positions must be done outside of compfc.m, with results placed in the proper columns of **fdata**. Program compfc.m assumes that thrust acts along the x body axis through the aircraft c.g.

12.8.5 *compmc.m*

Program compmc.m computes the nondimensional aerodynamic moment coefficients based on aircraft geometry and measured data arranged in standard **fdata** format. Program compmc.m uses angular accelerations from the **fdata** matrix if the data in the appropriate columns for those quantities are nonzero; otherwise, compmc.m computes the angular accelerations from the angular velocities using a local smoothed numerical differentiation (deriv.m).

Program compmc.m assumes that thrust acts along the x body axis and through the aircraft c.g.

A related program called compfmc.m uses both compfc.m and compmc.m to compute nondimensional aerodynamic force and moment coefficients, along with nondimensional rotational rates, then stores the results in the appropriate columns of the standard **fdata** matrix.

12.8.6 cmpsigs.m

Many aerodynamic models include terms which consist of a parameter multiplying a regressor or explanatory variable. Since the model also usually includes a bias term, an investigation of the modeling effectiveness of the regressors or explanatory variables should exclude both the scaling (accounted for by the model parameter for each term) and the bias (accounted for by the bias term in the model). Program cmpsigs.m strips away the bias and scaling of regressors or explanatory variables and makes a comparison plot. This allows an analyst to quickly determine the level of similarity of the waveforms for various regressors or explanatory variables. In a sense, the cmpsigs.m program provides a graphical depiction of the correlations among the regressors or explanatory variables. Pairwise correlation can be quantified mathematically using the correlation coefficient (see Chapter 5).

The vectors to be compared are columns of the input matrix **x**, and must have the same length as the time vector **t**. Scaling is always done relative to the first vector in the matrix **x**. For example, if `x=[x1,x2,x3]`, vectors **x2** and **x3** would be scaled to match the original scale of vector **x1**.

12.9 SIDPAC Programs

A list of all SIDPAC programs appears in the MATLAB® command window in response to typing `help sidpac` at the command prompt. Each name in the list is a hyperlink to the corresponding m-file header, which contains complete information on the inputs, outputs, functionality, and calling syntax of the m-file. Clicking on any of the hyperlinks displays the relevant m-file header. The m-files are grouped approximately by functionality, to familiarize the user with SIDPAC capabilities, and to help the user select appropriate SIDPAC tools.

References

[12.1] *MATLAB® Primer, R2015b*, The MathWorks, Inc., Natick, MA, 2015.

[12.2] Press, W.H., Teukolsky. S.A., Vettering, W.T., and Flannery, B.R., *Numerical Recipes in FORTRAN: The Art of Scientific Computing*, 2nd Ed., Cambridge University Press, New York, NY, 1992.

[12.3] Morelli, E.A., "System IDentification Programs for AirCraft (SIDPAC)," AIAA Paper AIAA-2002-4704, *AIAA Atmospheric Flight Mechanics Conference*, Monterey, CA, 2002.

[12.4] Morelli, E.A., "Practical Aspects of the Equation-Error Method for Aircraft Parameter Estimation," AIAA-2006-6144, *AIAA Atmospheric Flight Mechanics Conference*, Keystone, CO, 2006.

Appendix A

Mathematical Background

A.1 Linear Algebra

A.1.1 Basics

A matrix is a rectangular array. The matrix

$$\boldsymbol{A}=\begin{bmatrix} a_{11} & a_{12} & \cdots & a_{1n} \\ a_{21} & a_{22} & \cdots & a_{2n} \\ \vdots & \vdots & \vdots & \vdots \\ a_{m1} & a_{m2} & \cdots & a_{mn} \end{bmatrix}=\begin{bmatrix} a_{ij} \end{bmatrix} \quad \begin{matrix} i=1,2,\ldots,m \\ j=1,2,\ldots,n \end{matrix} \tag{A.1}$$

is called an $m \times n$ matrix, meaning there are m rows and n columns. The vector

$$\boldsymbol{x}=\begin{bmatrix} x_1 \\ x_2 \\ \vdots \\ x_n \end{bmatrix}$$

is an $n \times 1$ column vector. The vector

$$y=\begin{bmatrix} y_1 & y_2 & \cdots & y_m \end{bmatrix}$$

is an $1 \times m$ row vector. The dimension of a vector is the number of elements in the vector, e.g., the dimension of the vector $\boldsymbol{x}$ above is n. Matrices can be thought of as consisting of row vectors stacked vertically, or column vectors stacked horizontally.

The transpose of a matrix is the matrix obtained by switching the rows and columns. For the matrix $\boldsymbol{A}$ in Eq. (A.1), the transpose is an $n \times m$ matrix

$$\boldsymbol{A}^T=\begin{bmatrix} a_{11} & a_{21} & \cdots & a_{m1} \\ a_{12} & a_{22} & \cdots & a_{m2} \\ \vdots & \vdots & \vdots & \vdots \\ a_{1n} & a_{2n} & \cdots & a_{mn} \end{bmatrix} \tag{A.2}$$

Matrices with the same dimensions can be added by adding their corresponding elements,

$$A+B=\left[a_{ij}\right]+\left[b_{ij}\right]=\left[a_{ij}+b_{ij}\right] \qquad \begin{matrix} i=1,2,\ldots,m \\ j=1,2,\ldots,n \end{matrix}$$

Multiplying a matrix by a scalar is the same as multiplying each element in the matrix by that scalar,

$$kA=Ak=\left[ka_{ij}\right] \qquad \begin{matrix} i=1,2,\ldots,m \\ j=1,2,\ldots,n \end{matrix}$$

for any scalar k. Two matrices A and B can be multiplied in the order AB if the number of columns of A equals the number of rows of B. When this is true, the matrices A and B are said to be conformable or to have conformable dimensions. The product of an $m\times n$ matrix A with an $n\times p$ matrix B is an $m\times p$ matrix C, computed as

$$\begin{gathered} C=\left[c_{ij}\right]=AB \\ c_{ij}=\sum_{k=1}^{n} a_{ik}b_{kj} \qquad \begin{matrix} i=1,2,\ldots,m \\ j=1,2,\ldots,p \end{matrix} \end{gathered} \tag{A.3}$$

In general, matrix multiplication is not commutative, so $AB \neq BA$.

The inner product or dot product is a scalar defined for two vectors with the same dimension,

$$u\bullet v=u^T v=\sum_{i=1}^{n} u_i v_i$$

where

$$\begin{aligned} u^T &= \left[u_1 \;\; u_2 \;\; \ldots \;\; u_n\right] \\ v^T &= \left[v_1 \;\; v_2 \;\; \ldots \;\; v_n\right] \end{aligned}$$

Each element c_{ij} of the $m\times p$ product matrix C in Eq. (A.3) is the inner product of the ith row of A with the jth column of B.

The inner product of a vector with itself equals the sum of its squared elements, which is also the squared length of the vector,

$$\|u\|^2=u\bullet u=\sum_{i=1}^{n} u_i^2 \tag{A.4}$$

If the inner product of two vectors $\boldsymbol{u}$ and $\boldsymbol{v}$ is zero, the vectors are said to be orthogonal,

$$\boldsymbol{u}^T \boldsymbol{v} = 0 \Leftrightarrow \text{ vectors } \boldsymbol{u} \text{ and } \boldsymbol{v} \text{ are orthogonal} \tag{A.5}$$

Vectors that have unit length and are mutually orthogonal are called orthonormal vectors.

The outer product of an $n \times 1$ vector $\boldsymbol{u}$ with an $m \times 1$ vector $\boldsymbol{v}$ is an $n \times m$ matrix,

$$\boldsymbol{u}\,\boldsymbol{v}^T = \left[u_i \, v_j \right] \qquad \begin{array}{l} i = 1,2,\ldots,n \\ j = 1,2,\ldots,m \end{array}$$

A square matrix has the same number of rows and columns, $m = n$. A diagonal matrix is a square matrix with all zero values, except for the elements on the diagonal:

$$\boldsymbol{D} = \begin{bmatrix} d_{11} & 0 & 0 \\ 0 & d_{22} & 0 \\ 0 & 0 & d_{33} \end{bmatrix}$$

A diagonal matrix with ones on the diagonal is called the identity matrix,

$$\boldsymbol{I} = \begin{bmatrix} 1 & 0 & 0 \\ 0 & 1 & 0 \\ 0 & 0 & 1 \end{bmatrix}$$

A square matrix is called symmetric if

$$\boldsymbol{A} = \boldsymbol{A}^T$$

which means that the values on either side of the diagonal are the mirror image of the other side.

A matrix is Hermitian if it equals its complex conjugate transpose (see Section A.2),

$$\boldsymbol{A} = \boldsymbol{A}^\dagger$$

The trace of a square matrix is the sum of the diagonal elements,

$$Tr(A) = \sum_{i=1}^{n} a_{ii} \qquad \boldsymbol{A} = \left[a_{ij} \right] \quad i,j = 1,2,\ldots,n \tag{A.6}$$

For a square matrix $\boldsymbol{A}$, the matrix inverse $\boldsymbol{A}^{-1}$ is defined by

$$\boldsymbol{A}\boldsymbol{A}^{-1} = \boldsymbol{I} = \boldsymbol{A}^{-1}\boldsymbol{A} \tag{A.7}$$

when the inverse exists. A square matrix $\boldsymbol{A}$ is called orthogonal if

$$\boldsymbol{A}^T\boldsymbol{A} = \boldsymbol{I} = \boldsymbol{A}\boldsymbol{A}^T$$

so that

$$\boldsymbol{A}^T = \boldsymbol{A}^{-1} \tag{A.8}$$

The columns of an orthogonal matrix are orthonormal vectors, and similarly for the rows.

The determinant is a scalar function of a square matrix. For a 2×2 matrix

$$\boldsymbol{A} = \begin{bmatrix} a_{11} & a_{12} \\ a_{21} & a_{22} \end{bmatrix} \tag{A.9}$$

the determinant is defined as

$$det(\boldsymbol{A}) \equiv a_{11}a_{22} - a_{12}a_{21} = |\boldsymbol{A}| \tag{A.10}$$

For a 3×3 matrix

$$\boldsymbol{A} = \begin{bmatrix} a_{11} & a_{12} & a_{13} \\ a_{21} & a_{22} & a_{23} \\ a_{31} & a_{32} & a_{33} \end{bmatrix} \tag{A.11}$$

the determinant is defined as

$$\begin{aligned} det(\boldsymbol{A}) = |\boldsymbol{A}| = a_{11}a_{22}a_{33} + a_{12}a_{23}a_{31} + a_{13}a_{21}a_{32} \\ - a_{11}a_{23}a_{32} - a_{12}a_{21}a_{33} - a_{13}a_{22}a_{31} \end{aligned} \tag{A.12}$$

The cofactor of the element located in the ith row and j^{th} column of a square matrix is defined as $(-1)^{i+j}$ times the determinant of the minor determinant, which is the determinant of the square matrix remaining after the ith row and j^{th} column are removed. For example, the cofactor of element a_{12} in the matrix $\boldsymbol{A}$ of Eq. (A.11) is

$$C_{12} = -det\begin{bmatrix} a_{21} & a_{23} \\ a_{31} & a_{33} \end{bmatrix}$$

For any square matrix, the determinant can be found as the sum of all elements in a single row or column multiplied by its cofactor. For example, the determinant in Eq. (A.12) can be found by a sum of the products of each element in the second row of $\boldsymbol{A}$ with its cofactor,

$$det(\boldsymbol{A}) = -a_{21}\begin{vmatrix} a_{12} & a_{13} \\ a_{32} & a_{33} \end{vmatrix} + a_{22}\begin{vmatrix} a_{11} & a_{13} \\ a_{31} & a_{33} \end{vmatrix} - a_{23}\begin{vmatrix} a_{11} & a_{12} \\ a_{31} & a_{32} \end{vmatrix}$$

$$= -a_{21}a_{12}a_{33} + a_{21}a_{13}a_{32} + a_{22}a_{11}a_{33} - a_{22}a_{13}a_{31} - a_{23}a_{11}a_{32} + a_{23}a_{12}a_{31}$$

The same procedure can be used to find the determinant of a larger square matrix in terms of determinants of smaller square matrices. Using this idea repeatedly, the determinant of any square matrix can be reduced to a sum of 3×3 or 2×2 determinants, which can be calculated as shown above.

If two columns of a square matrix $\boldsymbol{A}$ are proportional, then the expression for the determinant resulting from expanding along each of these columns is the same except for multiplication by a constant. But the determinant must be the same regardless of which column is used for the expansion, so the determinant of a square matrix with proportional columns must be zero. The same logic applies for proportional rows. This can be easily verified using a 2×2 matrix,

$$\boldsymbol{A} = \begin{bmatrix} a_{11} & k\,a_{11} \\ a_{21} & k\,a_{21} \end{bmatrix}$$

$$det(\boldsymbol{A}) \equiv a_{11}ka_{21} - ka_{11}a_{21} = 0$$

When the determinant is nonzero, none of the columns of the matrix are proportional. For a square $n\times n$ matrix $\boldsymbol{A}$ comprised of n column vectors,

$$\boldsymbol{A} = [\boldsymbol{a}_1 \quad \boldsymbol{a}_2 \quad \dots \quad \boldsymbol{a}_n]$$

if $det(\boldsymbol{A}) \neq 0$, then the relation

$$c_1\boldsymbol{a}_1 + c_2\boldsymbol{a}_2 + \ldots + c_n\boldsymbol{a}_n = \boldsymbol{0} \qquad \text{(A.13)}$$

is only satisfied when $c_1 = c_2 = \ldots = c_n = 0$. When this is true, the columns of $\boldsymbol{A}$ are said to be linearly independent, and $\boldsymbol{A}$ is called a nonsingular matrix. A matrix $\boldsymbol{A}$ with n linearly independent columns is said to have rank n.

The inverse of a nonsingular square matrix can be computed from

$$\boldsymbol{A}^{-1} = \frac{[Cof(\boldsymbol{A})]^T}{|\boldsymbol{A}|} \qquad \text{(A.14)}$$

where $Cof(\boldsymbol{A})$ is the matrix of cofactors of $\boldsymbol{A}$

$$Cof(\boldsymbol{A}) = \begin{bmatrix} C_{11} & C_{12} & \cdots & C_{1n} \\ C_{21} & C_{22} & \cdots & C_{2n} \\ \vdots & \vdots & & \vdots \\ C_{n1} & C_{n2} & \cdots & C_{nn} \end{bmatrix}$$

This expression can be verified by

$$\boldsymbol{A}^{-1}\boldsymbol{A} = \frac{\left[Cof(\boldsymbol{A})\right]^T}{|\boldsymbol{A}|}\boldsymbol{A} = \frac{1}{|\boldsymbol{A}|}\sum_{k=1}^{n} C_{ki}\, a_{kj} = \begin{cases} 1 \text{ for } i = j \\ 0 \text{ for } i \neq j \end{cases} \quad i, j = 1, 2, \ldots n$$

$$= \boldsymbol{I}$$

since the summation is either an expansion for $|\boldsymbol{A}|$ along a column of $\boldsymbol{A}$, or an expansion for the determinant of a matrix with two identical columns, which is zero.

From Eq. (A.14), it follows that the inverse of a square matrix exists only when the determinant is nonzero. Equivalently, the matrix inverse exists only for a nonsingular matrix, which is a matrix with columns that are linearly independent.

A.1.2 Useful Matrix Relationships

The following is a list of useful matrix identities. In the expressions, $\boldsymbol{A}, \boldsymbol{B},$ and $\boldsymbol{C}$ are matrices with appropriate dimensions, k is a scalar, n is a positive integer, and $\boldsymbol{I}$ is the identity matrix with appropriate dimension.

$$\boldsymbol{A} + \boldsymbol{B} = \boldsymbol{B} + \boldsymbol{A} \qquad \text{In general, } \boldsymbol{AB} \neq \boldsymbol{BA} \tag{A.15}$$

$$(\boldsymbol{AB})^T = \boldsymbol{B}^T\boldsymbol{A}^T \qquad (\boldsymbol{AB})^{-1} = \boldsymbol{B}^{-1}\boldsymbol{A}^{-1} \qquad (\boldsymbol{A} + \boldsymbol{B})^T = \boldsymbol{A}^T + \boldsymbol{B}^T \tag{A.16}$$

$$\left(\boldsymbol{A}^n\right)^{-1} = \left(\boldsymbol{A}^{-1}\right)^n \qquad \left(\boldsymbol{A}^{-1}\right)^T = \left(\boldsymbol{A}^T\right)^{-1} \qquad \boldsymbol{IA} = \boldsymbol{AI} = \boldsymbol{A} \tag{A.17}$$

$$k(\boldsymbol{AB}) = (k\boldsymbol{A})\boldsymbol{B} = \boldsymbol{A}(k\boldsymbol{B}) \qquad \left(\boldsymbol{A}^T\right)^T = \boldsymbol{A} \tag{A.18}$$

$$\left(\boldsymbol{A}^{-1}\right)^{-1} = \boldsymbol{A} \qquad (k\boldsymbol{A})^{-1} = \frac{1}{k}\boldsymbol{A}^{-1} \qquad |k\boldsymbol{A}| = k^n|\boldsymbol{A}| \tag{A.19}$$

$$(\boldsymbol{A} = n \times n \text{ matrix})$$

$$|\boldsymbol{AB}| = |\boldsymbol{A}||\boldsymbol{B}| \qquad |\boldsymbol{A}| = \left|\boldsymbol{A}^T\right| \qquad \left|\boldsymbol{A}^{-1}\right| = \frac{1}{|\boldsymbol{A}|} \qquad |\boldsymbol{I}| = 1 \tag{A.20}$$

A.1.3 Matrix Inversion Lemma

Given square, nonsingular matrices $\boldsymbol{A}, \boldsymbol{C}$, and $\boldsymbol{A}+\boldsymbol{BCD}$ with appropriate dimensions,

$$(\boldsymbol{A}+\boldsymbol{BCD})^{-1} = \boldsymbol{A}^{-1} - \boldsymbol{A}^{-1}\boldsymbol{B}\left(\boldsymbol{C}^{-1} + \boldsymbol{DA}^{-1}\boldsymbol{B}\right)^{-1} \boldsymbol{DA}^{-1} \qquad \text{(A.21)}$$

This identity is called the matrix inversion lemma. To prove it, multiply both sides of the last equation by $\boldsymbol{A}+\boldsymbol{BCD}$,

$$\boldsymbol{I} = (\boldsymbol{A}+\boldsymbol{BCD})\left[\boldsymbol{A}^{-1} - \boldsymbol{A}^{-1}\boldsymbol{B}\left(\boldsymbol{C}^{-1} + \boldsymbol{DA}^{-1}\boldsymbol{B}\right)^{-1} \boldsymbol{DA}^{-1}\right]$$

By straightforward matrix manipulations,

$$\boldsymbol{I} = \boldsymbol{I} + \boldsymbol{BCDA}^{-1} - \boldsymbol{B}\left(\boldsymbol{C}^{-1} + \boldsymbol{DA}^{-1}\boldsymbol{B}\right)^{-1} \boldsymbol{DA}^{-1}$$

$$- \boldsymbol{BCDA}^{-1}\boldsymbol{B}\left(\boldsymbol{C}^{-1} + \boldsymbol{DA}^{-1}\boldsymbol{B}\right)^{-1} \boldsymbol{DA}^{-1}$$

$$\boldsymbol{I} = \boldsymbol{I} + \boldsymbol{BCDA}^{-1} - \boldsymbol{B}\left(\boldsymbol{I} + \boldsymbol{CDA}^{-1}\boldsymbol{B}\right)\left(\boldsymbol{C}^{-1} + \boldsymbol{DA}^{-1}\boldsymbol{B}\right)^{-1} \boldsymbol{DA}^{-1}$$

$$\boldsymbol{I} = \boldsymbol{I} + \boldsymbol{BCDA}^{-1} - \boldsymbol{BC}\left(\boldsymbol{C}^{-1} + \boldsymbol{DA}^{-1}\boldsymbol{B}\right)\left(\boldsymbol{C}^{-1} + \boldsymbol{DA}^{-1}\boldsymbol{B}\right)^{-1} \boldsymbol{DA}^{-1}$$

$$\boldsymbol{I} = \boldsymbol{I} + \boldsymbol{BCDA}^{-1} - \boldsymbol{BCDA}^{-1}$$

$$\boldsymbol{I} = \boldsymbol{I}$$

The original statement reduces to an identity, which proves the lemma.

A.1.4 Expressing Vectors in Different Reference Frames

An arbitrary three-dimensional vector can be expressed in two different reference frames with the same origin as follows

$$\text{Frame } A: \quad \boldsymbol{V} = \begin{bmatrix} v_{A_1} & v_{A_2} & v_{A_3} \end{bmatrix}^T$$

$$\text{Frame } B: \quad \boldsymbol{V} = \begin{bmatrix} v_{B_1} & v_{B_2} & v_{B_3} \end{bmatrix}^T$$

Figure A.1 illustrates the situation, showing only one axis from the *B* frame, for clarity.

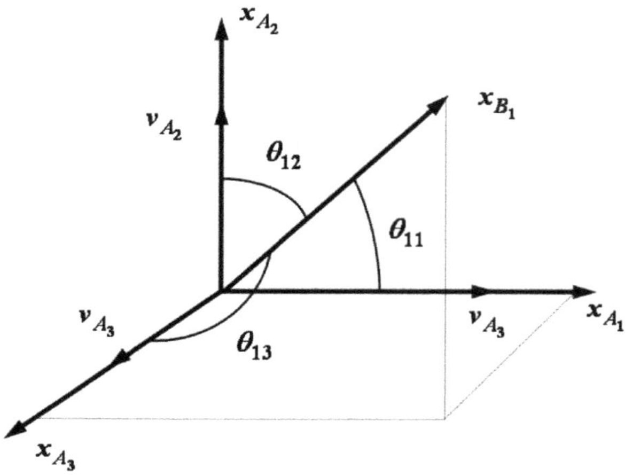

Figure A.1 **Vector components**

If θ_{ij} denotes the angle of the ith axis of the B frame relative to the jth axis of the A frame, then the components of V in the B frame can be expressed as

$$v_{B_i} = \sum_{j=1}^{3} v_{A_j} \cos\theta_{ij}$$

where

$$l_{ij} \equiv \cos\theta_{ij}$$

is called a direction cosine. If the direction cosines are arranged in a transformation matrix $\boldsymbol{L}_{BA}$ as

$$\boldsymbol{L}_{BA} \equiv \left[l_{ij}\right] \qquad i, j = 1,2,3$$

then

$$\boldsymbol{V}_B = \begin{bmatrix} v_{B_1} \\ v_{B_2} \\ v_{B_3} \end{bmatrix} = \boldsymbol{L}_{BA} \begin{bmatrix} v_{A_1} \\ v_{A_2} \\ v_{A_3} \end{bmatrix} = \boldsymbol{L}_{BA}\boldsymbol{V}_A$$

The last two equations are the required transformation for components of a vector from one reference frame to another with the same origin.

The vector $\boldsymbol{V}$ has the same magnitude, regardless of the reference frame,

$$|\boldsymbol{V}| = \boldsymbol{V}_A^T \boldsymbol{V}_A = \boldsymbol{V}_B^T \boldsymbol{V}_B = \boldsymbol{V}_A^T \boldsymbol{L}_{BA}^T \boldsymbol{L}_{BA} \boldsymbol{V}_A$$

It follows that

$$\boldsymbol{L}_{BA}^T \boldsymbol{L}_{BA} = \boldsymbol{I}$$

$$\boldsymbol{L}_{BA}^{-1} = \boldsymbol{L}_{BA}^T$$

$$\left|\boldsymbol{L}_{BA}^T \boldsymbol{L}_{BA}\right| = \left|\boldsymbol{L}_{BA}^T\right|\left|\boldsymbol{L}_{BA}\right| = \left|\boldsymbol{L}_{BA}\right|^2 = 1$$

The expressions above are orthogonality conditions, which show that the columns of $\boldsymbol{L}_{BA}$ are orthonormal, and that $\left|\boldsymbol{L}_{BA}\right| = 1$. It follows that the transformation is nonzero and does not alter the vector magnitude. Furthermore,

$$\boldsymbol{V}_B = \boldsymbol{L}_{BA} \boldsymbol{V}_A = \boldsymbol{L}_{BA} \boldsymbol{L}_{AB} \boldsymbol{V}_B$$

Combining the last two sets of expressions,

$$\boldsymbol{L}_{AB} = \boldsymbol{L}_{BA}^{-1} = \boldsymbol{L}_{BA}^T$$

The transformation matrix $\boldsymbol{L}_{BA}$ is therefore an orthogonal matrix.

The orthogonality conditions yield a set of 6 conditions on the columns of the transformation matrix,

$$\boldsymbol{L}_{BA} = \begin{bmatrix} \boldsymbol{l}_1 & \boldsymbol{l}_2 & \boldsymbol{l}_3 \end{bmatrix}$$

$$\boldsymbol{l}_i^T \boldsymbol{l}_j = \delta_{ij} = \begin{cases} 1 \text{ for } i = j \\ 0 \text{ for } i \neq j \end{cases} \qquad i, j = 1, 2, 3$$

This represents six relations among the nine elements of $\boldsymbol{L}_{BA}$, so only three of the nine elements of $\boldsymbol{L}_{BA}$ are independent. These three independent direction cosines specify the orientation of frame B relative to frame A.

As noted in Chapter 3, relative orientation of one reference frame to another for aircraft is typically done using single axis rotations. For such rotations, the transformation matrices are x-axis rotation (Figure A.2)

$$L_1(\theta_1) = \begin{bmatrix} 1 & 0 & 0 \\ 0 & \cos\theta_1 & \sin\theta_1 \\ 0 & -\sin\theta_1 & \cos\theta_1 \end{bmatrix}$$

$$V_B = L_1(\theta_1)V_A$$

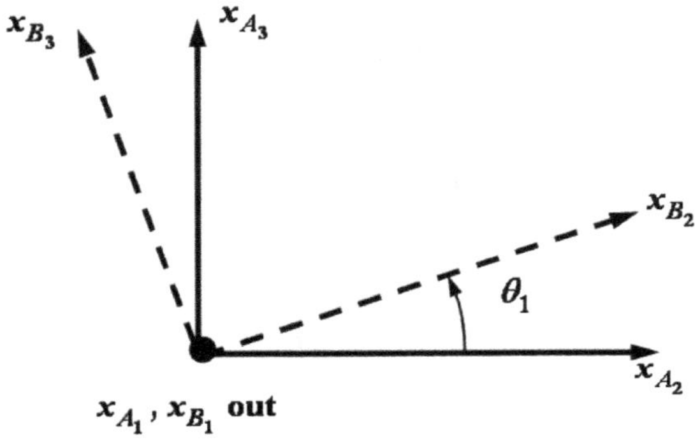

Figure A.2 ***x*-axis rotation**

y-axis rotation (Figure A.3)

$$L_1(\theta_1) = \begin{bmatrix} \cos\theta_2 & 0 & -\sin\theta_2 \\ 0 & 1 & 0 \\ \sin\theta_2 & 0 & \cos\theta_2 \end{bmatrix}$$

$$V_B = L_2(\theta_2)V_A$$

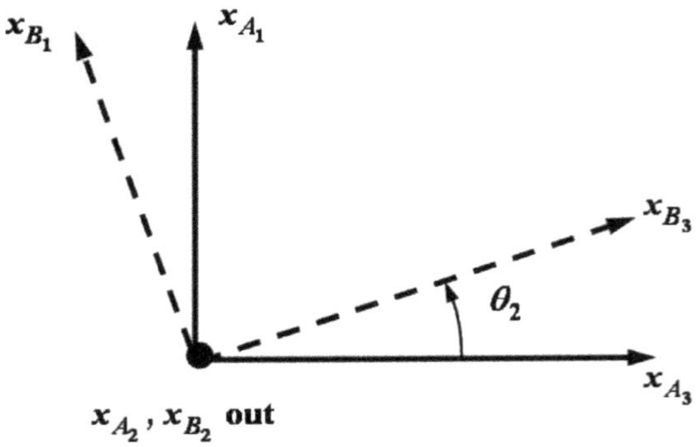

Figure A.3 ***y*-axis rotation**

z-axis rotation (Figure A.4)

$$L_3(\theta_3) = \begin{bmatrix} \cos\theta_3 & \sin\theta_3 & 0 \\ -\sin\theta_3 & \cos\theta_3 & 0 \\ 0 & 0 & 1 \end{bmatrix}$$

$$V_B = L_3(\theta_3) V_A$$

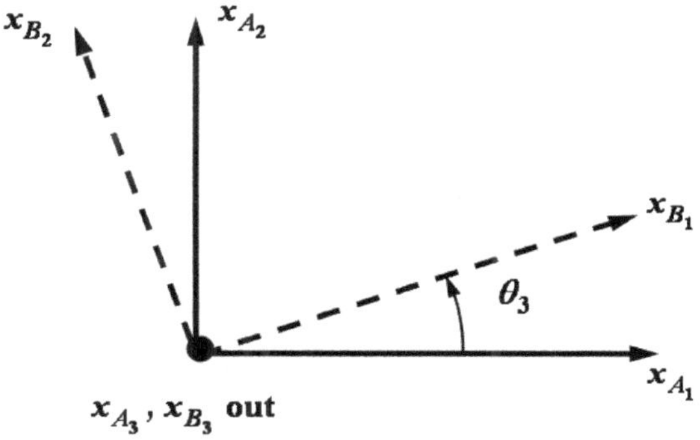

Figure A.4 **z-axis rotation**

Full specification of the relative orientation of one reference frame to another can be made with three single axis rotations. Common transformations for aircraft are from vehicle-carried earth axes to body axes using Euler angles,

$$L_{BV} = L_1(\phi) L_2(\theta) L_3(\psi) = L_{VB}^T$$

$$L_{BV} = \begin{bmatrix} 1 & 0 & 0 \\ 0 & \cos\phi & \sin\phi \\ 0 & -\sin\phi & \cos\phi \end{bmatrix} \begin{bmatrix} \cos\theta & 0 & -\sin\theta \\ 0 & 1 & 0 \\ \sin\theta & 0 & \cos\theta \end{bmatrix} \begin{bmatrix} \cos\psi & \sin\psi & 0 \\ -\sin\psi & \cos\psi & 0 \\ 0 & 0 & 1 \end{bmatrix}$$

$$L_{BV} = \begin{bmatrix} \cos\theta\cos\psi & \cos\theta\sin\psi & -\sin\theta \\ \sin\phi\sin\theta\cos\psi - \cos\phi\sin\psi & \sin\phi\sin\theta\sin\psi + \cos\phi\cos\psi & \sin\phi\cos\theta \\ \cos\phi\sin\theta\cos\psi + \sin\phi\sin\psi & \cos\phi\sin\theta\sin\psi - \sin\phi\cos\psi & \cos\phi\cos\theta \end{bmatrix}$$

and from wind axes to body axes using air flow angles,

$$L_{BW} = L_2(\alpha) L_3(-\beta) = L_{WB}^T$$

$$\boldsymbol{L}_{BW} = \begin{bmatrix} \cos\alpha & 0 & -\sin\alpha \\ 0 & 1 & 0 \\ \sin\alpha & 0 & \cos\alpha \end{bmatrix} \begin{bmatrix} \cos\beta & -\sin\beta & 0 \\ \sin\beta & \cos\beta & 0 \\ 0 & 0 & 1 \end{bmatrix}$$

$$\boldsymbol{L}_{BW} = \begin{bmatrix} \cos\alpha\cos\beta & -\cos\alpha\sin\beta & -\sin\alpha \\ \sin\beta & \cos\beta & 0 \\ \sin\alpha\cos\beta & -\sin\alpha\sin\beta & \cos\alpha \end{bmatrix}$$

A.2 Complex Numbers

A complex number s with real part σ and imaginary part ω can be written as $s = \sigma + j\omega$, and represented by a point in the complex plane, see Fig. A.5. The same point in the complex plane can also be represented as a vector with magnitude A and phase angle ϕ

$$s = \sigma + j\omega = A(\cos\phi + j\sin\phi) = A\,e^{j\phi} \tag{A.22}$$

where $\sigma, \omega, A,$ and ϕ are real numbers, j is the imaginary number, $j \equiv \sqrt{-1}$, and

$$A = |s| = \sqrt{\sigma^2 + \omega^2} \quad , \quad \phi = \tan^{-1}\left(\frac{\omega}{\sigma}\right) \tag{A.23}$$

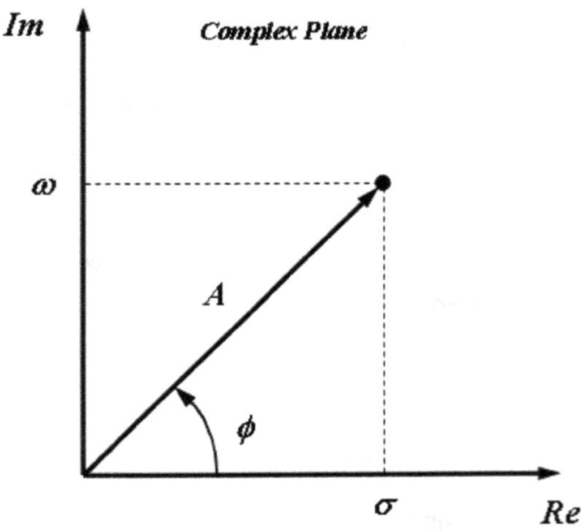

Figure A.5 **Complex number representations**

The last equality in Eq. (A.22) uses Euler's identity:

$$e^{j\phi} = \cos\phi + j\sin\phi \tag{A.24}$$

The representation of a complex number in terms of the amplitude A and the phase angle ϕ shown in Eq. (A.22), is called phasor notation.

For any two complex numbers,

$$s_1 = \sigma_1 + j\omega_1 = A_1\, e^{j\phi_1} \quad , \quad s_2 = \sigma_2 + j\omega_2 = A_2\, e^{j\phi_2} \tag{A.25}$$

$$s_1\, s_2 = A_1\, A_2\, e^{j(\phi_1+\phi_2)} \tag{A.26}$$

$$\frac{s_1}{s_2} = \frac{A_1}{A_2}\, e^{j(\phi_1-\phi_2)} \tag{A.27}$$

$$s_1 \pm s_2 = (\sigma_1 \pm \sigma_2) + j(\omega_1 \pm \omega_2) \tag{A.28}$$

If s is complex, then a function $G(s)$ is also complex, in general. Using Eq. (A.27), a vector with complex elements can be represented using relative magnitudes and phase angles, as in the following example for a complex vector with two elements:

$$\begin{bmatrix} \sigma_1 + j\omega_1 \\ \sigma_2 + j\omega_2 \end{bmatrix} = \begin{bmatrix} A_1 e^{j\phi_1} \\ A_2 e^{j\phi_2} \end{bmatrix} \Rightarrow \begin{bmatrix} 1 \\ \dfrac{A_2}{A_1} e^{j(\phi_2-\phi_1)} \end{bmatrix} \tag{A.29}$$

This technique is used to show relative magnitude and phasing of elements in complex vectors.

A.3 Calculus

A.3.1 Matrix Differentiation

For a square, nonsingular matrix $\boldsymbol{A}$,

$$\frac{d}{d\boldsymbol{A}} ln|\boldsymbol{A}| = \left(\boldsymbol{A}^{-1}\right)^T \tag{A.30a}$$

$$\frac{\partial}{\partial \boldsymbol{A}} Tr\left(\boldsymbol{A}^{-1}\boldsymbol{B}\right) = -\left[\boldsymbol{A}^{-1}\boldsymbol{B}\,\boldsymbol{A}^{-1}\right]^T \tag{A.30b}$$

If $\boldsymbol{A}$ is symmetric,

$$\frac{\partial}{\partial \boldsymbol{A}} Tr\left(\boldsymbol{B}\,\boldsymbol{A}\,\boldsymbol{B}^T\right) = 2\boldsymbol{B}\boldsymbol{A} \tag{A.30c}$$

When A is a function of a scalar θ,

$$\frac{d}{d\theta}\left(A^{-1}\right) = -A^{-1}\frac{dA}{d\theta}A^{-1} \tag{A.31}$$

For a quadratic form $Q = \boldsymbol{x}^T A\boldsymbol{x}$, where $\boldsymbol{x}$ is an $n\times 1$ vector and A is an $n\times n$ square matrix,

$$\nabla_x Q = \frac{\partial Q}{\partial \boldsymbol{x}} = 2A\boldsymbol{x} \tag{A.32}$$

When A and $\boldsymbol{x}$ are a function of a scalar θ,

$$\frac{dQ}{d\theta} = \left(\frac{\partial \boldsymbol{x}}{\partial \theta}\right)^T A\boldsymbol{x} + \boldsymbol{x}^T\frac{\partial A}{\partial \theta}\boldsymbol{x} + \boldsymbol{x}^T A\frac{\partial \boldsymbol{x}}{\partial \theta} \tag{A.33}$$

If A is symmetric, the last expression reduces to

$$\frac{dQ}{d\theta} = 2\left(\frac{\partial \boldsymbol{x}}{\partial \theta}\right)^T A\boldsymbol{x} + \boldsymbol{x}^T\frac{\partial A}{\partial \theta}\boldsymbol{x} \tag{A.34}$$

A.3.2 Laplace Transform

For a physical signal $x(t)$ of the type encountered in practical airplane flight, the Laplace transform is defined by

$$\mathcal{L}\left[x(t)\right] \equiv \int_0^\infty x(t)e^{-st}dt = \tilde{x}(s) \tag{A.35}$$

Since the Laplace transform is an integral transform, it is also a linear operator,

$$\mathcal{L}\left[ax(t)+by(t)\right] \equiv \int_0^\infty \left[ax(t)+by(t)\right]e^{-st}dt = a\tilde{x}(s)+b\tilde{y}(s) \tag{A.36}$$

The following formulae can be derived using the definition of the Laplace transform, along with integration by parts in some cases,

$$\mathcal{L}\left[\dot{x}(t)\right] = s\int_0^\infty x(t)e^{-st}dt - x(0) = s\tilde{x}(s) - x(0) \tag{A.37}$$

$$\mathcal{L}\left[\ddot{x}(t)\right] = s\int_0^\infty \dot{x}(t)e^{-st}dt - \dot{x}(0) = s\left[s\tilde{x}(s) - x(0)\right] - \dot{x}(0)$$

$$= s^2\tilde{x}(s) - sx(0) - \dot{x}(0)$$

$$\mathcal{L}\left[\frac{d^n x(t)}{dt^k}\right] = s^n \tilde{x}(s) - s^{n-1}x(0) - s^{n-2}\frac{dx}{dt}(0) - \ldots - \frac{d^{n-1}x}{dt^{n-1}}(0) \quad (A.38)$$

$$\mathcal{L}\left[\delta(t)\right] \equiv \int_0^{\infty} \delta(t) e^{-st}\, dt = 1 \quad (A.39)$$

$$\mathcal{L}\left[\mathbf{1}(t)\right] \equiv \int_0^{\infty} \mathbf{1}(t) e^{-st}\, dt = \frac{1}{s} \quad (A.40)$$

$$\mathcal{L}\left(e^{-at}\right) \equiv \int_0^{\infty} e^{-at}\, e^{-st}\, dt = \frac{1}{(s+a)} \quad (A.41)$$

$$\mathcal{L}\left(te^{-at}\right) \equiv \int_0^{\infty} te^{-at}\, e^{-st}\, dt = \frac{1}{(s+a)^2} \quad (A.42)$$

$$\mathcal{L}\left(\frac{t^k e^{-at}}{k!}\right) \equiv \int_0^{\infty} \frac{t^k e^{-at}}{k!} e^{-st}\, dt = \frac{1}{(s+a)^{k+1}} \quad (A.43)$$

A.3.3 Convolution Integral

For a scalar dynamic system, the output can be computed from the convolution integral

$$y(t) = \int_0^t h(t-\tau)u(\tau)d\tau = h(t) * u(t) \quad (A.44)$$

where $h(t)$ is the impulse response function, also known as the weighting function. Changing variables using $\xi = t - \tau$

$$\begin{aligned} y(t) = h(t)*u(t) &= \int_0^t h(t-\tau)u(\tau)d\tau = \int_t^0 h(\xi)u(t-\xi)(-d\xi) \\ &= \int_0^t u(t-\xi)h(\xi)d\xi = u(t)*h(t) \end{aligned} \quad (A.45)$$

which shows that the convolution operation is commutative. If the dynamic system is causal, meaning that effects always follow their cause, then

$$h(t-\tau) = 0 \qquad \text{for } \tau > t$$

The convolution integral can then be written as

$$y(t) = \int_0^{\infty} h(t-\tau)u(\tau)d\tau$$

Taking the Laplace transform,

$$\mathcal{L}[y(t)] = \mathcal{L}\left[\int_0^\infty h(t-\tau)u(\tau)d\tau\right] = \int_0^\infty e^{-st}\left[\int_0^\infty h(t-\tau)u(\tau)d\tau\right]dt$$

Changing the order of integration and substituting $\lambda = t - \tau$,

$$\mathcal{L}[y(t)] = \int_0^\infty \int_0^\infty h(t-\tau)e^{-st}dt\, u(\tau)d\tau = \int_0^\infty \int_0^\infty h(\lambda)e^{-s(\lambda+\tau)}d\lambda\; u(\tau)d\tau$$

$$\mathcal{L}[y(t)] = \int_0^\infty \int_0^\infty h(\lambda)e^{-s\lambda}d\lambda\; u(\tau)e^{-s\tau}d\tau = \int_0^\infty h(\lambda)e^{-s\lambda}d\lambda \int_0^\infty u(\tau)e^{-s\tau}d\tau$$

$$\tilde{y}(s) \equiv \mathcal{L}[y(t)] = \mathcal{L}[h(t)]\mathcal{L}[u(t)] = \tilde{h}(s)\tilde{u}(s) \tag{A.46}$$

The preceding development shows that convolution in the time domain is equivalent to multiplication in the Laplace domain. By changing the roles of the variables, the same development can be used to show that convolution in the Laplace domain is equivalent to multiplication in the time domain.

A.3.4 Fourier Transform

Replacing the Laplace transform variable s with $j\omega$ gives the Fourier transform,

$$\mathcal{F}[x(t)] = \tilde{x}(j\omega) = \int_0^\infty x(t)e^{-j\omega t}dt = \tilde{x}(s)\big|_{s=j\omega} \tag{A.47}$$

Setting $s = j\omega$ in the transfer function $\dfrac{\tilde{y}(s)}{\tilde{u}(s)} = H(s)$,

$$\frac{\tilde{y}(j\omega)}{\tilde{u}(j\omega)} = H(j\omega) \tag{A.48}$$

results in the frequency response $H(j\omega)$, which defines the steady-state response of the dynamic system to sinusoidal inputs.

A.4 Polynomial Splines

Spline functions are defined as piecewise polynomials functions of degree m in one or more independent variables. The term "piecewise" means that the polynomial is different for specific ranges of the independent variables. When continuity constraints are considered, the function values and derivatives agree at the points where the piecewise polynomials join. These points are called knots, and are defined as specific values of each independent variable. A polynomial spline $S_m(x)$ of degree m with continuous derivatives up to degree $m-1$, for a single independent variable $x \in [x_0, x_{max}]$, can be expressed as

$$S_m(x) = \sum_{r=1}^{m} C_r x^r + \sum_{i=1}^{k} D_i (x - x_i)_+^m \tag{A.49}$$

where

$$(x - x_i)_+^m = \begin{cases} (x - x_i)^m & x > x_i \\ 0 & x \le x_i \end{cases} \tag{A.50}$$

and the C_r and D_i are constants. The values $x_1, x_2, \ldots, x_k$ are knots which satisfy the condition

$$x_0 < x_1 < x_2 < \ldots < x_k < x_{max} \tag{A.51}$$

The special case of the polynomial spline for $m = 0$ (a spline of degree zero) represents an approximation by piecewise constants.

A polynomial spline in two independent variables x_1 and x_2 can be introduced to approximate a function of two independent variables over the range $x_1 \in [x_{10}, x_{1max}]$ and $x_2 \in [x_{20}, x_{2max}]$. Then, as in the one-dimensional case, the two ranges $[x_{10}, x_{1max}]$ and $[x_{20}, x_{2max}]$ are subdivided by sets of knots x_{1i} and x_{2i} where

$$x_{10} < x_{11} < x_{12} < \ldots < x_{1k_1} < x_{1max}$$

$$x_{20} < x_{21} < x_{22} < \ldots < x_{2k_2} < x_{2max}$$

The knots partition the rectangle defined by the full independent variable ranges into rectangular panels. A polynomial spline of degree m for x_1 and degree n for x_2, with continuous partial derivatives up to degree $(m-1)$ and $(n-1)$ respectively, on the rectangle defined by the intervals $[x_{10}, x_{1max}]$ and $[x_{20}, x_{2max}]$ can be formulated as

$$S_m(x_1,x_2)=\sum_{r=0}^{m}\sum_{s=0}^{n}C_{rs}x_1^r x_2^s+\sum_{i=1}^{k}P_i(x_2)(x_1-x_{1i})_+^m+\sum_{j=1}^{l}Q_j(x_1)(x_2-x_{2j})_+^n$$
$$+\sum_{i=1}^{k}\sum_{j=1}^{l}D_{ij}(x_1-x_{1i})_+^m(x_2-x_{2j})_+^n \tag{A.52}$$

where $P_i(x_2)$ and $Q_j(x_1)$ are polynomials of degree n and m, respectively, each with constant coefficients, and where C_{rs} and D_{ij} are constants.

Appendix B

Probability, Statistics, and Random Variables

B.1 Random Variables

A random variable X is a quantity that can take on values randomly according to a probability $\mathcal{P}(X \le x)$, where x is a selected value. The probability $\mathcal{P}(X \le x)$ gives a scalar value on the interval $[0,1]$ indicating the probability that the random variable X will take on a value less than or equal to x. Consequently, $\mathcal{P}(X \le x)$ depends on x. A probability value of zero corresponds to an impossibility, while a probability of 1 corresponds to certainty.

Random variables can be discrete or continuous. In the development given here, only continuous random variables will be considered, and each random variable is denoted by a capital letter. These conventions are not adhered to in the chapters.

B.1.1 Probability Distribution and Probability Density

The probability distribution function of a random variable X is defined by

$$P(x) \equiv \mathcal{P}(X \le x) \quad , \quad -\infty < x < \infty \tag{B.1}$$

where $P(x)$ is a scalar value on the interval $[0,1]$ indicating the probability that the random variable X will take a value less than or equal to x. Some properties of the probability distribution function are:

$$\begin{aligned} &P(x) \ge 0 \quad , \quad -\infty < x < \infty \\ &P(-\infty) = 0 \quad , \quad P(\infty) = 1 \\ &P(x_1) \le P(x_2) \quad \text{for } x_1 \le x_2 \end{aligned} \tag{B.2}$$

The probability density function of the random variable X, also known as the frequency function, is defined as

$$p(x) = \frac{dP(x)}{dx} \tag{B.3}$$

Then,

$$\int_{-\infty}^{\infty} p(x)\, dx = P(\infty) - P(-\infty) = 1 \tag{B.4}$$

and

$$\int_{x_1}^{x_2} p(x)\, dx = P(x_2) - P(x_1) = \mathcal{P}(x_1 \le X \le x_2) \tag{B.5}$$

B.1.2 Expected Value and Variance

The expected value, or the mean, of a random variable X is defined by

$$E(X) = \int_{-\infty}^{\infty} x\, p(x)\, dx \tag{B.6}$$

where $p(x)$ is the probability density function for X. If the same experiment is repeated N times, with each run producing a sample of the random variable X, the frequency interpretation of the expected value has the form

$$E(X) \approx \frac{1}{N} \sum_{i=1}^{N} x_i = \overline{x} \tag{B.7}$$

where the x_i, $i = 1, 2, \ldots, N$ are the sample values of the random variable X. In this case, the probability density for each sample is the constant $1/N$, and summation replaces the integral.

The most likely value of X is the constant x_m such that $p(x_m)$ has the maximum value. If $p(x)$ is even, then $p(x) = p(-x)$ and $E(X) = 0$. If $p(x)$ is symmetric about $x = a$, then $p(a - x) = p(a + x)$ and $E(X) = a$.

The variance of a random variable X with expected value $E(X) = \eta$ is defined by

$$\sigma^2 = E\left[(X - \eta)^2\right] \equiv \int_{-\infty}^{\infty} (x - \eta)^2\, p(x)\, dx \tag{B.8}$$

This definition can also be expressed as

$$\sigma^2 = E\left(X^2 - 2\eta X + \eta^2\right) = E\left(X^2\right) - 2\eta E(X) + \eta^2 = E\left(X^2\right) - \eta^2$$

or

$$\sigma^2 = E\left(X^2\right)-\left[E(X)\right]^2 \tag{B.9}$$

If X is a random variable with $E(X)=\eta$ and $E\left[(X-\eta)^2\right]=\sigma^2$, then $Y = cX$ for any constant c is also a random variable, with

$$E(Y)=cE[X]=c\eta \qquad Var(Y)=E\left[(cX-c\eta)^2\right]=c^2\sigma^2 \tag{B.10}$$

B.1.3 Two Random Variables

For two random variables X and Y, consider two sets of events defined by $(X \le x)$ and $(Y \le y)$, respectively. The probability distribution functions of X and Y are $P(x)=\mathcal{P}(X \le x)$ and $P(y)=\mathcal{P}(Y \le y)$, respectively. The joint probability distribution of the random variables X and Y is defined by

$$P(x,y) \equiv \mathcal{P}(X \le x, Y \le y) \tag{B.11}$$

The joint probability distribution has the following properties, which follow from the corresponding properties for a single random variable:

$$\begin{aligned} &P(x,y)\ge 0 \quad , \quad -\infty < x < \infty \quad , \quad -\infty < y < \infty \\ &P(-\infty,-\infty)=P(-\infty,y)=P(x,-\infty)=0 \\ &P(x,\infty)=P(x) \quad , \quad P(\infty,y)=P(y) \quad , \quad P(\infty,\infty)=1 \end{aligned} \tag{B.12}$$

The joint probability density function for the random variables X and Y is defined by

$$p(x,y)=\frac{\partial^2 P(x,y)}{\partial x \partial y} \tag{B.13}$$

The joint probability density $p(x,y)$ is related to the marginal probability densities $p(x)$ and $p(y)$ by

$$p(x)=\int_{-\infty}^{\infty} p(x,y)\,dy \tag{B.14}$$

$$p(y)=\int_{-\infty}^{\infty} p(x,y)\,dx \tag{B.15}$$

B.1.4 Uncorrelated and Independent Random Variables

Two random variables X and Y are called uncorrelated if

$$E(XY) = E(X)E(Y) \tag{B.16}$$

They are orthogonal if

$$E(XY) = 0 \tag{B.17}$$

and independent if

$$p(x,y) = p(x)\,p(y) \tag{B.18}$$

If two random variables X and Y are uncorrelated, then their covariance and correlation coefficient are zero:

$$\begin{aligned} Cov(X,Y) &= E\left[(X-\eta_x)(Y-\eta_y)\right] \\ &= E(XY) - \eta_x E(Y) - E(X)\eta_y + \eta_x\eta_y \\ &= E(X)E(Y) - \eta_x E(Y) - E(X)\eta_y + \eta_x\eta_y = 0 \end{aligned} \tag{B.19}$$

$$r = \frac{Cov(X,Y)}{\sigma_x\,\sigma_y} = 0 \tag{B.20}$$

where

$$\begin{aligned} \eta_x &= E(X) \quad , \quad \sigma_x^2 = E\left[(X-\eta_x)^2\right] \\ \eta_y &= E(Y) \quad , \quad \sigma_y^2 = E\left[(Y-\eta_y)^2\right] \end{aligned} \tag{B.21}$$

B.1.5 Functions of Two Random Variables

If the random variables X and Y are combined linearly, $aX + bY$, where a and b are known constants, then the mean value and variance are

$$E(aX + bY) = a\eta_x + b\eta_y \tag{B.22}$$

$$\begin{aligned} \sigma_{ax+by}^2 &= E\left\{\left[a(X-\eta_x) + b(Y-\eta_y)\right]^2\right\} \\ &= a^2\sigma_x^2 + b^2\sigma_y^2 + 2abr\sigma_x\sigma_y \end{aligned} \tag{B.23}$$

If random variables X and Y are combined nonlinearly, e.g., $f(X,Y)$, then

$$E\left[f(X,Y)\right] \approx f\left(\eta_x,\eta_y\right) \tag{B.24}$$

$$\begin{aligned}\sigma^2_{f(x,y)} &= E\left\{\left[f(X,Y)-f\left(\eta_x,\eta_y\right)\right]^2\right\} \\ &\approx \left(\frac{\partial f}{\partial x}\right)^2 \sigma_x^2 + \left(\frac{\partial f}{\partial y}\right)^2 \sigma_y^2 + 2\left(\frac{\partial f}{\partial x}\right)\left(\frac{\partial f}{\partial y}\right) r\,\sigma_x\sigma_y\end{aligned} \tag{B.25}$$

where the two approximate equalities follow from the definitions of the mean and variance, using the approximation

$$f(X,Y) \approx f\left(\eta_x,\eta_y\right) + \left(\frac{\partial f}{\partial x}\right)\left(X-\eta_x\right) + \left(\frac{\partial f}{\partial y}\right)\left(Y-\eta_y\right) \tag{B.26}$$

with the partial derivatives $\partial f/\partial x$ and $\partial f/\partial y$ evaluated at $x=\eta_x$, $y=\eta_y$.

B.1.6 Conditional Probability and Bayes's Rule

For two random variables X and Y, the conditional probability that $Y \le y$, given that $X \le x$, can be expressed as

$$\mathcal{P}\left(Y \le y \mid X \le x\right) = \frac{\mathcal{P}\left(Y \le y, X \le x\right)}{\mathcal{P}\left(X \le x\right)} = \frac{P(x,y)}{P(x)} \tag{B.27}$$

This conditional probability has a simple frequency interpretation as the number of events where both $Y \le y$ and $X \le x$ occur, divided by the number of events where $X \le x$ occurs. The analogous statement using probability density functions is

$$p\left(y \mid x\right) = \frac{p(x,y)}{p(x)} \tag{B.28}$$

Similarly, for the conditional probability density of x, given that $Y = y$,

$$p\left(x \mid y\right) = \frac{p(x,y)}{p(y)} \tag{B.29}$$

From the last two equations, it follows that

$$p(x,y) = p\left(x \mid y\right)p(y) = p\left(y \mid x\right)p(x) \tag{B.30}$$

or

$$p(x \mid y) = \frac{p(y \mid x)\,p(x)}{p(y)} \tag{B.31}$$

which is known as Bayes's rule. Its alternative form is

$$p(y \mid x) = \frac{p(x \mid y)\,p(y)}{p(x)} \tag{B.32}$$

B.1.7 Random Vector

A collection of N random variables $X_1, X_2, \ldots, X_N$ can be arranged as a random vector $\boldsymbol{X} = [X_1 \quad X_2 \quad \ldots \quad X_N]^T$. The joint probability distribution of $\boldsymbol{X}$ is defined by

$$P(x_1, x_2, \ldots, x_N) \equiv \mathcal{P}(X_1 \le x_1, X_2 \le x_2, \ldots, X_N \le x_N) \tag{B.33}$$

The joint probability density $p(x_1, x_2, \ldots, x_N)$ is obtained by differentiating the joint distribution with respect to $x_1, x_2, \ldots, x_N$.

The mean value and covariance matrix of the random vector $\boldsymbol{X}$ are defined by

$$E(\boldsymbol{X}) = \boldsymbol{\eta} \tag{B.34}$$

$$Cov(\boldsymbol{X}) = E\left[(\boldsymbol{X} - \boldsymbol{\eta})(\boldsymbol{X} - \boldsymbol{\eta})^T\right] = \boldsymbol{\Sigma} \tag{B.35}$$

If random vectors $\boldsymbol{X}_1, \boldsymbol{X}_2, \ldots, \boldsymbol{X}_N$ are mutually independent, then

$$p(\boldsymbol{X}_1, \boldsymbol{X}_2, \ldots \boldsymbol{X}_N) = p(\boldsymbol{X}_1)\,p(\boldsymbol{X}_2) \ldots p(\boldsymbol{X}_N) \tag{B.36}$$

where $p(\boldsymbol{X}_i)$, $i = 1, 2, \ldots, N$ are marginal densities. Random vectors $\boldsymbol{X}_1$ and $\boldsymbol{X}_2$ are called uncorrelated if

$$E\left(\boldsymbol{X}_1 \boldsymbol{X}_2^T\right) = E(\boldsymbol{X}_1)\,E\left(\boldsymbol{X}_2^T\right) = \boldsymbol{\eta}_1\,\boldsymbol{\eta}_2^T \tag{B.37}$$

Random vectors $\boldsymbol{X}_1$ and $\boldsymbol{X}_2$ are called orthogonal if their elements are mutually orthogonal, so that

$$E\left(\boldsymbol{X}_1 \boldsymbol{X}_2^T\right) = \boldsymbol{0} \tag{B.38}$$

If a random vector $\boldsymbol{X}$ is formed as

$$\boldsymbol{X} = \begin{bmatrix} \boldsymbol{X}_1 \\ \boldsymbol{X}_2 \end{bmatrix} \tag{B.39}$$

with

$$E(\boldsymbol{X}) = \begin{bmatrix} E(\boldsymbol{X}_1) \\ E(\boldsymbol{X}_2) \end{bmatrix} = \begin{bmatrix} \boldsymbol{\eta}_1 \\ \boldsymbol{\eta}_2 \end{bmatrix} = \boldsymbol{\eta} \tag{B.40}$$

$$Cov(\boldsymbol{X}) = E\left[(\boldsymbol{X}-\boldsymbol{\eta})(\boldsymbol{X}-\boldsymbol{\eta})^T\right] = \boldsymbol{\Sigma} = \begin{bmatrix} \boldsymbol{\Sigma}_{11} & \boldsymbol{\Sigma}_{12} \\ \boldsymbol{\Sigma}_{12} & \boldsymbol{\Sigma}_{22} \end{bmatrix} \tag{B.41}$$

Combining Eqs. (B.37), and (B.39)-(B.41), $\boldsymbol{\Sigma}_{12} = \boldsymbol{0}$ when $\boldsymbol{X}_1$ and $\boldsymbol{X}_2$ are uncorrelated.

B.2 Statistics

B.2.1 Gaussian (Normal) Distribution

A random variable is said to be normally distributed or Gaussian if its probability density function takes the form

$$p(x) = \frac{1}{\sigma\sqrt{2\pi}} exp\left[-\frac{(x-\eta)^2}{2\sigma^2}\right] \tag{B.42}$$

where η is the mean value of the random variable, and σ^2 is the variance. The normal probability density function is completely determined by the mean η and variance σ^2. The statement that a random variable X is normally distributed with mean η and variance σ^2 can be made as

$$X \text{ is } \mathbb{N}(\eta, \sigma^2) \tag{B.43}$$

The Gaussian probability density and probability distribution functions are sketched in Figs. B.1 and B.2, respectively. If σ is used as the unit on the abscissa, then the area under the $p(x)$ curve between $\eta - k\sigma$ and $\eta + k\sigma$ is given in Table B.1. The area under the $p(x)$ curve between selected values x_1 and x_2 corresponds to the probability that the random variable takes on a value in the range $[x_1, x_2]$.

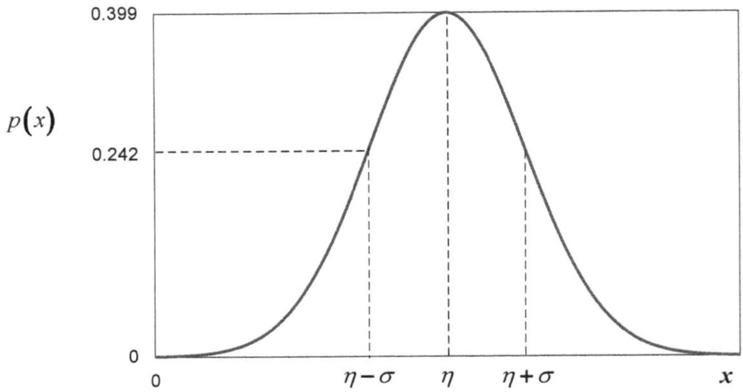

Figure B.1 **Gaussian probability density**

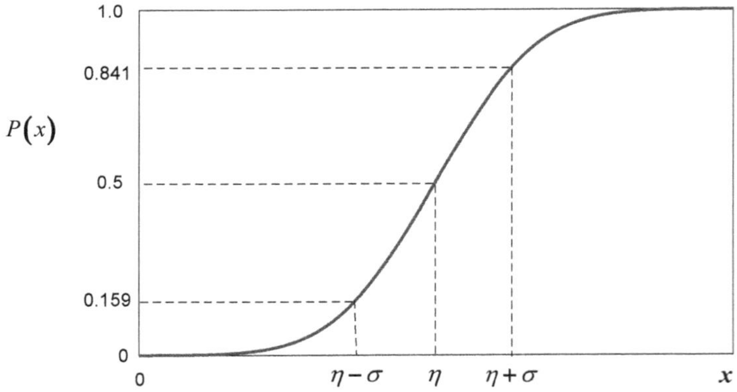

Figure B.2 **Gaussian probability distribution**

Table B.1 **Gaussian distribution probabilities**

k	0.6745	1.0	2.0	3.0
area $= \int_{\eta-k\sigma}^{\eta+k\sigma} p(x)\,dx$	0.50	0.683	0.955	0.997
$\mathcal{P}(\eta-k\sigma \le X \le \eta+k\sigma)$ (%)	50	68.3	95.5	99.7

Introducing the random variable $U = (X - \eta)/\sigma$, then

$$p(u) = \frac{1}{\sqrt{2\pi}} exp\left[-\frac{u^2}{2}\right] \tag{B.44}$$

which means that

$$U \text{ is } \mathbb{N}(0,1) \tag{B.45}$$

Values for the area under the curve $p(u)$ from 0 to various selected values of u are tabulated in numerous textbooks on statistics.

An N-dimensional random vector $\boldsymbol{X}$ with probability density function $p(\boldsymbol{x})$, $\boldsymbol{x} = [x_1 \;\; x_2 \;\; \dots \;\; x_N]^T$, is said to be normally distributed if $p(\boldsymbol{x})$ is of the form

$$p(\boldsymbol{x}) = \frac{1}{(2\pi)^{N/2} |\Sigma|^{1/2}} exp\left[-\frac{(\boldsymbol{X} - \boldsymbol{\eta})^T \Sigma^{-1} (\boldsymbol{X} - \boldsymbol{\eta})}{2}\right] \tag{B.46}$$

$$E(\boldsymbol{X}) = \boldsymbol{\eta} \tag{B.47}$$

$$Cov(\boldsymbol{X}) = E\left[(\boldsymbol{X} - \boldsymbol{\eta})(\boldsymbol{X} - \boldsymbol{\eta})^T\right] = \boldsymbol{\Sigma} \tag{B.48}$$

A normally distributed random vector with mean $\boldsymbol{\eta}$ and covariance $\boldsymbol{\Sigma}$ is denoted as

$$\boldsymbol{X} \text{ is } \mathbb{N}(\boldsymbol{\eta}, \boldsymbol{\Sigma}) \tag{B.49}$$

B.2.2 t Distribution

The mean $\overline{x}$ of N stochastically independent random variables taken from an ensemble of random variables with mean η and variance σ^2 is normally distributed with mean η and variance σ^2/N, i.e.,

$$u = \frac{\overline{x} - \eta}{\sigma/\sqrt{N}} \quad \text{is} \quad \mathbb{N}(0,1) \tag{B.50}$$

where

$$\overline{x} = \frac{1}{N}\sum_{i=1}^{N} x_i \tag{B.51a}$$

Replacing σ^2 by the estimate

$$s^2 = \frac{1}{N-1}\sum_{i=1}^{N}(x_i - \overline{x})^2 \tag{B.51b}$$

the u variable is changed to a new variable

$$t = \frac{\overline{x} - \eta}{s/\sqrt{N-1}} \tag{B.52}$$

The t distribution is independent of both parameters η and σ. It depends only on the number of statistical degrees of freedom f for s^2, i.e., $f = N-1$. The number of degrees of freedom indicates how many independent pieces of information are necessary to compute a given statistic. The s^2 statistic computed from Eq. (B.51b) has $N-1$ degrees of freedom, because $\sum_{i=1}^{N}(x_i - \overline{x}) = 0$ from (B.51a), which means that only $N-1$ of the x_i are independent. As $N \to \infty$ the t distribution approaches the distribution of u, which is $\mathbb{N}(0,1)$.

B.2.3 F Distribution

If the random variables $u_1, u_2, \ldots, u_f$ are each $\mathbb{N}(0,1)$, then

$$\chi^2 = \sum_{i=1}^{f} u_i^2 \tag{B.53}$$

is called a chi-squared random variable with f degrees of freedom. The ratio of two chi-squared random variables is a new random variable F, defined as

$$F = \frac{\chi_1^2/f_1}{\chi_2^2/f_2} \tag{B.54}$$

with f_1 and f_2 degrees of freedom. The random variable F can take only positive values and has two parameters, f_1 and f_2. The F distribution is tabulated in many statistics textbooks as a function of f_1 and f_2.

B.2.4 Partitioning Sum of Squares

The relationship among the quantities SS_T, SS_R, and SS_E given in Chapter 5 is

$$SS_T = SS_R + SS_E \tag{B.55}$$

To derive this relationship, start with the identity

$$z(i) - \bar{z} = \left[\hat{y}(i) - \bar{z}\right] + \left[z(i) - \hat{y}(i)\right] \tag{B.56}$$

then square both sides and sum over N data points,

$$\begin{aligned}\sum_{i=1}^{N}\left[z(i)-\bar{z}\right]^2 &= \sum_{i=1}^{N}\left[\hat{y}(i)-\bar{z}\right]^2 + \sum_{i=1}^{N}\left[z(i)-\hat{y}(i)\right]^2 \\ &\quad + 2\sum_{i=1}^{N}\left[\hat{y}(i)-\bar{z}\right]\left[z(i)-\hat{y}(i)\right]\end{aligned} \tag{B.57}$$

or

$$\begin{aligned}\sum_{i=1}^{N}\left[z(i)-\bar{z}\right]^2 &= \sum_{i=1}^{N}\left[\hat{y}(i)-\bar{z}\right]^2 + \sum_{i=1}^{N}\left[z(i)-\hat{y}(i)\right]^2 \\ &\quad + 2\sum_{i=1}^{N}\hat{y}(i)\left[z(i)-\hat{y}(i)\right] - 2\bar{z}\sum_{i=1}^{N}\left[z(i)-\hat{y}(i)\right]\end{aligned} \tag{B.58}$$

The normal equations (5.9c) show that the inner product of each regressor with the residual vector $\boldsymbol{z} - \hat{\boldsymbol{y}}$ is zero. Since the first regressor is a vector of ones, the sum of the residuals must be zero, and the last summation on the right side of the equation above is zero. The estimated output vector $\hat{\boldsymbol{y}} = \boldsymbol{X}\hat{\boldsymbol{\theta}}$ is a linear sum of the regressors, so the inner product of $\hat{\boldsymbol{y}}$ with the residual vector is also zero. Therefore, the third summation on the right is also zero. With these simplifications,

$$\sum_{i=1}^{N}\left[z(i)-\bar{z}\right]^2 = \sum_{i=1}^{N}\left[\hat{y}(i)-\bar{z}\right]^2 + \sum_{i=1}^{N}\left[z(i)-\hat{y}(i)\right]^2 \tag{B.59}$$

or, using the definitions in Eqs. (5.25)-(5.28),

$$SS_T = SS_R + SS_E \tag{B.60}$$

B.2.5 Alternate Expression for the Fisher Information Matrix

The Fisher information matrix $\boldsymbol{M}$ is defined as

$$\boldsymbol{M} \equiv E\left[\left(\frac{\partial \ln \mathbb{L}}{\partial \boldsymbol{\theta}}\right)\left(\frac{\partial \ln \mathbb{L}}{\partial \boldsymbol{\theta}}\right)^T\right] \tag{B.61}$$

where $\mathbb{L}$ is the likelihood function, equal to the probability density function of z given $\boldsymbol{\theta}$, i.e.,

$$\mathbb{L}(z;\boldsymbol{\theta}) \equiv p(z\,|\,\boldsymbol{\theta}) \tag{B.62}$$

so that

$$\boldsymbol{M} \equiv E\left\{\left[\frac{\partial \ln p(z\,|\,\boldsymbol{\theta})}{\partial \boldsymbol{\theta}}\right]\left[\frac{\partial \ln p(z\,|\,\boldsymbol{\theta})}{\partial \boldsymbol{\theta}}\right]^T\right\} \tag{B.63}$$

An alternate expression for $\boldsymbol{M}$ can be derived by starting with the identity

$$\int_{-\infty}^{\infty} p(z\,|\,\boldsymbol{\theta})\, dz = 1 \tag{B.64}$$

which follows from the definition of conditional probability density functions. Assuming that $\ln p(z\,|\,\boldsymbol{\theta})$ is sufficiently smooth, the gradient with respect to $\boldsymbol{\theta}$ is

$$\frac{\partial \ln p(z\,|\,\boldsymbol{\theta})}{\partial \boldsymbol{\theta}} = \frac{1}{p(z\,|\,\boldsymbol{\theta})}\left[\frac{\partial p(z\,|\,\boldsymbol{\theta})}{\partial \boldsymbol{\theta}}\right] \tag{B.65}$$

which can be rearranged to

$$\frac{\partial p(z\,|\,\boldsymbol{\theta})}{\partial \boldsymbol{\theta}} = p(z\,|\,\boldsymbol{\theta})\frac{\partial \ln p(z\,|\,\boldsymbol{\theta})}{\partial \boldsymbol{\theta}} \tag{B.66}$$

Differentiating Eq. (B.64) twice with respect to $\boldsymbol{\theta}$, and using Eq. (B.66),

$$\int_{-\infty}^{\infty}\left[\frac{\partial \ln p(z\,|\,\boldsymbol{\theta})}{\partial \boldsymbol{\theta}}\frac{\partial \ln p(z\,|\,\boldsymbol{\theta})}{\partial \boldsymbol{\theta}^T} + \frac{\partial^2 \ln p(z\,|\,\boldsymbol{\theta})}{\partial \boldsymbol{\theta}\,\partial \boldsymbol{\theta}^T}\right] p(z\,|\,\boldsymbol{\theta})\, dz = \boldsymbol{0} \tag{B.67}$$

Introducing expectation operator notation, it must be that

$$E\left\{\left[\frac{\partial \ln p(z\,|\,\boldsymbol{\theta})}{\partial \boldsymbol{\theta}}\right]\left[\frac{\partial \ln p(z\,|\,\boldsymbol{\theta})}{\partial \boldsymbol{\theta}}\right]^T + \frac{\partial^2 \ln p(z\,|\,\boldsymbol{\theta})}{\partial \boldsymbol{\theta}\,\partial \boldsymbol{\theta}^T}\right\} = \boldsymbol{0}$$

or

$$E\left\{\left[\frac{\partial \ln p(z\,|\,\boldsymbol{\theta})}{\partial \boldsymbol{\theta}}\right]\left[\frac{\partial \ln p(z\,|\,\boldsymbol{\theta})}{\partial \boldsymbol{\theta}}\right]^T\right\} = -E\left[\frac{\partial^2 \ln p(z\,|\,\boldsymbol{\theta})}{\partial \boldsymbol{\theta}\,\partial \boldsymbol{\theta}^T}\right] \tag{B.68}$$

Combining Eqs. (B.61), (B.62), and (B.68),

$$\boldsymbol{M} \equiv E\left[\left(\frac{\partial \ln \mathbb{L}}{\partial \boldsymbol{\theta}}\right)\left(\frac{\partial \ln \mathbb{L}}{\partial \boldsymbol{\theta}}\right)^T\right] = -E\left(\frac{\partial^2 \ln \mathbb{L}}{\partial \boldsymbol{\theta}\,\partial \boldsymbol{\theta}^T}\right) \tag{B.69}$$

The term on the far right of Eq. (B.69) is the alternate expression for the Fisher information matrix $\boldsymbol{M}$.

B.2.6 Derivation of the Cramér-Rao Lower Bound

The theoretical minimum of the estimated parameter covariance matrix for an unbiased estimator, called the Cramér-Rao lower bound, equals the inverse of the Fisher information matrix. This means that the estimated parameter covariance matrix satisfies the inequality

$$Cov\left(\hat{\boldsymbol{\theta}}\right) \geq \boldsymbol{M}^{-1} \tag{B.70}$$

To derive this expression, start with the following statement that is true by the definition of an unbiased estimator,

$$E\left(\hat{\boldsymbol{\theta}} - \boldsymbol{\theta}\right) = \int_{-\infty}^{\infty}\left(\hat{\boldsymbol{\theta}} - \boldsymbol{\theta}\right) p(z\,|\,\boldsymbol{\theta})\, dz = \boldsymbol{0} \tag{B.71}$$

Differentiating with respect to $\boldsymbol{\theta}$ gives

$$\int_{-\infty}^{\infty} -\boldsymbol{I}\; p(z\,|\,\boldsymbol{\theta})\, dz + \int_{-\infty}^{\infty}\left(\hat{\boldsymbol{\theta}} - \boldsymbol{\theta}\right)\left[\frac{\partial p(z\,|\,\boldsymbol{\theta})}{\partial \boldsymbol{\theta}}\right]^T dz = \boldsymbol{0} \tag{B.72}$$

where $\boldsymbol{I}$ is an identity matrix. Since $\int_{-\infty}^{\infty} p(z\,|\,\boldsymbol{\theta})\, dz = 1$,

$$\int_{-\infty}^{\infty}\left(\hat{\boldsymbol{\theta}} - \boldsymbol{\theta}\right)\left[\frac{\partial p(z\,|\,\boldsymbol{\theta})}{\partial \boldsymbol{\theta}}\right]^T dz = \boldsymbol{I} \tag{B.73}$$

Introducing expectation operator notation, and using Eq. (B.66),

$$E\left\{\left(\hat{\boldsymbol{\theta}} - \boldsymbol{\theta}\right)\left[\frac{\partial \ln p(z\,|\,\boldsymbol{\theta})}{\partial \boldsymbol{\theta}}\right]^T\right\} = \boldsymbol{I} \tag{B.74}$$

The following lemma is now required (Ref. [B.1]).

Lemma:

Let $\boldsymbol{x}$ and $\boldsymbol{y}$ be two random vectors with the same dimension. Then,

$$E\left(\boldsymbol{x}\boldsymbol{x}^T\right) \geq E\left(\boldsymbol{x}\boldsymbol{y}^T\right)\left[E\left(\boldsymbol{y}\boldsymbol{y}^T\right)\right]^{-1} E\left(\boldsymbol{y}\boldsymbol{x}^T\right) \tag{B.75}$$

Proof:

Let $\boldsymbol{Q}$ be an arbitrary nonrandom matrix with conformable dimensions. Then, since any covariance matrix must be nonnegative,

$$E\left[(\boldsymbol{x}-\boldsymbol{Q}\boldsymbol{y})(\boldsymbol{x}-\boldsymbol{Q}\boldsymbol{y})^T\right] \geq \boldsymbol{0} \tag{B.76}$$

Expanding the left side of the above equation,

$$E\left(\boldsymbol{x}\boldsymbol{x}^T\right) - \boldsymbol{Q}E\left(\boldsymbol{y}\boldsymbol{x}^T\right) - E\left(\boldsymbol{x}\boldsymbol{y}^T\right)\boldsymbol{Q}^T + \boldsymbol{Q}E\left(\boldsymbol{y}\boldsymbol{y}^T\right)\boldsymbol{Q}^T \geq \boldsymbol{0}$$

$$E\left(\boldsymbol{x}\boldsymbol{x}^T\right) \geq \boldsymbol{Q}E\left(\boldsymbol{y}\boldsymbol{x}^T\right) + E\left(\boldsymbol{x}\boldsymbol{y}^T\right)\boldsymbol{Q}^T - \boldsymbol{Q}E\left(\boldsymbol{y}\boldsymbol{y}^T\right)\boldsymbol{Q}^T \tag{B.77}$$

Now choose

$$\boldsymbol{Q} = E\left(\boldsymbol{x}\boldsymbol{y}^T\right)\left[E\left(\boldsymbol{y}\boldsymbol{y}^T\right)\right]^{-1} \tag{B.78}$$

so that the two terms on the far right of Eq. (B.77) cancel. Substituting Eq. (B.78) into Eq. (B.77),

$$E\left(\boldsymbol{x}\boldsymbol{x}^T\right) \geq E\left(\boldsymbol{x}\boldsymbol{y}^T\right)\left[E\left(\boldsymbol{y}\boldsymbol{y}^T\right)\right]^{-1} E\left(\boldsymbol{y}\boldsymbol{x}^T\right) \tag{B.79}$$

This ends the proof of the lemma.

Using Eq. (B.79) with

$$\boldsymbol{x} \equiv \left(\hat{\boldsymbol{\theta}} - \boldsymbol{\theta}\right) \qquad \boldsymbol{y} \equiv \frac{\partial \ln p(\boldsymbol{z}\,|\,\boldsymbol{\theta})}{\partial \boldsymbol{\theta}}$$

and invoking Eq. (B.73) results in

$$E\left[\left(\hat{\boldsymbol{\theta}} - \boldsymbol{\theta}\right)\left(\hat{\boldsymbol{\theta}} - \boldsymbol{\theta}\right)^T\right] \geq \boldsymbol{I}\left\{E\left[\frac{\partial \ln p(\boldsymbol{z}\,|\,\boldsymbol{\theta})}{\partial \boldsymbol{\theta}} \frac{\partial \ln p(\boldsymbol{z}\,|\,\boldsymbol{\theta})}{\partial \boldsymbol{\theta}^T}\right]\right\}^{-1} \boldsymbol{I} \tag{B.80}$$

Combining Eq. (B.80) with Eq. (B.63) gives

$$Cov(\hat{\theta}) \equiv E\left[(\hat{\theta}-\theta)(\hat{\theta}-\theta)^T\right] \geq M^{-1} \tag{B.81}$$

which is the Cramér-Rao inequality stated in Eq. (B.70).

B.3 Random Process Theory

B.3.1 Random Process

A random process, also called a stochastic process, is a collection of random variables $X(t)$ indexed by the parameter t from a set T. A random process is said to be continuous if T is a connected set, and discrete if T is a finite set, i.e., $T = (t_1, t_2, \ldots, t_N)$. A realization or sample of a random process is an assignment of specific values to $X(t)$ for each t of T, from among the possible values of $X(t)$ at each t. A complete description of a general random process would require specification of all possible joint probability density functions $p\left[X(t_1), X(t_2), \ldots, X(t_N)\right]$.

B.3.2 Stationary Random Process

A random process $\boldsymbol{X}(t)$ is said to be a stationary random process if, for each $(t_1, t_2, \ldots, t_N)$ the joint probability distribution of $\left[X(t_1+\tau), X(t_2+\tau), \ldots, X(t_N+\tau)\right]$ is the same for any τ. The properties of a stationary process are:

1) The mean value $\eta = E\left[\boldsymbol{X}(t)\right]$ is a constant, independent of *t*.

2) The covariance matrix
$Cov(\boldsymbol{X}(t_1), \boldsymbol{X}(t_2)) = E\left\{\left[\boldsymbol{X}(t_1)-\eta\right]\left[\boldsymbol{X}(t_2)-\eta\right]^T\right\}$ exists.

3) The covariance matrix $Cov\left[\boldsymbol{X}(t_1), \boldsymbol{X}(t_2)\right]$ depends only on $t_1 - t_2$ and not on the magnitudes of t_1 or t_2.

B.3.3 White Noise Process

A discrete random process $\boldsymbol{X}(t)$ is said to be a white noise process if for $E\left[\boldsymbol{X}(t)\right]=0$, the covariance matrix $Cov\left[\boldsymbol{X}(t_1),\boldsymbol{X}(t_2)\right]$ can be expressed as

$$Cov\left[\boldsymbol{X}(t_1),\boldsymbol{X}(t_2)\right]=E\left[\boldsymbol{X}(t_1),\boldsymbol{X}^T(t_2)\right]=\begin{cases}\boldsymbol{Q}(t_1) & t_1=t_2\\ 0 & t_1\neq t_2\end{cases} \tag{B.82}$$

where $\boldsymbol{Q}(t_1)$ is a positive semi-definite matrix. The above condition implies that $\boldsymbol{X}(t_1)$ and $\boldsymbol{X}(t_2)$ are uncorrelated for any $t_1 \neq t_2$.

B.3.4 Correlation Functions

For stationary random processes $x(t)$ and $y(t)$, the mean values are constant, independent of t, i.e.,

$$\eta_x = E\left[x(t)\right] \qquad \text{and} \qquad \eta_y = E\left[y(t)\right] \tag{B.83}$$

The covariance functions are also independent of t, and depend only on the time shift τ. For arbitrary t and τ they can be expressed as

$$\begin{aligned}\Sigma_x(\tau)&=E\left\{\left[x(t)-\eta_x\right]\left[x(t+\tau)-\eta_x\right]\right\}=E\left[x(t)x(t+\tau)\right]-\eta_x^2\\ \Sigma_x(\tau)&=E\left\{\left[y(t)-\eta_y\right]\left[y(t+\tau)-\eta_y\right]\right\}=E\left[y(t)y(t+\tau)\right]-\eta_y^2\\ \Sigma_x(\tau)&=E\left\{\left[x(t)-\eta_x\right]\left[y(t+\tau)-\eta_y\right]\right\}=E\left[x(t)y(t+\tau)\right]-\eta_x\eta_y\end{aligned} \tag{B.84}$$

where

$$\begin{aligned}E\left[x(t)x(t+\tau)\right]&\equiv\mathcal{R}_{xx}(\tau)\\ E\left[y(t)y(t+\tau)\right]&\equiv\mathcal{R}_{yy}(\tau)\\ E\left[x(t)y(t+\tau)\right]&\equiv\mathcal{R}_{xy}(\tau)\end{aligned} \tag{B.85}$$

$\mathcal{R}_{xx}(\tau)$ and $R_{yy}(\tau)$ are known as the autocorrelation functions of $x(t)$ and $y(t)$ respectively, and $R_{xy}(\tau)$ is called the cross-correlation function of $x(t)$ and $y(t)$. It follows that for a zero-mean stationary random process $x(t)$,

$$\Sigma_x(0)=E\left[x(t)x(t)\right]=Var\left[x(t)\right]=\mathcal{R}_{xx}(0) \tag{B.86}$$

The autocorrelation functions are even functions,

$$\mathcal{R}_{xx}(-\tau) = \mathcal{R}_{xx}(\tau)$$
$$\mathcal{R}_{yy}(-\tau) = \mathcal{R}_{yy}(\tau) \tag{B.87}$$

whereas the cross-correlation functions satisfy

$$\mathcal{R}_{xy}(-\tau) = \mathcal{R}_{yx}(\tau) \tag{B.88}$$

The autocorrelation function for white noise is a Dirac delta function at $\tau = 0$, indicating that values at each time t are uncorrelated with values at other times. In the general case, when a random process has an autocorrelation function that is some even function of τ, the random process is called colored noise.

B.3.5 Power Spectral Density Functions

The stationary random processes $x(t)$ and $y(t)$ have autocorrelation functions $\mathcal{R}_{xx}(\tau)$ and $\mathcal{R}_{yy}(\tau)$ respectively, and cross-correlation function $\mathcal{R}_{xy}(\tau)$. The Fourier transforms of the these functions are defined as the power spectral density functions,

$$S_{xx}(\omega) = \int_{-\infty}^{\infty} \mathcal{R}_{xx}(\tau)\, e^{-j\omega\tau}\, d\tau$$
$$S_{yy}(\omega) = \int_{-\infty}^{\infty} \mathcal{R}_{yy}(\tau)\, e^{-j\omega\tau}\, d\tau \tag{B.89}$$
$$S_{xy}(\omega) = \int_{-\infty}^{\infty} \mathcal{R}_{xy}(\tau)\, e^{-j\omega\tau}\, d\tau$$

$S_{xx}(\omega)$ and $S_{yy}(\omega)$ are called two-sided power spectral density functions (or spectral densities) of the random processes $x(t)$ and $y(t)$ respectively, and $S_{xy}(\omega)$ is called the two-sided cross spectral density function (or cross spectral density) between $x(t)$ and $y(t)$. The frequency ω in rad/s varies over the range $(-\infty, \infty)$.

The inverse Fourier transform of the auto spectral density is

$$R_{xx}(\tau) = \frac{1}{2\pi}\int_{-\infty}^{\infty} S_{xx}(\omega)\, e^{j\omega\tau}\, d\omega \tag{B.90}$$

When $\tau = 0$, the autocorrelation function equals the mean square magnitude of the signal in the time domain,

$$R_{xx}(0) = E\left[x^2(t)\right] = \frac{1}{2\pi}\int_{-\infty}^{\infty} S_{xx}(\omega)\,d\omega \tag{B.91}$$

Equation (B.91) is the basis for the name "power spectral density". Historically, electrical engineers considered the mean square of a current or voltage signal to be proportional to power using the resistance,

$$\text{power} = (\text{current})^2 * \text{resistance} = (\text{voltage})^2 / \text{resistance} \tag{B.92}$$

The terminology has persisted, and is applied in cases where the signals have nothing to do with power.

From the symmetry of the autocorrelation function,

$$\begin{aligned} \mathcal{R}_{uu}(\tau) &= \mathcal{R}_{uu}(-\tau) \\ \mathcal{R}_{yy}(\tau) &= \mathcal{R}_{yy}(-\tau) \\ \mathcal{R}_{uy}(\tau) &= \mathcal{R}_{yu}(-\tau) \end{aligned} \tag{B.93}$$

It follows that

$$\begin{aligned} S_{uu}(-\omega) &= S_{uu}(\omega) \\ S_{yy}(-\omega) &= S_{yy}(\omega) \\ S_{uy}(-\omega) &= S_{uy}^*(\omega) = S_{yu}(\omega) \end{aligned} \tag{B.94}$$

For the input $u(t)$ and output $y(t)$ of a linear, time-invariant dynamic system, the cross correlation can be expressed as

$$\begin{aligned} \mathcal{R}_{uy}(\tau) &= \lim_{T\to\infty} \frac{1}{T}\int_0^T u(t)\,y(t+\tau)\,dt \\ &= \lim_{T\to\infty} \frac{1}{T}\int_0^T \int_0^\infty H(t+\tau-\lambda)\; u(\lambda)\,d\lambda\, u(t)\,dt \\ &= \lim_{T\to\infty} \frac{1}{T}\int_0^T \int_0^\infty H(\mu)\; u(t+\tau-\mu)\,d\mu\, u(t)\,dt \\ &= \int_0^\infty H(\mu) \lim_{T\to\infty} \frac{1}{T}\int_0^T u(t)\,u(t+\tau-\mu)\,dt\,d\mu \end{aligned} \tag{B.95}$$

Combining Eq. (B.95) with the expression for the autocorrelation $\mathcal{R}_{uu}(\tau)$,

$$\mathcal{R}_{uu}(\tau) = \lim_{T\to\infty} \frac{1}{T}\int_0^T u(t)\,u(t+\tau)\,dt \tag{B.96}$$

gives

$$\mathcal{R}_{yu}(\tau) = \int_0^\infty H(\mu)\,\mathcal{R}_{uu}(\tau-\mu)\,d\mu \tag{B.97}$$

For practical work, it is more convenient to use spectral densities for positive frequencies only. The associated spectral densities are called one-sided spectral densities, defined as

$$G_{uu}(\omega) = \begin{cases} 2\,S_{uu} = 2\int_{-\infty}^{\infty} \mathcal{R}_{uu}(\tau)\, e^{-j\omega\tau}\, d\tau & \omega > 0 \\ S_{uu} = \int_{-\infty}^{\infty} \mathcal{R}_{uu}(\tau)\; d\tau & \omega = 0 \\ 0 & \omega < 0 \end{cases} \tag{B.98a}$$

$$G_{yy}(\omega) = \begin{cases} 2\,S_{yy} = 2\int_{-\infty}^{\infty} \mathcal{R}_{yy}(\tau)\, e^{-j\omega\tau}\, d\tau & \omega > 0 \\ S_{yy} = \int_{-\infty}^{\infty} \mathcal{R}_{yy}(\tau)\; d\tau & \omega = 0 \\ 0 & \omega < 0 \end{cases} \tag{B.98b}$$

$$G_{uy}(\omega) = \begin{cases} 2\,S_{uy} = 2\int_{-\infty}^{\infty} \mathcal{R}_{uy}(\tau)\, e^{-j\omega\tau}\, d\tau & \omega > 0 \\ S_{uy} = \int_{-\infty}^{\infty} \mathcal{R}_{uy}(\tau)\; d\tau & \omega = 0 \\ 0 & \omega < 0 \end{cases} \tag{B.98c}$$

To standardize results by removing weather differences among different days, a standard set of atmospheric conditions has been defined, called the standard atmosphere. Using the standard atmosphere, comparisons can be properly made using the same atmospheric conditions. Table C.1 shows conditions for the 1976 U.S. standard atmosphere, Ref. [C.1]. The data in Table C.1 shows that the air density, static pressure, and temperature all decrease with increasing altitude. Warmer, denser, higher pressure air lies at the bottom of the atmosphere.

Table C.1 **U.S. Standard Atmosphere, 1976**

Alt., ft	T, °R	p, lb/ft^2	ρ/ρ_0	ρ, lb-s^2/ft^4	a, ft/s
0	518.67	2116.2	1.00000	0.0023769	1116.44
1,000	515.10	2040.8	0.97107	0.0023082	1112.60
2,000	511.54	1967.7	0.94278	0.0022409	1108.76
3,000	507.97	1896.7	0.91513	0.0021751	1104.89
4,000	504.41	1827.7	0.88811	0.0021109	1100.98
5,000	500.84	1760.9	0.86170	0.0020482	1097.08
6,000	497.28	1696.0	0.83590	0.0019869	1093.18
7,000	493.71	1633.1	0.81070	0.0019270	1089.27
8,000	490.15	1572.0	0.78609	0.0018684	1085.33
9,000	486.59	1512.9	0.76206	0.0018113	1081.36
10,000	483.02	1455.6	0.73859	0.0017555	1077.40
11,000	479.46	1400.0	0.71568	0.0017011	1073.43
12,000	475.88	1346.2	0.69333	0.0016480	1069.42
13,000	472.34	1294.1	0.67151	0.0015961	1065.42
14,000	468.78	1243.6	0.65022	0.0015455	1061.38
15,000	465.22	1194.8	0.62946	0.0014962	1057.35
16,000	461.66	1147.5	0.60921	0.0014480	1053.31
17,000	458.09	1101.7	0.58946	0.0014011	1049.25
18,000	454.53	1057.5	0.57021	0.0013553	1045.14
19,000	450.97	1014.7	0.55144	0.0013107	1041.04
20,000	447.42	973.3	0.53317	0.0012673	1036.94
21,000	443.86	933.2	0.51534	0.0012249	1032.81
22,000	440.30	894.6	0.49798	0.0011837	1028.64
23,000	436.74	857.2	0.48108	0.0011435	1024.48
24,000	433.18	821.2	0.46462	0.0011044	1020.31
25,000	429.62	786.3	0.44859	0.0010663	1016.11

Table C.1 **U.S. Standard Atmosphere, 1976 (continued)**

Alt., ft	T, °R	p, lb/ft^2	ρ/ρ_0	ρ, lb-s^2/ft^4	a, ft/s
26,000	426.07	752.7	0.43300	0.0010292	1011.88
27,000	422.51	720.1	0.41782	0.0009931	1007.64
28,000	418.95	688.9	0.40305	0.0009580	1003.41
29,000	415.40	658.8	0.38869	0.0009239	999.15
30,000	411.84	629.7	0.37473	0.0008907	994.85
31,000	408.28	601.6	0.36115	0.0008584	990.55
32,000	404.73	574.6	0.34795	0.0008270	986.22
33,000	401.17	548.5	0.33513	0.0007966	981.89
34,000	397.62	523.5	0.32267	0.0007670	977.53
35,000	394.06	499.3	0.31058	0.0007382	973.13
36,000	390.51	476.1	0.29883	0.0007103	968.73
37,000	389.97	453.9	0.28525	0.0006780	968.08
38,000	389.97	432.6	0.27191	0.0006463	968.08
39,000	389.97	412.4	0.25920	0.0006161	968.08
40,000	389.97	393.1	0.24708	0.0005873	968.08
42,000	389.97	357.2	0.22452	0.0005336	968.08
44,000	389.97	324.6	0.20402	0.0004849	968.08
46,000	389.97	295.0	0.18540	0.0004407	968.08
48,000	389.97	268.1	0.16848	0.0004005	968.08
50,000	389.97	243.6	0.15311	0.0003639	968.08
52,000	389.97	221.4	0.13914	0.0003307	968.08
54,000	389.97	201.2	0.12645	0.0003006	968.08
56,000	389.97	182.8	0.11492	0.0002731	968.08
58,000	389.97	166.2	0.10444	0.0002482	968.08
60,000	389.97	151.0	0.09492	0.0002256	968.08
65,000	389.97	118.9	0.07475	0.0001777	968.08
70,000	392.25	93.7	0.05857	0.0001392	970.90
75,000	394.97	74.0	0.04591	0.0001091	974.28
80,000	397.69	58.5	0.03606	0.0000857	977.62
85,000	400.42	46.3	0.02837	0.0000674	980.94
90,000	403.14	36.8	0.02236	0.0000531	984.28
95,000	405.85	29.2	0.01765	0.0000419	987.60
100,000	408.57	23.3	0.01396	0.0000332	990.91

C.2 Elementary Aerodynamics

Components of the aerodynamic force and moment acting on an aircraft can be written in terms of nondimensional coefficients as follows:

Forces

Body Axes	*Stability Axes*	
$X = \overline{q} S C_X$	$D = \overline{q} S C_D = -X_S$	(C.5)
$Z = \overline{q} S C_Z$	$L = \overline{q} S C_L = -Z_S$	(C.6)
$Y = \overline{q} S C_Y$	$Y = \overline{q} S C_Y = Y_S$	(C.7)

Moments

$$L = \overline{q} b S C_l \tag{C.8}$$

$$M = \overline{q} \overline{c} S C_m \tag{C.9}$$

$$N = \overline{q} b S C_n \tag{C.10}$$

where

$\overline{q} =$ dynamic pressure $= \frac{1}{2}\rho V^2$ lbf/ft^2

$\rho =$ air density $= \rho(P,T) = 0.002377$ slug/ft^3 at sea level

$V =$ airspeed, ft/sec

$S =$ wing reference area, ft^2

$b =$ wing span, ft

$\overline{c} =$ mean aerodynamic chord (MAC), ft

$$\overline{c} \equiv \frac{1}{S}\int_{-b/2}^{b/2} \left[c(y)\right]^2 dy \tag{C.11}$$

Thrust can also be written in terms of a nondimensional thrust coefficient,

$$T = \overline{q} S C_T \tag{C.12}$$

When the thrust force comes from a jet engine, the aerodynamic nondimensionalization in Eq. (C.12) can be inadequate. In that case, it is more common to characterize the thrust force directly. For propeller aircraft, the characterization in Eq. (C.12) is appropriate because the propulsion force arises from rotating airfoils that comprise the propeller blades.

Reference geometry definitions for a typical airplane are shown in Fig. C.1. Wing chord c is the distance from the leading edge of the wing airfoil section to the trailing edge. The straight line connecting the leading edge and the trailing edge of the airfoil is called the chord line. Wing span b is the distance between wing tips. When the wing is not rectangular, the chord changes along the wing span, as in Fig. C.1. The mean aerodynamic chord $\bar{c}$ is computed from Eq. (C.11). The mean aerodynamic chord is used as a reference length for nondimensionalization and also as a reference to specify longitudinal location of the c.g. Wing reference area S normally includes all the shaded area shown in Fig. C.1.

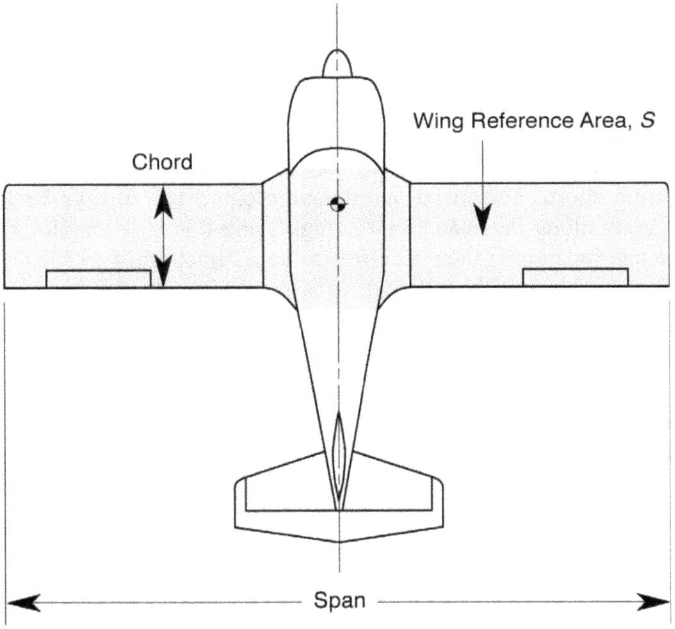

Figure C.1 **Airplane geometry definitions**

The defining relations in Eqs. (C.5)-(C.12) make all the coefficients nondimensional. Consequently, values of the nondimensional coefficients are tied to the geometric quantities used in the nondimensionalization. For example, if the reference area S is changed, the nondimensional force and moment coefficients change, but of course the aerodynamic forces and moments do not.

The nondimensional aerodynamic coefficients (and the corresponding aerodynamic forces and moments) are classified as longitudinal and lateral as follows:

$$\begin{aligned} &\text{Longitudinal:} \quad C_X, C_Z, C_m \ \text{ or } \ C_D, C_L, C_m \\ &\text{Lateral:} \quad C_Y, C_l, C_n \end{aligned} \tag{C.13}$$

The above classification arises from the fact that the longitudinal forces and moments affect the airplane motion in the body-axis *x-z* plane defined at the start of a maneuver, and therefore relate to longitudinal motion. The lateral forces and moments affect lateral airplane motion outside the body-axis *x-z* plane defined at the start of the maneuver. Longitudinal motion is also called symmetric flight, and lateral motion is called asymmetric flight.

C.3 Mass / Inertia Properties

Aircraft mass and longitudinal c.g. position are easily found by weighing the aircraft using scales at each point where the aircraft contacts the ground. Longitudinal c.g. location is often specified as a location along the mean aerodynamic chord. The mean aerodynamic chord can always be located at some location along the span of the wing. Using this location along the wing span, the c.g. location is then specified as some fraction of the distance along the mean aerodynamic chord, e.g. $0.35\bar{c}$, which would correspond to a location 35 percent of the distance from the leading edge to the trailing edge of the mean aerodynamic chord. It is also common practice to locate the c.g. by specifying right-handed Cartesian coordinates relative to a datum. Normally, the coordinates are called fuselage station (+*x* aft), butt line (+*y* out the right wing), and water line (+*z* up).

There are two ways to find the inertial properties for an aircraft, which are quantified by the elements of the inertia tensor. The first method, which might be called the analytical method, involves treating the airplane as a sum of individual parts, such as the fuselage, wings, vertical tail, etc. Inertia tensor elements for each individual part are summed together using the parallel axis theorem,

$$I_p = I + md^2 \tag{C.15}$$

where I is the moment of inertia about a given axis, m is the mass, and I_p is the moment of inertia about a parallel axis displaced a perpendicular distance d away. Often the shapes of the individual parts are approximated by simple shapes whose inertial properties are known, assuming constant mass density. An example would be approximating the fuselage as a cylinder. This method is also used to modify known inertial properties for added equipment, stores, instrumentation, cargo, or crew/passengers.

The second method, which might be called the experimental method, involves a ground test where the airplane is set up to oscillate about one of the body axes with negligible damping and a known spring constant. The measured time for the period of the oscillations can be used to calculate the mass moment of inertia about the axis of rotation for the oscillation. The expression for calculating the moment of inertia about the rotation axis is

$$I = K(m_a + m)T^2 + I_a \tag{C.16}$$

where subscript a denotes properties of the measurement apparatus, K is the spring constant of the apparatus, m is the aircraft mass, and T is the period. For small airplane models, this can be done using a pendulum apparatus, carrying out the experiment for each body axis.

To obtain the cross-product of inertia I_{xz}, one additional test is done to determine the mass moment of inertia about an intermediate axis in the body-axis Oxz plane, at an angle ξ to the body x axis. Then I_{xz} is computed from

$$I_{xz} = \frac{I_\xi - I_x \cos^2 \xi - I_z \sin^2 \xi}{\sin 2\xi} \tag{C.17}$$

where I_ξ denotes the moment of inertia about the intermediate axis.

The cross product of inertia I_{xz} arises mostly from the vertical tail and therefore has a much lower magnitude than the diagonal elements of the inertia matrix I_x, I_y, and I_z. This can be seen in the inertia values for the F-16 simulation, given in Appendix D, Table D.1.

C.4 Greek Alphabet and Conversion Factors

Table C.2 **Greek Alphabet**

Letter	Lower Case	Upper Case
Alpha	α	A
Beta	β	B
Gamma	γ	Γ
Delta	δ	Δ
Epsilon	ε	E
Zeta	ζ	Z
Eta	η	H
Theta	θ	Θ
Iota	ι	I
Kappa	κ	K
Lambda	λ	Λ
Mu	μ	M
Nu	ν	N
Xi	ξ	Ξ
Omicron	o	O
Pi	π	Π
Rho	ρ	P
Sigma	σ	Σ
Tau	τ	T
Upsilon	υ	Y
Phi	ϕ	Φ
Chi	χ	X
Psi	ψ	Ψ
Omega	ω	Ω

Conversion Factors

1 ft = 0.3048 m

1 nautical mile = 1.152 mi = 6080.2 ft

1 kt = 1.152 mph = 1.689 ft/s = 0.5148 m/s

1 slug = 32.174 lbm = 14.59 kg

1 slug/ft^3 = 515.2 kg/m^3

1 lbf = 4.448 N

1 atm = 2116.2 lbf/ft^2

1 hp = 550 ft-lbf/s

1 deg = 0.01745 rad

g = 32.174 ft/s^2 = 9.806 m/s^2 at sea level

References

[C.1] *U.S. Standard Atmosphere: 1976*, National Oceanic and Atmospheric Administration, NASA, and U.S. Air Force, Washington, DC, 1976.

Appendix D

F-16 Nonlinear Simulation

This appendix provides a detailed description of a nonlinear simulation for the F-16 fighter aircraft. The simulation is written completely in MATLAB®, and is included as part of the software package associated with this textbook. It is essentially a MATLAB® version of the FORTRAN simulation given in Appendix A of Ref. [D.1]. The simulation is based on information in Ref. [D.2], which includes a wind tunnel aerodynamic database for a 16% scale model of the F-16 aircraft, and a ground test database for engine thrust.

The following sections provide details for various aspects of the F-16 nonlinear simulation, including the F-16 aircraft, equations of motion, aerodynamic model, engine model, and analysis tools. The F-16 nonlinear simulation documented here is one of several included in Ref. [D.3].

D.1 F-16 Aircraft

The F-16 is a single-seat, multirole fighter with a blended wing / body and a cropped delta wing planform with leading edge sweep of 40°. Figure D.1 shows a three-view drawing of the F-16. The wing is fitted with leading edge flaps and trailing edge flaperons (flaps / ailerons). Tail surfaces are swept and cantilevered. The horizontal stabilator is composed of two all-moving tail plane halves, while the vertical tail is fitted with a trailing edge rudder. Thrust is provided by one General Electric F110-GE-100 or Pratt & Whitney F100-PW-220 afterburning turbofan engine mounted in the rear fuselage.

The aircraft was modeled with controls for throttle δ_{th}, stabilator δ_s, aileron δ_a, and rudder δ_r. Speed brake and flaps were assumed fixed at zero deflection. Throttle deflection was limited to the range $0 \le \delta_{th} \le 1$, stabilator deflection was limited to $-25^{\text{o}} \le \delta_s \le 25^{\text{o}}$, aileron deflection was limited to $-21.5^{\text{o}} \le \delta_a \le 21.5^{\text{o}}$, and rudder deflection was limited to $-30^{\text{o}} \le \delta_r \le 30^{\text{o}}$. These limits represent the physicals stops.

D.2 Equations of Motion

The aircraft was assumed rigid with constant mass density and symmetry about the Oxz plane in body axes. Thrust was assumed to act along the x body axis and through the center of gravity. A stationary atmosphere was

Differentiating the equations (D.4) with respect to time gives

$$\dot{V} = \frac{u\dot{u} + v\dot{v} + w\dot{w}}{V}$$

$$\dot{\alpha} = \frac{u\dot{w} - w\dot{u}}{u^2 + w^2} \tag{D.6}$$

$$\dot{\beta} = \frac{V\dot{v} - v\dot{V}}{V^2\sqrt{1-\left(\frac{v}{V}\right)^2}}$$

The translational states for the nonlinear aircraft simulations are V, α, and β. The values of u, v, and w are computed from Eqs. (D.5), using current values of V, α, and β. Time derivatives $\dot{u}, \dot{v}$, and $\dot{w}$ are computed from Eqs. (D.1), then Eqs. (D.6) are used to obtain $\dot{V}, \dot{\alpha}$, and $\dot{\beta}$ for use in the numerical integration routines. This approach avoids the lengthy nonlinear calculations associated with the equations of motion written in wind axes, and allows aircraft nondimensional aerodynamic coefficient data to be implemented in the body-axis coordinate system, with no need for converting the data to wind axes components.

For rotational motion, straightforward algebraic manipulation transforms the equations (D.2) into a form suitable for numerical integration, with only one time derivative on the left side of each equation,

$$\dot{p} = \left(c_1 r + c_2 p - c_4 h_p\right)q + \bar{q}Sb\left(c_3 C_l + c_4 C_n\right)$$

$$\dot{q} = \left(c_5 p + c_7 h_p\right)r - c_6\left(p^2 - r^2\right) + \bar{q}S\bar{c}\,c_7 C_m \tag{D.7}$$

$$\dot{r} = \left(c_8 p - c_2 r - c_9 h_p\right)q + \bar{q}Sb\left(c_4 C_l + c_9 C_n\right)$$

where

$$c_1 = \frac{\left(I_y - I_z\right)I_z - I_{xz}^2}{I_x I_z - I_{xz}^2} \qquad c_2 = \frac{\left(I_x - I_y + I_z\right)I_{xz}}{I_x I_z - I_{xz}^2} \qquad c_3 = \frac{I_z}{I_x I_z - I_{xz}^2}$$

$$c_4 = \frac{I_{xz}}{I_x I_z - I_{xz}^2} \qquad c_5 = \frac{I_z - I_x}{I_y} \qquad c_6 = \frac{I_{xz}}{I_y} \tag{D.8}$$

$$c_7 = \frac{1}{I_y} \qquad c_8 = \frac{\left(I_x - I_y\right)I_x - I_{xz}^2}{I_x I_z - I_{xz}^2} \qquad c_9 = \frac{I_x}{I_x I_z - I_{xz}^2}$$

The nonlinear aircraft simulation also includes rotational kinematic equations and navigation equations. The rotational kinematic equations, which relate Euler angular rates to body-axis angular rates, are given by:

$$\dot{\phi} = p + \tan\theta\left(q\sin\phi + r\cos\phi\right)$$

$$\dot{\theta} = q\cos\phi - r\sin\phi \tag{D.9}$$

$$\dot{\psi} = \frac{q\sin\phi + r\cos\phi}{\cos\theta}$$

Equations (D.9) are nonlinear state equations for the Euler angles $\phi, \theta,$ and ψ, in a form suitable for numerical integration. The rotational kinematic equations describe the time evolution of the aircraft attitude angles, which are required to properly resolve the gravity force along the aircraft body axes in Eqs. (D.1). In the simulation code, quaternions are used to compute Euler angle derivatives $\dot{\psi}$ and $\dot{\phi}$, to avoid the singularities in the rotational kinematic state equations at $\theta = \pm 90^{\circ}$.

The navigation equations relate aircraft translational velocity components in body axes to earth-axis components, neglecting wind. These differential equations describe the time evolution of the position of the aircraft c.g. relative to earth axes,

$$\begin{aligned}
\dot{x}_E &= u\cos\psi\cos\theta + v\left(\cos\psi\sin\theta\sin\phi - \sin\psi\cos\phi\right) \\
&\quad + w\left(\cos\psi\sin\theta\cos\phi + \sin\psi\sin\phi\right) \\
\dot{y}_E &= u\sin\psi\cos\theta + v\left(\sin\psi\sin\theta\sin\phi + \cos\psi\cos\phi\right) \\
&\quad + w\left(\sin\psi\sin\theta\cos\phi - \cos\psi\sin\phi\right) \\
\dot{h} &= u\sin\theta - v\cos\theta\sin\phi - w\cos\theta\cos\phi
\end{aligned} \tag{D.10}$$

Assuming thrust acts along the x body axis, the body-axis accelerations a_x, a_y, and a_z in g units are calculated from

$$a_x = \frac{\bar{q}SC_X + T}{mg}$$

$$a_y = \frac{\bar{q}SC_Y}{mg} \tag{D.11}$$

$$a_z = \frac{\bar{q}SC_Z}{mg}$$

Eqs. (D.1) through (D.10) are the nonlinear aircraft equations of motion implemented in the F-16 nonlinear simulation.

D.3 Engine Model

The F-16 is powered by a single afterburning turbofan jet engine, which was modeled taking into account throttle gearing and engine power level lag.

The engine power dynamic response was modeled with an additional state equation, as a simple first order lag in the actual power level response to commanded power level:

$$\dot{P}_a = \frac{1}{\tau_{eng}}\left(P_c - P_a\right) \tag{D.12}$$

Commanded power level was computed as a function of throttle position,

$$P_c = P_c\left(\delta_{th}\right) \tag{D.13}$$

Throttle gearing in Eq. (D.13) is implemented in the file tgear.m, which translates the throttle deflection δ_{th} in the interval [0,1] to commanded power level in the interval [0,100]. Commanded power P_c as a function of δ_{th} is

$$P_c\left(\delta_{th}\right) = \begin{cases} 64.94\delta_{th} & \text{if } \delta_{th} \le 0.77 \\ 217.38\delta_{th} - 117.38 & \text{if } \delta_{th} > 0.77 \end{cases} \tag{D.14}$$

The routine rtau.m computes the engine power time constant τ_{eng}, and pdot.m implements the engine thrust dynamics in Eq. (D.12).

Engine thrust force is computed by linear interpolation of an engine thrust database, as a function of actual power level P_a, altitude h, and Mach number M,

$$T = T\left(P_a, h, M\right) \tag{D.15}$$

For the F-16, the engine thrust is given in tabular form as a function of altitude and Mach number for idle, military, and maximum power. The data table for each power level includes values for altitude and Mach number in the ranges $0 \text{ ft} \le h \le 50{,}000 \text{ ft}$ and $0 \le M \le 1$. Engine data tables are defined in the file f16_engine_setup.m. Linear interpolation and thrust calculation is performed in the module f16_engine.m. Thrust is computed from

$$T = \begin{cases} T_{idle} + \left(T_{mil} - T_{idle}\right)\left(\dfrac{P_a}{50}\right) & \text{if } P_a < 50 \\ T_{mil} + \left(T_{max} - T_{mil}\right)\left(\dfrac{P_a - 50}{50}\right) & \text{if } P_a \geq 50 \end{cases} \tag{D.16}$$

The engine angular momentum h_p is assumed to act along the aircraft x body axis with a fixed value of 160 slug-ft^2/s.

D.4 Aerodynamic Model

Nondimensional aerodynamic force and moment coefficient data were derived from a low-speed static wind tunnel test and a dynamic forced oscillation wind tunnel test, both conducted on a 16% scale model of the F-16. The aerodynamic database applies to the F-16 flown out of ground effect, with landing gear retracted, and no external stores, see Refs. [D.1] and [D.2]. Static aerodynamic data are in tabular form as a function of angle of attack and sideslip angle over the ranges $-10^{\circ} \leq \alpha \leq 45^{\circ}$ and $-30^{\circ} \leq \beta \leq 30^{\circ}$, respectively. Dynamic data is provided in tabular form at zero sideslip angle over the angle of attack range $-10^{\circ} \leq \alpha \leq 45^{\circ}$. Dependence of the nondimensional coefficients on $\dot{\alpha}$ is included in the q dependencies, due to the manner in which the data was collected in the wind tunnel.

Each nondimensional aerodynamic force and moment coefficient is built up from a set of component functions, where the value of each component function is determined by a table look-up in the wind tunnel database. The aerodynamic database was simplified slightly by dropping second-order dependencies, e.g., dependence of longitudinal aerodynamic forces on sideslip angle. For the table look-ups, angle of attack, sideslip angle, and control surface deflections are in degrees. Aircraft angular velocities are in radians per second. The value for each component function is found by linear interpolation, using current values of the states and controls. For values of states and controls outside the range of the available data, the interpolation routines extrapolate linearly using the nearest data points. Aerodynamic coefficients are referenced to a center of gravity location at $0.35\bar{c}$, so $x_{cg_{ref}} = 0.35$. Corrections to the flight center of gravity position are made in the coefficient build-up equations.

The nondimensional aerodynamic force and moment coefficients for the F-16 vary with air flow angles (α, β), aircraft angular velocities (p, q, r), and control surface deflections $(\delta_e, \delta_a, \delta_r)$. Moment coefficients C_m and C_n include a correction for the center of gravity position. The coefficients are computed as follows:

$$C_X = C_{X_0}(\alpha, \delta_e) + \left(\frac{q\bar{c}}{2V}\right) C_{X_q}(\alpha)$$

$$C_Y = -0.02\beta + 0.021\left(\frac{\delta_a}{20}\right) + 0.086\left(\frac{\delta_r}{30}\right) + \left(\frac{b}{2V}\right)\left[C_{Y_p}(\alpha)p + C_{Y_r}(\alpha)r\right]$$

$$C_Z = C_{Z_0}(\alpha)\left[1 - \left(\frac{\beta\pi}{180}\right)^2\right] - 0.19\left(\frac{\delta_e}{25}\right) + \left(\frac{q\bar{c}}{2V}\right) C_{Z_q}(\alpha)$$

$$C_l = C_{l_0}(\alpha, \beta) + \Delta C_{l_{\delta_a}}(\alpha, \beta)\left(\frac{\delta_a}{20}\right) + \Delta C_{l_{\delta_r}}(\alpha, \beta)\left(\frac{\delta_r}{30}\right)$$
$$+ \left(\frac{b}{2V}\right)\left[C_{l_p}(\alpha)p + C_{l_r}(\alpha)r\right]$$

$$C_m = C_{m_0}(\alpha, \delta_e) + \left(\frac{q\bar{c}}{2V}\right) C_{m_q}(\alpha) + \left(x_{cg_{ref}} - x_{cg}\right) C_Z$$

$$C_n = C_{n_0}(\alpha, \beta) + \Delta C_{n_{\delta_a}}(\alpha, \beta)\left(\frac{\delta_a}{20}\right) + \Delta C_{n_{\delta_r}}(\alpha, \beta)\left(\frac{\delta_r}{30}\right)$$
$$+ \left(\frac{b}{2V}\right)\left[C_{n_p}(\alpha)p + C_{n_r}(\alpha)r\right] - \left(\frac{\bar{c}}{b}\right)\left(x_{cg_{ref}} - x_{cg}\right) C_Y$$

where the symbols with arguments, e.g., $C_{l_0}(\alpha, \beta)$, represent values obtained from table look-ups.

Data tables for the component functions of the nondimensional coefficients are defined in f16_aero_setup.m. The nondimensional aerodynamic force and moment coefficients are computed in f16_aero.m, using linear interpolation of the data tables.

D.5 Atmosphere Model

Air density and the speed of sound are calculated using relations modeling the 1976 U.S. Standard Atmosphere, see Appendix C. Quantities that depend on these atmospheric properties, namely Mach number M, and dynamic pressure $\overline{q}$, are also calculated. The relationships are:

$$T^* = 1 - 0.703 \times 10^{-5}\, h$$

$$\rho = 0.002377\left(T^*\right)^{4.14}$$

$$M = \begin{cases} \dfrac{V}{\sqrt{1.4(1716.3)390}} & h \geq 35{,}000 \text{ ft} \\[2ex] \dfrac{V}{\sqrt{1.4(1716.3)\left(519T^*\right)}} & h < 35{,}000 \text{ ft} \end{cases}$$

$$\overline{q} = \frac{1}{2}\rho V^2$$

The function atm.m is used to compute these quantities, given altitude h and airspeed V .

D.6 Mass / Inertia Properties

The mass properties of the F-16 are given in the file f16_massprop.m. These properties include the weight of the aircraft, longitudinal c.g. position x_{cg} , and moments of inertia $I_x, I_y, I_z,$ and I_{xz} . This module also calculates the constants $c_1, c_2, \ldots, c_9$ for use in the body-axis moment equations. The longitudinal center of gravity location x_{cg} is given as a fraction of mean aerodynamic chord, and can be adjusted within this module to account for different longitudinal center of gravity locations. Table D.1 lists the mass properties used in the F-16 nonlinear simulation. The default longitudinal c.g. position is set at a forward location, so that the aircraft can be flown open-loop, without stability augmentation feedback control.

Table D.1 **Mass properties of the simulated F-16**

Parameter	Value
Weight (lbf)	20,500
x_{cg}	0.25
I_x (slug-ft^2)	9,496
I_y (slug-ft^2)	55,814
I_z (slug-ft^2)	63,100
I_{xz} (slug-ft^2)	982

D.7 Analysis Tools

D.7.1 Aircraft States and Controls

For the F-16 nonlinear simulation, the state vector is

$$\boldsymbol{x} = \begin{bmatrix} V & \alpha & \beta & \phi & \theta & \psi & p & q & r & x_E & y_E & h & P_a \end{bmatrix}^T \tag{D.17}$$

and the control vector is

$$\boldsymbol{u} = \begin{bmatrix} \delta_{th} & \delta_s & \delta_a & \delta_r \end{bmatrix}^T \tag{D.18}$$

The set of coupled, nonlinear, first-order ordinary differential equations that comprise the simulation model can be represented by the vector differential equation

$$\dot{\boldsymbol{x}} = \boldsymbol{f}\left(\boldsymbol{x}, \boldsymbol{u}\right) \tag{D.19}$$

The output equations can be represented by the vector equation

$$\boldsymbol{y} = \boldsymbol{h}\left(\boldsymbol{x}, \boldsymbol{u}\right) \tag{D.20}$$

The nonlinear equations of motion are implemented in f16_deq.m. This routine implements Eq. (D.19), which computes the state vector derivatives, given the states and controls. The function f16.m is the main routine for the F-16 nonlinear simulation. This routine initializes the data tables, carries out the numerical integration of Eq. (D.19), and implements Eq. (D.20).

D.7.2 Trim

A steady flight condition with state derivatives that are constant or zero is called a trimmed flight condition. Finding the states and controls associated with these conditions is useful for initializing the nonlinear simulation, stability analysis, and defining a reference condition for linearization. The equations defining a trim condition are Eqs. (D.19) with the state derivatives set to a constant vector c_{trim},

$$c_{trim} = \boldsymbol{f}(\boldsymbol{x}, \boldsymbol{u}) \tag{D.21}$$

This constitutes a set of coupled nonlinear algebraic equations that can be solved to find trim values of the states and controls. Note that the state equations related to the position variables x_E and y_E are not included in the equations used for trim, because they are not relevant for dynamics and control analysis. The values of the state and control vectors which satisfy the trim equations and any added constraint equations define the trim condition. Constraint equations for common steady flight conditions are described below.

<u>*Trim Constraint for Steady Translational Flight*</u>

For a steady translational flight condition,

$$\phi = p = q = r = 0 \tag{D.22}$$

The earth position states $(x_E \text{ and } y_E)$ are dropped, since their values matter only for navigation. Altitude h will be specified for a desired flight condition. Heading angle ψ is arbitrary, so the remaining states are $V, \alpha, \beta, \theta,$ and P_a. For a given aircraft wing loading, V and α are related, so only one of these can be chosen for the desired flight condition. Sideslip angle β is used to balance lateral asymmetry and side force from lateral control deflections. Controls $\delta_s, \delta_a,$ and δ_r balance aerodynamic moments. The engine power P_a in steady state is a function of throttle δ_{th} via the throttle gearing. Engine power combined with altitude and airspeed determine the thrust, which must balance drag.

When airspeed V and flight path angle $\gamma \equiv \theta - \alpha$ are specified for the steady flight condition, then the kinematic relationship between the velocity expressed in wind axes and in earth axes gives

$$\underbrace{\begin{bmatrix} * \\ * \\ -V \sin\gamma \end{bmatrix}}_{\text{earth axes}} = L(\psi)^T \, L(\theta)^T \, L(\phi)^T \, L_{WB}(\alpha, \beta)^T \underbrace{\begin{bmatrix} V \\ 0 \\ 0 \end{bmatrix}}_{\text{wind axes}}$$

Equating z components on both sides,

$$sin\gamma = a\,sin\theta - b\,cos\theta$$

where

$$a = cos\,\alpha\,\,cos\,\beta$$

$$b = sin\phi\,\,sin\,\beta + cos\,\phi\,\,sin\alpha\,\,cos\,\beta$$

Solving for $tan\theta$,

$$tan\theta = \frac{ab + sin\gamma\,\sqrt{a^2 - sin^2\gamma + b^2}}{a^2 - sin^2\gamma} \quad ; \quad \theta \neq \pm\frac{\pi}{2} \tag{D.23}$$

The expression above is the rate of climb constraint, which can be appended to the trim equations when computing a trim solution.

Trim Constraint for Steady Turning Flight

A steady turn can be specified by turn rate $\dot{\psi}$, or centripetal acceleration G in g units, since

$$G = \frac{1}{g}\frac{V^2}{R} = \frac{\dot{\psi}V}{g} \tag{D.24}$$

When the turn is coordinated, the aerodynamic forces are balanced and there is no acceleration in the body-axis y direction. The body-axis y force equation becomes

$$0 = pw - ru + g\,cos\theta\,sin\phi \tag{D.25}$$

which means that the gravity force is balanced by inertial terms. In a steady turn, $\dot{\phi} = \dot{\theta} = 0$, so Eq. (3.25) reduces to

$$\begin{aligned} p &= -\dot{\psi}\,sin\theta \\ q &= \dot{\psi}\,cos\theta\,sin\phi \\ r &= \dot{\psi}\,cos\theta\,cos\phi \end{aligned} \tag{D.26}$$

Combining Eqs. (D.24)-(D.26), and substituting for u, v, and w in terms of V, α, and β from Eqs. (D.5),

$$sin\phi = G\,cos\,\beta\left(sin\alpha\,\,tan\theta + cos\,\alpha\,\,cos\,\phi\right) \tag{D.27}$$

Equation (D.27) is the coordinated turn constraint, which can be appended to the trim equations when computing a trim solution.

Combining the rate of climb constraint (D.23) and the coordinated turn constraint (D.27) by substituting for $tan\theta$ in the coordinated turn constraint gives

$$tan\phi = G\frac{cos\,\beta}{cos\,\alpha}\,\frac{\left(a-b^2\right)+b\,tan\alpha\,\sqrt{c\left(1-b^2\right)+G^2\,sin^2\,\beta}}{a^2-b^2\left(1+c\,tan^2\,\alpha\right)} \tag{D.28}$$

where

$$a = 1 - G\,tan\alpha\,sin\,\beta$$

$$b = \frac{sin\gamma}{cos\,\beta} \tag{D.29}$$

$$c = 1 + G^2\,cos^2\,\beta$$

For $\gamma = 0$, Eq. (D.29) becomes

$$tan\phi = \frac{G\,cos\,\beta}{cos\,\alpha - G\,sin\alpha\,sin\,\beta} \tag{D.30}$$

The F-16 nonlinear simulation can include the rate of climb constraint Eq. (D.23) and/or the turn constraint Eq. (D.27) in f16_trm.m to calculate a trimmed flight condition.

D.7.3 Linearization

The nonlinear equations of motion can be linearized using finite differences, as discussed in Chapter 3. The function lnze.m does this, using the function f16_deq.m to compute values of the right side of Eqs. (D.19) for nominal and perturbed states and controls.

D.7.4 Solving Nonlinear Aircraft Equations of Motion

The nonlinear equations of motion, with experimental data for the aerodynamics and thrust, and arbitrary control surface input time series, require numerical integration to obtain the state and output time histories. This is an initial value problem, which is: given the initial state at $t = 0$ and the control for $t \geq 0$, find the state trajectory for $t > 0$. A standard algorithm for continuous systems is fourth-order Runge-Kutta, where the state equations are integrated repeatedly over one time step Δt , as follows:

$$\boldsymbol{x}(t+\Delta t) = \boldsymbol{x}(t) + \Delta t\left(\frac{k_1}{6}+\frac{k_2}{3}+\frac{k_3}{3}+\frac{k_4}{6}\right)$$

where

$$\Delta t = \text{integration time step}$$

$$k_1 = \boldsymbol{f}(\boldsymbol{x},\boldsymbol{u},t)$$

$$k_2 = \boldsymbol{f}(\boldsymbol{x}^*,\boldsymbol{u},t^*)$$

$$k_3 = \boldsymbol{f}(\boldsymbol{x}^{**},\boldsymbol{u},t^{**})$$

$$k_4 = \boldsymbol{f}(\boldsymbol{x}^{***},\boldsymbol{u},t^{***})$$

and

$$\boldsymbol{x}^* = \boldsymbol{x}(t) + \frac{\Delta t}{2}\boldsymbol{f}(\boldsymbol{x},\boldsymbol{u},t) \quad ; \quad t^* = t + \frac{\Delta t}{2}$$

$$\boldsymbol{x}^{**} = \boldsymbol{x}(t) + \frac{\Delta t}{2}\boldsymbol{f}(\boldsymbol{x}^*,\boldsymbol{u},t^*) \quad ; \quad t^{**} = t + \frac{\Delta t}{2}$$

$$\boldsymbol{x}^{***} = \boldsymbol{x}(t) + \Delta t\, \boldsymbol{f}(\boldsymbol{x}^{**},\boldsymbol{u},t^{**}) \quad ; \quad t^{***} = t + \Delta t$$

Second-order Runge-Kutta is less accurate (depending on the size of Δt, the control surface inputs, and the aircraft dynamics), but executes twice as fast because half as many state derivative calculations are required:

$$\boldsymbol{x}(t+\Delta t) = \boldsymbol{x}(t) + \Delta t\, k_2$$

where

$$\Delta t = \text{integration time step}$$

$$k_2 = \boldsymbol{f}(\boldsymbol{x}^*,\boldsymbol{u},t^*)$$

and

$$\boldsymbol{x}^* = \boldsymbol{x}(t) + \frac{\Delta t}{2}\boldsymbol{f}(\boldsymbol{x},\boldsymbol{u},t) \quad ; \quad t^* = t + \frac{\Delta t}{2}$$

A flow chart for nonlinear simulation is shown in Fig. D.2.

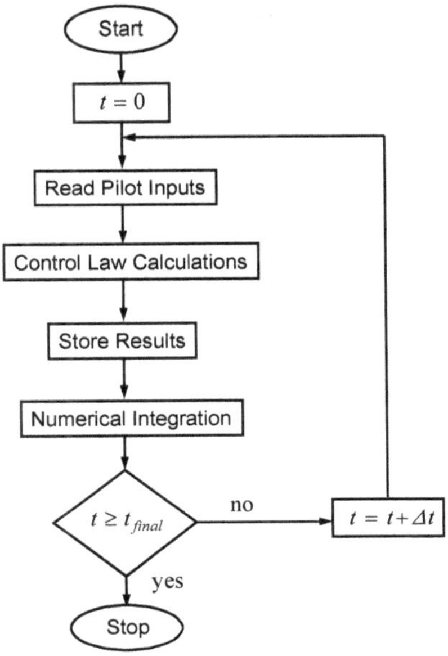

Figure D.2 **Flow chart for nonlinear simulation**

Note that the F-16 nonlinear simulation does not include a feedback control law, so the control law calculations are simply scaling the pilot inputs to commanded control surface deflections.

D.7.5 Real-Time Piloted Simulation

The F-16 nonlinear simulation can be flown with a joystick or computer mouse in real time, using commercially-available software called Aviator Visual Design Simulator (AVDS), available from RasSimTech, Ltd. at http://www.rassimtech.com. The interface file f16_avds_matlab_sim.m contains comments in the header that include detailed instructions on how to use this capability. The F-16 real-time piloted simulation can be run using the script fly_f16.m. AVDS reads pilot inputs and displays a real-time image of the F-16 while the simulation is running in MATLAB®. Full details of this interface are available in Ref. [D.3].

Supporting Materials

SIDPAC is available by request from NASA Langley at:

https://software.nasa.gov/software/LAR-16100-1

Full contact information and organizational affiliation will be collected from all requestors, and all requestors must pass NASA security screening before receiving SIDPAC.

Information about SIDPAC and its development is available at:

http://sunflyte.com/SIDBook_SIDPAC.htm

Information about this textbook is available at:

http://sunflyte.com/SIDBook.html

www.ingramcontent.com/pod-product-compliance
Lightning Source LLC
Chambersburg PA
CBHW050452190326
41458CB00005B/1252
9780997430615